Advances in Architectural Acoustics

Advances in Architectural Acoustics

Editors

Nikolaos M. Papadakis
Massimo Garai
Georgios E. Stavroulakis

MDPI • Basel • Beijing • Wuhan • Barcelona • Belgrade • Manchester • Tokyo • Cluj • Tianjin

Editors

Nikolaos M. Papadakis
Technical University of Crete
Greece

Massimo Garai
University of Bologna
Italy

Georgios E. Stavroulakis
Technical University of Crete
Greece

Editorial Office
MDPI
St. Alban-Anlage 66
4052 Basel, Switzerland

This is a reprint of articles from the Special Issue published online in the open access journal *Applied Sciences* (ISSN 2076-3417) (available at: https://www.mdpi.com/journal/applsci/special_issues/Architectural_Acoustics).

For citation purposes, cite each article independently as indicated on the article page online and as indicated below:

LastName, A.A.; LastName, B.B.; LastName, C.C. Article Title. *Journal Name* **Year**, *Volume Number*, Page Range.

ISBN 978-3-0365-4295-9 (Hbk)
ISBN 978-3-0365-4296-6 (PDF)

Cover image courtesy of the Ministry of Culture and Sports of Greece
Cover image (Neoria Monument, Chania, Crete, Greece)
Cover image shows the interior of Neoria monument in Chania during acoustic measurements, and is courtesy of the Ministry of Culture and Sports of Greece.

© 2022 by the authors. Articles in this book are Open Access and distributed under the Creative Commons Attribution (CC BY) license, which allows users to download, copy and build upon published articles, as long as the author and publisher are properly credited, which ensures maximum dissemination and a wider impact of our publications.

The book as a whole is distributed by MDPI under the terms and conditions of the Creative Commons license CC BY-NC-ND.

Contents

About the Editors vii

Preface to "Advances in Architectural Acoustics" ix

Nikolaos M. Papadakis, Massimo Garai and Georgios E. Stavroulakis
Special Issue: Advances in Architectural Acoustics
Reprinted from: *Appl. Sci.* **2022**, *12*, 1728, doi:10.3390/app12031728 1

Dario D'Orazio
Italian-Style Opera Houses: A Historical Review
Reprinted from: *Appl. Sci.* **2020**, *10*, 4613, doi:10.3390/app10134613 7

Nikolaos M. Papadakis and Georgios E. Stavroulakis
Review of Acoustic Sources Alternatives to a Dodecahedron Speaker
Reprinted from: *Appl. Sci.* **2019**, *9*, 3705, doi:10.3390/app9183705 35

Arianna Astolfi, Elena Bo, Francesco Aletta and Louena Shtrepi
Measurements of Acoustical Parameters in the Ancient Open-Air Theatre of Tyndaris
(Sicily, Italy)
Reprinted from: *Appl. Sci.* **2020**, *10*, 5680, doi:10.3390/app10165680 67

Enedina Alberdi, Miguel Galindo and Ángel L. León-Rodríguez
Evolutionary Analysis of the Acoustics of the Baroque Church of San Luis de los
Franceses (Seville)
Reprinted from: *Appl. Sci.* **2021**, *11*, 1402, doi:10.3390/app11041402 85

**Hanna Autio, Mathias Barbagallo, Carolina Ask, Delphine Bard, Eva Lindqvist Sandgren
and Karin Strinnholm Lagergren**
Historically Based Room Acoustic Analysis and Auralization of a Church in the 1470s
Reprinted from: *Appl. Sci.* **2021**, *11*, 1586, doi:10.3390/app11041586 105

Takeshi Okuzono, M Shadi Mohamed and Kimihiro Sakagami
Potential of Room Acoustic Solver with Plane-Wave Enriched Finite Element Method
Reprinted from: *Appl. Sci.* **2020**, *10*, 1969, doi:10.3390/app10061969 131

Takumi Yoshida, Takeshi Okuzono and Kimihiro Sakagami
Time Domain Room Acoustic Solver with Fourth-Order Explicit FEM Using Modified Time
Integration
Reprinted from: *Appl. Sci.* **2020**, *10*, 3750, doi:10.3390/app10113750 149

Erling Nilsson and Emma Arvidsson
An Energy Model for the Calculation of Room Acoustic Parameters in Rectangular Rooms with
Absorbent Ceilings
Reprinted from: *Appl. Sci.* **2021**, *11*, 6607, doi:10.3390/app11146607 173

Hequn Min and Ke Xu
Coherent Image Source Modeling of Sound Fields in Long Spaces with a
Sound-Absorbing Ceiling
Reprinted from: *Appl. Sci.* **2021**, *11*, 6743, doi:10.3390/app11156743 201

Amy Bastine, Thushara D. Abhayapala and Jihui Zhang
Power Response and Modal Decay Estimation of Room Reflections from Spherical Microphone
Array Measurements Using Eigenbeam Spatial Correlation Model
Reprinted from: *Appl. Sci.* **2021**, *11*, 7688, doi:10.3390/app11167688 213

Juan Óscar García Gómez, Oliver Wright, Bertie van den Braak, Javier Sanz, Liam Kemp and Thomas Hulland
On the Sequence of Unmasked Reflections in Shoebox Concert Halls
Reprinted from: *Appl. Sci.* **2021**, *11*, 7798, doi:10.3390/app11177798 229

Valtteri Hongisto and Jukka Keränen
Comfort Distance—A Single-Number Quantity Describing Spatial Attenuation in Open-Plan Offices
Reprinted from: *Appl. Sci.* **2021**, *11*, 4596, doi:10.3390/app11104596 243

Matteo Cingolani, Giulia Fratoni, Luca Barbaresi, Dario D'Orazio, Brian Hamilton and Massimo Garai
A Trial Acoustic Improvement in a Lecture Hall with MPP Sound Absorbers and FDTD Acoustic Simulations
Reprinted from: *Appl. Sci.* **2021**, *11*, 2445, doi:10.3390/app11062445 253

Hequn Min and Yitian Liao
Mechanism Analysis of the Influence of Seat Attributes on the Seat Dip Effect in Music Halls
Reprinted from: *Appl. Sci.* **2021**, *11*, 9768, doi:10.3390/app11209768 273

Lili Pan and Francesco Martellotta
A Parametric Study of the Acoustic Performance of Resonant Absorbers Made of Micro-perforated Membranes and Perforated Panels
Reprinted from: *Appl. Sci.* **2020**, *10*, 1581, doi:10.3390/app10051581 283

Louena Shtrepi, Sonja Di Blasio and Arianna Astolfi
Listeners Sensitivity to Different Locations of Diffusive Surfaces in Performance Spaces: The Case of a Shoebox Concert Hall
Reprinted from: *Appl. Sci.* **2020**, *10*, 4370, doi:10.3390/app10124370 299

Yaw-Shyan TSAY and Chiu-Yu YEH
A Machine Learning Based Prediction Model for the Sound Absorption Coefficient of Micro-Expanded Metal Mesh (MEMM)
Reprinted from: *Appl. Sci.* **2020**, *10*, 7612, doi:10.3390/app10217612 323

About the Editors

Nikolaos M. Papadakis

Nikolaos M. Papadakis is a researcher in the Institute of Computational Mechanics and Optimization (Co.Mec.O) at the Technical University of Crete and an adjunct lecturer in the Department of Music Technology and Acoustics in the Hellenic Mediterranean University. He is also Visiting Lecturer at the National and Kapodistrian University of Athens. He holds a Ph.D. in Architectural and Computational Acoustics from Technical University of Crete, MSc in Audio Acoustics from University of Salford, UK, and a BSc in Physics from the National and Kapodistrian University of Athens. His work focuses on architectural acoustics, computational acoustics, psychoacoustics, and environmental acoustics. He is currently working on the acoustics of monuments, modeling of noise barriers, audiovisual interaction, and soundscapes. Additionally, he participates as an acoustic consultant in various projects of public and private interest. He is a recipient of several research grants.

Massimo Garai

Massimo Garai is Full Professor at the Department of Industrial Engineering of the University of Bologna and head of the Applied Acoustics Lab. His work is focused on architectural acoustics, building acoustics, environmental acoustics, new measuring techniques in acoustics, properties of materials, and digital signal processing. He is currently working on modeling new metamaterials for acoustic applications, environmental sustainability, and applications of machine learning in acoustics. He is President of the Acoustics and Vibration commission of the Italian standardization body (UNI), convenor of two CEN working groups, and an ISO Expert for acoustics. He is a level 3 (max) expert in acoustics as certified by CICPND. He is one of the most cited authors of Applied Acoustics (2002–2005) and won best paper awards at the international congresses IBPC 2015, ISTD 2011, and ISTD 2009. He served as general chairman of the international congress ISTD 2017.

Georgios E. Stavroulakis

Georgios E. Stavroulakis is Professor and Director of the Institute for Computational Mechanics and Optimization at the Technical University of Crete. His work is focused on applications of optimization in mechanics, either for modeling of novel effects, such as unilateral contact problems, using nonsmooth optimization, or for optimal design and parameter identification tasks. He is currently working on modeling and identification of monuments, design of shunted piezocomposites with applications on vibration suppression, and application of artificial neural networks for the solution of engineering problems. He holds the Privatdozent title from the Technical University of Braunschweig and is an Honorary Professor at Jordan University of Science and Technology. He serves as managing board member of the Greek Association for Computational Mechanics (past president) and the Hellenic Society for Theoretical and Applied Mechanics (president).

Preface to "Advances in Architectural Acoustics"

Satisfactory acoustics is crucial for the ability of spaces, such as auditoriums and lecture rooms, to perform their primary function. The acoustics of dwellings and offices greatly affects our life quality, since we are all consciously or subconsciously aware of the sounds to which we are subjected to daily. Architectural acoustics, which encompasses room and building acoustics, is the scientific field that deals with these topics and can be defined as the study of the generation, propagation, and effects of sound in enclosures. Modeling techniques, as well as related acoustic theories for accurately calculating the sound field, have been the center of many of the major new developments. In addition, the image conveyed by a purely physical description of sound would be incomplete without regarding human perception; hence, the interrelation between objective stimuli and subjective sensations is a field of important studies.

A holistic approach in terms of research and practice is the optimum way for solving the perplexing problems that arise in the design or refurbishment of spaces, since current trends in contemporary architecture, such as transparency, openness, and preference for bare sound-reflecting surfaces, are continually pushing the very limits of functional acoustics. All recent advances in architectural acoustics are gathered in this Special Issue, and we hope to inspire researchers and acousticians to explore new directions in this age of scientific convergence.

Nikolaos M. Papadakis, Massimo Garai, and Georgios E. Stavroulakis
Editors

Editorial

Special Issue: Advances in Architectural Acoustics

Nikolaos M. Papadakis [1,2,*], Massimo Garai [3] and Georgios E. Stavroulakis [1]

1. Institute of Computational Mechanics and Optimization (Co.Mec.O), School of Production Engineering and Management, Technical University of Crete, 73100 Chania, Greece; gestavr@dpem.tuc.gr
2. Department of Music Technology and Acoustics, Hellenic Mediterranean University, 74100 Rethymno, Greece
3. Department of Industrial Engineering, University of Bologna, 40136 Bologna, Italy; massimo.garai@unibo.it
* Correspondence: nikpapadakis@isc.tuc.gr

Introduction

Satisfactory acoustics is crucial for the ability of spaces such as auditoriums and lecture rooms to perform their primary function. The acoustics of dwellings and offices greatly affects the quality of our life since we are all consciously or subconsciously aware of the sounds to which we are daily subjected. The aim of this special issue was to gather advances in architectural acoustics that hopefully could inspire researchers and acousticians to explore new directions in this age of scientific convergence and multidisciplinary cooperation.

The special issue was an exciting journey for us in which we had the opportunity to communicate with many people in the field from all over the world. Authors from Australia, Belgium, China, Finland, Greece, Italy, Japan, New Zealand, Spain, Sweden, Taiwan and the UK—including 19 universities and 4 acoustic firms and corporations—participated in this special issue.

Among the numerous submissions, 17 successfully passed the review process. For better presentation in this introductory text, the papers have been categorized as follows:

- Review studies;
- Historical Acoustics;
- Computational Acoustics;
- Design of concert or conference halls and open-plan offices;
- Miscellaneous (sound absorbers, listeners perception, machine learning).

Review Studies

1. Italian-Style Opera Houses: A Historical Review

In his work [1], D'Orazio investigated the historical development of Italian-style opera houses from the 16th century to the present day. Called "Italian" due to their origin, operas developed thanks to the mutual influence of the genre and the building characteristics. The acoustics of historical opera houses is now considered as intangible cultural heritage. The paper addresses the state-of-the-art literature—most of which is available in Italian—which can be driven easily by the sharing of historical and contemporary knowledge.

2. Review of Acoustic Sources: Alternatives to a Dodecahedron Speaker

In this study by Papadakis and Stavroulakis [2], fifteen acoustic sources alternative to a dodecahedron speaker are presented. Emphasis is placed on features such as omnidirectionality, repeatability, adequate sound pressure levels, even frequency response, accuracy in the measurement of acoustic parameters and the fulfillment of ISO 3382-1 requirements for sound sources. The collected data from this review can be used in many areas for the appropriate selection of an acoustic source according to the expected use.

Historical Acoustics

3. Measurements of Acoustical Parameters in the Ancient Open-Air Theatre of Tyndaris (Sicily, Italy)

The outcomes of a measurement campaign of acoustical parameters in the ancient theatre of Tyndaris (Sicily) are presented in this paper by Astolfi et al. [3]. The results show that the reverberation time and sound strength values were relatively low when compared with other theatres because of the lack of the original architectural element of the scaenae frons. When combining this effect with the obvious condition of an unroofed space, issues emerge in terms of applicability of the protocols recommended in the ISO standard.

4. Evolutionary Analysis of the Acoustics of the Baroque Church of San Luis De Los Franceses (Seville)

The church of San Luis de los Franceses, built by the Jesuits for their novitiate in Seville (Spain), is an example of a Baroque church with a central floor plan. The acoustics of this church were studied by Alberdi et al. [4] through in situ measurements and virtual models. The main objective was to analyze the evolution and perception of its sound field from the 18th to 21st centuries, considering the different audience distributions and sound sources as well as the modifications in furniture and coatings.

5. Historically Based Room Acoustic Analysis and Auralization of a Church in the 1470s

This paper by Autio et al. [5] describes the historical acoustics of an important abbey church in Sweden in the 1470s. A digital historical reconstruction was developed, liturgical material specific to this location was recorded and auralized and a room acoustic analysis was performed. The analysis was guided by the liturgical practices in the church and the monastic order connected to it.

Computational Acoustics

6. Potential of Room Acoustic Solver with the Plane-Wave Enriched Finite Element Method

A preliminary study on the partition of unity finite element method (PUFEM) as a room acoustic solver was presented by Okuzono et al. [6]. The PUFEM performance was examined against a standard FEM in a single room and a coupled room, including frequency-dependent complex impedance boundaries of Helmholtz resonator-type sound absorbers and porous sound absorbers. The results demonstrated that the PUFEM could accurately predict wideband frequency responses under a single coarse mesh with considerably fewer degrees of freedom than the standard FEM.

7. Time Domain Room Acoustic Solver with a Fourth-Order Explicit FEM Using Modified Time Integration

This paper by Yoshida et al. [7] presents a proposal of a time domain room acoustic solver using a novel fourth-order accurate explicit time domain finite element method (TD-FEM). The proposed method could use irregularly shaped elements whilst maintaining fourth-order accuracy in time without an additional computational complexity compared with the conventional method. The practicality of the method at kilohertz frequencies was presented via two numerical examples of acoustic simulations in a rectangular sound field, including complex sound diffusers and in a concert hall with a complex shape.

8. An Energy Model for the Calculation of Room Acoustic Parameters in Rectangular Rooms with Absorbent Ceilings

In this paper by Nilsson and Arvidsson [8], a statistical energy analysis (SEA) model was derived where a non-isotropic sound field was considered. The sound field was subdivided into a grazing and a non-grazing part, where the grazing part referred to waves propagating almost parallel to the suspended ceiling. A comparison with measurements was performed for a classroom configuration, which revealed that the new model agreed better with the measurements than the classical one (diffuse field).

9. Coherent Image Source Modeling of Sound Fields in Long Spaces with a Sound-Absorbing Ceiling

This paper by Min and Xu [9] presents a coherent image source model for a simple yet accurate prediction of the sound field in long enclosures with a sound-absorbing ceiling. In the proposed model, the reflections on the absorbent boundary were separated from those on the reflective ones when evaluating the reflection coefficients. The model was compared with the classic wave theory, an existing coherent image source model and a scale model experiment.

10. Power Response and Modal Decay Estimation of the Room Reflections from Spherical Microphone Array Measurements Using Eigenbeam Spatial Correlation Model

The application of the eigenbeam spatial correlation method in estimating the time frequency-dependent directional reflection powers and model decay times was presented by Bastine et al. [10]. The experimental results evaluated the application of the proposed technique for two rooms with distinct environments using their room impulse response measurements recorded by a spherical microphone array. The experimental observations proved that the proposed model is a promising tool in characterizing early and late reflections, which is beneficial in controlling the perceptual factors of room acoustics.

Design of Concert or Conference Halls and Open-Plan Offices

11. On the Sequence of Unmasked Reflections in Shoebox Concert Halls

This work by García Gómez et al. [11] is a tribute to the 90th anniversary of Sir Harold Marshall and his early innovative ideas of the acoustic signature of a hall. By analyzing the cross-sections of three concert halls, this study quantified the potential links between the architectural form of a hall, the resultant skeletal reflections and the properties of its acoustic signature. Whilst doing so, this study identified potential masking reflections through a visual and an analytical assessment of the skeletal reflections.

12. Comfort Distance—A Single-Number Quantity Describing Spatial Attenuation in Open-Plan Offices

This study by Hongisto and Keränen [12] introduces a new single-number quantity, the comfort distance r_C, that integrates the quantities of the A-weighted sound pressure level of speech, $L_{p,A,S,4m}$, and the spatial decay rate of speech, $D_{2,S}$, used in ISO 3382-3. The new quantity describes the distance from an omnidirectional loudspeaker where the A-weighted sound pressure level of normal speech falls below 45 dB. The study explains why the comfort criterion level is set to 45 dB and explores the comfort distances in 185 offices reported in previous studies.

13. A Trial Acoustic Improvement in a Lecture Hall with MPP Sound Absorbers and FDTD Acoustic Simulations

The feasibility and performance of micro-perforated panels (MPPs) when used as an acoustic treatment in lecture rooms were investigated in this work by Cingolani et al. [13]. Three different micro-perforated steel specimens were first designed following existing predictive models and then physically manufactured through 3D additive metal printing. Numerical simulations were carried out using a full-spectrum wave-based method: a finite-difference time-domain (FDTD) code was chosen to better handle time-dependent signals as the verbal communication. The outcomes of the process showed the influence of the acoustic treatment in terms of the reverberation time (T_{30}) and sound clarity (C_{50}).

14. Mechanism Analysis of the Influence of Seat Attributes on the Seat Dip Effect in Music Halls

In this paper [14], Min and Liao performed numerical simulations on the basis of the finite element method to study the influence of seat attributes (seat height, seat spacing and seat absorption) on the seat dip effect (SDE) and the corresponding mechanism. The mapping of the sound spatial distribution related to the SDE was employed to observe the behavior of sound between the seats. A mechanism analysis revealed that the SDE was highly associated with standing waves inside the seat gaps and with the "diffusion" effect on the grazing incident waves by the energy flow vortexes around the top surfaces of the seats.

Miscellaneous (Sound Absorbers, Listeners Perception, Machine Learning)

15. A Parametric Study of the Acoustic Performance of Resonant Absorbers Made of Micro-perforated Membranes and Perforated Panels

The paper by Pan and Martellotta [15] first investigated the reliability of prediction models for perforated and micro-perforated panels by a comparison with measured data. Subsequently, whilst taking advantage of a parametric optimization algorithm, it was shown how to design an absorber covering three octave bands from 500 Hz to 2 kHz with an average sound absorption coefficient of approximately 0.8. Such a solution might be conveniently realized whilst using optically transparent panels, which might offer extra value as they could ensure visual contact whilst remaining neutral in terms of the design.

16. Listeners Sensitivity to Different Locations of Diffusive Surfaces in Performance Spaces: The Case of a Shoebox Concert Hall

The effects of diffusive surfaces on the acoustic design parameters in a real shoebox concert hall with variable acoustics (Espace de Projection, IRCAM, Paris, France) were investigated by Shtrepi et al. [16]. Acoustic measurements were performed in six hall configurations by varying the location of the diffusive surfaces over the front, mid and rear part of the lateral walls whilst the other surfaces were maintained absorptive or reflective. Conventional ISO 3382 objective acoustic parameters were evaluated along with a subjective investigation performed by using the ABX method with auralization at two listening positions.

17. A Machine Learning Based Prediction Model for the Sound Absorption Coefficient of a Micro-Expanded Metal Mesh (MEMM)

The objective of this study by Tsay and Yeh [17] was to develop a prediction model for a MEMM via a machine learning approach. An experiment including 14 types of MEMM was first performed in a reverberation room measured according to ISO 354. To predict the sound absorption coefficient of the MEMM, the capability of three conventional models and three machine learning (ML) models of the supervised learning method were studied for the development of the prediction model.

Conclusions

The diversity and richness of the presented papers demonstrates the liveliness of the research in the field of architectural acoustics. In this exciting context, it is easy to guess that more novel ideas are yet to appear along with further research and new applications. We hope that this collection will serve as an inspiration for our fellow acousticians, especially the young ones, to explore new ways in architectural acoustics in the future.

Acknowledgments: We are grateful to all contributors who made this special issue a success. First of all, to the authors who trusted their work to us. Thank you all, one by one. We express our gratitude to the *Applied Sciences* editorial team for their effective communication and tireless work. We would like to personally thank our assistant editor, Enoch Li, for unlimited help, kindness and understanding. Enoch, thank you so much for your help and positive spirit, even after numerous exchanges of emails to resolve small or big issues. We also thank our initial assistant editor, Luca Shao. Finally, we extend our thanks to all reviewers, who made it possible to evaluate and select the excellent works presented in this special issue.

Conflicts of Interest: The authors declare no conflict of interest.

References

1. D'Orazio, D. Italian-Style Opera Houses: A Historical Review. *Appl. Sci.* **2020**, *10*, 4613. [CrossRef]
2. Papadakis, N.M.; Stavroulakis, G.E. Review of Acoustic Sources Alternatives to a Dodecahedron Speaker. *Appl. Sci.* **2019**, *9*, 3705. [CrossRef]
3. Astolfi, A.; Bo, E.; Aletta, F.; Shtrepi, L. Measurements of Acoustical Parameters in the Ancient Open-Air Theatre of Tyndaris (Sicily, Italy). *Appl. Sci.* **2020**, *10*, 5680. [CrossRef]
4. Alberdi, E.; Galindo, M.; León-Rodríguez, Á.L. Evolutionary Analysis of the Acoustics of the Baroque Church of San Luis de los Franceses (Seville). *Appl. Sci.* **2021**, *11*, 1402. [CrossRef]
5. Autio, H.; Barbagallo, M.; Ask, C.; Bard Hagberg, D.; Lindqvist Sandgren, E.; Strinnholm Lagergren, K. Historically Based Room Acoustic Analysis and Auralization of a Church in the 1470s. *Appl. Sci.* **2021**, *11*, 1586. [CrossRef]
6. Okuzono, T.; Mohamed, M.S.; Sakagami, K. Potential of room acoustic solver with plane-wave enriched finite element method. *Appl. Sci.* **2020**, *10*, 1969. [CrossRef]
7. Yoshida, T.; Okuzono, T.; Sakagami, K. Time domain room acoustic solver with fourth-order explicit FEM using modified time integration. *Appl. Sci.* **2020**, *10*, 3750. [CrossRef]
8. Nilsson, E.; Arvidsson, E. An energy model for the calculation of room acoustic parameters in rectangular rooms with absorbent ceilings. *Appl. Sci.* **2021**, *11*, 6607. [CrossRef]
9. Min, H.; Xu, K. Coherent Image Source Modeling of Sound Fields in Long Spaces with a Sound-Absorbing Ceiling. *Appl. Sci.* **2021**, *11*, 6743. [CrossRef]
10. Bastine, A.; Abhayapala, T.D.; Zhang, J.A. Power Response and Modal Decay Estimation of Room Reflections from Spherical Microphone Array Measurements Using Eigenbeam Spatial Correlation Model. *Appl. Sci.* **2021**, *11*, 7688. [CrossRef]
11. Gómez, J.Ó.G.; Wright, O.; van den Braak, B.; Sanz, J.; Kemp, L.; Hulland, T. On the Sequence of Unmasked Reflections in Shoebox Concert Halls. *Appl. Sci.* **2021**, *11*, 7798. [CrossRef]
12. Hongisto, V.; Keränen, J. Comfort Distance—A Single-Number Quantity Describing Spatial Attenuation in Open-Plan Offices. *Appl. Sci.* **2021**, *11*, 4596. [CrossRef]
13. Cingolani, M.; Fratoni, G.; Barbaresi, L.; D'orazio, D.; Hamilton, B.; Garai, M. A Trial Acoustic Improvement in a Lecture Hall with MPP Sound Absorbers and FDTD Acoustic Simulations. *Appl. Sci.* **2021**, *11*, 2445. [CrossRef]
14. Min, H.; Liao, Y. Mechanism Analysis of the Influence of Seat Attributes on the Seat Dip Effect in Music Halls. *Appl. Sci.* **2021**, *11*, 9768. [CrossRef]
15. Pan, L.; Martellotta, F. A parametric study of the acoustic performance of resonant absorbers made of micro-perforated membranes and perforated panels. *Appl. Sci.* **2020**, *10*, 1581. [CrossRef]
16. Shtrepi, L.; Di Blasio, S.; Astolfi, A. Listeners sensitivity to different locations of diffusive surfaces in performance spaces: The case of a shoebox concert hall. *Appl. Sci.* **2020**, *10*, 4370. [CrossRef]
17. Tsay, Y.-S.; Yeh, C.-Y. A Machine Learning Based Prediction Model for the Sound Absorption Coefficient of Micro-Expanded Metal Mesh (MEMM). *Appl. Sci.* **2020**, *10*, 7612. [CrossRef]

Review

Italian-Style Opera Houses: A Historical Review

Dario D'Orazio [†]

Department of Industrial Engineering, University of Bologna, 40136 Bologna, Italy; dario.dorazio@unibo.it; Tel.: +39-051-2090549
† Current address: Viale Risorgimento 2, 40136 Bologna, Italy.

Received: 18 May 2020; Accepted: 24 June 2020; Published: 3 July 2020

Abstract: Attending an opera involves a multi-sensory evaluation (acoustical, visual, and more), cultural background and other emotional parameters. The present work aims to investigate the historical development of Italian-style opera houses, from the 16th century until today. Called "Italian" due to their origin, they developed thanks to the mutual influence of the genre and the building characteristics. Furthermore, the acoustics of historical opera houses is now considered as intangible cultural heritage, so it should be known and preserved. The paper addressed the state-of-the-art literature—most of which was proposed in Italian—which can be driven easily by the sharing of historical and contemporary knowledge.

Keywords: acoustics; opera house; intangible cultural heritage

1. The Social-Historical Context

1.1. The Birth of the Genre: The Early Age of Opera House

The word Opera in Latin is the plural of opus, which means "act, performance". Thus, opera means the simultaneous act of a performer—including their voice and gestures—and music. Florence and its 16th-Century cultural influences are commonly accepted as the birthplace of Opera. This kind of performance acquired a semi-public dimension, such as the Teatro Mediceo in Florence (1586). After Florence, some North-Italian courts hosted opera composers and independent architectures were opened to the ruler and the court [1]. In the Venetian area, the Teatro Olimpico, Vicenza (1585) [2] were designed by A. Palladio (1508–1580); in the courts of Milan, the Teatro all'antica, Sabbioneta (1590) by V. Scamozzi (1548–1616) [3]; in Parma, the Teatro Farnese (1618, damaged during WWII and rebuilt) [4,5] by G. B. Aleotti (1546–1636); by the same architect, in Ferrara, the Teatro degli Intrepidi (1604, burned in 1679) [6]. In these early spaces designed for melodrama, one of the most significant aspects is the structural and typological background, deriving from the form of the Roman architecture, including the Roman theatre [7]. This latter was called 'Ancient theatre' until the 18th Century [8,9], in contrast with the 'new theatre' for opera.

The turning point in the opera-house history was 6 March 1637, when all social classes attended the inauguration of the San Cassiano theatre in Venice. The paying audience led to a redefinition of the theatre shape: it made it possible to plan the theatrical seasons and the related investments, thus building permanent structures. During some years, several similar theatres were built in Venice, often named by the nearest church: Ss. Giovanni e Paolo, 1638; Novissimo 1640; San Moise 1640; San Giovanni Grisostomo (now Malibran) 1678 [10]. These latter cited opera houses allowed the development of opera as we know it today. Indeed, these halls hosted the second generation of opera composers, F. Cavalli (1602–1676), A. Cesti (1623–1669), who exported the opera outside of Italy, respectively in France and in Austria; in the early 18th Century the operas of A. Vivaldi (1678–1741) and then the evolution of the so-called Neapolitan School—among others, D. Scarlatti (1685–1757), G.B. Pergolesi (1710–1736), G. Paisiello (1740–1816), D. Cimarosa (1749–1801). Before the demolition, some of these theatres were used until the 19-Century hosting the representation of G. Rossini (1792–1868) and G. Donizetti (1797–1848). In other words, they were used from the birth of melodramma in the early 17th Century to its standardisation in the 19th Century. On one hand, this means that the geometrical form was useful to visual and acoustic needs of the audience [11]. The Basilica form of the early court-theatres did not have satisfactory acoustic requirements for this novel kind of performance. On the other hand, these typologies of theatres could justify by hosting enough paying spectators the affordability of new developments/constructions for the opera manager. The role of the 17th-Century Venetian managers in reviewing the acoustics of opera should be acknowledged [12]: thanks to commercial documents is possible to know, today, all the details about orchestras, audience, representations [13,14].

The theatre of Ss Giovanni e Paolo (1638) was probably the earliest one hosting audience in the boxes [10,15] (see Figure 1). As a matter of fact, the early court theatres did not need a large attendance because they hosted the court only. Instead, the public Venetian theatres needed to increment the seating capacity and improve the visual conditions of the attending public. At the same time, the upper classes began to claim independent and private spaces, leading to the building of wooden partitions on the different tiers, creating the so-called "boxes". This spatial division reflected the social subdivision into classes. As cited by Venetian chronicler C. Ivanovich (1620–1689) [16]:

"Giravano d'intorno cinque file di loggie l'una sovraposta all'altra con parapetti avanti a balaustri di marmo ... Le due piú alte, e piú lontane file [di logge] erano ripiene di cittadinanza, nella terza sedevano i signori Scolari, e i nobili stranieri, il secondo come luogo piú degno era dei Sig. Rettori e de' Nobili veneti, e nel primo se ne stavano le gentildonne, e i principali gentilhuomini della cittá". (Five tiers of boxes went round, one on top of the other, with marble balustrades. The upmost and furthest two tiers [of boxes] were stuffed with ordinary citizens, in the third one sat Scholars and foreign nobility, the second was reserved to Rectors and noblemen of the Venetian region, and in the first sat the gentlewomen, and main gentlemen from the city).

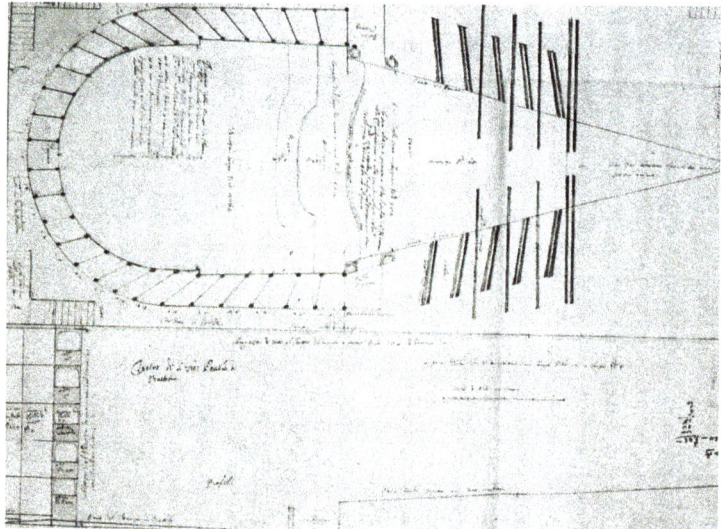

Figure 1. Plan of theatre of Ss Giovanni e Paolo, in Venice (1638).

1.2. The Influence of Italian Opera House Outside of Italy

In 1637, the opera Andromeda by F. Mannelli (1595–1697) marked the birth of Libretto. This was a booklet which included the score of the opera, the parts of the soloist and a very detailed description of the scene and of the scenic actions. This allowed to reply the same opera in different places. By this way, the opera was exported outside of Italy. In 1645, the first opera in Paris, La Finta Pazza, was presented. Some Italian composers, such as the aforementioned Cavalli or Giovanni Battista Lulli (1632–1687)—also know as Jean-Baptiste Lully—influenced the early French opera composers, such J.-P. Rameau (1683–1764).

As well as the contents, i.e., the opera, Italy exported also the container, i.e., the opera house. As a paradigmatic example, the Galli-Bibiena family of architects moved from Italy to European Courts, working as scenic artists and opera-house architects [17]. They designed around the Europe: Grosse Hoftheater, Wien (1708); Great Theatre at Nancy (1709); Teatro Filarmonico, Verona (1719, burned in 1749); Teatro Alibert, Rome (1720); Mannheim Opernhaus (1719, burned in 1795); Royal Theatre, Mantua (1731, burned in 1781); Markgräfliches Opernhaus, Bayreuth (1748) [18]; Dresden opera (1750, burned in 1849); Ópera do Tejo, Lisbon (1752, ruined in 1755); Teatro Rossini, Lugo (1760) [19,20]; Teatro Comunale, Bologna (1763) [21–23]; Teatro de Quattro Cavalieri, Pavia (now Fraschini, 1773) [24,25]; Scientific theatre, Mantua (1775, now Bibiena), and many others [26].

The development of the opera house building was closely related to the development of the opera and the society. In Italy, between the late 17th and the 19th Century, there was a debate on the form of the opera house [27]. The first essay dealing with the so-called "Italian theatre" was written by the architect F. Carini Motta (death 1699) [28]. The author recognised the different typologies: the 16th-Century Court theatre with steps, with boxes or galleries, with boxes joined or not joined with the proscenium. Regarding this latter one, he proposed two different models of the horseshoe plan typology. In order to increase the width of the proscenium and increase the visibility of the stage, some authors—as F. Algarotti (1712–1764) [29], F. Riccati (1718–1790) [30], A. Memmo (1729–1793) [31]—proposed the bell-shape plan, the one used by the already mentioned Galli-Bibbiena architects in all they works. An alternative proposal was the elliptic shape, proposed by theorists [32], but also used by C. Morelli (1732–1812) [33]—one of the most influential architect of the Papal State—for building the theatre of Imola (1780, now Stignani Theatre), or by A. Petrocchi for the theatre of Lugo (1758–1760, now Rossini Theatre). The instances of the Enlightenment [34] were followed by theorists Enea Arnaldi (1716–1794) [35] and Francesco Milizia (1725–1798) [36]. Influenced by some of the French designs [37,38], they published on the topic of the "ideal theatre". Milizia exalted the semi-circular ancient theatre, where all the spectators were able to see and hear properly, while the modern theatre with boxes did not allow satisfying visuals to the whole public [39]. T. C. Beccega (17??–18??) tried to match the Graeco-Roman architecture to the construction of the Modern Italian Theatre [40]. He followed the Vitruvius's theory and, at the same time, he tried to develop a theatre standard, whose shape was a semi-circle with elongated extremities. Unfortunately, none of these ideal designs were built. A virtual simulation of V. Ferrarese's theatre, based on Milizia's ideal theatre, has been proposed and discussed in recent years [41].

Until the 20th Century, the theatres were built by wooden structures, with very few exceptions built with masonry—such as the aforementioned Teatro Comunale in Bologna (1763). As seen, theatres were periodically ruined by fires. This might be the reason that triggered the evolutionary process of the theatrical form [42]. In the early 19th Century, the form of the opera house should be considered already evolved, with the works of Piermarini—who designed La Scala theatre in Milan—and Niccolini [43]—who rebuilt the San Carlo theatre in Naples. This form was defined around some fixed points: the materials, the horse-shoe shape, the dimension (four tiers of boxes and a gallery with very few exceptions, a cavea volume of about 6000–10,000 m^2) [44].

Figure 2 shows the evolution of theatres in Italy: after the initial period of court theatres (1585–1637), the evolution of the form as it has been discussed, and the 19th century provided the dissemination of opera house. In order to host the main representations, in each city a mid- or large-sized theatre was built. According to Prodi et al. [44], a 'mid-sized' group means a volume of the main hall between 3000 m^3 and 9000 m^3, corresponding to a current occupancy of 800–1000 people. A 'large-sized' group means a volume larger than 9000 m^3, corresponding to more than 1300 seats.

As an example, in Venice the small 17th century theatres were replaced by the La Fenice theatre, which was built for the first time in 1792 [31] and rebuilt in 1837 [45]. Technical essays of this period seems to confirm a well-consolidated technical knowledge [46,47]. The authors revealed the knowledge of the European discussion on the opera-house acoustics. To confirm this, the design of the last large 19th century opera house, the Vittorio Emanuele Theatre in Palermo (now Teatro Massimo, 1897), shows the influences of the contemporary great opera houses of the time [48].

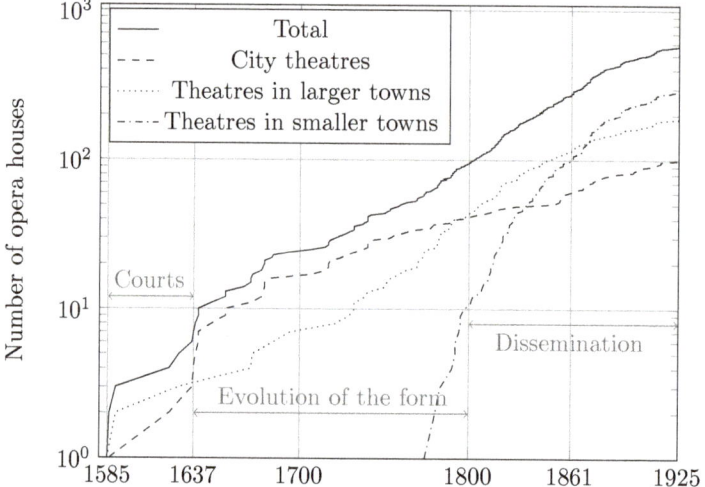

Figure 2. Progression of opera houses in Italy from 1585 to 1925.

In the United Kingdom, the early opera developed based on Elizabethan theatre, through composers such as H. Purcell (1659–1695) and G. F. Händel (1685–1759). Nevertheless, the English opera did not have a 19th century development comparable the other European tradition. Until the middle of 19th century the Covent Garden [49] theatre hosted performances in baroque style. There was a coexistence of English, French, and Italian tradition. Concerning the design of theatres, it should be remarked Saunders's work [50], who surveyed the dimensions of the ancient and the modern theatres, founding the most advantageous form for the voice and sight. His works was rediscussed in a contemporary way in several papers on Pre-Sabinian design criteria [51–53]. After the birth of architectural acoustics as a science, the engineer H. Bagenal (1888–1979) and the physicist A. Wood (1879–1950) collected the first pioneristic acoustic survey on theatres and opera houses [54].

In the aforementioned French tradition, after the first one lead by Rameau, the new National-opera, called Comédie Francaise and the Italian-style opera, the Comédie-Italienne, coexisted. The Enlightenment debate and, then the French Grand Opéra, moved away from Italian tradition, both in opera-style and in buildings. About the first point, the late 18th-Century theorists proposed designs where the architecture reflected the social ideals of equity, in contrast with the social in-equities of Italian opera houses. Although there were many designs, the only built theatre was the Musical Theatre of Besançon by C-J. Ledoux (1736–1806) [37]. About the second point, the Grand-Opéra—which was usually dated back to La Muette au Portici (1828) written by D. F. Auber (1782–1871)—needs larger opera houses. The prototype of this new kind of opera house was the new opera house in Paris (1875), designed by C. Garnier (1825–1898). The architect declared intentionally—and probably provocatively declared— to do not follow any acoustic instance in his work [55]:

"je n'ai adopté aucun principe [...] je ne me suis basé sur aucune théorie [...] c'est du hasard seul que j'attends ou l'insuccés ou la réussite" (I have not adopted any principle [...] I did not base myself on any theory [...], I wait for either failure or success by mere chance).

He designed tiers of gallieries instead of tiers of boxes, aiming to democratise the opera attendance. The fact remains that, after the Paris Opéra (now Opéra Garnier)—and after the almost contemporary Bayreuther Festspielhaus (1876) [56,57], the gallieries gradually replaced the boxes in opera houses, allowing a better visual and a larger attendance of public.

In the Austrian–German culture, the 18th-Century opera was characterised by C W. Gluck (1714–1787) and then W.A. Mozart (1756–1791), both in a dialectical relationship between a new opera in German and the Italian opera, especially the Neapolitan school. For instance, after a century and a half, during the years 1783–1786, the Performance Calendar for the National Court Theatres in Vienna provided only operas wrote in Italian [58]. The 'national' German opera was developed in the 19th Century, thanks to the work of C. M. von Weber (1786–1826) and, then, R. Wagner (1813–1883). This new German opera needed the building of new opera houses, such as the Dresden oper, by G. Semper (1803–1879) and the already mentioned Bayreuth Festspielhaus. It should be noted that some features, such as the stage inclination and the orchestra pit position—lower than the audience in the stalls—were previously proposed by theorists [40] and by architects—such as in the aforementioned Ledoux's Besançon theatre (1784, burned in 1958).

In the late 19th century the opera was exported to the Americas. The most relevant opera houses were the Academy of Music (1854, demolished in 1926) or the Metropolitan Opera (also known as MET—1883, rebuilt in 1966) in New York City, and the Teatro Colon (1908) in Buenos Aires [59]. These opera houses were built as a synthesis of what was done during the previous two centuries. For instance, in the old MET, there were three tiers of private boxes, where New York's powerful new industrial families—which also provided for building this new theatre—could display their wealth and establish their social prominence (see Figure 3). The interest of American people for the opera led some composers, such as G. Puccini (1858–1924), to write original operas inspired by the American tradition, such as Madama Butterfly (1904), Fanciulla del West (1914). The first representations of the latter one and the following Gianni Schicchi (1918) took place at Metropolitan Theatre in New York City. At the beginning of the 20th Century, the USA needed new theatres, and in one of these new opera house took place the first acoustical correction for such kind of spaces. It was made by W. C. Sabine (1868–1919), recognised as the founder of the architectural acoustics, who proposed to place a canopy on the vault in order to modify the reflections from ceiling [60]. Despite this, the theatre was demolished, due to its poor acoustics.

Figure 3. A concert by pianist Josef Hofmann in the old Metropolitan Opera House in 1936. National Archives at College Park/Public domain.

1.3. The 20th Century

The 19th century was also the age of the new houses of parliament, due to the new political asset of European Monarchies. Indeed the word "Parliament" appeared at the time of the French constitutional monarchy of 1830–1848: the French parliament house (Palais Bourbon) was renovated in 1828, the English one (Palace of Westminster) was rebuilt in 1847–52, the Italian one (Palazzo Montecitorio) was built in 1871 and renovated in 1903 by E. Basile (1857–1932)—who was the architect of the already mentioned Teatro Massimo in Palermo. The intelligibility of a single speaker in a hall occupied by about one thousand people became a new instance for the acousticians of this period [47].

The 20th Century marked also the birth of Cinema, and—thanks to the 'new' electronic features—the gradual replacement of a real orchestra by a reproduced music. Acousticians seemed to curb their interesting in opera house. Moreover, the sound of singers and orchestra started to be enhanced by electro-acoustics [61].

Opera survived, but was in need of some changes. On one hand, the performances required more spaces—this was one of the reason of the rebuilt of 'old' MET into the new hall at Lincoln Center in the 1960s [62] (see the timeline in Figure 4). On the other hand, the opera houses became 'iconic' buildings, to the partial disadvantage of their acoustic function. This was the largely debated case of Sydney Opera house, whose long building process (1957–1973) forced the acoustic to deal with new instances in architectural acoustics [63].

During 1950s and early 1960s a comparative analysis between subjective and objective parameter were used by L. L. Beranek [64] as a preliminary study in order to design the acoustics of the new Philharmonic Hall (rebuilt during the 1970s as Avery Fisher Hall, from 2015 renamed as David Geffen Hall), in the complex of Lincoln Center. The acoustic design was optimised for the acoustic intimacy, which seemed to be the most important subjective attribute in the previous survey. As a matter of fact, the acoustics of the new Philharmonic hall was inadequate and in the hall was renovated. The failure was due to several reasons, most of which are not imputable to Beranek's design.

In the early 1990s, L. Beranek—who left the acoustic design activity after the Philarmonic Hall failure [65]—was asked by Hidaka to collaborate for the acoustic design of the new Tokyo Opera House [66]. Several international opera houses were surveyed again [67], and new acoustic criteria were taken into account [68].

In Italy, in 1992 the fire destroyed Petruzzelli theatre in Bari. The same fate, in 1996, for La Fenice theatre in Venice. After these unfortunate events, Italian scholars were committed to recognise the acoustics of Italian Historical Opera House as intangible cultural heritage [69–72].

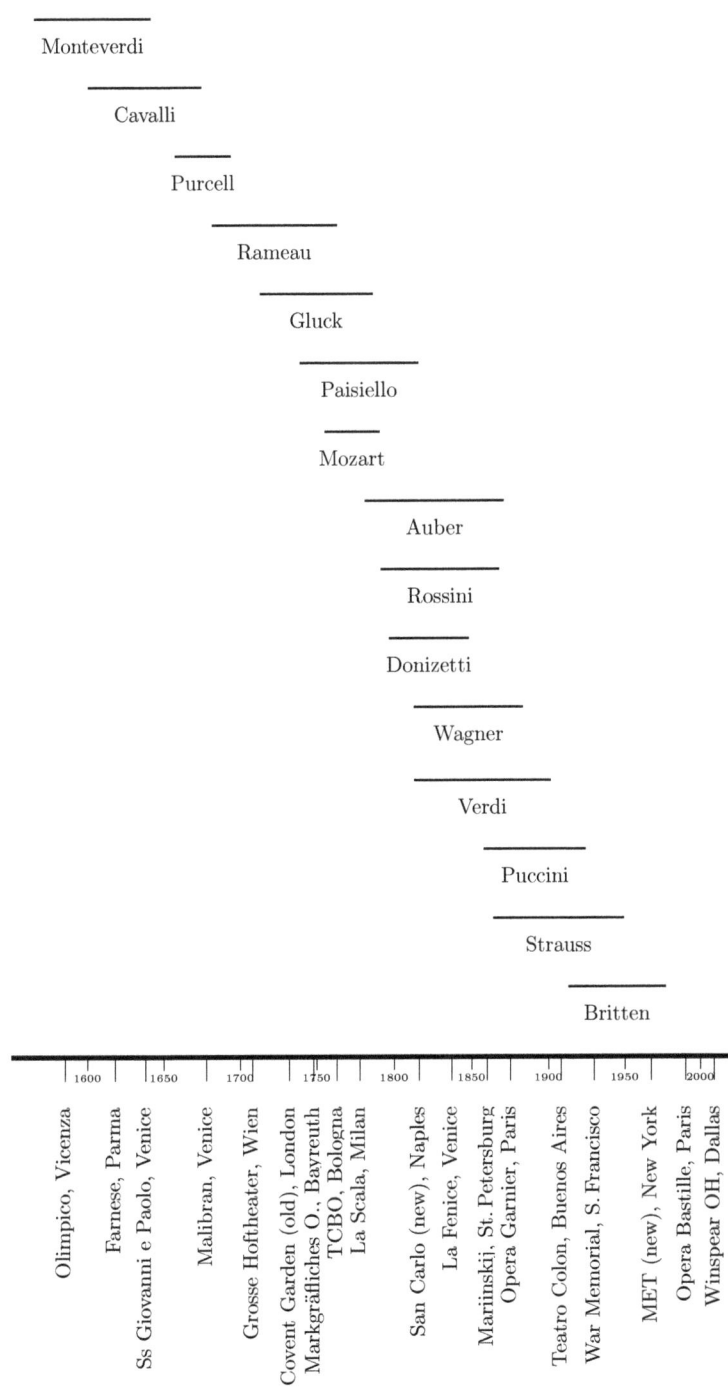

Figure 4. Timeline of some of mentioned opera houses of and life span of some of mentioned composers.

2. Italian-Style Opera Houses: A Chronotaxis

2.1. Dissemination

Depending on the construction year, Italian opera houses can be divided into four categories:

1585–1637 Court theatres of Northern Italy, unopened to the public. They host the early operas, played by instruments and voices different from the today ones. Despite this, these early-age theatres get almost all the peculiarities of the future opera house. For this reason, they were exhaustively measured and simulated by scholars [5].

1637–1800 Public theatres, opened earlier in Venice and later in the main cities. This period marked the evolution of the form. The scholars discussed and wrote on the best shape, the right dimension and the proper materials [27]. Opera houses were built and quickly demolished, often due to fires. This may be viewed as a genetic selection: only the best-sounding theatres were rebuilt using the same materials and techniques, otherwise different ways to build were tested [42]. The interest of people in opera increased during this period, leading to the need for larger opera houses. The goal of theatre-designers was to assure the best stage-visibility for a largest number of people. This means having the widest proscenium arch with respect to the hall dimensions. The horse-shoe shape emerged as the best compromise between these two reasons. Most of the opera houses in the largest cities were build in this period: la Pergola in Florence (1656) [73], Argentina in Roma (1732), San Carlo in Naples (1737, burned in 1816 and rebuilt in 1817) [74,75], Regio in Torino (1740, burned in 1936), Comunale in Bologna (1763), [21,23,73,76,77], La Scala in Milan (1778) [73,78,79], La Fenice in Venice (1792, burned in 1836 and rebuilt in 1837, burned again in 1996) [80].

1800–1925 Until 1861, Italy was still divided into pre-unitarian states. In this period the theatrical form was well defined, and it was replied in order to have a opera house in each town. As shown in Figure 5, this dissemination process varied depending on pre-unitarian state, earlier in the Kingdom of Lombardy–Venetia, then in the Grand Duchy of Tuscany and in the Papal State, later in the Kingdom of Two-Sicilies and in the Kingdom of Sardinia. There were social and cultural reasons to this temporal misalignment, but after the Italian unification the differences were diminished. In Italy, opera played a role of unification between people who spoke slightly different languages and had somewhat different cultures. For instance, during the Risorgimento process, the sentence "W V.E.R.D.I." was used also as acronym, meaning "W Vittorio Emanuele, King of Italy". Furthermore, some G. Verdi's operas, such as *Nabucco* (1842) or *I Lombardi alla Prima Crociata* (1843), hidden the revolutionary instances of the Lombard people which was under the Austrian government until 1861. It should be noted the case of peripheral regions of the Papal State—corresponding today to Umbria, Marche and Romagna regions—in which each city, large or little town was provided by its own opera house. A huge number of these small-sized theatres still exist. Some of them were surveyed by local studies, such as in the case of Campania [81], Romagna [20], Marche [82], Puglia [83], and Veneto [84]. Due to the well-defined form, architects repeated the same project with few variations. Indeed, theatres built in the same geographical area and in the same years often shared the same workers and same building techniques. Due to the large number of small/mid-sized opera house, a "regional" typology can be identified.

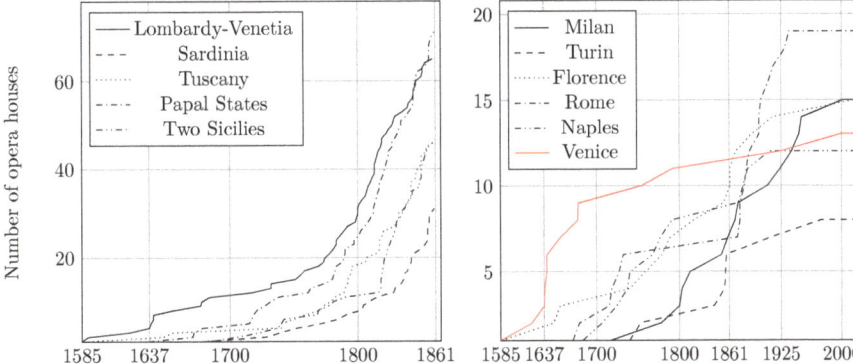

Figure 5. Dissemination of opera houses in the pre-unitarian states (**left** figure) and in the corresponding most important cities (**right** figure).

Moreover, several large-sized theatre were built at the end of the 19th century in Rome (from 1871) and in the South of Italy. Among others, they should be mentioned the Costanzi theatre in the new Rome (1880, renowned in 1928) [85], the Bellini theatre in Catania (1890) [86], the Massimo theatre in Palermo (1897) which is still today the largest Italian opera house—and the Petruzzelli theatre in Bari (1902, burned in 1992). The design of these latter opera houses was influenced by the instances of most recent European opera houses. There were both boxes and galleries—two tiers of galleries instead of the Loggione. The scenic arch was much wider than previously, as in the case of Massimo theatre in Figure 6. Indeed, the Verdian orchestra was larger than the previous tradition [87,88] an the soloists needed to increase their sound strength. In order to reach the proper balance, architects extended the stage, through the so-called proscenium, below the arch. This latter was tilted for two reasons: to reflect the singer's voice on the audience and avoid the flutter echo effects for the orchestra. It should be noted that in some cases, such as the Costanzi and Massimo theatres, the orchestra floor can be also lowered, as "the most recent German theatres" [48].

Figure 6. On the left: Proscenium arch of Massimo Theatre in Palermo; Reproduced with permission from Teatro Massimo, Copyright Franco Lannino, 2020. On the right: plan of Massimo theatre in Palermo A: proscenium, B: orchestra; $btsd$: proscenium arch; x, y: stage boxes.

1925–today Opera houses were replaced by cinemas. Often, opera houses were converted to cinemas and then progressively abandoned. Sometimes, such in the case of the Romagna in the 1970–1980s [89], the refurbishing of a group of these abandoned opera houses was managed by local government. Otherwise, the restoration has been based on a public debate [90]. Several opera houses, damaged by bombs during the WWII, had to be rebuilt. Most of the remaining historical theatres needed refurbishments due to age, or adjustments to the needs of new kind of performances. These two points involved the acoustician debate, and will be treated in the next sections.

2.2. Refurbishments

Due to reasons which will be discussed in a next section, during the 20th Century almost all opera houses were modified in order to create the orchestra pit. In some cases, the orchestra pit replaced the proscenium area, so the stage was reduced. This choice influenced the acoustics of the opera house, because the soloist is moved back and the strength of his/her voice decreases [91]. Such a kind of intervention was usually done without an acoustic consultancy.

To the author's knowledge, the early acoustic consultancy in this field was done by G. Sacerdote and C. Bordone, concerning the renovation of De La Sena theatre of Feltre [15]. Afterwards, case studies were documented by Cocchi et al. [92,93]; Tronchin and Farolfi—theatre of Gradisca d'Isonzo [94]; Pisani [95]; Facondini [96]—Rossini theatre of Pesaro; Fausti et al. [97]—Valli theatre of Reggio Emilia. It was with the burning of the La Fenice theatre [98] that the interest in this field increased. The acoustics of historical opera house was recognized as intangible culturale heritage [71,72].

When some materials are changed or removed, the acoustic quality can dramatically change. This was the case of wooden acoustic cavities in the stalls of Alighieri theatre in Ravenna, removed during refurbishment works [99]. A more recent works in the same theatre was focused on the replacement of the wooden stage [91]. Both of these interventions have influenced the acoustic quality of the theatre, reducing the reverberation.

2.3. New Buildings

During the 19th Century, as shown in Figure 2, opera houses were disseminated. This means that operas were represented in larger cities as well as in little towns. Nowadays, it is still the same. Table 1 shows the statistics of season 2018–2019 for the lyrical performances collected for country/city for various national contexts. In some countries—e.g., USA, Russia and France—the performances are concentrated in one or two main cities. In other countries, e.g., Germany or Italy, the attendance is more distributed [100]. As a consequence, opera houses were still designed and built, or re-built, in the last decades in Italy.

Table 1. Opera performances vs. Cities. in the season 2018–2019. Data from operabase.com.

	USA		Russia		Germany		France		Italy	
1	New York	277	Moscow	616	Berlin	625	Paris	429	Milan	170
2	San Francisco	86	St Petersburg	539	Hamburg	419	Lyon	87	Rome	154
3	Philadelphia	70	Ekaterinburg	94	München	396	Marseille	60	Venice	153
4	Chicago	69	Novosibirsk	71	Dresden	357	Strasbourg	55	Trieste	137
5	Houston	63	Samara	69	Erfurt	194	Toulouse	46	Florence	132
...	...		...		...		...		...	
10	Seattle	45	Rostov-on-Don	41	Hannover	161	Nice	31	Bologna	90

Some opera houses were burned during the 20th Century [101], such as the aforementioned cases of Regio theatre in Turin (1936) [102], Petruzzelli in Bari (1992), La Fenice in Venice (1996). The Regio theatre was rebuilt in 1973, with the new design of C. Mollino and acoustic consultancy of G. Sacerdote [103]. La Fenice theatre and Petruzzelli theatres were rebuilt keeping the historical design [98]. They were reopened, respectively in 2004 [104] and 2009 [105].

Other opera houses were damaged during WWII: this was the case for the Carlo Felice theatre in Genua, damaged in 1943 and rebuilt with a new design of architect A. Rossi in 1991; the Teatro delle Muse in Ancona, damaged in 1943 and rebuilt in 2002, acoustic consultant: A. Cocchi [106]; the Galli theatre in Rimini was damaged in 1943 and reopened after a long debate, using the original L. Poletti's design, in 2018 [107]. The reconstruction of this latter one was followed by a large group of acoustic consultants—including the author of the present work [108].

Moreover, new opera houses were built: the Teatro Lirico opened in 1993, acoustic consultant: R. Pompoli [109]; the Arcimboldi theatre in Milan, inaugurated in 2001—architect V. Gregotti and acoustic consultant D. Commins [110]; and the new Teatro del Maggio Musicale Fiorentino in Florence (2011), whose acoustic consultant was J. Reinhold of MüllerBBM [111].

Finally, it should be mentioned the renovation of La Scala theatre in Milan during the years 2002–2004. The architect of the new fly tower was M. Botta, the acoustic consultant was H. Arau [78].

3. A Taxonomy of Surveys, and Some Unresolved Instances

Early measurements in Italian opera houses were done by Faggiani in 1930s [112,113]. After the WWII, it should be mentioned the pioneer work of G. Sacerdote, C. Bordone and the workgroup of National Institute of Standards and Technology "Galileo Ferraris" in Turin [15]. On the academic front, many research practices were based at the University of Bologna between 1970s and 1980s [114,115]. The amount of acoustic measures intensified in the 1990s, involving further research groups [86,116–118].

After a successful season of meetings in the 1990s, the community of scholars found standard procedures to qualify the acoustics of historical opera house [119,120], collected in the so-called *Ferrara charter*. For instance, the Ferrara charter stated the use of dodecadrical sound source [121] instead of alternative sound sources [122] used in some early surveys [73]. Moreover, the charter stated how to place the source on the stage and in the pit [123], and how many receivers must be measured in the stalls and in the boxes [124]. By this way, the measurement campaigns done in the years between 2000 and 2010 returned comparable results, which were collected by Prodi et al. [44]. This work is also the source for measurement results and theathers' data, such as volume or occupation.

Tables 2–4 report a literature review of measurement campaigns, aiming to collect surveys on Italian-style opera houses (Tables 2 and 3), including the ones built outside of Italy (Table 4). The aforementioned classification of theatre by Prodi et al. [44] was used, based on volume of the main hall: small- ($V < 3000 \, \text{m}^3$), mid- ($3000 < V < 9000 \, \text{m}^3$), and large-sized theatres ($V > 9000 \, \text{m}^3$).

Table 2. Surveyed Italian-style opera houses built in Pre-unitarian Italian States (1585–1860). The "measurement" column specifies which kind of measurements were performed: T means reverberation time, early decay time, and more; C means, e.g., early-to-energy ratio, sound clarity, definition; G means Sound Strength, G_E, G_L using a calibrated sound source; S means spaciousness criteria, such as Inter-Aural Crosscorrelation Coefficient ($IACC$), Cosine-Lateral fraction ($LFC80$), and more.

Theatre(s)	Year	Size	Measurement	Repository
Olimpico, Vicenza	1585	mid	TCS [4] TCGS [5]	CAD, IRs [125]
All'antica, Sabbioneta	1590	small	TCS [4] TCGS [5]	CAD, IRs [125]
Farnese, Parma	1628	large	TCS [4] TCGS [5]	CAD, IRs [125]
La Pergola, Florence	1656	mid	TCS [73]	
Malibran, Venice	1678	mid	TG [126]	
del Pavone, Perugia	1717	mid	TCGS [44]	
Filarmonico, Verona	1732	mid	TC [44]	
San Carlo, Napoli	1737	large	TCG [44,74,75]	
Verdi, Padova	1751	mid	TC [127]	
Accademico, Castelfranco V.	1758	small	TC [44,128]	
Rossini, Lugo	1761	small	TCGS [20,129]	
Comunale, Bologna	1763	mid	TCGS [21,73] TCGS [22,44,77] TCGS [23]	CAD [130]
Court, Caserta	1769	small	TC [131]	
Fraschini, Pavia	1773	mid	TCGS [25,44]	
Mercadante, Napoli	1777	mid	TC [26,44]	
La Scala, Milano	1778	large	TCGS [68] TCS [73] T [78] TCGS [44,79]	
Morlacchi, Perugia	1781	mid	TCGS [44]	
Stignani, Imola	1782	small	TCGS [20,129]	
Zandonai, Rovereto	1786	mid	T [44]	
Masini, Faenza	1788	small	TCGS [20,129]	
dell'Aquila, Fermo	1790	mid	TS [77]	
Abbado, Ferrara	1797	mid	TCGS [44,132]	
Verdi, Trieste	1801	mid	TCS [44,84,133]	
Grande, Brescia	1810	mid	TCG [44,134]	
Sociale, Como	1813	mid	TC [25,44]	
Rossini, Pesaro	1818	mid	TC [96]	
Sociale, Trento	1819	mid	TC [127]	
Verdi, S. Severo	1819	mid	TCGS [83]	
Garibaldi, Gallipoli	1825	small	TCGS [83]	
Regio, Parma	1829	large	TGCS [44,135]	
Bonci, Cesena	1846	mid	TC [136] TCG [44,77] TCGS [20,129]	
Petrarca, Arezzo	1833	mid	TC [137]	
Milanollo, Savigliano	1836	small	TC [138]	
Marrucino, Chieti	1818	small	TC [139]	
La Fenice, Venezia	1837	mid	TCGS [80]	
Garibaldi, Lucera	1837	small	TCGS [83]	
Traetta, Bitonto	1838	small	TCGS [83]	
Civico, Tortona	1838	small	TC [138]	
Dragoni, Meldola	1838	small	TCGS [20,129]	
Fenaroli, Lanciano	1841	small	TC [139]	
Pavarotti, Modena	1841	mid	[44]	
di Bartolo, Buti	1842	small	T [44,140]	
Goldoni, Bagnacavallo	1845	small	TCGS [20,129]	
Alighieri, Ravenna	1852	mid	TCGS [20,91,129]	
V. Emanuele, Messina	1852	mid	[94,141]	
Verdi, Florence	1854	large	TCS [73]	
Piccinini, Bari	1854	mid	TCGS [44,83,142]	
Baudi, Selve in Vigone	1855	small	TC [138]	
Valli, Reggio Emilia	1857	mid	TCS [44,97] [143]	
Alfieri, Asti	1860	mid	TCG [44,144]	
Chiari, Cervia	1860	small	TCGS [20,129]	

Table 3. Surveyed Italian-style opera houses built in the unitarian Italy (1861–2020). Rebuildings are indicated by brackets. The "measurement" column specify which kind of measurements were performed: T means reverberation time, early decay time, and more; C means, e.g., early-to-energy ratio, sound clarity, definition; G means Sound Strength, G_E, G_L using a calibrated sound source; S means spaciousness criteria, such as $IACC$, $LFC80$, and more.

Theatre(s)	Year	Size	Measurement	Repository
Guerrini, Benevento	1862	small	TC [44,127]	
Menotti, Spoleto	1864	mid	TC [93]	
Comunale, Cesenatico	1865	small	TCGS [20,129]	
Verdi, Pisa	1867	mid	T [44,140]	
Mercadante, Cerignola	1868	mid	TCGS [83]	
Verdi, Busseto	1868	small	T [114]	
Del Monaco, Treviso	1869	mid	TCS [44,84,94]	
Verdi, Salerno	1872	mid	[44,75,145]	
Curci, Barletta	1872	small	TCGS [83]	
Paisiello, Lecce	1872	small	TCGS [44,83,142]	
R. Margherita, Caltanissetta	1875	small	TC [127]	
Rossetti, Trieste	1878	mid	TC [127]	
Comunale, Russi	1887	small	TCGS [20,129]	
Bellini, Catania	1890	large	TC [86,127]	
Van Vesterhout, Mola	1896	small	TCGS [83]	
Massimo, Palermo	1897	large	TC [127]	
Sociale, Rovigo	1904	mid	TCS [44,84]	
Comunale, Nardó	1908	small	TCGS [83]	
Civico, Schio	1909	small	TC [44,84]	
Opera, Roma	(1928)	large	TCG [44,85]	
Comunale, Adria	1935	mid	TCS [44,84]	
Duse, Bologna	(1943)	mid	TCG [146]	CAD, IRs [147]
TEA, Bologna	1975	large	TCGS [73]	
Carlo Felice, Genova	(1991)	large	TC [44]	
Lirico, Cagliari	1993	large	TCG [44,109]	
Arcimboldi, Milano	2001	large	TCGS [44,110]	
La Fenice, Venezia	(2003)	mid	TCS [44,84,133]	
Garibaldi, Bisceglie	(2003)	mid	TCGS [83]	
Verdi, Brindisi	(2006)	large	TCGS [83]	
Comunale, Gradisca d'Isonzo	(2009)	mid	TC [44,148]	
Petruzzelli, Bari	(2009)	large	TC [105]	
Maggio Musicale Fiorentino	2011	large	TC [111]	
Eschilo, Gela	(2013)	small	TC [149]	
Civico, Schio	(2014)	small	TC [90]	
Galli, Rimini	(2018)	mid	TC [107,108]	

However, there are still some unresolved instances. The first one concerns the setup of the fly tower during the measurements. It is well known that the presence or absence of draperies on the stage can influence significantly the sound behaviour at listeners' position [150,151]. Moreover, the sound source position in a coupled space can influence the sound energy decay [129,146,152]. The Ferrara-charter procedure states two positions of the sound source on the stage—the one in the fore-stage, the other one in the back stage—in order to 'average' the effects of coupling between the fly-tower and the cavea. Moreover, the Ferrara charter states a minimum amount of drapes in the fly tower—about 400 m^2 in case of mid-sized opera house—in order to have a 'acoustic' scene comparable to the opera performance. Indeed, when the amount of drapes increases, the sound energy in the main hall decreases and the opposite one. Some scholars tried to measure the fly tower and the main hall when they are separated by the firedoor, some others measure the audience by closing the scene curtains [96].

The second unresolved point concerns the measure of spaciousness. Indeed, the result of spatial measurement depends on the technologies used in the measurements: different microphones can provide different results [153]. Moreover, the spaciousness varies considerably if it measured in the center of the stalls, or near the side walls; if the receiver is in the boxes, it may vary if it is placed close to the edge or inside the box. Spatial measurements [154,155] could become a helpful tool to resolve this ambiguity [156].

Table 4. Surveyed Italian-style opera houses built outside Italy.

Ref.	Theatre(s)	Year	Country
[18]	Bayreuth Markgräfliches	1750	Germany
[18]	L'Opéra Royal Versailles	1770	France
[157]	Grand Theatre, Bordeaux	1780	France
[158]	S. Carlos, Lisbon	1793	Portugal
[159]	Palace Theatre, Archangelskoje	1818	Russia
[160]	Bolshoi, Moscow	1825	Russia
[161]	Teatro Principal, Valencia	1832	Spain
[162]	Opera House, Wroclaw	1841	Poland
[158]	Donna Maria II, Lisbon	1846	Portugal
[160]	Mariinsky, St. Petersburg	1860	Russia
[160]	Opera House, Saratov	1860	Russia
[158]	Trindade, Lisbon	1867	Portugal
[160]	Opera House, Voronezh	1870	Russia
[158]	Sá de Miranda, Viana do Castelo	1879	Portugal
[68]	Hungarian State Opera House	1884	Hungary
[158]	Coliseu dos Recreios, Lisbon	1890	Portugal
[158]	Garcia Resende, Evora	1892	Portugal
[158]	S. Luiz, Lisbon	1894	Portugal
[158]	Viriato, Viseu	19th Century	Portugal
[59,68]	Colon, Buenos Aires	1908	Argentina
[163]	Grand Theatre, Poznan	1910	Poland
[160]	Opera House, Ekaterinburg	1912	Russia
[164]	Municipal Theatre, Lima	1920	Perú
[158]	S. Joao, Oporto	1920	Portugal
[160]	Music Hall, St. Petersburg	1928	Russia
[160]	Maliy, St. Petersburg	1944	Russia
[165]	Teatro Argentino, La Plata	1999	Argentina
[166]	Opera House, Astana	2013	Kazakhstan

The third point concerns the sound strength. This criterion was still rarely measured, according to Tables 2 and 3. Moreover, a theoretical model for sound energy distribution in an opera house is needed. It must take into account the size of aperture, the coupling effect between the two sub-rooms [146,167–170]—the fly tower and the main hall—and the sound source position [129]. Furthermore, due to the subjective relevance of the balance and listener envelopment, the SISO (single-input-single-output) model of sound strength could be extended. An attempt of MIMO (multiple-input-multiple-output) measurements were done in Comunale Theatre in Bologna. A loudspeaker orchestra was adapted for the pit layout, taking into account the loudspeaker directivity and the sound power level of each source [171].

Virtual acoustics can be a useful tool for improving the knowledge of opera houses [172]. Simulated impulse responses allow us to extend the virtual experience of an opera house, including the occupation and other variables, such as the scene configuration. Virtual acoustics allows also to enjoy a no-longer existing space [91,173]. While scale model were widely used in the design and analysis of opera houses until recent years [110,166,174,175], numerical simulation of sound field was introduced, in the context of Italian opera houses, during the 1990s [148,176,177]. However, many models have been developed for acoustic consultancies and they are not available on a repository. As a consequence, there is a limited literature on the materials and acoustic peculiarities of Italian opera houses.

Instead of this, the data sharing could provide many benefits. Neal et al. [178] shared measurements, made by a 32-element spherical microphone array, and both omnidirectional and a directional sound source, of American and European concert halls. Concerning the opera houses, as of today there are few examples of free-available data. Büttner et al. [125] shared their work on Italian court theatres, including measured and simulated impulse responses. More models were provided by the author's workgroup for Bayreuther Festspielhaus [179], Alighieri Theatre in Ravenna [91], Comunale of Bologna [130]. In the author's hope, this approach could be shared by the scholars' community, in order to increase the knowledge on this complex topic.

4. About Performers and Music inside of Opera Houses

4.1. The Evolution of the Orchestra, and Orchestra Pit

In the early 17th Century, musicians were hidden behind the stage, as sentenced by G. B. Doni [180]

"Tutta questa moltitudine d'instrumenti [...] rende cosi poco suono, che appena si ode da' più vicini alla scena" (All this multitude of instruments [...] yields so little sound, that can only just be heard from closer to the scene).

Some years later, still in the court–theatre context, Monteverdi placed the musicians in front of the stage [181]. In the Venetian public theatre with boxes, the orchestra was placed on the stalls floor [182]. This kind of performance needed few musicians (see the St. Moise theatre case in 1720, Table 5) [183]. With the first large-sized theatres, such as the S. Carlo in Naples in 1741 (Table 5), the orchestra increased. Its composition was characterised by a large number of violins and more double-basses than cellos. It should be noted that the sound power level of historical instruments might differ from the one today, and generally historical instruments sound lower. In case of strings it is due to the string materials, in case of woodwinds it depends on the keys. An accurate measurement of these differences was recently made by Weinzierl et al. [184]. Mozart set the balance between the woodwind section, fixing the ratio between flutes, oboes, clarinets, bassons and horns as 2:2:2:2:4 [185]. This ratio was keep during the evolution of opera. Indeed, it was increased to base "at-three"—i.e., three flutes, three oboes, three clarinets and three bassoons—in the Verdian orchestra [87,88] and, then, to base at-four" by the German composers R. Wagner [186] and R. Strauss.

As noted in severals works [187,188], the balance between voice and orchestra plays a predominant role in opera-house acoustics. This is due to the fact that vocal and orchestral signals need different boundary conditions of the sound field, and a well-designed opera house should satisfy both. In an opera, the voice needs more intelligibility and more "focusing" on the soloist [189,190]. This means a lower reverberance, and a lower Apparent Source Width, using ISO 3382 subjective categories. In terms of objective criteria, this corresponds to low EDT values and high Inter-Aural Crosscorrelation Coefficient ($IACC$) values. Instead, music needs more sustain. This means more reverberance and more envelopment, which corresponds to higher EDT values and low $IACC$ values.

The development of the opera house from an acoustic standpoint is essentially aimed at the fulfillment of these needs. In order to assure these conditions the orchestra was placed in the Italian opera house in front of the proscenium. This provides a lot of reflections from the side walls, and scattering from boxes. If the geometry is well-balanced, a very low value of $IACC$ is achieved. Moreover, the energy decay is quite regular, meaning an EDT/T ratio of about one [20]. The singers were placed instead in the fly tower, which provides less reflections due to side curtains, achieving a high $IACC$ value. Moreover, the fly tower is coupled with the main hall through the scenic arch. The effect of this coupling is often a double-slope decay, then an EDT/T ratio is lower than one [191].

Table 5. Orchestral configuration during the years. *vli*: violins—if data are available, they are shown as first parts + second (+ third in Strauss's Salome only); *vla*: violas; *cl*: cellos; *db*: double-basses; *fl*: flutes + piccolo; *ob*: oboes + english horn; *cla*: clarinets + bass clarinet; *bas*: bassons + contrabasson; *ho*: french horns; *tp*: trumpets + bass trumpets; *tb*: trombones + cimbasso/Wagner tubas; *perc*: percussions. Data taken from [87,88,183,185,186].

Year	Opera House	Performance	Strings				Woodwinds				Brasses			Others	
			vli	vla	cl	db	fl	ob	cla	bas	ho	tp	tb	perc	harp
1720	St. Moise, Venice		2	1	1	1		2		2	2				
1741	San Carlo, Naples		38	8	2	4		6		4	3	2			
1758	Argentina, Rome		16	4	2	3		2		2	3				
1770	Ducale, Milan	Mozart: *Mitridate Re di Ponto*	14+14	6	2	6	2	2	2	2	4	2		yes	
1778	Comunale, Bologna		25	8	2	4	2	3		2	4	4	4	yes	
1778	La Scala, Milan (opening)	Salieri: *Europa riconosciuta*	30	8	5	8	2	3	3	2	4	4			
1797	San Carlo, Naples		25	4	2	6		2	2	4	4				
1818	Covent Garden, London		8	1	2		1	2		2	1	2	1	yes	
1825	La Scala, Milan		30	8	6	10	2	2	2	2	2	1	1		
1858	La Scala, Milan	Verdi: *Un ballo in maschera*	28	8	8	8	2+1	2+1	3	3	5	3	3+1	yes	1
1876	Bayreuther Festspielhaus (opening)	Wagner: *Ring*	16+16	12	12	8	4	3+1	3+1	3+1	8	4	4+5	yes	4
1909	Dresden Oper	Strauss: *Elektra*	8+8+8	6+6+6	6+6	8	3+1	3+1	5+3	3+1	8	6+1	2+3	yes	2

The evolution of the opera during the 19th Century needed more instruments. This was one of the reasons for the original Wagnerian idea of mystic gulf, conceived for the aforementioned Bayreuth Festspielhaus (1872) [57]. However, that was not the only reason: the dramaturgy was increasing in complexity. More complex characters meant more focusing on the soloist. The presence of the orchestra pit increases the envelopment and increases the reverberance (EDT/T ratio higher than one). The pit provides the perceptual effect of moving the orchestra to the background, while the soloist are still well 'focused' on the stage. This was the reason why some authors compared the Wagnerian opera—and the Festspielhaus—to the birth of Cinema: the soloist is the actor, the orchestra is the background [192]. During the 20th Century, orchestra pits were progressively introduced in Italian theatres [23,91]. It should be remarked that orchestra pits require further adjustments, to increase the communication to hall and stage [76,193], and to ensure safe working conditions to musicians [194,195].

Furthermore, the music of the 18th-century opera is mainly based on strings and (few) woodwinds and brasses. In the 19th century the number of woodwinds and brasses increases, as shown by orchestral configuration in Table 5.

4.2. The Repertory and the Availability of Anechoic Recordings

The opera-attending experience involves many factors: multi-sensorial, cultural, or emotional. Which kind of anechoic recordings should be available to scholars? A choice criterion could concern the contemporary use of opera houses. Indeed, until the advent of cinema, opera-house seasons mixed new operas and opera from the so-called repertory. Nowadays, besides rare cases, opera seasons are based mostly only on this repertory, which means a selection of compositions of few authors. Table 6 resumes the most represented performances during ten seasons of the last decade, confirming this trend. As an example, the most represented opera (Verdi's *Traviata*, which has 6843 representations over 1373 productions) alone collects almost the same number collected by all Wagner's operas (7729 representations over 1866 productions). The statistics show that some languages—i.e., the Italian—and some styles—i.e., from late 18th century to the early 20th century—are preferred by opera listeners. This can be viewed as an evolutionary selection of the 'software', as the horse-shoe shape was the results of a natural selection of the 'hardware'.

It is interesting to compare the repertory statistics to the anechoic material available for researchers. The earliest anechoic excerpts were recorded in the BBC anechoic chamber [196], by a small orchestra through a single microphone. These recordings had a low dynamics and sounded bit lo-fi, if are compared to today's practices. In 1988 Hidaka et al. [197] recorded a large orchestra in a damped concert hall, surrounding the stage with an acoustically absorptive enclosure. Both previous recording sets include an opera ouverture, but no vocal parts.

In 1998 Farina et al. [198] recorded three excerpts—two arias and a *Romanza*—for soprano and accompanying piano. Patÿnen et al. [199] recorded symphonic and opera music by using a 22 microphone array. Several musicians and a soloist recorded individually an aria from Mozart's Don Giovanni. Speech excerpts were recorded by the TU-Berlin group [125,200]. In more recent years, other opera excerpts were recorded using the Patÿnen-Vigeánt workflow [201], taking into account the statistics of one-year representations in the most important opera houses [202].

Table 7 collects the previously released anechoic recordings.

Table 6. Statistics of most represented performances in the decade 2010–2019. In the upper table, the ten most represented composer are sort, then are also shown the most represented Czech composer (Janaceck), the most represented composers of 18th century (Gluck, after Mozart), 17th century (Purcell), and 16th century (Monteverdi). In the lower table, the ten most represented operas are sort, and also shown are the two most represented German operas written by Wagner and Mozart, and the most represented operas in Czech and Russian, which are, respectively, Dvorak's Rusalka and Tchaikovsky's Pikovaya Dama. Data from operabase.com.

By Composer/period				
Composer	Nationality	Century	Performances	Productions
Verdi	Italian	19th	27,194	5911
Puccini	Italian	19–20th	20,297	3805
Mozart	Austrian	18th	19,860	4207
Rossini	Italian	19th	8674	1898
Donizetti	Italian	19th	7716	1756
Wagner	German	19th	7229	1866
Bizet	French	19th	6517	1303
Tchaikovsky	Russian	19th	4314	1041
R. Strauss	German	20th	3970	832
Haendel	German/English	17th	2852	504
Britten	English	20th	2368	513
...				
Janacek	Czech	19–20th	2071	423
...				
Gluck	German	18th	1435	298
...				
Purcell	English	17th	1102	280
...				
Monteverdi	Italian	16–17th	981	250
By Opera/language				
Composer	Title	Language	Performances	Productions
Verdi	La traviata	Italian	6843	1373
Mozart	Die Zauberflöte	German (Singspiel)	5839	900
Bizet	Carmen	French	5728	1110
Puccini	La boheme	Italian	5316	1005
Puccini	Tosca	Italian	4611	989
Rossini	Il barbiere di Siviglia	Italian	4236	918
Puccini	Madama Butterfly	Italian	4230	929
Mozart	Le nozze di Figaro	Italian	4143	815
Verdi	Rigoletto	Italian	3952	866
Mozart	Don Giovanni	Italian	3949	755
...				
Wagner	Der fliegende Holländer	German	1593	332
...				
Mozart	Die Entführung aus dem Serail	German	1413	269
...				
Dvorak	Rusalka	Czech	1133	193
...				
Tchaikovsky	Pikovaya Dama	Russian	816	209

Table 7. Previously-released anechoic recordings of opera excerpts. "Solo" means that performer(s) was (were) recorded each-at-a-time, "Ensemble" means that performers were recorded together. "Multi-Tracks" means that recordings are available as multiple tracks with individual instruments/voices.

Ref.	Year	Type	Location	Music Materials	Multi-Tracks
[196]	1969	ensemble	BBC	Wagner: *Siegfried Idyll*	no
[197]	1988	ensemble	Osaka	Mozart, *Le Nozze De Figaro* (Ouverture) Glinka: *Ruslan And Lyudmila* (Ouverture) Verdi: *La Traviata* (Preludio)	no
[198]	1998	solo [1]	Parma	Haendel: "Lascia Ch'io Pianga" from *Almira* Tosti: "Ricordi ancora il di' che c'incontrammo" Mozart: "In uomini, in soldati" from *Così Fan Tutte*	yes
[199]	2008	solo	Aalto	Mozart: "Mi tradí quell'alma ingrata" from *Don Giovanni*	yes
[5,125,200]	2015–2019	solo	TU-Berlin	Speech from Sofocle's *Edipo tiranno* Speech from Cicero's *Catiline* Oration	(yes)
[201,202]	2016–2020	solo	Bologna	Donizetti: "O mio Paride vezzoso" from *L'elisir d'amore* Verdi: "Di tale amor che dirsi" from *Il trovatore* Puccini: "O mio babbino caro" from *Gianni Schicchi*	yes

[1] The orchestral parts were perfomed by a piano accompaniment.

5. Final Considerations

Acoustic concerns were taken into account during the development of the form of the Italian-style opera house. This is the reason why the historical context and the relative technical literature were widely reviewed. Moreover, cultural instances take a relevant place in the acoustic analysis of opera houses.

Many opera houses were surveyed in the past. The results of these measurement campaigns were shared mostly within the Italian acoustic community. However, these contributions led to important results, the most relevant of which is the recognition of opera house acoustics as an intangible cultural heritage.

The cultural background of opera houses was discussed in the present work by several points of view. In the author's hope, this review could fill some gaps, pushing forward the emerging approaches in this research fields.

Funding: This research received no external funding.

Acknowledgments: The author gratefully acknowledge Alessia Milo, who provided useful comments and revisions of the text. Thanks go to Massimo Garai, who helped the author in the final proof reading. Finally, the author wish to thank Nicola Prodi for sharing his experience and material.

Conflicts of Interest: The author declares no conflict of interest.

Abbreviations

EDT	Early Decay Time
G	Sound Strength
G_E	Sound Strength integrated over the direct field and the early reflections
G_L	Sound Strength integrated the late reverberated part of the impulse response $G_L = G - G_E$
$IACC$	Inter-Aural Crosscorrelation Coefficient
LFC_{80}	Cosine-Lateral fraction
T	Reverberation time

References

1. Barbieri, P. The state of architectural acoustics in the late renaissance. In *Architettura e Musica Nella Venezia del Rinascimento*; Howard, D., Moretti, L., Eds.; B. Mondadori: Milan, Italy, 2006.
2. Sanvito, P.; Weinzierl, S. The Acoustics of the Teatro Olimpico in Vicenza. *Odeo Olimp.* **2013**, *23*, 463–492. (In Italian)
3. Scamozzi, V. *L'idea Della Architettura Universale (About of Universal Architecture)*; Expensis Auctoris: Venezia, Italy, 1615. (In Italian)
4. Prodi, N.; Pompoli, R. The acoustics of three Italian historical theatres: The early days of modern performance spaces. In Proceedings of the Tecniacustica 2000, Madrid, Spain, 16–20 October 2000.
5. Weinzierl, S.; Sanvito, P.; Schultz, F.; Büttner, C. The acoustics of Renaissance Theatres in Italy. *Acta Acust. United Acust.* **2015**, *101*, 632–641. [CrossRef]
6. Torello–Hill, G. The exegesis of Vitruvius and the creation of theatrical spaces in Renaissance Ferrara. *Renaiss. Stud.* **2014**, *29*, 227–246. [CrossRef]
7. Girón, S.; Álvarez-Corbacho, A.; Zamarreno, T. Exploring the Acoustics of Ancient Open-Air Theatres. *Arch. Acoust.* **2020**, *45*, 181–208. [CrossRef]
8. Bondin, N. *Teatro Antico. Ragionamento Sopra la Forma e la Struttura (Ancient Theatre. Reasoning on Form and Structure)*; (n.p.): Venezia, Italy, 1746. (In Italian)
9. Damun, J. *Prospectus du Noveau Theatre Tracé sur les Principes des Grecs et des Romains (Project of a New Theatre Based on Greek and Roman Principles)*; Lambert: Paris, France, 1772. (In French)
10. Johnson, E.J. *Inventing the Opera House: Theater Architecture in Renaissance and Baroque Italy*; Cambridge University Press: Cambridge, UK; New York, NY, USA, 2018.
11. Jeon, J.Y.; Kim, Y.H.; Cabrera, D.; Bassett, J. The effect of visual and auditory cues on seat preference in an opera theater. *J. Acoust. Soc. Am.* **2008**, *123*, 4272–4282. [CrossRef]

12. Glixon, B.; Glixon, J. *Inventing the Business of Opera. The Impresario and His Work in Seventeenth-Century*; Oxford University Press: Oxford, UK, 2006.
13. Bianconi, L.; Walker, T. Production, consumption and political function of seventeenth-century Italian opera. *Early Music Hist.* **1984**, *4*, 209–296. [CrossRef]
14. Rosand, E. *Opera in Seventeenth-Century Venice: The Creation of a Genre*; University of California Press: Berkley, CA, USA, 2007.
15. Bordone, C.; Sacerdote, G. Acoustic problems of boxes of opera houses. In Proceedings of the 2nd International Conference on Acoustics and Musical Research (CIARM95), Ferrara, Italy, 19–21 May 1995; pp. 245–250. (In Italian)
16. Ivanovich, C. *Memorie Teatrali di Venezia (Theatrical memories of Venice)*; N. Pezzana: Venezia, Italy, 1688.
17. Galli Bibiena, F. *L'architettura Civile Preparata su la Geometria e Ridotta alle Prospettive (Civil Architecture: Geometry and Perspective)*; Paolo Monti: Parma, Italy, 1711. (In Italian)
18. Bassuet, A. Acoustics of a selection of famous 18th-Century opera houses: Versailles, Markgräfliches, Drottningholm, Schweitzingen. *J. Acoust. Soc. Am.* **2008**, *123*, 3192. [CrossRef]
19. Cervellati, P.L. *Il Rossini di Lugo: Sul Restauro di un Celebre Teatro (The Rossini in Lugo: On the Restoration of a Famous Theatre)*; Nuova Alfa Editoriale: Bologna, Italy, 1986. (In Italian)
20. Garai, M.; Morandi, F.; D'Orazio, D.; De Cesaris, S.; Loreti, L. Acoustic measurements in eleven Italian opera houses: Correlations between room criteria and considerations on the local evolution of a typology. *Build. Environ.* **2015**, *94*, 900–912. [CrossRef]
21. Cocchi, A.; Garai, M.; Tavernelli, C. Boxes and sound quality in an Italian opera house. *J. Sound Vib.* **2000**, *232*, 171–191. [CrossRef]
22. Tronchin, L.; Shimokura, R.; Tarabusi, V. Spatial sound characteristics in the Theatre Comunale in Bologna, Italy. In Proceedings of the 9th Western Pacific Acoustics Conference, Seoul, Korea, 26–28 June 2006.
23. D'Orazio, D.; Fratoni, G.; Rovigatti, A.; Garai, M. A virtual orchestra to qualify the acoustics of historical opera houses. *Build. Acoust.* **2020**. [CrossRef]
24. Giordano, L. The "Teatro dei Quattro Cavalieri" and the work of A. Galli Bibiena in Pavia. *Bollettino D'arte* **1975**, *60*, 88–102. (In Italian)
25. Magrini, A.; Ricciardi, P. Acoustic characterization of the Teatro Sociale di Como e del Teatro Fraschini. In Proceedings of the Opera houses of the Unity of Italy, Venice, Italy, 23 November 2011. (In Italian)
26. Iannace, G.; Maffei, L. Acoustic characterization of the Teatro Mercadante. In Proceedings of the Opera Houses of the Unity of Italy, Venice, Italy, 23 November 2011. (In Italian)
27. D'Orazio, D.; Nannini, S. Towards Italian Opera Houses: A review of acoustic design in pre-Sabine scholars. *Acoustics* **2019**, *1*, 252–280. [CrossRef]
28. Carini Motta, F. *Treaty about the Structure of the Theatres and of the Scenes*; Giavazzi: Guastalla, Italy, 1676; republished in Il Polifilo: Milano, Italy, 1972. (In Italian)
29. Algarotti, F. *Saggio Sopra L'Opera in Musica (Essay on Opera)*, 2nd ed.; Marco Coltellini: Livorno, Italy, 1764. (In Italian)
30. Riccati, G. *Della Costruzione De' Teatri Secondo il Costume D'Italia: Vale a Dire Divisi in Piccole Logge (About the Construction of Theatres According to the Italian Style, i.e., with Small Boxes)*; Remondini from Venice: Bassano, Italy, 1790. (In Italian)
31. Memmo, A. *Semplici Lumi Tendenti a Render Cauti I Soli Interessati nel Teatro da Erigersi Nella Parocchia di S Fantino in Venezia [...] (Simple Hints for Those Who Are Interested in the Theatre to Build in the Parish of S. Fantino in Venice)*; (n.p.): Venice, Italy, 1790. (In Italian)
32. Patte, P. *Essai sur L'architecture Théatrale, ou, De L'ordonnance la Plus Avantageuse à Une Salle de Spectacles, Relativement aux Principes de L'optique et de L'acoustique: Avec un Examen des Principaux Téatres de l'Europe, et une Analyse des écrits les plus Importans sur Cette Matiere (Essay on Theatre Architecture, or on the Most Advantageous Design of a Performance Hall, and an Analysis of the Most Important Writings on This Topic)*; Moutard: Paris, France, 1782. (In French)
33. Morelli, C. *Pianta e Spaccato del Nuovo Teatro d'Imola (Plan and Section of the New Theatre of Imola)*; Casaletti: Roma, Italy, 1780. (In Italian)
34. Braham, A. *The Architecture of the French Enlightenment*; Thames and Hudson: London, UK, 1980.

35. Arnaldi, E. *Idea di un Teatro Nelle Principali Sue Parti Simile A'teatri Antichi All'uso Moderno Accomodato (Idea of a Theatre in Its Principal Parts Similar to Ancient Theatres Arranged for the Modern Use)*; Veronese: Vicenza, Italy, 1762. (In Italian)
36. Milizia, F. *Trattato Completo, Formale e Materiale del Teatro (About the Theatre)*; Stamperia Pasquali: Venezia, Italy, 1794. (In Italian)
37. Ledoux, C.N. *L'architecture Considerée sour le Rapport de l'Art, des Moeurs et de la Legislation ... (The Architecture Considered through the Connections with Art, Traditions and Law)*; Perroneau: Paris, France, 1804. (In French)
38. Boullée, E.L. *Architecture, Essai sur l'art (Architecture, Essay on Art)*; Rosenau, H., Ed.; Tiranti: London, UK, 1953. (In French)
39. Barbieri, P.; Tronchin, L. L'impostazione acustica dei teatri nei progetti del primo neoclassicismo Italiano (1762–1772), (The Acoustical Structure of Theatres in the First Italian Neoclassical Projects). In *Francesco Milizia e il Teatro del suo Tempo Architettura, Musica, Scena, Acustica*; Collana Studi e Ricerche n. 2; Russo: Trento, Italy, 2011; pp. 137–161, ISBN 978-88-8443-396-1. (In Italian)
40. Beccega, T. *Sull'architettura Greco-Romana Applicata Alla Costruzione del Teatro Moderno Italiano e Sulle Macchine Teatrali (On the Application of the Graeco-Roman Architecture to the Construction of the Modern Italian Theatre and to the Theatrical Machines)*; Alvisopoli: Venezia, Italy, 1817. (In Italian)
41. Tronchin, L. Francesco Milizia (1725–1798) and the Acoustics of his Teatro Ideale (1773). *Acta Acust. United Acust.* **2013**, *99*, 91–97. [CrossRef]
42. Forsyth, M. *Buildings for Music*; The Press Syndicate: Cambridge, UK, 1985.
43. Niccolini, A. *Alcune Idee Sulla Risonanza del Teatro (Some Ideas on the Resonance of the Theatre)*; Masi: Napoli, Italy, 1816. (In Italian)
44. Prodi, N.; Pompoli, R.; Martellotta, F.; Sato, S. Acoustics of Italian historical opera houses. *J. Acoust. Soc. Am.* **2015**, *138*, 769–781. [CrossRef] [PubMed]
45. Meduna, T.; Meduna, G. *Il Teatro La Fenice in Venezia...(On the La Fenice Theatre...)*; Antonelli: Venezia, Italy, 1849. (In Italian)
46. de Cesare, F. *La Scienza dell'Architettura Applicata alla Costruzione, alla Distribuzione, alla Decorazione Degli Edifici Civili (The Science of Architecture on the Construction, Distribution and Decoration of Civil Buildings)*; Pellizzone: Napoli, Italy, 1885. (In Italian)
47. Favaro, A. *L'Acustica Applicata alla Costruzione delle sale per Spettacoli e Pubbliche Adunanze (Acoustics Applied to the Construction of Performance Spaces and Public Meetings)*; Camilla e Bertolero: Torino, Italy, 1882. (In Italian)
48. Basile, G.B.F. *Sulla costruzione del Teatro Massimo Vittorio Emanuele (On the Construction of Teatro Massimo Vittorio Emanuele)*; Tip. Dello Statuto: Palermo, Italy, 1883. (In Italian)
49. Newton, J.P. Room Acoustics Measurements at the Royal Opera House, London. In Proceedings of the 17th International Congress on Acoustics (ICA), Rome, Italy, 2–7 September 2001.
50. Saunders, G. *Treatise on Theaters*; I. and J. Taylor: London, UK, 1790.
51. Postma, B. A history of the use of time intervals after the direct sound in concert hall design before the reverberation formula of Sabine became generally accepted. *Build. Acoust.* **2013**, *20*, 157–176. [CrossRef]
52. Postma, B.N.; Jouan, S.; Katz, B.F. Pre-Sabine room acoustic design guidelines based on human voice directivity. *J. Acoust. Soc. Am.* **2018**, *143*, 2428–2437. [CrossRef]
53. Postma, B.N.; Katz, B.F. Pre-Sabine room acoustic assumptions on reverberation and their influence on room acoustic design. *J. Acoust. Soc. Am.* **2020**, *147*, 2478–2487. [CrossRef] [PubMed]
54. Bagenal, H.; Wood, A. *Planning for Good Acoustics*; Methuen & Co.: London, UK, 1931.
55. Garnier, C. *Le Théâtre (The Theatre)*; Librairie Hachette: Paris, France, 1871; Chapitre IX.
56. Müller, K. Festspielhaus Bayreuth—The unique acoustic situation. *J. Acoust. Soc. Am.* **1999**, *105*, 929. [CrossRef]
57. D'Orazio, D.; De Cesaris, S.; Morandi, F.; Garai, M. The aesthetics of the Bayreuth Festspielhaus explained by means of acoustic measurements and simulations. *J. Cult. Herit.* **2018**, *34*, 151–158. [CrossRef]
58. Link, D. *The National Court Theater in Mozart's Vienna: Sources and Documents 1783–1792*; Oxford University Press: New York, NY, USA, 1998,
59. Basso, G. Acoustical evaluation of the Teatro Colón of Buenos Aires. *Proc. Meet. Acoust.* **2017**, *30*, 015014.
60. Sabine, W.C. Theater Acoustics. *Am. Archit.* **1913**, *104*, 257.
61. Burris-Meyer, H. Sound in the theater. *J. Acoust. Soc. Am.* **1940**, *11*, 346. [CrossRef]

62. Hutchins-Viroux, R. Opera and Society in the English-speaking World. The American Opera Boom of the 1950s and 1960s: History and Stylistic Analysis. *Lit. Hist. Ideas Images Soc. Engl.-Speak. World* **2004**, *2*, 145–163.
63. Jordan, V.L. Acoustical design considerations of the Sydney Opera House. *J. Proc. R. Soc. N. S. W.* **1973**, *106*, 33–53.
64. Beranek, L.L. *Music, Acoustics & Architecture*; John Wiley & Sons: New York, NY, USA, 1962.
65. Beranek, L.L. *Riding the Waves*; The MIT Press: Cambridge, MA, USA, 2008.
66. Beranek, L.L.; Hidaka, T.; Masuda, S. Acoustical design of the opera house of the New National Theatre, Tokyo, Japan. *J. Acoust. Soc. Am.* **2000**, *107*, 355–367. [CrossRef] [PubMed]
67. Beranek, L.L. *Concert and Opera Halls. How They Sound*; American Institute of Physics: Woodbury, NY, USA, 1996.
68. Hidaka, T.; Beranek, L.L. Objective and subjective evaluations of twenty-three opera houses in Europe, Japan, and the Americas. *J. Acoust. Soc. Am.* **2000**, *107*, 368–383. [CrossRef]
69. Fausti, P.; Pompoli, R.; Prodi, N. Acoustics of opera houses: A cultural heritage. *J. Acoust. Soc. Am.* **1999**, *105*, 929. [CrossRef]
70. Farina, A.; Ayalon, R. Recording concert hall acoustics for posterity. In Proceedings of the 24th AES Conference on Multichannel Audio, Banff, AB, Canada, 26–28 June 2003.
71. Prodi, N.; Pompoli, R. Acoustics in the restoration of Italian historical opera houses: A review. *J. Cult. Herit.* **2016**, *21*, 915–921. [CrossRef]
72. Prodi, N. From tangible to intangible heritage inside Italian historical opera houses. *Heritage* **2019**, *2*, 826–835. [CrossRef]
73. Farina, A. Acoustic quality of theatres: Correlations between experimental measures and subjective evaluations. *Appl. Acoust.* **2001**, *62*, 899–916. [CrossRef]
74. Iannace, G.; Ianniello, C.; Maffei, L.; Romano, R. Objective measurement of the listening condition in the old Italian opera house Teatro di San Carlo. *J. Sound Vib.* **2000**, *232*, 239–249. [CrossRef]
75. Dragonetti, R.; Ianniello, C.; Mercogliano, F.; Romano, R.A. The Teatro di San Carlo in Naples and its smaller clone Teatro Verdi in Salerno. In Proceedings of the Acoustics 08, Paris, France, 30 June–4 July 2008; pp. 1367–1372.
76. Facondini, M.; Bignozzi, L. Variable sound orchestra pit for the "Teatro Comunale" of Bologna. In Proceedings of the 17th International Congress on Acoustics (ICA), Rome, Italy, 2–7 September 2001.
77. Tronchin, L.; Tarabusi, V. Acoustic characterization of the Teatro Comunale in Bologna and the Teatro Bonci in Cesena. In Proceedings of the Opera Houses of the Unity of Italy, Venice, Italy, 23 November 2011. (In Italian)
78. Arau, H. Renovating Teatro alla Scala Milano for the 21st century, Part I and II. *J. Acoust. Soc. Am.* **2005**, *117*, 2522. [CrossRef]
79. Farina, A.; Capra, A.; Armelloni, E.; Varani, C.; Amendola, A. Acoustic survey of the Teatro alla Scala in Milan. In Proceedings of the Opera Houses of the Unity of Italy, Venice, Italy, 23 November 2011. (In Italian)
80. Tronchin, L.; Farina, A. Acoustics of the former Teatro La Fenice in Venice. *J. Audio Eng. Soc.* **1997**, *45*, 1051–1062.
81. Ceniccola, G. Architetture in Scena. Teatri Storici in Campania tra XVIII e XX Secolo: Conoscenza e Nodi Critici nel Progetto di Conservazione (Performance Architecture. Historical Theatres in Campania between 18th and 20th Century: Knowledge and Criticalities in the Conservation Project). Ph.D. Thesis, University of Naples "Federico II", Naples, Italy, 2011. (In Italian)
82. Quagliarini, E. *Costruzioni in Legno nei Teatri All'Italiana del '700 e '800: Il Patrimonio Nascosto Dell'architettura Teatrale Marchigiana (Wooden Buildings in Italian Historical Opera Houses: The Hidden Heritage of Theatrical Architecture)*; Allinea: Firenze, Italy, 2008. (In Italian)
83. Martellotta, F. Acoustics of Apulian Historical Theatres: Further developments of the research. In Proceedings of the Opera Houses of the Unity of Italy, Venice, Italy, 23 November 2011. (In Italian)
84. Zecchin, R.; Di Bella, A.; Boniotto, E.; Boscolo, S.; Bovo, M.E.; Granzotto, N.; Rinaldi, C. Survey on opera houses in Triveneto: Analysis of the main acoustic parameters. In Proceedings of the 33rd National Meeting of Italian Acoustic Society, Ischia, Italy, 10–12 May 2006. (In Italian)
85. Iannace, G.; Ianniello, C.; Maffei, L.; Romano, R. Acoustics measurements in the Teatro dell'Opera in Rome. In Proceedings of the 17th International Congress on Acoustics (ICA), Rome, Italy, 2–7 September 2001.

86. Cammarata, G.; Fichera, A.; Marletta, L. Indici di qualità acustica del Teatro Bellini di Catania (Acoustical quality indexes of the Bellini Theatre in Catania). In Proceedings of the 27th National Meeting of Italian Acoustic Society, Padua, Italy, 31 March–2 April 1993. (In Italian)
87. Harwood, G.W. Verdi's Reform of the Italian Opera Orchestra. *19th-Century Music* **1986**, *10*, 108–134. [CrossRef]
88. Giordani, E.; Salvarani, M. *Orchestral staff in 19th century Opera Houses, Introductory Issues for a Psycho-Acoustic Evaluation*; Studi Verdiani 16; Mattioli 1885: Parma, Italy, 2002. (In Italian)
89. Bortolotti, L.; Masetti, L. *Teatri Storici. Dal Restauro allo Spettacolo (Historical Theatres. From Renovation to Performance)*; Nardini: Fiesole, Italy, 1977. (In Italian)
90. Silingardi, V.; Rinaldi, C.; Granzotto, N.; Barbaresi, L.; di Bella, A. A restoration based on the result of a public debate: The case of Civic Theatre of Schio. *Riv. Ital. Acust.* **2017**, *41*, 1–14.
91. D'Orazio, D.; Rovigatti, A.; Garai, M. The Proscenium of Opera Houses as a disappeared intangible heritage: A virtual reconstruction of the 1840s original design of the Alighieri Theatre in Ravenna. *Acoustics* **2019**, *1*, 694–710. [CrossRef]
92. Cocchi, A.; Farina, A.; Tronchin, L. Computer assisted methods and acoustic quality: Recent restoration case histories. In Proceedings of the MCHA95, Kirishima, Japan, 15–18 May 1995.
93. Cocchi, A.; Consumi, M.C.; Shimokura, R. Considerations about the acoustic properties of Teatro Nuovo in Spoleto after the restoration works. In Proceedings of the Acoustics 08, Paris, France, 30 June–4 July 2008; pp. 1391–1394.
94. Tronchin, L. The design of acoustical enhancements and diffusion in the opera house of Treviso, Italy. In Proceedings of the 6th International Conference on Auditorium Acoustics, IOA, Copenhagen, Denmark, 5–7 May 2006.
95. Pisani, R.; Duretto, F. The restoration and the acoustic issues of historical theatres. In Proceedings of the 27th National Meeting of Italian Acoustic Society, Genua, Italy, 26–28 May 1999. (In Italian)
96. Facondini, M. Acoustic restoration of the Teatro Comunale Gioacchino Rossini in Pesaro. In Proceedings of the 17th International Congress on Acoustics (ICA), Rome, Italy, 2–7 September 2001.
97. Fausti, P.; Prodi, N. On the testing of renovations inside historical opera houses. *J. Sound Vib.* **2002**, *258*, 563–575. [CrossRef]
98. Strada, M.; Pompoli, R. Acoustic program in the competition for the reconstruction of the "La Fenice" opera house after the fire of 29 January 1996. *J. Sound Vib.* **2000**, *232*, 9–15 [CrossRef]
99. Cocchi, A.; Garai, M.; Tronchin, L. Influenza di cavità risonanti poste sotto la fossa orchestrale: Il caso del teatro Alighieri di Ravenna (The Influence of Resonating Cavities under the Orchestra Pit: The Case of the Alighieri Theatre in Ravenna). In *Teatri Storici. Dal Restauro Allo Spettacolo*; Nardini Editore: Firenze, Italy, 1997; pp. 67–84, 135–153. (In Italian)
100. Cancellieri, G.; Turrini, A. The Phantom of Modern Opera: How Economics and Politics Affect the Programming Strategies of Opera Houses. *Int. J. Arts Manag.* **2016**, *18*, 25–36.
101. Bernardini, G. Application to a Case Study: Fire Safety in Historical Theaters. In *Fire Safety of Historical Buildings*; Springer Briefs in Applied Sciences and Technology; Springer: Berlin, Germany, 2017.
102. Sacerdote, G.G.; Pisani, R. The new Teatro Regio of Torino. *Acustica* **1975**, *32*, 138–146
103. Guglielmino, D. Il Regio e la Sua Acustica (The Regio Opera House and Its Acoustics). Master's Thesis, Politecnico di Torino, Turin, Italy, 2007.
104. Di Bella, A.; Zecchin, R. Acoustic characteristics of Teatro La Fenice. In Proceedings of the Opera Houses of the Unity of Italy, Venice, Italy, 23 November 2011. (In Italian)
105. Facondini, M.; Ponteggia, D. Acoustics of the Restored Petruzzelli Theater. In Proceedings of the 128th AES Convention, London, UK, 22–25 May 2010; Convention Paper 8024.
106. Cocchi, A.; Ando, Y. The Le Muse Theatre in Ancona: Recent Developments. In Proceedings of the Auditorium Acoustics, London, UK, 19–21 July 2002.
107. Tronchin, L. The reconstruction of the Teatro Galli in Rimini: The acoustic design. In Proceedings of the International Symposium on Room Acoustics (ISRA), Melbourne, Australia, 29–31 August 2010.
108. Tronchin, L.; Facondini, M.; D'Orazio, D.; Farnetani, A. The acoustic quality of the Teatro Galli in Rimini after the restoration. In Proceedings of the46th National Meeting of Italian Acoustic Society, Pesaro, Italy, 29–31 May 2019. (In Italian)

109. Pompoli, R.; Farina, A.; Fausti, P. The acoustics of the Nuovo Teatro Comunale in Cagliari International. In Proceedings of the Institute of Acoustics, London, UK, 10–12 February 1995.
110. Commins, D.; Pompoli, R.; Farina, A.; Fausti, P.; Prodi, N. Acoustics of Teatro degli Arcimboldi in Milano design, computer and scale models, details, results. In Proceedings of the Institute of Acoustics (IOA2002), London, UK, 19–21 July 2002.
111. Reinhold, J.; Conta, S. The acoustics of the new opera house in Florence: Innovative choices in a quasi-classical theatre. In Proceedings of the 39th National Meeting of Italian Acoustic Society, Rome, Italy, 4–6 July 2012. (In Italian)
112. Faggiani, D. Sul tempo di circonsonanza delle sale di spettacolo (About the reverberation time of performance halls). *Atti del Regio Istituto Lombardo* **1936**, *16–20*, 1–8.
113. Faggiani, D. La reazione del palcoscenico sulla dinamica sonora delle sale (The effects of stage acoustic response on the sound of the halls). *Il Politecnico* **1937**, *85*, 3–10.
114. Cocchi, A. The Giuseppe Verdi theatre in Busseto. In Proceedings of the 1st National Meeting of Italian Acoustic Society, Roma, Italy, 1 May 1973; pp. 9–23. (In Italian)
115. Cocchi, A. Riflessioni in merito all'acustica dei teatri (Meditations about the acoustics of theaters). *Riv. Ital. Acust.* **1984**, *8*, 69. (In Italian)
116. Ianniello, C. The acoustics of the Italian-style Opera House. In Proceedings of the 17th International Congress on Acoustics (ICA), Rome, Italy, 2–7 September 2001.
117. Ianniello, C. Some notes on historical theatres for opera. *Riv. Ital. Acust.* **2002**, *26*, 45–62. (In Italian)
118. Cammarata, G.; Fichera, A.; Rizzo, G. Analysis of objective acoustic parameters in some Italian theatres (influence of the diffuse reflection coefficient). In Proceedings of the 27th National Meeting of Italian Acoustic Society, Genoa, Italy, 26–28 May 1999; pp. 78–81. (In Italian)
119. Prodi, N.; Pompoli, R. Guidelines for acoustic measurements inside historical opera houses. Procedures and validation *J. Sound Vib.* **2000**, *232*, 281–301.
120. Fausti, P.; Farina, A. Acoustic measurements in opera houses: A comparison between different techniques and equipment. *J. Sound Vib.* **2000**, *232*, 213–229. [CrossRef]
121. D'Orazio, D.; De Cesaris, S.; Guidorzi, P.; Barbaresi, L.; Garai, M.; Magalotti, R. Room acoustic measurements using a high SPL dodecahedron. In Proceedings of the 140th AES Convention, Paris, France, 4–7 June 2016.
122. Papadakis, N.M.; Stavroulakis, G.E. Review of Acoustic Sources Alternatives to a Dodecahedron Speaker. *Appl. Sci.* **2019**, *9*, 3705. [CrossRef]
123. Pompoli, R.; Prodi, N. A study on balance inside an historical opera house. In Proceedings of the 17th International Congress on Acoustics (ICA), Rome, Italy, 2–7 September 2001.
124. Pelorson, X.; Vian, J.P.; Polack, J.D. On the variability of room acoustical parameters: Reproducibility and statistical validity. *Appl. Acoust.* **1992**, *37*, 175–198. [CrossRef]
125. Büttner, C.; Schultz, F.; Weinzierl, S. Room Acoustical Measurements and Simulations of Italian Renaissance Theatres. 2014. Available online: http://dx.doi.org/10.14279/depositonce-32.2 (accessed on 3 July 2020).
126. Strada, M.; Romagnoni, P.; Carbonari, A. Sound level measurements and assessments on internal acoustics in a historic Venetian theater: The Malibran. In Proceedings of the National Meeting of Italian Acoustic Society, Venice, Italy, 5–7 May 2004. (In Italian)
127. Cammarata, G.; Fichera, A.; Pagano, A.; Rizzo, G. Acoustical prediction in some Italian theatres. *Acoust. Res. Lett. Online* **2001**, *2*, 61–66. [CrossRef]
128. Bonsi, D.; Stanzial, D. Using orchestra shell for acoustic optimization of the Teatro Accademico in Castelfranco Veneto. In Proceedings of the 29th National Conference of Italian Acoustic Association, Ferrara, Italy, 12–14 June 2002; pp. 601–604. (In Italian)
129. Garai, M.; De Cesaris, S.; Morandi, F.; D'Orazio, D. Sound energy distribution in Italian opera houses. *Proc. Meet. Acoust.* **2016**, *28*, 015019.
130. D'Orazio, D.; Fratoni, G. TCBO CAD and Measured IRs, Mendeley Data. 2019. Available online: http://dx.doi.org/10.17632/ggty3v22cx (accessed on 3 July 2020).
131. Iannace, G.; Ianniello, C.; Maffei, L.; Romano, R. The acoustic of the Teatro di Corte Reggia in Caserta. In Proceedings of the 17th International Congress on Acoustics (ICA), Rome, Italy, 2–7 September 2001.
132. Fausti, P.; Pompoli, R. The acoustic of the Comunale Theatre in Ferrara. In Proceedings of the Acoustics of Historical Theatres: A Cultural Heritage, Ferrara, Italy, 4 November 1998; pp. 53–62. (In Italian)

133. Di Bella, A.; Zecchin, R. Acoustic characteristics of Teatro Verdi in Trieste. In Proceedings of the Opera Houses of the Unity of Italy, Venice, Italy, 23 November 2011. (In Italian)
134. Masoero, M.; Astolfi, A.; Bottalico, P.; Pisani, R. Acoustics of the Teatro Grande in Brescia. In Proceedings of the Opera Houses of the Unity of Italy, Venice, Italy, 23 November 2011. (In Italian)
135. Farina, A.; Capra, A.; Armelloni, E.; Varani, C.; Amendola, A. Acoustic survey of the Teatro Regio in Parma. In Proceedings of the Opera Houses of the Unity of Italy, Venice, Italy, 23 November 2011. (In Italian)
136. Facondini, M. *The Sound of the Stage*; Il Ponte Vecchio: Cesena, Italy, 1999; ISBN 88-8312-036-1. (In Italian)
137. Mazzarella, L.; Cairoli, M. Petrarca Theatre: A case study to identify the acoustic parameters trends and their sensitivity in a horseshoe shape opera house. *Appl. Acoust.* **2018**, *136*, 61–75. [CrossRef]
138. Astolfi, A.; Bortolotto, A.; Filippi, A.; Masoero, M. Acoustical characterisation of small Italian opera house. In Proceedings of the Forum Acusticum, Budapest, Hungary, 29 August–2 September 2005.
139. Lori, V.; Serpilli, F.; Cesini, G.; Costanzo, E.; Montelpare, S.; Mataloni, G. Acoustics survey of some historical opera houses of Abruzzo region. In Proceedings of the 43rd National Meeting of Italian Acoustic Society, Alghero, Italy, 25–27 May 2016. (In Italian)
140. Bartoli, C. Acoustics of theatres and auditoria in the province of Pisa: First results. In Proceedings of the 33rd National Conference of Italian Acoustic Association, Ischia, Italy, 10–12 May 2006; pp. 533–536. (In Italian)
141. Tronchin, L. Acoustical design of diffusing panels in the Theatre Vittorio Emanuele, Messina, Italy. In Proceedings of the Room Acoustics: Design and Science (RADS), Awaji, Japan, 11–13 April 2004.
142. Cirillo, E.; dell'Alba, M.; Martellotta, F. Acoustic survey of historical Apulian theatres. In Proceedings of the 38th National Meeting of Italian Acoustic Society, Rimini, Italy, 8–10 June 2011. (In Italian)
143. Farina, A.; Capra, A.; Armelloni, E.; Varani, C.; Amendola, A. Acoustic survey of the Teatro Valli in Reggio Emilia. In Proceedings of the Opera Houses of the Unity of Italy, Venice, Italy, 23 November 2011. (In Italian)
144. Masoero, M.; Astolfi, A.; Bottalico, P.; Pisani, R. Acoustics of the Teatro Alfieri in Asti. In Proceedings of the Opera Houses of the Unity of Italy, Venice, Italy, 23 November 2011. (In Italian)
145. Iannace, G.; Maffei, L. Acoustic characterization of the Teatro Verdi in Salerno. In Proceedings of the Opera Houses of the Unity of Italy, Venice, Italy, 23 November 2011. (In Italian)
146. D'Orazio, D.; Fratoni, G.; Garai, M. Overhead stage canopies in a coupled volume theatre: Effects on the sound energy distribution and on the secondary reverberation. *J. Acoust. Soc. Am.* **2019**, *146*, 2802. [CrossRef]
147. D'Orazio, D. Duse Theatre CAD and Measured IRs, Mendeley Data. 2019. Available online: http://dx.doi.org/10.17632/br9x8hp52m (accessed on 3 July 2020).
148. Tronchin, L.; Farolfi, G. Acoustic design through measurements and computer simulations: The Teatro Comunale of Gradisca d'Isonzo. *Riv. Ital. Acust.* **1996**, *20*, 17–27. (In Italian)
149. Tronchin, L.; Tarabusi, V. The acoustic design of the Teatro Eschilo, Gela (Italy). In Proceedings of the International Symposium on Room Acoustics (ISRA), Melbourne, Australia, 29–31 August 2010.
150. Qandil, E.; Ianniello, C.; Iannace, G. The effect of stage scenery on the acoustics of an Italian opera house. In Proceedings of the 17th International Congress on Acoustics (ICA), Rome, Italy, 2–7 September 2001.
151. Jeon, J.Y.; Kim, J.H.; Ryu, J.K. The effects of stage absorption on reverberation times in opera house seating areas. *J. Acoust. Soc. Am.* **2015**, *137*, 1099–1107. [CrossRef]
152. Bradley, D.T.; Wang, L.M. The effects of simple coupled volume geometry on the objective and subjective results from nonexponential decay. *J. Acoust. Soc. Am.* **2005**, *118*, 1480–1490. [CrossRef]
153. Vigeant, M.C.; Giacomoni, C.B.; Scherma, A.C. Repeatability of spatial measures using figure-of-eight microphones. *Appl. Acoust.* **2013**, *74*, 1076–1084. [CrossRef]
154. Farina, A.; Tronchin, L. 3D sound characterisation in theatres employing microphone arrays. *Acta Acust. United Acust.* **2013**, *99*, 118–125. [CrossRef]
155. Martellotta, F. On the use of microphone arrays to visualize spatial sound field information. *Appl. Acoust.* **2013**, *74*, 987–1000. [CrossRef]
156. Dick, D.A.; Vigeant, M.C. An investigation of listener envelopment utilizing a spherical microphone array and third-order ambisonics reproduction *J. Acoust. Soc. Am.* **2019**, *145*, 2795–2809. [CrossRef]
157. Semidor, C.; Barlet, A. Objective and subjective surveys of opera house acoustics: Example of the Grand Theatre de Bordeaux. *J. Sound Vib.* **2000**, *232*, 251–261. [CrossRef]
158. Santiago, F. Portuguese Theatres and Concert Halls Acoustics. Master's Thesis, Engineer & Arquitecture La Salle Arquitactural and Environmental Acoustics Master, Barcelona, Spain, 15 March 2007.

159. Lannie, M. Acoustics of Gonzago Theatre in the Palace and Park Museum of Archangelskoje. *Appl. Acoust.* **1993**, *40*, 347–353. [CrossRef]
160. Lannie, M.; Makrinenko, L. Acoustics of Russian classical opera houses. In Proceedings of the 102nd Audio Engineering Society Convention, Munich, Germany, 22–25 March 1997; Preprint 4427 (D8).
161. Cerdá, S.; Giménez, A.; Romero, J.; Cibrián, R.M.; Miralles, J.L. Room acoustical parameters: A factor analysis approach. *Appl. Acoust.* **2009**, *70*, 97–109. [CrossRef]
162. Rudno-Rudziński, K.; Dziechciński, P. Reverberation time of Wrocław opera house after restoration. *Arch. Acoust.* **2006**, *31*, 247–252.
163. Sygulska, A. The influence of the stage layout on the acoustics of the auditorium of the Grand Theatre in Poznan. In Proceedings of the Acoustics 08, Paris, France, 29 June–4 July 2008.
164. Dianderas, C.J. Acoustical evalutation of the Municipal Theatre of Lima. *Appl. Acoust.* **1992**, *35*, 153–166. [CrossRef]
165. Baggio, M. Measurements of Teatro Argentino de La Plata. In *Acoustical Instruments and Measurements*; Universidad Nacional de Tres de Febrero: Buenos Aires, Argentina, 2013.
166. Cairoli, M.; Moretti, E.; Gade, A.C. New opera house in Astana: A recent opportunity to use a room acoustic scale model. *Proc. Meet. Acoust.* **2013**, *19*, 015131. [CrossRef]
167. Ermann, M. Coupled Volumes: Aperture size and the double-sloped decay of concert halls. *Build. Acoust.* **2005**, *12*, 1–14. [CrossRef]
168. Cappello, D.; Prodi, N.; Pompoli, R. A case history of coupled volumes in an historical opera house. In Proceedings of the ISRA 2007, Seville, Spain, 10–12 September 2007.
169. D'Orazio, D.; Fratoni, G.; Garai, M. Acoustics of a chamber music hall inside a former church by means of sound energy distribution. *Can. Acoust.* **2017**, *45*, 7–17.
170. Stanzial, D.; Bonsi, D.; Prodi, N. Measurement of new energetic parameters for the objective characterisation of an opera house. *J. Sound Vib.* **2000**, *232*, 193–211. [CrossRef]
171. D'Orazio, D.; Barbaresi, L.; Garai, M. A loudspeaker orchestra for opera houses studies. *J. Acoust. Soc. Am.* **2017**, *141*, 3998. [CrossRef]
172. Lokki, T.; Pätynen, J. Applying anechoic recordings in auralization. In Proceedings of the EAA Symposium on Auralization, Espoo, Finland, 15–17 June 2009; pp. 15–17.
173. Fabbri, P.; Farina, A.; Fausti, P.; Pompoli, R. The second life of the Teatro degli Intrepidi by G. B. Aleotti through the new techniques of virtual acoustics. In Proceedings of the 2nd International Conference on Acoustics and Musical Research (CIARM95), Ferrara, Italy, 19–21 May 1995. (In Italian)
174. Cocchi, A.; Farina, A.; Rocco, L. Reliability of scale-model researches: A concert hall case. *Appl. Acoust.* **1990**, *30*, 1–13. [CrossRef]
175. Ryu, J.K.; Jeon, J.Y. Subjective and objective evaluations of a scattered sound field in a scale model opera house. *J. Acoust. Soc. Am.* **2008**, *124*, 1538–1549. [CrossRef] [PubMed]
176. Cammarata, G.; Fichera, A.; Graziani, S. A Virtual Instrument for the Analysis of Objective Acoustic Parameters. In Proceedings of the 2nd International Conference on Acoustics and Musical Research (CIARM95), Ferrara, Italy, 19–21 May 1995.
177. Farina, A. Aurora listens to the traces of pyramid power. *Noise Vibr. Worldw.* **1995**, *26*, 6–9.
178. Neal, M.T.; Vigeant, M.C. A measurement database of US and European concert halls for realistic auralization and study of individual preference. *Proc. Inst. Acoust.* **2018**, *40*, 27–38
179. D'Orazio, D. Bayreuth Festspielhaus CAD and Virtual BRIRs, Mendeley Data. 2019. Available online: http://dx.doi.org/10.17632/d85cbd4d6y (accessed on 3 July 2020).
180. Doni, G.B.; Gori, A.F.; Passeri, G.B. *Lyra Barbarina Amphichordos: De' Trattati Di Mvsica*; Caesar Stamperia Imperiale: Florence, Italy, 1763. (In Italian)
181. Barbieri, P. The acoustics of Italian opera houses and auditoriums. *Recercare* **1998**, *10*, 263–328.
182. Bjurstrom, P. *Giacomo Torelli and Baroque Stage Design*; Almqvist & Wiksell: Stockholm, Sweden, 1961.
183. Zaslaw, N.; Spitzer, J. *The Birth of the Orchestra: History of an Institution*; Oxford University Press: New York, NY, USA, 2004.
184. Weinzierl, S.; Lepa, S.; Schultz, F.; Detzner, E.; von Coler, H.; Behler, G. Sound power and timbre as cues for the dynamic strength of orchestral instruments. *J. Acoust. Soc. Am.* **2018**, *144*, 1347–1355. [CrossRef] [PubMed]

185. Halpin, P.W. The Wind Band in Mozart's Opera Orchestra: Origins, Function, and Legacy. Available online: https://opencommons.uconn.edu/dissertations/AAI3265776 (accessed on 3 July 2020).
186. Bebbington, W.A. The Orchestral Conducting Practice of Richard Wagner. Ph.D. Thesis, Musicology, City Univ. of New York, New York, NY, USA, 1984.
187. Prodi, N.; Velecka, S. A scale value for the balance inside an historical opera house. *J. Acoust. Soc. Am.* **2005**, *117*, 771–779. [CrossRef]
188. Sato, S.; Prodi, N. On the subjective evaluation of the perceived balance between a singer and a piano inside different theatres. *Acta Acust. United Acust.* **2009**, *95*, 532–539. [CrossRef]
189. D'Orazio, D.; De Cesaris, S.; Garai, M. A comparison of methods to compute the "effective duration" of the autocorrelation function and an alternative proposal. *J. Acoust. Soc. Am.* **2011**, *130*, 1954–1961. [CrossRef] [PubMed]
190. D'Orazio, D.; Garai, M. The autocorrelation-based analysis as a tool of sound perception in a reverberant field. *Riv. Estet.* **2017**, *66*, 133–147. [CrossRef]
191. De Cesaris, S.; Morandi, F.; Loreti, L.; D'Orazio, D.; Garai, M. Notes about the early to late transition in Italian theatres. In Proceedings of the ICSV22, Florence, Italy, 12–16 July 2015.
192. Polack, J.-D.; Retbi, M. Wagner and the Cinema: A Cognitive Approach to the Acoustics of the Bayreuth Festival Theatre. In Proceedings of the Institute of Acoustics, Dublin, Ireland, 20–22 May 2011.
193. Prodi, N.; Pompoli, R.; Parati, L. The acoustics of the municipal theatre in Modena. In Proceedings of the Forum Acusticum 2002, Seville, Spain, 3–7 September 2002.
194. Iannace, G.; Iannello, C.; Maffei, L.; Romano, R. Room acoustic conditions of performers in an old opera house. *J. Sound Vib.* **2000**, *232*, 17–26. [CrossRef]
195. D.Lgs. 81/2008 Art. 98 Worker Safety: Guide Lines to Musicians and Perfomances. 2012. Available online: https://www.lavoro.gov.it/documenti-e-norme/studi-e-statistiche/Documents/Testo%20Unico%20sulla%20Salute%20e%20Sicurezza%20sul%20Lavoro/Testo-Unico-81-08-Edizione-Giugno%202016.pdf (accessed on 3 July 2020).
196. Burd, A.N. *Non-Reverberant Music for Acoustic Studies*; The British Broadcasting Corporation Report No. 1969/17; British Broadcasting Corporation: London, UK, 1969.
197. Hidaka, T.; Kageyama, K.; Masuda, S. Recording of anechoic orchestral music and measurement of its physical characteristics based on the auto-correlation function. *Acustica* **1988**, *67*, 68–70.
198. Farina, A.; Bigi, L.; Tronchin, L. Anechoic Recordings. Available online: http://www.angelofarina.it/Public/Anecoic/ (accessed on 3 July 2020).
199. Pätynen, J.; Pulkki, V.; Lokki, T. Anechoic recording system for symphony orchestra. *Acta Acust. United Acust.* **2008**, *94*, 856–865. [CrossRef]
200. Böhm, C.; Fiedler, F.; Weinzierl, S.; Holter, E.; Muth, S.; Schaefer, U.; Schwesinger, S. An Anechoic Recording of Ciceros 3rd Cataline Oration: Italian, Latin and German. 2019. Available online: http://dx.doi.org/10.14279/depositonce-8536 (accessed on 3 July 2020).
201. D'Orazio, D. Anechoic recordings of Italian opera played by orchestra, choir, and soloists. *J. Acoust. Soc. Am.* **2020**, *147*, 157–163. [CrossRef] [PubMed]
202. D'Orazio, D.; De Cesaris, S.; Garai, M. Recordings of Italian opera orchestra and soloists in a silent room. *Proc. Meet. Acoust.* **2016**, *28*, 015014. [CrossRef]

© 2020 by the authors. Licensee MDPI, Basel, Switzerland. This article is an open access article distributed under the terms and conditions of the Creative Commons Attribution (CC BY) license (http://creativecommons.org/licenses/by/4.0/).

Review

Review of Acoustic Sources Alternatives to a Dodecahedron Speaker

Nikolaos M. Papadakis [1,2,*] and Georgios E. Stavroulakis [1]

1. Institute of Computational Mechanics and Optimization (Co.Mec.O), School of Production Engineering and Management, Technical University of Crete, 73100 Chania, Greece
2. Department of Music Technology and Acoustics, Hellenic Mediterranean University, 74100 Rethymno, Greece
* Correspondence: nikpapadakis@isc.tuc.gr

Received: 18 July 2019; Accepted: 2 September 2019; Published: 6 September 2019

Abstract: An omnidirectional source is required in many acoustic measurements. Commonly a dodecahedron speaker is used but due to various factors (e.g., high cost, transportation difficulties) other acoustic sources are sometimes preferred. In this review, fifteen acoustic source alternatives to a dodecahedron speaker are presented while emphasis is placed on features such as omnidirectionality, repeatability, adequate sound pressure levels, even frequency response, accuracy in measurement of acoustic parameters and fulfillment of ISO 3382-1 source requirements. Some of the alternative acoustic sources have the appropriate features to provide usable results for acoustic measurements, some have acoustic characteristics better than a dodecahedron speaker (e.g., omnidirectionality in the high-frequency range), while some can potentially fulfill the ISO 3382-1 source requirements. Collected data from this review can be used in many areas (e.g., ISO measurements, head-related transfer functions measurements) for the appropriate selection of an acoustic source according to the expected use. Finally, suggestions for uses and future work are given aimed at achieving further advances in this field.

Keywords: acoustic measurements; impulse response measurements; omnidirectional source; dodecahedron; acoustic parameters; sound source; reverberation time; ISO 3382; auralization

1. Introduction

From the point of view of acoustics, a source is a region of space, in contact with the fluid medium where new acoustic energy is being generated, to be radiated outward as sound waves [1]. Source mechanisms, according to Fahy [2], may be broadly placed in one of the following three general categories on a phenomenological basis: fluctuating volume/mass displacement or injection, accelerating/fluctuating force on fluid and fluctuating fluid shear stress. Enlightening examples and description of the generation of sound from each category can be found in [3]. The most common category is the first one (fluctuating volume/mass displacement or injection) with sources such as loudspeakers, handclaps and vibration surfaces. The fluctuating volume/mass displacement or injection is the rate of change of the rate of fluid volume displacement (i.e., the volume acceleration) which determines the strength of the sound generated. It has to be noted that vibrating surfaces could also exert fluctuating forces on a contiguous fluid as byproducts of the fluid displacement activity [2]. Categorization of sources according to the generation of sound is also presented by Kurze [4].

Insights into the behavior of many practical acoustic sources can be obtained by considering elementary, idealized sources. A source that is concentrated at a point and produces an omnidirectional sound field is called a simple source or a monopole source [5]. The conceptually simplest sound source with finite extension is the spherical source, often referred to as 'pulsating' or 'breathing sphere' [6], which falls into the category of fluctuating volume/mass displacement or injection. A sphere, pulsating harmonically at any frequency at which its circumference is very much less than an acoustic wavelength, generates a sound field close to that of an ideal point monopole, except in the near field [7].

1.1. Acoustic Measurements with Omnidirectional Sources

A practical implementation of a pulsating sphere producing an omnidirectional field is useful in many fields of acoustics. An omnidirectional sound source is required in many acoustic measurements set by the International Organization for Standardization (ISO) as well as many standards set by national organizations, such as the American National Standards Institute (ANSI), the German Institute for Standardization (DIN-Deutsches Institut für Normung) and the British Standards Institution (BSI).

Due to the large number of standards, we will only mention standards for acoustic measurements by the ISO where a source with omnidirectional characteristics is required, such as ISO 3382-1 [8], ISO 3382-2 [9], ISO 3382-3 [10], ISO 354 [11], ISO 17497-1 [12], ISO 17497-2 [13], ISO 16283-1 [14], ISO 16283-3 [15] and ISO 10140-2 [16].

ISO 3382-1 describes the appropriate measurement of the impulse response and acoustic parameters of performance spaces, ISO 3382-2 of ordinary rooms and ISO 3382-3 of open-plan offices. An impulse response is the temporal evolution of the sound pressure observed at a point in a room as a result of the emission of a Dirac impulse at another point in the room. Acoustic parameters can be obtained directly from the impulse response and used to assess the acoustic quality of a space and provide guidance for possible improvements. ISO 354 describes the measurement of the sound absorption coefficient of materials performed in a reverberation chamber. ISO 17497-1 describes the measurement of the random-incidence scattering coefficient in a reverberation room (part two in a free field). ISO 16283-1 describes the measurement of the airborne sound insulation in buildings (part three for façade sound insulation) and ISO 10140-2 describes the laboratory measurement of sound insulation of building elements.

An acoustic source is also necessary for the measurement of head-related transfer functions (HRTFs) which are important for auralization purposes in many fields, such as virtual and augmented reality. HRTFs are defined as the free-field transfer functions, from a point sound source to each of the two ears on a fixed head. In practice HRTFs, describe the overall filtering effect imposed by anatomical structures. As stated by Xie [17], "in far-field HRTF measurements, the point sound source needed can be approximated by a common, small, loudspeaker system, where measurement errors caused by the size and directivity of the loudspeaker, as well as the multiple scattering between subject and loudspeaker, are negligible". However, in near-field measurements, the size of the source and scattering between subject and source is a key element for accurate measurements [18]. Under such conditions, a commonly used loudspeaker can no longer be regarded as a point source. Therefore the appropriate sound source is necessary for the measurement of the HRTFs. In relevant research [19], a dodecahedral sound source with a small radius (0.035 m) is proposed.

1.2. Requirements for Omnidirectional Sources

Requirements for an omnidirectional sound source aimed for acoustic measurements are provided in ISO 3382-1 [8], ISO 16283-1 [14], ISO 10140-5 [20] and ISO 140-3 [21]. However, the most referred ones are the requirements on ISO 3382-1. Those are:

"The sound source shall be as close to omnidirectional as possible. A maximum deviation of directivity of source in decibels for excitation with octave bands of pink noise and measured in free field is expected (Table 1)".

"The sound source shall produce a sound pressure level sufficient to provide decay curves with the required minimum dynamic range, without contamination by background noise. In the case of measurements of impulse responses using pseudo-random sequences (e.g., maximum-length sequence (MLS) [22]), the required sound pressure level might be quite low because a strong improvement of the signal-to-noise ratio by means of synchronous averaging is possible. In the case of measurements which do not use a synchronous averaging (or other) technique to augment the decay range, a source level will be required that gives at least 45 dB above the background level in the corresponding frequency band".

Table 1. The maximum deviation of directivity of source in decibels for excitation with octave bands of pink noise and measured in free field (ISO 3382-1).

Frequency (Hz)	125	250	500	1000	2000	4000
Maximum deviation (dB)	±1	±1	±1	±3	±5	±6

Less stringent requirements for a maximum deviation of directivity are described in ISO 16283-1 [14] and ISO 10140-5 [20] (Table 2). Specifications are also described in ISO 140-3 [21] which is now withdrawn. ISO 10140-5 states that: "Uniform omnidirectional radiation can be assumed if the directivity index (DI) values are within the limits of ±2 dB in the frequency range of 100 Hz to 630 Hz. In the range of 630 Hz to 1000 Hz, the limits increase linearly from ±2 dB to ±8 dB. They are 8 dB for frequencies of 1000 Hz to 5 000 Hz". It should be noted that the specifications are exactly the same for ISO 16283-1 beside a small difference. The ISO 16283-1 requires ±5 dB for 800 Hz a difference which is indistinguishable between the standards.

Table 2. The maximum deviation of directivity of source in decibels for excitation with octave bands of pink noise and measured in free field (ISO 16283-1, ISO 10140-5 and ISO 140-3 (withdrawn)).

Frequency (Hz)	100	630	1000	5000
Maximum deviation (dB)	±2	±2	±8	±8

Figure 1 presents the limits for the maximum deviation of directivity for all the standards. ISO 140-3 is also presented since it is still sometimes referred in some dodecahedron speaker specifications. Directions for the measurement of directivity can be found in some standards such as ISO 3382-1, ISO 10140-5, ISO 16283-1 and ISO 16283-2. The maximum acceptable deviations from omnidirectionality are measured when averaged over 'gliding' 30° arcs in a free sound field. In case a turntable cannot be used, measurements per 5° should be performed, followed by 'gliding' averages, each covering six neighboring points. The reference value shall be determined from a 360° energetic average in the measurement plane. The minimum distance between source and microphone shall be 1.5 m during these measurements. However some concerns have been expressed about measuring the directivity according to those standards [23–25]. Also, a new descriptor for measuring the directivity of dodecahedron speakers and omnidirectional sound sources has been proposed [26].

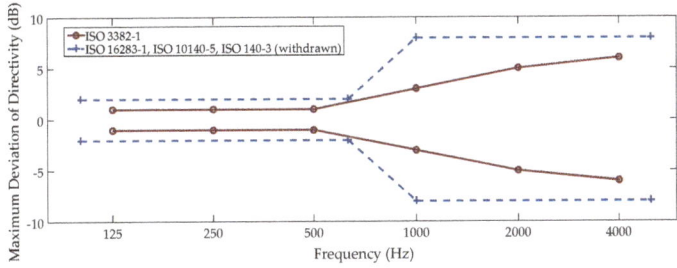

Figure 1. The maximum deviation of directivity of source in decibels for excitation with octave bands of pink noise and measured in free field for ISO 3382-1, ISO 16283-1, ISO 10140-5 and ISO 140-3 (withdrawn).

1.3. Dodecahedron Speakers

The most common practical implementation of a pulsating sphere producing an omnidirectional field is a dodecahedron speaker. As Kuttruff states [27]: "its radiation characteristics (pulsating sphere) can be approximated within certain limits by a regular dodecahedron or icosahedron composed of 12 or 20 regular polygons, respectively, each of them fitted out with a loudspeaker in its center". However, commercial implementations exist mainly for the dodecahedron speakers and less for the icosahedron speakers.

Dodecahedron speakers are omnidirectional sources widely used for room acoustics measurements. Commercially available dodecahedron speakers are manufactured and tested in order to meet the ISO 3382-1 sound source requirements. The omnidirectional directivity is approached by placing 12 electrodynamic loudspeakers (direct radiator type) in a regular 12-face polyhedron. The dodecahedron speaker, in essence, is a "dodecahedron arrangement of drivers to approximate the omnidirectional sound radiation characteristics of a monopole" [28]. However, other polyhedron loudspeakers can also be used which in some cases can be viewed as equally omnidirectional [29].

In practice, since dodecahedron speakers cannot emit sufficient acoustic power if an impulsive signal is applied directly, alternative excitation signals are used. The most common ones are maximum-length sequence (MLS) [22] and exponential sine sweep (ESS) [30] which are described in Annex A and B of ISO 18233 [31], respectively. Application of these methods ensures that the loudspeaker can emit a large amount of energy with longer duration without challenging its limited peak power capability, while impulse responses with a high time resolution can still be obtained afterward through post-processing. These deterministic excitation signals can be accurately reproduced and thereby enhance the repeatability of the measurements. Since the dodecahedron speaker utilizes MLS or ESS signals, enhancement of the signal-to-noise ratio by 20 dB to 30 dB or more compared to the classical method may be obtainable, as ISO 18233 [31] states. However, as ISO 18233 also states, the use of loudspeakers typically introduces non-linear distortion in the system which increases with the excitation level and violates the requirement for linearity, hence appropriate signal levels should be chosen. Optimum signal-to-noise ratios can be found in Stan et al. [32], while comparison of the two methods can be found in [33]. The appropriate choice of excitation signal for acoustic measurements can be depended on the background noise [34].

Drawbacks of Dodecahedron Speakers

Despite the widespread use of dodecahedron speakers, there are certain drawbacks associated with them. Namely, we can say deviation from omnidirectionality, low-frequency performance and practical reasons such as high cost.

The dodecahedron speaker fails to be an exact approximation of a monopole source since there is a deviation from omnidirectionality. The directivity of a dodecahedral loudspeaker can be considered uniform in the low-frequency range, while at higher frequencies (namely above 1 kHz), sound radiation shows greater deviation [26]. Because of the finite difference in distance between the loudspeaker diaphragms as well as the fact that these usually have a conical shape and are not a continuous part of the spherical surface, the radiation pattern is not ideally spherical at frequencies where the distance between loudspeakers or the depth of the cones are larger than a small fraction of the wavelength. Typically, the deviations start to become large when $ka > 3$, where a is the radius of the sphere and k is the wavenumber [35]. These deviations seem to increase, the smaller the measuring distance from the sound source inside the critical distance [36]. It has also been shown that constructive interference of the pressure field across the spherical baffle surface and not individual loudspeaker pistonic radiation characteristics is the most significant factor with respect to deviations from omnidirectional radiation [37]. Another reason for the deviation from omnidirectionality is that the impulse response of a dodecahedron speaker will feature contributions due to edge diffraction. Correspondingly, the frequency response of the loudspeaker will feature frequency response irregularities [35]. It has to be noted that these deviations from omnidirectionality have an effect on the measurement of the impulse response and acoustic parameters of a space [24,25].

However, stepwise rotation of a dodecahedron sound source can be employed to improve the accuracy of room acoustic measurements [38]. Also, three-way measuring loudspeakers with omnidirectional characteristics have been proposed as a possible solution [39]. There are also commercial implementations in which the twelve drivers of the dodecahedron speaker are placed in a sphere in order to avoid edge diffraction.

Another drawback, especially for dodecahedron speakers of smaller size, is their output in the low-frequency range. Since dodecahedron speakers utilize electrodynamic loudspeakers (direct radiator type) their performance at low frequencies depends on the size of the drivers [40,41]. This means that smaller-size dodecahedron speakers usually have lower output at low frequencies. However, subwoofers can be used in combination with a dodecahedron speaker in order to increase the sound pressure levels at low frequencies [39,42].

There are also other drawbacks of dodecahedron speakers, mainly due to practical reasons. Their high cost makes them probably the most expensive equipment in an acoustic measurement setup. Also, their heavy weight combined with their large volume makes transportation difficult e.g., transfers to airports. There are also cases where their use is required in places where there is no electricity supply. An external generator or an appropriate dodecahedron speaker with an internal generator can be used, which further increases the cost.

1.4. Aim of This Review

The authors would like to express the two main reasons which motivated the writing of this paper. Firstly, in order to provide the acoustic community with an organized overview of the alternative acoustic sources to a dodecahedron speaker and to present them in a way that emphasizes their important elements. There is a considerable amount of literature on the subject. However, there are no surveys presenting and examining all of these sources. Collective reference for audio sources can be found in the introduction of some publications [26,43].

Secondly, there seems to be a need for alternative sound sources to a dodecahedron speaker for acoustic measurements. Numerous examples can be found in the literature where alternative sound sources were used. Some examples and the sources that were used are: The measurement of impulse responses in open-air theatres (firecracker) [44,45], in churches (pistol shots and balloons) [46], in Buddhist temples (balloons) [47], measuring the acoustics of catacombs (balloons and firecrackers) [48,49], measurements in Stonehenge (balloons) [50], measurements in the Notre-Dame cathedral (balloons) [51], measurements in the Hagia Sofia (balloons) [52], measurements in urban environments (pistol shots) [53], green roofs absorption (pistol shots) [54], measurements in subway stations (firecrackers) [55], the acoustic of caves (balloons) [56,57], room acoustics (handclap) [58], barrier attenuation (shotshell primer) [59] and classroom acoustics [60] (wooden clapper). The reason that prompted the use of these alternative sources will be presented in the following related chapters.

It is worth noting that this review refers to sound-source alternatives to a dodecahedron speaker. This review does not include sources for the determination of sound power levels as stated by ISO 6926 [61] and used in various measurements such as ISO 3741 [62], ISO 3743-1 [63] and by the survey methods described in ISO 3747 [64]. Also this review does not include sources described in IEC 60268-16 [65] which specifies objective methods for rating the transmission quality of speech with respect to intelligibility (the standard requires a mouth simulator having similar directivity characteristics to those of the human head/mouth).

This paper has been organized in the following way: Section 2 presents the acoustic sources along with the relevant studies. The third section presents a discussion of the aforementioned studies and a critique of the significant findings and identifies areas for further research. Finally, the conclusion section gives a brief summary and contextualizes the research.

2. Acoustic Source Alternatives to a Dodecahedron Speaker

This chapter will present fifteen acoustic sources alternatives to a dodecahedron speaker (Table 3). The list includes only sources for which relevant publications were found in the literature. The most common alternative acoustic sources with the most references and practical applications are the balloon, gunshot, firecracker, handclap and inverse horn design. These sources are going to be presented first. The next sources in the list (wooden clapper, shotshell primer, rotation of directional speaker, ultrasound piezoelectric transducer, ring radiator, explosive mixture of acetylene gas with air and compressor nozzle hiss) have a smaller number of references in the literature. Directional speakers which are next in the list are not alternative sources to a dodecahedron speaker since they are not omnidirectional. However, they are used as such and therefore have been included in the list. There is a considerable amount of literature concerning the application of lasers in acoustics. However, a laser-induced breakdown is a relevant new acoustic source with potential for practical applications. Finally, electric spark sources have been mainly used for the simulation of acoustic phenomena using scale models. However, there are some publications where they are used as an alternative source for acoustic measurements and they are therefore included in this review.

Table 3. List of acoustic source alternatives to a dodecahedron speaker.

1. Balloon	7
2. Gunshot	9
3. Firecracker	10
4. Handclap	11
5. Inverse horn design	13
6. Wooden clapper	14
7. Shotshell primer	15
8. Rotation of a directional speaker	15
9. Ultrasound piezoelectric transducer (spherical distribution)	16
10. Ring radiator	17
11. Explosive mixture of acetylene gas with air	18
12. Compressor nozzle hiss	18
13. Directional speaker	19
14. Laser-induced air breakdown	20
15. Electric spark source	21

Emphasis will be placed on features such as omnidirectionality, repeatability, frequency response, adequate sound power, accuracy in measurement of acoustic parameters and fulfillment of ISO 3382-1 sound source requirements. The reason why these features were chosen to be presented in is because they are the most commonly studied features to be found in the literature and they also define the acoustic behavior of acoustic sources.

Source omnidirectionality is one of the two sound source characteristics required from the ISO 3382-1 standard. Omnidirectionality ensures uniform space excitation necessary for correct impulse response measurement. The directivity of the source influence the measurement of impulse responses and acoustic parameters of spaces [24,25]. For each source, we have collected various research that have investigated the directivity of the source in accordance with ISO 3382-1 or with more general criteria.

Source repeatability ensures that the same sound filed is produced from the sound source for each measurement and hence similar impulse response and acoustic parameters are measured. Source repeatability is considered to be given for acoustics sources that utilize electrodynamic speakers (e.g., dodecahedron speaker). However, for sources of the impulsive type (e.g., handclaps and balloons) repeatability is not certain and may involve large variations. For each source we have collected relevant data from the literature.

Adequate sound pressure levels (according to ISO 3382-1) ensure that there will be no contamination in the acoustic measurements by background noise. Sound pressure levels should be studied in conjunction with a frequency response of the sound source since ISO 3382-1 requires source level at least 45 dB above the background level in the corresponding frequency band (if synchronous averaging is not available). For each source, total sound pressure levels have been collected from the literature and for each frequency band if available.

Even frequency response (or relatively even frequency response) ensures correct impulse response measurement which is important for auralization purposes through the process of convolution [66]. Although there are no specific restrictions in ISO 3382-1 about the source even frequency response, beside the following: "the source and associated equipment should be adequate to radiate a sufficient signal level in all of the octave bands for 125 Hz to 4000 Hz", however, restrictions can be found in ISO 16283-3: "The sound field generated by the loudspeaker shall be steady and have a continuous spectrum in the frequency range considered. The differences between the sound power levels in the one-third octave bands that define the octave bands shall not be greater than 6 dB in the 125 Hz octave band, 5 dB in the 250 Hz octave band and 4 dB in octave bands with higher center frequencies". Relevant information considering the frequency response has been collected from the literature for each source.

Accuracy in measurements of acoustics parameters depends on omnidirectionality, frequency response and sound pressure levels of the acoustic source. Research from the literature is presented in acoustic parameters that were measured with the use of alternative acoustic sources. Emphasis is given where comparisons have been made with dodecahedron speakers.

Finally, ISO 3382-1 source requirements as stated in the introduction involve omnidirectionality and adequate sound pressure levels. Concluding remarks if the acoustic source fulfills ISO 3382-1 source requirements or potentially fulfills the requirements are going to be presented.

2.1. Balloon

The balloon is an affordable solution commonly used as an alternative source for acoustic measurements. The impulsive nature of the popping explosion is the fundamental property of interest [67]. However, the impulse response of a balloon burst is not ideal. Measurements can be found in many kinds of research and for different balloon sizes [67,68]. "When a balloon is burst perfectly, the resulting acoustic disturbance should have the shape of the letter N" [69]. In reality, there are deviations in the expected N shape of the acoustic disturbance which results in an N wave spectrum containing nulls [70]. As Horvat et al. [68] states, "balloons require more time to release the acoustic energy (compared to other sources) due to the certain amount of time required for cracking the balloon wall".

According to the research presented in the introduction, the justifications that prompted its use were mainly the lack of electric supply [47,56,57], affordability and ease of use [52]. Referring to the research that utilized the balloon as a sound source: "The difficulty of operating in a cumbersome environment prevented the use of a sound source like a dodecahedron loudspeaker with a power amplifier" (Acoustic of caves [57]), "However, the site is in condition of repairs; hence, it is not possible to use such measurement techniques (dodecahedron speaker)" (Buddhist temple [47]), "Balloons are inexpensive and easy use" (Hagia Sophia [52]), "Due to the impossibility to connect to the electricity grid, the use of the whole equipment was not possible as recommended by ISO 3382" (Acoustic of caves [56]).

Considering omnidirectionality, the balloon as a sound source does not fulfill the ISO 3382-1 standard, especially for lower frequency bands. A study from Pätynen et al. [67] showed that for different balloon types, the magnitude of deviations for directivity below the 500 Hz octave band is on the order of 6–9 dB, well above the standard limits. However, directivity at higher frequencies fulfilled the omnidirectional source conditions. The degree of omnidirectionality improved with balloon size for midrange frequencies and larger balloons were close to the standard in the 1 kHz band. The balloon was found to radiate mostly toward the direction of needle impact. Also in the same study there seemed to be evidence that there are variations in actual directivity as a function of inflation levels. Similar results about the omnidirectionality have been reported by Griesinger [71] and Cheenne et al. [72]. Schlieren imaging of a balloon burst [73] reveals that the shock front is not quite spherical. However, Vernon et al. [74] showed that hydrogen-oxygen balloons seem to show better omnidirectionality with much less directional deviation. The maximum deviations fit within the ISO 3382-1 limits at the high frequencies and nearly fit within the limits at the low frequencies.

Concerning the question of repeatability, the balloon does not satisfy this requirement, but only under certain conditions. Studies by Griesinger [71] and Horvat et al. [68] have shown that the balloon has poor repeatability. Also, Topa et al. [75], in order to evaluate the repeatability of measurements and the effect this has on the acoustic parameters, performed measurements with balloons for the same source and microphone positions. Results for Clarity (C_{80}) revealed differences for the whole frequency range. However, Pätynen et al. [67] have reported that balloon directivity patterns are stable over repetitions if certain criteria are met. Spectra and radiated sound from a balloon were quite constant for a given balloon type with consistent inflation and performing the same popping method. Consistent inflations levels were evaluated by measuring the maximum width diameter with a 1 cm margin of error. Also a study by Cheenne et al. [72], where anechoic recordings of balloon bursts were systematically acquired for various conditions of balloon diameters, puncture location and inflation pressure, reports that the results are quite consistent when averaged over one-third octave bands. However it seems both the results found in the studies of Griesinger [71] and Cheenne et al. [72] are difficult to replicate in real-life measurements in order to achieve similar results.

Relating to sound pressure levels, Horvat et al. [68] measured differed sized balloons and found levels ranging from 133 to 138 dB. Results from Pätynen et al. [67] ranged from 121 to 137.5 dB. Highest sound pressure levels were found, as expected, for the largest balloons.

Regarding frequency response, the balloon does not offer adequate excitation at the low frequencies [68]. The spectral content of balloon bursts clearly indicates the direct relation between balloon size and the overall spectrum. Specifically, the largest balloons were found to provide the highest amount of excitation, especially at low $1/3$-octave (or octave) bands, where proper excitation always represents a problem. For smaller balloon size, the excitation at low bands decreased. This is expected to cause variations in acoustic measurements especially where the background noise is high. Results also indicate that frequency responses have two emphasized frequencies which depend on balloon size and inflation level [67]. However, exploding hydrogen-oxygen balloons produce primarily low-frequency content, with characteristic frequencies on the order of 100–200 Hz [74].

In regard to the measurement of acoustic parameters, there are several studies where this matter is addressed [68,75–78]. A study by Jambrosic et al. [76] showed that for measurements of reverberation time (RT) with the use of a balloon, results will deviate in the low-frequency range in rooms, compared with studies using standard techniques (dodecahedron speaker). However, if the measured room is large and reverberant, smaller deviations are to be expected. A study by Topa et al. [75], revealed similar deviation in the low-frequency range for measurements of RT and early decay time (EDT). Also measurements of clarity (C_{80}) and center time (T_s) suffered from great deviations in the whole frequency spectrum. Griesinger [71] also stated that directionality of the source (balloon) is important for measurements of speech intelligibility. However, a technique by Abel et al. [70] can improve the measured results by converting recorded balloon pops into full audio bandwidth impulse responses. The technique is synthesizing the impulse response of the balloon pop according to the echo density and frequency band energies estimated in running windows.

In conclusion, the balloon does not meet the ISO 3382-1 sound source requirements since it cannot be considered an omnidirectional source and it also has a low sound level in the low-frequency range.

2.2. Gunshot

A gunshot is produced by a firearm which can be characterized as a "heat engine that converts stored chemical energy into kinetic energy" [79]. The sound from a firearm discharge consists of multiple acoustic events: the ballistic shockwave, internal gas leaks or ejections, the muzzle blast and reflections. The primary sound is the muzzle blast which is an explosive shockwave in air produced by propellant gasses under extremely high pressure that expands rapidly once the bullet exits the muzzle. The impulse response and the waveforms from each acoustic events which it is composed are presented in a study by Beck [80]. Impulse responses from different firearms can be found in the literature [81].

Traditionally, acousticians have used a firearm with blank cartridges as a sound source for acoustic measurements since it is an impulsive source that is lightweight and small enough to be easily transported. A gunshot is explicitly specified in the standard ISO 354 [11] as a possible alternative sound source. As the ISO states: "It is impossible in practice to create and radiate true Dirac delta functions, but short transient sounds (e.g., from shots) may offer close enough approximations for practical measurements". It is important to notice that a lot of the research for firearms has been conducted for forensic analysis or gunshot detection systems [81–84].

According to research presented in the introduction, the justifications that prompted its use as a sound source were emission of high sound levels [54] and ease of use [53]. Referring to the research: "Use of a starter pistol as an excitation source is a potentially useful impulse response recording method, especially in situations where use of the equipment required for sine sweep measurement is impractical or inappropriate" (Measurements in urban environments [53]), "A main advantage of such a device is the emission of high sound levels which makes a shot easily identifiable even at locations with high background noise levels" (Green roofs absorption measurements [54]).

Considering omnidirectionality, the gunshot appears to have directional characteristics. As presented in ISO 17201 [85], the muzzle blast is directional with sound levels on-axis ahead of the muzzle higher than levels directly behind the muzzle by up to 20 dB. Freytag et al. [79] state that "one explanation for the directivity of muzzle blasts is that the sound source is rapidly accelerating at muzzle discharge". Griesinger [71] states that the pistol appears to be directional at low frequencies, where there is a rise in energy from the side. Similar results about the directional characteristics of the handgun and for large-caliber weapons can be found in [86,87]. Recordings of different handguns as a function of azimuth can be found in [81]. In a study by Settles et al. [88] utilizing shadowgraphy, shockwaves of gunshots are depicted and variation in directionality is evident among different guns. However, a study of different handguns by Lamothe and Brandley [89] indicated that a 0.38 caliber gun has the best omnidirectional characteristics.

Concerning the question of repeatability, Dezelak et al. [90] state that differences in the noise characteristics between individual cartridges for the same gun are usually small, so the impulsive source can be replicated to a high degree. However, a study by Maher and Routh [81] comparing the on-axis peak pressure levels for ten shots for different handguns revealed that variability was observed among different guns. Griesinger [71] states that handguns have poor repeatability.

Regarding frequency response, several studies have shown that energy falls rapidly in the low-frequency range for handguns [71,79,89–91]. The frequency spectrum of a gun typically has a peak energy output in the 1 kHz to 2 kHz region [71]. Below this frequency the energy falls off rapidly (~14 dB/octave on a 1/3 octave analyzer). However the 0.38 caliber gun exhibits a significantly flatter frequency response [89]. For that reason, Bradley [92] used a 0.38 caliber pistol firing black powder in his research for auditorium acoustic. On the contrary, high caliber weapons have higher energy in the lower frequency range [87].

Relating to sound pressure levels, measurements from Beck et al. [80] ranged from 151 to 161 dB and measurements from Jambrosic et al. [91] measured variations from 148 to 168 dB. As expected, the handgun emits one of the highest sound pressure levels among sound sources.

In regard to the measurement of acoustic parameters, research by Fausti and Farina [77] revealed differences between RT measurements with a gun and techniques utilizing a dodecahedron speaker especially in the low-frequency range (more than 0.3 s difference in the 125 Hz octave band). A study by Jambrosic et al. [76] revealed differences in the measurement of RT between pistol shots of different calibers. In a study by Bradley [92], comparisons between measured values and calculated values (RT, C_{80}, C_{50} and C_{35}) based on ideal exponential decays showed that such predictions are not particularly accurate but give an indication of the mean trends. In a study by Dezelac et al. [90], a gunshot was used for sound insulation measurements. Similar results were found in a comparison between the apparent sound reduction index of a common partition obtained by the gunshot and conventional methods (utilizing a dodecahedron speaker). An advantage that a gunshot has as an impulsive sound source is that it offers the possibility of removing the flanking transmission, since a gun is much more decoupled with a floor, than a heavy loudspeaker [90].

In conclusion, the gunshot does not meet the ISO 3382-1 sound source requirements since it cannot be considered an omnidirectional source. However, it produces high sound pressure levels and therefore it is used in practice as a source in various measurements.

2.3. Firecracker

Firecrackers are small explosive devices primarily designed to produce a large amount of noise. They are wrapped in a cardboard or plastic, and usually in cylindrical cartridges casing to contain the explosive compound. The propellant inside is a kind of powder which can be a mixture of substances such as flash powder, cordite, smokeless powder, black powder, sulfur, charcoal, potassium nitrate, etc. Firecrackers, similar to other impulsive acoustic sources, have a typical N-pattern sound wave. As Horvat et al. states [68] "The explosion of a firecracker filled with explosive charge will provide a typical N-pattern sound wave, due to the fact that the burning speed of the explosive charge is high enough to enable almost instantaneous combustion of the whole quantity of explosive, thereby releasing a high amount of energy in a very short period of time". Impulse responses of firecrackers can be found in [26,93] while comparisons of different firecracker impulse responses can be found in [94].

According to research presented in the introduction, the main justification that prompted its use was to maximize signal to (background) noise ratio (SNR) for outdoor measurements [44,95]. Referring to the research: "Firecrackers were used in S1 and S2 (source positions), in order to overcome the problem of the low signal-to-noise ratio" (Open-air theatre [44]), "they (firecrackers) could maximize the signal-to-background noise ratio (SNR), ... dodecahedral source could have a limited sound power for open-air conditions" (Open-air theatre [95]). It is important to notice that in both these cases the source was used for open theater measurements. Firecrackers were also used in the research due to the lack of electricity [45,49].

Considering omnidirectionality, firecrackers seem to be among the few acoustic sources that do not exhibit directional characteristics. Arana et al. [93] performed directivity measurements with 16 microphones in a sphere of a 1.65 m diameter. For the measurements, different combinations were used, and microphones were placed at the vertical and horizontal planes and also on different meridians and parallels. Results presented in a polar shape (microphones were put on the equatorial circumference every 22.5°) showed that the directivity index of the sound levels obtained from a 20-explosion sample is less than 1 dB for all third-octave bands from 125 Hz to 8 kHz. The omnidirectional directivity of a firecracker can also be seen in a study by Settles et al. [88], exhibiting a shadowgram sequence of an explosion of 1 g of triacetone triperoxide in a cardboard cylinder.

Concerning repeatability, Arana et al. [93] presented temporary forms of acoustic signals from firecrackers revealing small but noticeable differences. In the same study, dispersion is evident in spectral power measurements and directivity diagrams. In a study by San Martin et al. [26] the standard deviation (STD) of the effective decay range (EDR) between measurements with a dodecahedron and a firecracker was assessed. The EDR (in dB) is a parameter which describes the available decay curve range to obtain the RT. It was found that due to the different sound power from each explosion, firecrackers have greater dispersion, especially in the high and medium frequencies compared to measurement with a dodecahedron speaker. The study concludes that, concerning repeatability, the firecrackers do not guarantee the extraordinary repeatability of techniques based on deterministic signals (MLS and ESS). In order to check the repeatability of measurements with firecrackers and the effect this has on the acoustic parameters, Topa et al. [75] performed measurements for the same source and microphone positions. Results for C_{80} revealed differences for the lower frequency band.

Regarding sound pressure levels, Horvat et al. [68] measured differed sized firecrackers and found levels ranging from 156 to 166 dB according to the size of the firecracker. Furthermore, the sound-to-noise ratio (SNR) between 1 and 8 kHz was found in a study [26] to be higher compared with measurements with dodecahedron speaker utilizing MLS and ESS signals. In an extensive work by Sharma et al. [96], sound pressure levels were measured for many different categories of firecrackers (ground shots, aerial shots and garland type) in an anechoic chamber and in a test site. There are also many kinds of research that have been conducted regarding sound pressure levels of firecrackers [94,96–98], related to the effect these levels have on hearing. It should be noted that because firecrackers cause high-pressure levels, care must be taken for the selection of a suitable microphone in acoustic measurements as well as appropriate precautions for the risk of hearing loss.

Relating to frequency response, results presented in various research [68,93,94] indicate that most energy emitted is concentrated in the frequency range from 500 Hz to 2 kHz with a roll-off below 1 kHz. San Martin et al. [26] stated that the biggest drawback of firecrackers is the low sound-to-noise ratio obtained for low frequencies. The same was denoted by Arana et al. [93], that the spectral power of a pseudo impulsive source (firecracker) is not sufficient in the low-frequency range. However, studies by Flamme et al. [94] and Horvat et al. [68] presented frequency spectra for different firecrackers showed that the maximum of energy is shifted towards lower frequencies with the increase of amount of explosive. The largest firecracker measured by Horvat et al. [68] exceeds 100 dB in the low-frequency range. This observation leads to the conclusion that in order to satisfy the ISO 3382-1 sound pressure level requirements (45 dB SNR) at low frequencies, the selection of the appropriate firecracker is necessary. Hence the utilization of firecrackers as an acoustic source in the low-frequency range should be exercised with caution and in conjunction with knowledge of the background noise levels for the optimal firecracker selection.

In regard to the measurement of acoustic parameters, San Martin et al. [26] present the most comprehensive research. Impulse response measurements across different halls performed with firecrackers (ten explosions) and two commercial dodecahedron speakers. The authors state that if both techniques are applied correctly, similar values are obtained for the acoustic parameters. In the same research, the STD for T_{30}, C_{80}, and G was also compared for the two sources (firecracker, dodecahedron speaker). The STD was found to be greater for the firecracker in the low-frequency range for T_{30} measurements across different enclosures. However the STD is remarkably higher for the dodecahedron speaker for C_{80} and source strength (G) in the high-frequency range across different enclosures. In a study by Arana et al. [93] acoustic measurements were performed with a firecracker. In the study it is stated that "average values for EDT were practically identical to those obtained with two other recent techniques" (however the techniques were not identified in the study). Also in a study by Topa et al. [75] measurements of acoustic parameters were performed with firecrackers, but they were only compared with measurements performed with balloons.

In conclusion, the firecracker as a sound source can potentially meet the ISO 3382-1 requirements concerning omnidirectionality. Research needs to be carried out according the ISO 3382-1 measurement procedure to certify that the firecracker can cover this requirement. Concerning the sound pressure levels requirements of ISO 3382-1, as stated in the introduction, they need to be at least 45 dB above the background level in the corresponding frequency band. Whether this condition is met depends on the firecracker used and the sound level that it creates in the low-frequency range, as well as the background noise in this frequency range.

2.4. Handclap

Handclap, the production of sound by striking the hands together, is an attractive source for acoustic measurements. It is common for acousticians to excite a room with a handclap in order to assess the quality of the space. It can also be used to detect unwanted acoustic phenomena such as echoes and flutter-echoes [26].

Sound generation by a handclap was the focus of some studies [99,100]. Research by Fletcher et al. [99] examined the underlying physics in order to explain the sound variations from a dull thud through a low-frequency pulse to a sharp high-frequency snap. Possible simplified geometries of a flat impact and a domed impact were examined. It was shown that a shock wave is generated for configurations of the hands that produce a loud sharp sound. The addition of a Helmholtz-type resonance is involved in the case of domed impacts. An impulse response of a handclap is presented in the same research. Schlieren imaging was used in a study [73] in order to depict shock waves from a handclap. In a study by Repp [100], the different sounds of a handclap were categorized in eight clapping modes according to hand configuration. A description and a figure for each configuration can be found in the research. The author states that additional variations may derive from such factors as hand curvature, stiffness, fleshiness of the palms, tightness of the fingers, precision and striking force. Relevant research by Peltola et al. [101] presents an analysis for synthesizing hand clapping sounds.

Some research utilized the handclap for acoustic measurements through the use of a smartphone as a sound recorder [102–104]. Measurements with a handclap were used mainly as a survey method for these studies to monitor room acoustics.

Considering omnidirectionality, a study by Griesinger [71] showed that a handclap has directional characteristics. Measurements in different directions revealed differences of more than 15 dB in certain frequency ranges. Therefore the handclap does not fulfill ISO 3382-1 requirements for omnidirectionality as expected.

Concerning repeatability, the handclap may have one of the lowest among acoustic sources. The reason being that the slightest variation of hand configuration may alter the generated impulse and spectral characteristics [100]. Also, there is considerable variability in spectral shapes of handclaps across individuals. In the same research, the amplitude standard deviations ranged from 0.7 to 5.2 dB across subjects. However it was found that it is possible to improve the repeatability of a handclap if it is produced by the same individual utilizing the exact same hand configuration every time.

Regarding sound pressure levels, a study measured the handclap to be 75.8 dB [102]. In a study by Seetharaman and Tarzia [103] the handclap averaged 26.4 dB above the background level (in a concert hall) with a standard deviation of 4.4 dB across measurements. In the same study however it was noted that signal processing steps can be introduced in order to improve its performance as a sound source. However, the handclap as a sound source does not fulfill ISO 3382-1 requirements for sound pressure levels.

Relating to frequency response, the handclap can be generated in different ways which result in different frequency responses. As stated before, eight clapping modes according to hand configuration were presented in a study by Repp [100]. It can be seen in all frequency responses that a roll-off is evident below 500 Hz. The same is reported in a similar study [102] and is in accordance with Griensiger [71] which states "handclaps suffer from poor low-frequency content". In a study by Fletcher [99], measurements of the sound of nominally flat and cupped natural handclaps were made in an anechoic environment. The author states that "a flat clap produces broad-band sound that typically extends to about 10 kHz while the spectrum of a domed clap usually has a subsidiary maximum somewhere below 1 kHz and then declines with frequency more rapidly than does the flat clap".

In regard to measurements of acoustic parameters, research utilized a handclap as a sound source and a smartphone as a sound recorder [103,104]. The handclap was used mainly as a survey method for these studies to monitor room acoustics. A study by Seetharamn and Tarzia [103] performed measurements of RT. Octave bands measurements above 250 Hz gave consistent results with small deviation, while lower frequencies were unreliable. However, the measurements have to be considered with caution since they were compared with measurements made with a balloon source. Another study by Huang [104] measured acoustic parameters (RT and T_s) in the Bayreuth Festspielhaus. Acoustic parameters were compared with results that were found in the literature and deviations were found that were attributed to reasons such as low SNR and the performance of micro electro-mechanical systems (MEMS) microphones found in smartphones.

In conclusion, the handclap as a sound source does not meet the ISO 3382-1 requirements since it cannot be considered an omnidirectional source and it has low sound pressure levels.

2.5. Inverse Horn Design

An inverse horn design sound source is formed by fitting the diaphragm of a conventional loudspeaker to an aperture hole through an inverted horn for concentrating the acoustical energy. The design was proposed by Polack et al. [105] and is based on the idea that a source of small dimensions is a good approximation of a point source with omnidirectional characteristics. The performance of such a point source depends on the interaction between the loudspeaker and the inverted horn. The careful design makes it possible to construct a very compact omnidirectional sound source satisfying the international standards. The original design has a very irregular frequency response due to the resonances of air generated inside the inverted horn. Cobo et al. [106] proposed applying inverse filtering in order to improve the frequency response of the source. Impulse responses of the source with and without inverse filtering are also presented in this research. It is worth noting that there are commercial implementations of the inverse horn design that fulfill the ISO 3382-1. Also construction of an inverse horn design source through 3-d printing is possible [107].

A similar approach to the inverse horn design is the loudspeaker-pipe sound source [108]. A point source can be formed if a loudspeaker is connected with a waveguide whose exit is small compared with the wavelength. The source is included in this section of the review since it is based on the same idea as the inverse horn design, even if it is not used as much as an alternative to a dodecahedron speaker. The major difficulties that hinder its use are its non-flat frequency response and the reflections between the pipe exit and its interior, but a digitally synthesized input signal can be used to overcome these drawbacks [109]. It has been used for the study of nozzle transmission characteristics [110] and to control internal noise propagation from aircraft engines [108]. It is stated in the ANSI S1.18 standard [111] for determining the acoustic impedance of ground surfaces, that a loudspeaker-pipe source can be utilized. Application for in situ ground impedance measurements can be found in the literature [112].

Considering omnidirectionality, various research and implementations showed that the inverse horn design can fulfill ISO 3382-1 directivity requirements [105–107,113]. In some of the research [106,107,113] inverse filtering was applied for flattening the frequency response but it also had a positive effect on the omnidirectionality of the sound source. Excellent results are presented in research by Cobo et al. [106], where the source, when equalized by inverse filtering, deviates less than ±1 dB in various frequencies. However, it has to be noted that the omnidirectionality of the source depends on the implementation.

Concerning repeatability, since the sound source is utilizing a conventional loudspeaker, it can be assumed that the same sound field is generated for each measurement. Therefore there will be no differences for measured impulse responses of a space, common for measurements conducted with an impulsive sound source.

Regarding frequency response, strong resonances can be seen in figures from various publications [105,106,113]. The amplitude of the resonances deviates more than 20 dB from the flat part of the frequency response [106]. As Ortiz et al. [113] states "it is governed (the frequency response) by the strong influence created by the Helmholtz resonance due to volume compliance and the mass of air in the aperture opening that affords regularly spaced peaks". The frequency of the resonant peaks and harmonics is determined by the volume of the cone. However, to equalize such irregular frequency response arising from resonances of air inside the inverted cone, as mentioned before, Cobo et al. [106] proposed applying inverse filtering. The technique pre-emphasizes the MLS signal driving the sound source so that zero phase or minimum-phase cosine magnitudes are radiated. Applying the proposed inverse filtering has many advantages: it flattens the frequency response, shortens the time response making the signal of the source more adequate for acoustical measurements and improves omnidirectionality. Inverse filtering was also utilized for other inverse horn design implementations [107,113].

Relating to sound pressure levels, 85 dB in each one-third-octave band between 100 Hz and 5 kHz and 102 dB full band was measured [105]. A little higher sound pressure levels can be found in a commercial implementation [114]. The inverse horn design has lower levels than traditional omnidirectional sources (dodecahedron speakers). However, since signals such as MLS [106,113] are usually utilized for acoustic measurements with the inverse horn design, synchronous averaging can be applied, thus enhancing the signal-to-noise ratio.

In regard to measurements of acoustic parameters, research where the inverse horn design was utilized as an alternative source and results obtained were compared with measurements with a dodecahedron speaker or another source could not be found in the literature. Since the source fulfills ISO 3382-1 requirements then it should provide similar results for the measurement of acoustic parameters compared with dodecahedron speakers. However, the inverse horn design was used as an omnidirectional source according to the ANSI S1.18 standard [115] for the measurement of ground impedance in a study by Cobo et al. [106].

In conclusion, the inverse horn design as a sound source meets the ISO 3382-1 criteria since it fulfills the requirements for omnidirectionality and emits acceptable sound pressure levels. Considerations should be taken in order to avoid contamination by background noise in acoustic measurements, especially in the low-frequency range.

2.6. Wooden Clapper

An impulsive sound can be created if two plates of wood are struck against each other. A study by Sumarac-Pavlovic et al. [116] created and measured a specially designed wooden clapper intended to operate as a source for acoustic measurements. The clapper consists of two identical parts connected by a hinge, allowing the two parts to revolve around the same axis. However, the shape of the waveform of the wooden clapper does not have the typical N-shape found in other impulsive sounds. It is more complex and represents a short intrinsic reverberation process lasting some few tenths of milliseconds. According to the classifications found in the literature, a clapper impulse can be considered as reverberant impact pulse waveform with B-duration of about 15 ms [117].

According to a research, the justifications that prompted its use as a sound source was the experimental estimations of impulse responses of old wood churches located at remote places (no road, no electricity) where none of the standard sound sources could be used [118]. The source was also used for the measurement of classroom acoustics [60]. Also, a wooden clapper can be used as an acoustic source for survey acoustic measurements through the use of a smartphone app [102].

Considering omnidirectionality, measurements performed in an anechoic environment showed that the proposed source does not fulfill the requirements of the ISO 3382-1. The polar diagrams presented in the study indicate that the deviations from omnidirectionality at the lower octave bands at central frequencies of 125, 250 and 500 Hz in both planes are within ±3 dB, above the ISO 3382-1 limits. However, the radiation of the wooden clapper is within the standard limits at higher frequencies.

Concerning repeatability, the variations of the impulse levels are relatively small if the clapper is operated by a trained person. They are within the limits of ±1–2 dB in all octave bands.

Regarding sound pressure levels, octave spectrum levels of the clapper impulse level at 1 m distance are ranging from (about) 72 dB (low-frequency range) to 100 dB. As the authors' state "measurements performed in rooms located in urban environments, where the ambient noise level was higher, showed that the dynamic range in the octave band at 125 Hz might be insufficient and that the measurement results have to be considered valid for the octave band at 250 Hz and higher".

Relating to frequency response, the authors present only the octave spectrum levels, in which the source seems to have a roll-off below 250 Hz. Similar roll-off was found for a wooden clapper in another study [102].

In regard to measurements of acoustic parameters, research where the wooden clapper was utilized as an alternative source and results obtained were compared with measurements with a dodecahedron speaker or another source could not be found in the literature.

In conclusion, the wooden clapper as a sound source does not meet the ISO 3382-1 sound source requirements since it cannot be considered an omnidirectional source and also does not have sufficient sound power in the low-frequency range.

2.7. Shotshell Primer

Shotshell primers contain a small amount of explosive, which is ignited inside of a tube by compressing its back section, crushing the explosive mixture. The tube directs the hot gas away from the holder region, the effective sound source being at the end of the tube where the gas expands out thus creating an impulsive sound. Research by Don et al. [119] has investigated the properties and design of a shotshell primer as an impulsive source. The source has also been used for measuring the effects of moisture content on soil impedance [120] and for the measurement of sound barrier attenuation [59].

Considering omnidirectionality, according to a polar graph presented in the research by Don et al. [119], the source appears to not fulfill the recommendations of ISO 3382-1. It is found experimentally that the sound field is conically symmetric around the axis formed by the tube.

Concerning repeatability, due to differences in the primers, small variations were measured in the peak levels typically by ±1 dB. Hence the source appears to have acceptable repeatability.

Regarding sound pressure levels, changes in the level tube can produce peak levels from 140 to 150 dB approximately. Increasing the tube length from 0.5 m to 2 m caused the peak level to drop by 10 dB. A number of different brands of primers have been tested with the peak level characteristics vary slightly, depending on the amount of explosive.

Relating to the frequency response, a normalized frequency spectra presented by the authors contains energies at frequencies between 100 Hz and 10 kHz, with the maximum intensity around 1 kHz. However, there is a steep roll-off in the low- and the high-frequency range.

In regard to measurements of acoustic parameters, the accuracy of measurements with a shotshell primer as an acoustic source has not been compared with measurements with a source that fulfills the ISO 3382-1 standard. However, as mentioned before the source have been used for measuring the effects of moisture content on soil impedance [120] and for the measurement of sound barrier attenuation [59]. For the measurements of sound barrier attenuation, the experimental results were compared with predictions from a variety of approximate and exact diffraction theories which allowed the attenuation to be determined with an accuracy of 1 dB. For the measurement of real and imaginary components of soil impedance, good agreement was obtained between measured and calculated values.

In conclusion, the shotshell primer as a sound source does not meet the ISO 3382-1 sound source requirements since it cannot be considered an omnidirectional source and also does not have sufficient sound power in the low-frequency range.

2.8. Rotation of a Directional Speaker

A sound source utilizing a common directional speaker through rotation and measurements in different placements was proposed and evaluated [43,121]. The source mimics the sound field emitted by a dodecahedron speaker by breaking it down in twelve different sound fields created from a single speaker for twelve different placements of the speaker (similar to the twelve positions of the faces of a dodecahedron speaker) [121]. Measurements were performed for every placement of the speaker and as a final step, the measurements were superimposed (the impulse responses) creating a single impulse response. In the following study [43], different common directional loudspeakers were used for creating an omnidirectional sound field for impulse response measurements and different placements of the loudspeakers were performed (different rotations of the loudspeakers for a total sum of twelve, twenty-six and fourteen positions).

Considering omnidirectionality, evidence from the research [43] indicated that utilization of different directional speakers results in different directional characteristics of the created sound field. Two-way design speakers that were used in the study employing a tweeter driver have a higher directivity in the high-frequency range compared with typical driver speakers. Hence, differences in the sound field created were detected which resulted in variations in the measurement of acoustic parameters. Another aspect of the results showed that a higher number of speaker placements possibly resulted in better omnidirectionality in the high-frequency range. The results indicated that the design of an appropriate directional speaker can potentially form a sound field with omnidirectional characteristics that fulfill the ISO 3382-1 sound source requirements.

Concerning repeatability, since a conventional speaker is utilized, it can be assumed that the same sound field can be generated for every set of measurements if the exact same placements of the speaker are performed. Therefore, there will be no differences for measured impulse responses of a space, common for measurements conducted with an impulsive sound source.

Regarding sound pressure levels, the suggested source can reach high levels that depend on the directional speaker that can be used. Also, since signals such as MLS and ESS can be utilized for acoustic measurements with the rotation of directional speakers, synchronous averaging can be applied thus enhancing the signal-to-noise ratio.

Relating to frequency response, signal energy tends to be evenly distributed across a wide range of frequencies in the case of a directional speaker of good quality. Hence, if a common directional speaker with a flat frequency response is utilized for the proposed method, then the emitted sound could presumably also approach a flat frequency response.

In regard to the measurement of acoustic parameters, measurements with a dodecahedron speaker and the proposed source were assessed in order to quantify the results. Evidence from these studies [43,121] point toward the idea that RT and EDT can be measured with the proposed source with excellent accuracy compared with measurements with a dodecahedron speaker. RT measurements showed a mean absolute error of less than 0.08 s, compared with measurements with a dodecahedron speaker. Results for C_{80} and definition (D_{50}) showed also a satisfactory accuracy but not as much as the results for the RT and EDT. A possible explanation might be that the differences in the sound fields created between the dodecahedron speaker and the proposed method are more profound in the early stages of the measured impulse responses for each case. Hence, since C_{80} and D_{50} parameters are an indication of early-to-late arriving sound energy, these differences affect the measurement of C_{80} and D_{50} more than the measurement of RT and EDT.

In conclusion, the rotation of an appropriate directional speaker can potentially form a sound field with omnidirectional characteristics that fulfill the ISO 3382-1 sound source requirements. However, that still remains to be verified.

2.9. Ultrasound Piezoelectric Transducer (Spherical Distribution)

An omnidirectional acoustic source can be realized consisting of a spherical distribution of hundreds of small ultrasound transducers [23,122]. The transducers emit audible sound thanks to the parametric acoustic array phenomenon which was first proposed by Westervelt [123]. In the parametric acoustic array phenomenon, an ultrasonic wave is emitted from a piezoelectric transducer, consisting of an ultrasonic carrier modulated with the desired audible signal. Thanks to nonlinear propagation effects in air, the primary field gets naturally demodulated, resulting in a strongly focused beam of audible sound.

Most utilizations of the parametric acoustic array phenomenon have been devoted to exploiting the highly focused beams that can be generated through planar arrays of piezoelectric transducer such as the audio spotlight applications [124]. However, it was not until the work of Sayin et al. [122] that it was used for the generation of omnidirectional sound fields instead of focused ones. A continuation and improvement of the work were provided by Arnela et al. [23]. Results from these two studies and implementations are going to be presented in the following paragraphs.

Considering omnidirectionality, measurements in an anechoic chamber with the ultrasound piezoelectric transducer (spherical distribution) [23] fulfilled the ISO 3382-1 requirements above 1 kHz and performed better than a dodecahedron speaker in the high-frequency range above 2 kHz. However, the directivity index deviations revealed that the source below 800 Hz does not fulfill the ISO 3382-1 sound source requirements. Similar results for omnidirectionality were reported for the prototype ultrasound piezoelectric transducer (spherical distribution) [122].

Concerning repeatability, since the source is using ultrasound piezoelectric transducers, it can be assumed that the same sound field can be generated for every measurement. Therefore there will be no differences for measured impulse responses of a space, common for measurements conducted with an impulsive sound source.

Regarding sound pressure levels, the source prototype [122] reached 90 dB SPL. Measurements presented for generation of pure tones [23] showed levels ranging from 46 to 59 dB for the ultrasound piezoelectric transducer for different octave bands. The sound source clearly has lower SPLs than the dodecahedron for all the frequencies. However, the authors state that in future works, broadband signals such as sine sweeps will be evaluated to get higher sound pressure levels and to characterize the source according to current regulations.

Regarding frequency response, measurements performed in octave bands [23] and 1/3 octave bands [122] indicate that the source experiences difficulties to generate strong sound pressure levels for the low-frequency range.

In regard to measurements of acoustic parameters, there are no results displayed, however the authors state [23] that in the future "the behavior of the source when performing acoustic tests in buildings, such as the RT of a room or the airborne sound insulation of a partition, will also be examined".

In conclusion, the ultrasound piezoelectric transducer (spherical distribution) does not meet the ISO 3382-1 omnidirectionally requirements in the low-frequency range. Also, considerations should be taken in order to avoid contamination of background noise in acoustic measurements, especially in the low-frequency area. However, the omnidirectionality of the source in the high-frequency range can possibly be utilized in conjunction with other acoustic sources.

2.10. Ring Radiator

An omnidirectional sound source consisting of two electrodynamic drivers, positioned face-to-face at a short distance (10 to 40 mm) has been proposed by Kruse et al. [125,126]. The drivers are equipped with smooth conical cabinets which are filled with damping material. Sound is emitted from the edge of the cavity which has a ring-shaped form. Different cabinet design, driver size and driver arrangements have been considered for the sound source. Results from the aforementioned studies are presented in the following paragraphs.

Considering omnidirectionality, measurements of directivity (250 to 12000 Hz) were performed for the proposed sound source in an anechoic chamber [125]. The directivity patterns were determined by playing back white noise and calculating the 3rd-octave spectra, recorded at 3 m distance with 2° resolution. Results are presented for a source with two 5 cm drivers and a source with two 12 cm drivers both at a distance of 20 mm. As the authors state "a closer inspection of the data reveals a maximum variation of ±4 dB up to 8 kHz for the 5 cm drivers" [125]. Greater variation is observed for the 12 cm drivers. Results for directivity can also be found in [126] for 5 cm drivers. The aforementioned results for the ring radiator seem very promising. However, it is not clear if the source fulfills the ISO 3382-1 omnidirectionality requirements. A study should be conducted according to the specifications of ISO 3382-1 directivity measurements and also in more planes.

Concerning repeatability, since the sound source is utilizing conventional electrodynamic drivers, it can be assumed that the same sound field is generated for each measurement. Therefore there will be no differences for measured impulse responses of a space, common for measurements conducted with an impulsive sound source.

Regarding sound pressure levels, a sound pressure of 107 dB at 1 m distance was measured [126] when driving the loudspeakers at their maximal specified power with pink noise between 150 Hz and 6 kHz. Self-built sources by the authors were able to generate 100 dB (small source) and 106 dB (medium source) [125].

Relating to frequency response, figures presented at [125,126] demonstrate strong peaks at 2.5 kHz. As authors state: "At about 2.5 kHz, a resonance can be observed associated with the diameter of the cavity enclosed by the loudspeakers". Experimentation showed that this resonance can be reduced by placing a sheet of damping material in the cavity. Authors state that inverted filtering could be applied to smooth the frequency response. Also a roll of below 150 Hz is observed, which may be justified by the frequency response of the drivers.

In regard to measurements of acoustic parameters, there are no results displayed in the research. However a problem concerning the impulse response that could possibly affect the measurement of acoustic parameters was identified by the authors [125]: "A problem was the impulse response, which in case of the medium (size) source indicated decay times in excess of 30 ms at 10 kHz. At high frequencies, the distance between the two drivers is no longer small compared to the wavelength, and waves emitted from one chassis will be reflected at the other on". The authors provided a solution by inserting a small piece of acoustic foam into the space between the two drivers, thus dampening axial wave propagation and lowering the decay times to a maximum of 12 ms at 2.5 kHz.

In conclusion, the ring radiator seems to have omnidirectional characteristics that can potentially fulfill the ISO 3382-1 requirements. However, that still remains to be verified. Also, the source can produce adequate sound pressure levels for acoustic measurements.

2.11. Explosive Mixture of Acetylene Gas with Air

An explosive air–gas mixture has been investigated as an impulse source by Jambrosic et al. [91]. The reaction of calcium carbide with water produces acetylene as the explosive gas and calcium hydroxide as the byproduct. In the study, the air–gas mixture was stored in a container with a soft ball as sealant.

Relating to sound pressure levels, the source reached 164.9 dB at 1 m. Measurements were performed outdoors and not in an anechoic chamber due to safety reasons.

Concerning frequency response, the source produced more than sufficient sound pressure levels in the entire frequency range of interest, with a superior low-frequency spectral content as well. The explosive mixture of acetylene gas had higher sound pressure levels in the low-frequency range than firecrackers, gunshots or balloons measured in the same study.

Regarding omnidirectionality, repeatability or measurements of acoustic parameters, the authors did not provide any results.

Finally, concerning the fulfillment of ISO 3382-1 requirements, the explosive mixture of acetylene gas with air seems to create a satisfactory sound pressure level in the whole spectrum and especially in the low-frequency range but there are no data concerning the omnidirectionality of the source. However, due to the fact that the source is of the impulsive type it is possible that an omnidirectional sound field can be created. However, the applicability of the source in real life measurements is doubtful due to practicality and safety reasons.

2.12. Compressor Nozzle Hiss

In a study by Szlapa et al. [86], a compressor with a small-diameter nozzle producing a hiss noise was used as a sound source. The compressor nozzle hiss was employed as a broad-band noise for the interrupted noise method, described in ISO 354 [11].

Considering omnidirectionality, four microphones were arranged symmetrically around the source at 2 m distance. As the authors state [86], relative ranges of the maximum sound levels revealed a mean value of around 2.7%. However, measurements were performed in a room and not in an anechoic chamber.

Concerning repeatability, the authors state that the general features generated by the source remained unchanged.

Relating to frequency response, measurements performed in 1/3 octave bands, up to 40 kHz in different rooms. Results reveal that the maximum may lie in the 40 kHz band or even at a higher frequency. The characteristic of the frequency response of the source which extends well beyond the human hearing range, and can possibly be used for animal studies. Also, the authors state that the maxima of the source are very likely to be distributed bimodally, with the second maximum occurring at 125–160 Hz with values being significantly lower than that at 40 kHz. However, the source has significant low levels in the low-frequency range.

Regarding sound pressure levels, measurements for the compressor nozzle hiss revealed 92 dB (A-filtered) while for the same study 133 and 126 dB were measured for a gunshot and a balloon burst respectively for the same configuration. Again the measurements were performed in a room.

In regard to the measurement of acoustic parameters, RT presented by the authors shows higher deviation in the low-frequency range for the nozzle compared to measurements with balloons and gunshots. However, in the mid and high and frequency range, the results appear to be similar.

In conclusion, there is not enough evidence concerning the omnidirectionality of the compressor nozzle hiss fulfilling the ISO 3382-1 requirements. Also, the source does not have sufficient sound pressure level in the low-frequency range.

2.13. Directional Speaker

Directional, electrodynamic (direct-radiator type) speakers have been used as a source for acoustic measurements [75,127–129]. An electrodynamic or moving-coil loudspeaker is an electromagnetic transducer for converting electrical signals into sounds [130]. There are two principal types of loudspeakers: Those in which the vibrating surface (called the diaphragm) radiates sound directly into the air (direct-radiator type), and those in which a horn is interposed between the diaphragm and the air (horn type). The direct-radiator type is the most common one and it is used in home and car entertainment, mobile devices and in public-address systems.

Directional speakers are not alternative sources to a dodecahedron speaker since it is well documented that they are not omnidirectional [131]. However, they have been used as a sound source for acoustic measurements due to low cost, convenience and also for speech intelligibility measurements. Therefore they have been included in this review. Referring to the research that utilized the source due to low cost and convenience: "simplification in the measurements due to technical limitations" (Acoustics of Orthodox Churches [127]) and "this paper objective assessment of the room's acoustics, using simple low-cost equipment and available software" (Experimental methods [75]).

Directional speakers have also been used for the determination of speech intelligibility. The IEC 60268-16 [65] (International Electrotechnical Commission) specifies objective methods for rating the transmission quality of speech with respect to intelligibility and requires a mouth simulator having similar directivity characteristics to those of the human head/mouth. Commercial implementations are available. A study by Soeta et al. [128] utilized an electrodynamic speaker in order to investigate the effects of the style of the liturgy on acoustic parameters. In the study, it was stated "a directional sound source might be a better approximation than an omnidirectional sound source for the purpose of the present research". The same approach was followed by Brezina [129] for the measurement of intelligibility and clarity of speech in Romanesque churches and by Dordevic et al. [132] for an Orthodox church.

Considering omnidirectionality, directional speakers do not fulfill ISO 3382-1 directivity requirements for the sound source. Directivity patterns can be found in Beranek and Mellow [131] or in specifications of commercial implementation. As Beranek and Mellow state: "above the frequency where $ka = 2$ (k is the wavenumber and a is the radius of the diaphragm of the speaker, usually between 800 and 2000 Hz), a direct-radiator speaker can be expected to radiate less and less power". The rate at which the radiated power would decrease, if the cone were a rigid piston, is between 6 and 12 dB for each doubling of frequency. This decrease in power output is not as apparent directly in front of the loudspeaker as at the sides because of directivity. That is to say, at high frequencies, the cone directs a larger proportion of the power along the axis than in other directions.

Concerning repeatability, it can be assumed that the same sound field can be generated for every set of measurements. Therefore there will be no differences for measured impulse responses of a space, common for measurements conducted with an impulsive sound source.

Regarding sound pressure levels, the suggested source can reach high levels that depend on the speaker that will be used. Also, since signals such as MLS and ESS can be utilized for acoustic measurements, synchronous averaging can be applied thus enhancing the signal-to-noise ratio.

Relating to frequency response, signal energy tends to be evenly distributed across a wide range of frequencies in the case of a directional speaker of good quality. Hence, if a common directional speaker with a flat frequency response is utilized for the proposed method, then the emitted sound could presumably also approach a flat frequency response.

In regard to the measurement of acoustic parameters, both a directional speaker and a dodecahedron speaker were used in a number of studies [133–135] for comparison purposes. In a study by Jambrosic et al. [133], speakers of different directivity and size were used (near-field monitor, sound reinforcement loudspeaker, active subwoofer). RT measurements were performed in three acoustically different spaces (acoustically treated listening room, reverberant corridor and theatre) and compared with measurements with an omnidirectional source. Deviations can be observed in the results with the largest measured in the reverberant corridor. The authors state: "if the room has an irregular shape or its surfaces are very reflective, the results are again significantly different from the ones obtained using the omnidirectional speaker as a referent source". In a study by Wallace and Harvie-Clark [135], RT measurements were performed in a multipurpose hall with a directional speaker and an omnidirectional one. Significant differences were measured if the directional speaker was used in one orientation only, compared with the omnidirectional speaker. However, measurements averaged made with the directional speaker pointing both horizontally and vertically, more accurately represent the results obtained with an omnidirectional speaker. In another study by the National Physical Laboratory (NPL) [134], a large recording studio and a small control room were given to participants to perform acoustic measurements. Among others, the effect of sound source was evaluated. Different sources were used (omnidirectional source, array speakers and directional speaker). Little variations were found in the high-frequency range. However in the low-frequency range, larger variations were found which is unexpected since directional speakers have better omnidirectionality in this range.

In conclusion, a directional speaker does not meet the ISO 3382-1 requirements since it cannot be considered an omnidirectional source.

2.14. Laser-Induced Air Breakdown

One of the many uses of pulsed lasers is to generate acoustic pulses in the air beside solid and liquid media [136,137] which can be used as a source for impulse response measurements within the audible bandwidth [138,139]. An acoustic point source can be generated by focusing a pulsed laser beam to rapidly heat the air at a focal point which produces a small expanding plasma ball. Plasma is formatted "through the cascade process caused by electrons emitted from atoms and molecules that have absorbed multiple photons through a multi-photon process when a laser beam is focused in a gas. A portion of this plasma energy is used to create a shock wave, which is the sound source generated by laser-induced breakdown" [139].

Referring to research, in a study by Bolanos et al. [138], the source was used for the measurement of the impulse response of a room. The accuracy of the results was assessed by comparison with impulse responses measured with a custom-made spherical loudspeaker and a directional loudspeaker. In a study by Hosoya et al. [139], the source was validated by measuring the resonant frequencies (from the impulse response) of a very small space and comparing them with those computed by a theoretical model. The source is also proposed for scale model work [140,141].

Considering omnidirectionality, measurements were performed in an anechoic chamber for directivity in octave bands at 1 m (10° resolution) [140]. The central frequencies were 75 kHz, 37.5 kHz, 18.75 kHz, 9.375 kHz, 4.6875 kHz and frequencies below 3.125 kHz. The results present an omnidirectional pattern for all frequency bands. The maximum difference in energy level is less than 1 dB within the range from 0° up to 90° for all frequency bands. In a subsequent study by the author [138], the magnitude response of the pressure pulse measured along 0°, 45° and 90° directions deviates by less than 0.5 dB at frequencies above 10 kHz, and no noticeable differences were found at frequencies less than 10 kHz. In a study by Hosoya et al. [139], the point sound source was generated in an anechoic box were twelve microphones (30° differences) were placed 80 mm from the sound source. The average power spectra measured at each point was presented showing small differences.

Concerning repeatability, a study by Bolanos et al. [140] measured twenty consecutive pulses at a distance of 0.8767 m. The waveform shape was conserved between emissions and only small sound level differences were evident. This was also true for other source–receiver distances. The standard deviation of the peak values of 20 repetitions was less than 0.3 dB for all measurement distances. In subsequent research [138], the standard deviation for the magnitude response of the pressure pulse at 1 m averaged over 100 measurements was less than 0.8 dB for all frequencies. The authors state: "The repeatability of laser-induced breakdown (LIB) depends mainly on the laser configuration, but other parameters, e.g., dust, may affect the generation of LIB and consequently produce deviations in the waveform from pulse to pulse". However, in another study [139] it is stated that: "as the laser pulse energy increases, the sound pressure of the point sound source generated by LIB increases, also the reproducibility decreases and the fluctuations become apparent".

Regarding sound pressure levels, the peak pressure value of the LIB was measured in a study [140] at 0.8 m and was approximately 360 Pa (145 dB). In another study [139] three levels of laser pulse energy were used: 335.9 mJ, 798.2 mJ and 990.9 mJ. As the laser pulse energy increased, the sound pressure of the point sound source generated by LIB increased. Different time responses of sound pressure generated by LIB ranged from to 480.3 Pa (147.6 dB) to 698.9 Pa (150.87 dB). It was also found that the amplitude of the sound pressure varies depending on the spot radius.

Relating to frequency response, differences were found among the research [139,140], possibly due to different implementations. In a study by Bolanos et al. [140], the magnitude response has its maximum at 20 kHz, a rising edge of 6 dB/octave for frequencies below 10 kHz and a decay edge of approximately 10 dB/octave for frequencies between 30 and 120 kHz. In a study by Hosoya et al. [139], the frequency response appears to be more even.

In regard to measurements of acoustic parameters, results and comparisons with other sources such as dodecahedron speakers were not found in the literature. However, a comparison between the impulse responses obtained with a dodecahedron speaker, a directional speaker and an LIB can be found in [138]. The impulse response obtained with LIB shows distinct reflections in contrast with measurements with the spherical loudspeaker. The authors justify this fact due to interference with reflections produced by the different driver elements in the spherical loudspeaker. The authors state: "The LIB presents characteristics close to an ideal point source thus providing an accurate measurement of the impulse response of the acoustic system". Therefore differences in the impulse response in regard to measurements with a dodecahedron speaker are to be expected according to the research. The results of these differences in acoustic parameter measurements have not yet been explored.

In conclusion, a LIB can form a sound field that can potentially have omnidirectional characteristics that fulfill the ISO 3382-1 requirements. However, that still remains to be verified. The LIB appears to be a very promising acoustic source for practical acoustic measurement applications.

2.15. Electric Spark Source

An electric spark discharge may be used as an impulse sound source for acoustic measurements [142,143]. The principle is based on the generation of an electric discharge by applying a high voltage between two electrodes. On the electrical breakdown of the air gap, heat is generated in the spark, causing a rapid expansion of the core gas [142]. This expansion results in the propagation of a shock wave which is the primary source of sound [144]. Following the initial shock, the air in the region of the core is raised in temperature above the ambient and the cooling of this air results in a secondary wave of lower frequency and intensity. Steel electrodes with variable radius can be used combined with a variable gap.

Electric spark sources have been mainly used as a source for scale models for the simulation of acoustic phenomena [145]. The most common scale model application being auditorium acoustics [146] while it has also been used for the study of sound propagation in urban areas [147]. Electric spark sources have also been used for the acoustical spectrometry of sound propagation through air-filled porous materials [148]. In addition, the acoustic properties of an absorbent material have been measured with a high voltage spark discharge as an impulse source [149]. Finally, high energy spark discharges have been used in studies for the measurement of impulse responses and acoustic parameters in spaces [150,151].

Considering omnidirectionality, a study by Ayrault et al. [145] showed that the behavior of spark discharges and their acoustic radiation depends greatly on the electrodes gap. Polar plots measured in an anechoic chamber for different electrode gaps showed that the overpressure (maximum pressure of the impulse) is reduced 2.5 dB at 90° (compared to 0°) and 4.4 dB for electrode gaps of 5 mm and 20 mm respectively. The authors did not present polar plots for smaller electrode gaps. In a similar study, polar plots presented by Hidaka et al. [152] showed a sufficient omnidirectionality. However, the authors showed results only for 5, 10 and 20 kHz without presenting details how the measurements were performed. It has to be noted that the research was implemented for scale models which require higher frequencies. Finally, a study by Shibayama et al. [153] presented high correlation between theoretical and experimental directivity patters of a spark discharge. However, results are presented for frequencies above 20 kHz. The study also presents high correlation between the estimated sound pulse waveforms and the measured ones for different directions.

Concerning repeatability, in a study by Hidaka et al. [152] the sound pressure waveforms of the spark discharge sound source (superposed 64 times) were presented with fairly good results. However, in the study, it is stated that "a waveform with high repeatability and big sound energy are incompatible conditions". In a study by Ayrault et al. [145], in order to establish repeatability, measurements were performed 40 times. The standard deviation of the pressure measurements was found not to decrease significantly with a greater measurements number. Cabot et al. [150,151] in his studies for the measurement of impulse responses in rooms, states that "an electric spark discharge provides the best combination of intensity, repeatability, reset time and portability". In a study by Picaut and Simon [147], a good reproducibility was obtained as small deviations were found in the spectrum for different discharges. However Qin and Attenborough [136] state about the variation of electric spark discharges that: "first, there can be significant variations in the spark waveforms from spark to spark under the same testing conditions. This is the result of temporal variations in the breakdown potential, randomness of the air breakdown channel between the electrodes, and vibration of the electrodes". In the same study, the relative variation of the peak pressure, defined as the standard deviation divided by the average pressure, was found less than 3% for laser generated acoustic shocks but about 9% for the acoustic signals from the electric sparks.

Regarding sound pressure levels, in a study by Ayrault et al. [145] up to 140 dB SPL at 1 m from the source were measured. In another study and implementation by Wyber [142], sound pressure levels reached 133 dB.

Relating to frequency response, figures are provided in Shibayama et al. [153], Latham [154], Hidaka et al. [152], Cabot et al. [151] and Picaut and Simon [147]. Spectra reveal that spark sources tend to be lacking in low-frequency energy. Also, the shape of the spectrum seems to have a roll-off starting at about 10 kHz.

In regard to measurements of acoustic parameters, while systems utilizing electric spark charges for measurements of acoustic parameters were proposed and implemented [150,151], measurements and comparison were not presented.

In conclusion, results so far about the electric spark source show that it is not omnidirectional in the audible frequency range, thus it does not fulfill the ISO 3382-1 requirements. Also, the source does not provide sufficient sound pressure levels in the low-frequency range.

3. Discussion

Fifteen sources have been identified in the literature which can be used as alternative sources to a dodecahedron speaker for acoustic measurements. The majority of the sound sources (9) are of the impulsive type: balloons, guns, firecrackers, handclaps, wooden clappers, shotshell primers, an explosive mixture of acetylene gas with air, laser-induced air breakdowns and electric spark sources. Four acoustic sources utilize an electrodynamic loudspeaker: Inverse horn design, rotation of directional loudspeakers, ring radiators and directional loudspeakers. One study utilizes piezoelectric transducers through spherical distribution. Finally, one study used a compressor nozzle hiss. A categorization of the sources according to whether they are impulsive or continuous is presented in Table 4.

Table 4. Categorization of the sources.

Source Categories	Sources
Impulsive	Balloon, gun, firecracker, handclap, wooden clapper, shotshell primer, an explosive mixture of acetylene gas with air, laser-induced air breakdown, electric spark source
Continuous	Inverse horn design, rotation of directional loudspeaker, ring radiator, directional loudspeaker, piezoelectric transducers through spherical distribution, compressor nozzle hiss

According to ISO 354 [11] and ISO 3382-1 [8], the methods for measuring RT and acoustic parameters can be classified to the interrupted noise method and the integrated impulse response method. A further classification presented in ISO 354 separates the integrated impulse response method into the direct and the indirect method. An extensive description of the methods can be found in the aforementioned standards. All the impulsive sources can be utilized in the direct integrated impulse response method. All the sources which employ an electrodynamic loudspeaker and the source of piezoelectric transducers through spherical distribution can be utilized in the indirect integrated impulse response method. Those sources can also be utilized in the interrupted noise method. The compressor nozzle hiss can only be employed in the interrupted noise method.

A summary of the features examined for the sources (omnidirectionality, repeatability, adequate sound pressure levels, even frequency response, accuracy in measurement of acoustic parameters and fulfillment of ISO 3382-1 sound source requirements) is presented in Table 5. A blank in the table implies that this feature has not been studied or that there is insufficient data to draw conclusions. The table has been created to present a general overview of the characteristics of the sources. However, the data in the table should be evaluated with caution and in conjunction with the corresponding research. Determining whether a source can be utilized and the expected accuracy that can be achieved in the measurements requires deeper knowledge. For example, in the case of the balloon as a sound source, we note that RT can be measured (except in the low-frequency range). This holds true but the accuracy of the measurements will be affected by the background noise in each frequency band as well as the shape and characteristics of the measuring space as it is stated in the presented research.

Table 5. Features of acoustic sources.

	Sources	Omnidirectionality	Repeatability	Even Frequency Response	Adequate Sound Pressure Levels	Accuracy in Measurement of Acoustic Parameters	Fulfillment of ISO 3382	Featured References
1	Balloon	No	No (yes under conditions)	No	Yes (no in the low-frequency range)	No (yes for RT except low-frequency range)	No	[47,52,55,56,67–78]
2	Gunshot	No	Yes (under conditions)	No	Yes (no in the low-frequency range)	No (yes for RT except low-frequency range)	No	[53,54,71,76,77,79–92]
3	Firecracker	Possibly Yes	Yes (slight deviations)	No	Yes (no in the low-frequency range)	Possibly yes	Possibly yes	[26,44,45,49,68,75,88,93–98]
4	Handclap	No	No	No	No	No	No	[71,73,99–104]
5	Inverse horn design	Yes	Yes	Yes (through inverse filtering)	Yes	Yes	Yes	[105–115]
6	Wooden clapper	No	Yes (under conditions)	No	Yes (no in the low-frequency range)	-	No	[60,102,116–118]
7	Shotshell primer	No	Yes	No	Yes (except low-frequency range)	-	No	[59,119,120]
8	Rotation of directional speaker	-	Yes	Yes	Yes	Yes (for RT)	-	[43,121]
9	Ultrasound Piezoelectric Transducer	No	Yes	No	No	-	No	[23,122–124]
10	Ring radiator	Possibly yes (under conditions)	Yes	No	Yes	-	-	[125,126]
11	Explosive mixture of acetylene gas with air	-	-	Yes	Yes	-	-	[91]
12	Compressor Nozzle Hiss	-	-	No	No	-	-	[86]
13	Directional Loudspeaker	No	Yes	Yes	Yes	No (yes for RT under conditions)	No	[75,127–135]
14	Laser-induced air breakdown	Possibly yes	Yes	No	Yes (no in the low-frequency range)	-	-	[13b–141]
15	Electric spark source	No (depends on the implementation)	Yes	Yes (depends on the implementation)	No (depends on the implementation)	-	No	[142–152]
-	Dodecahedron	Yes	Yes	Yes (depends on the implementation)	Yes (depends on the implementation for low-frequency range)	Yes	Yes	[26–12]

Data from this review indicate that besides dodecahedron speakers, other sound sources that fulfill ISO 3382-1 requirements are inverse horn designs and possibly firecrackers. However, for firecrackers there is no research that has been carried out according to the specifications of ISO 3382-1 for the measurement of omnidirectionality. Other sound sources that can potentially fulfill ISO 3382-1 source requirements are the rotation of a directional speaker, ring radiator and laser-induced breakdown.

Uses and Future Work

The primary use of this review is to present the acoustic sources alternatives to a dodecahedron speaker, so that the reader can obtain a broad perspective of the literature and the available choices as well as their most important features. Hopefully the review will help the researcher, the acoustic consultant and the sound engineer for the appropriate choice of acoustic source according to the acoustic measurement.

The review may also be useful for selecting a suitable source for performing acoustic measurements with minimum cost. As stated in the introduction, a dodecahedron speaker is the most expensive equipment in an acoustic measurement setup. Some of the low-cost sources that were presented are the balloon, firecracker, handclap and rotation of directional speaker. The reader can be informed about the aspects of each method, the accuracy that can be achieved, as well as details of the appropriate application.

Another possible use of this review is to help identify research gaps. As presented in Table 5, there are gaps in the description of characteristics for many acoustic sources. Also, some features of the sources have not been explored in the best possible way. For example, a common feature not studied in all sources is omnidirectionality according to the specifications described in ISO 3382-1. Finally some sources such as the rotation of a directional speaker, ring radiator and laser-induced air breakdown are promising and justify further research.

An option for future research is also the combination of sources for acoustic measurements. For example, research has shown that some sources have better omnidirectionality at high frequencies (e.g., firecracker) and could be combined with sources that have better omnidirectionality and higher-pressure levels at low frequencies (e.g., dodecahedron speaker) for a better excitation of a space.

Finally, we hope that this review will be useful in relevant areas where acoustics measurements are necessary. As mentioned in the introduction, an acoustic source is essential for the measurement of HRTFs which are important for auralization purpose in many fields such as virtual and augmented reality. This review can possibly be useful for researchers working in those fields.

4. Conclusions

Relevant studies have been presented in this review concerning acoustic sources alternatives to a dodecahedron speaker. Fifteen sources were identified in the literature. The majority of them are of the impulsive type: balloons, guns, firecrackers, handclaps, wooden clappers, shotshell primers, an explosive mixture of acetylene gas with air, laser-induced air breakdowns and electric spark sources. Four acoustic sources utilize an electrodynamic loudspeaker: inverse horn design, rotation of directional loudspeakers, ring radiators and directional loudspeakers. The two final sources are a compressor nozzle hiss and piezoelectric transducers through spherical distribution. Emphasis was placed on features such as omnidirectionality, repeatability, adequate sound pressure levels, even frequency response, accuracy in measurement of acoustic parameters and fulfillment of ISO 3382-1 sound source requirements. Elements about the generation of sound and the impulse response were also presented where they were available.

Results from this review have shown that besides dodecahedron speakers, other sound sources that fulfill ISO 3382-1 requirements are inverse horn designs and possibly firecrackers. Sources that can potentially fulfill ISO 3382-1 source requirements are the rotation of a directional speaker, ring radiator and laser-induced breakdown. Reviewing the literature has led us to conclude that there are alternative sound sources that in some occasions can provide usable results concerning the measurements of acoustic parameters. This study has identified research gaps which can be a fruitful area of future work. A possible way for future research could be the combination of different sources for acoustic measurements. Finally, we hope that this review will be useful in relevant areas such as the measurement of HRTFs which are important for auralization purposes.

Author Contributions: N.M.P. analyzed the data and wrote the manuscript; G.E.S. provided suggestions and guidance for the work, reviewed and edited the manuscript.

Funding: This research received no external funding.

Conflicts of Interest: The authors declare no conflict of interest.

Abbreviations

The following abbreviations are used in this manuscript:

ANSI	American National Standards Institute
C_{80}	Clarity
D_{50}	Definition
DI	Directivity Index
DIN	Deutsches Institut für Normung
EDR	Effective Decay Range
EDT	Early Decay Time
ESS	Exponential Sine Sweep
G	Source Strength
HRTF's	Head-Related Transfer Functions
IEC	International Electrotechnical Commission
ISO	International Organization for Standardization
LIB	Laser-Induced Breakdown
MLS	Maximum Length Sequence
NPL	National Physics Laboratory
RT	Reverberation Time
SNR	Sound to Noise Ratio
STD	Standard Deviation
T_s	Centre Time

References

1. Morse, P.M.; Ingard, K.U. *Theoretical Acoustics*; Princeton university press: Princeton, NJ, USA, 1986; p. 306.
2. Fahy, F.J. *Foundations of Engineering Acoustics*; Elsevier: Amsterdam, The Netherlands, 2000; p. 96.
3. Fahy, F. *Air: The Excellent Canopy*; Horwood Publishing: Amsterdam, The Netherlands, 2009; p. 68.
4. Müller, G.; Möser, M. *Handbook of Engineering Acoustics*; Springer Science & Business Media: Berlin/Heidelberg, Germany, 2012; p. 353.
5. Crocker, M.J. *Encyclopedia of Acoustics*; John Wiley: Hoboken, NJ, USA, 1997; p. 107.
6. Kuttruff, H. *Room Acoustics*; CRC Press: Boca Raton, FL, USA, 2016; p. 84.
7. Fahy, F.J. *Foundations of Engineering Acoustics*; Elsevier: Amsterdam, The Netherlands, 2000; p. 116.
8. ISO. *ISO 3382-1: 2009. Measurement of Room Acoustic Parameters-Part 1: Performance Spaces*; ISO: Geneva, Switzerland, 2009.
9. ISO. *ISO 3382-2: 2008. Measurement of Room Acoustic Parameters-Part 2: Reverberation Time in Ordinary Rooms*; ISO: Geneva, Switzerland, 2009.
10. ISO. *ISO 3382-3: 2012. Measurement of Room Acoustic Parameters-Part 3: Open Plan Offices Rooms*; ISO: Geneva, Switzerland, 2009.
11. ISO. *ISO 354: 2003. Acoustics–Measurement of Sound Absorption in a Reverberation Room*; ISO: Geneva, Switzerland, 2003.
12. ISO. *ISO 17497-1:2004. Acoustics—Sound-Scattering Properties of Surfaces, Part 1: Measurement of the Random-Incidence Scattering Coefficient in a Reverberation Room*; ISO: Geneva, Switzerland, 2004.

13. ISO. *ISO 17497-2:2004. Acoustics—Sound-Scattering Properties of Surfaces-Part 2: Measurement of the Directional Diffusion Coefficient in a Free Field*; ISO: Geneva, Switzerland, 2004.
14. ISO. *ISO 16283-1:2014. Acoustics—Field Measurement OF Sound Insulation IN Buildings AND OF Building Elements-Part 1: Airborne Sound Insulation*; ISO: Geneva, Switzerland, 2014.
15. ISO. *ISO 16283-3:2014. Acoustics—Field Measurement of Sound Insulation in Buildings and of Building Elements-Part 3: Façade Sound Insulation*; ISO: Geneva, Switzerland, 2014.
16. ISO. *ISO 10140-2 Acoustics—Laboratory Measurement of Sound Insulation of Building Elements. Part 2: Measurement of Airborne Sound Insulation*; ISO: Geneva, Switzerland, 2014.
17. Xie, B. *Head-Related Transfer Function and Virtual Auditory Display*; J. Ross Publishing: Richmond, VA, USA, 2013; p. 68.
18. Guang-Zheng, Y.; Bo-Sun, X.; Dan, R. Effect of sound source scattering on measurement of near-field head-related transfer functions. *Chin. Phys. Lett.* **2008**, *25*, 2926. [CrossRef]
19. Yu, G.-Z.; Xie, B.-S.; Rao, D. Directivity of spherical polyhedron sound source used in near-field HRTF measurements. *Chin. Phys. Lett.* **2010**, *27*, 124302. [CrossRef]
20. ISO. *ISO 10140-5.Acoustics—Laboratory Measurement of sound Insulation of Building Elements—Part 5: Requirements for Test Facilities and Equipment*; ISO: Geneva, Switzerland, 2014.
21. ISO. *ISO 140-3: 1995. Acoustics —Measurement of Sound Insulation in Buildings and of Building Elements—Part 3: Laboratory Measurement of Airborne Sound Insulation of Building Elements*; ISO: Geneva, Switzerland, 1995.
22. Schroeder, M.R. Integrated impulse method measuring sound decay without using impulses. *J. Acoust. Soc. Am.* **1979**, *66*, 497–500. [CrossRef]
23. Arnela, M.; Guasch, O.; Sánchez-Martín, P.; Camps, J.; Alsina-Pagès, R.; Martínez-Suquía, C. Construction of an Omnidirectional Parametric Loudspeaker Consisting in a Spherical Distribution of Ultrasound Transducers. *Sensors* **2018**, *18*, 4317. [CrossRef]
24. San Martín, R.; Arana, M. Uncertainties caused by source directivity in room-acoustic investigations. *The J. Acoust. Soc. Am.* **2008**, *123*, 133–138. [CrossRef] [PubMed]
25. Knüttel, T.; Witew, I.B.; Vorländer, M. Influence of "omnidirectional" loudspeaker directivity on measured room impulse responses. *J. Acoust. Soc. Am.* **2013**, *134*, 3654–3662. [CrossRef] [PubMed]
26. San Martín, R.; Arana, M.; Machín, J.; Arregui, A. Impulse source versus dodecahedral loudspeaker for measuring parameters derived from the impulse response in room acoustics. *J. Acoust. Soc. Am.* **2013**, *134*, 275–284. [CrossRef]
27. Kuttruff, H. *Room Acoustics*; CRC Press: Boca Raton, FL, USA, 2016; p. 85.
28. Kleiner, M. *Acoustics and Audio Technology*; J. Ross Publishing: Richmond, VA, USA, 2011; p. 8.
29. Leishman, T.W.; Rollins, S.; Smith, H.M. An experimental evaluation of regular polyhedron loudspeakers as omnidirectional sources of sound. *J. Acoust. Soc. Am.* **2006**, *120*, 1411–1422. [CrossRef]
30. Farina, A. Simultaneous measurement of impulse response and distortion with a swept-sine technique. In Proceedings of the Audio Engineering Society Convention 108, Paris, France, 19–22 February 2000.
31. ISO. *ISO 18233-2006 Acoustics—Application of New Measurement Methods in Building and Room Acoustics*; ISO: Geneva, Switzerland, 2006.
32. Stan, G.-B.; Embrechts, J.-J.; Archambeau, D. Comparison of different impulse response measurement techniques. *J. Audio Eng. Soc.* **2002**, *50*, 249–262.
33. Guidorzi, P.; Barbaresi, L.; D'Orazio, D.; Garai, M. Impulse responses measured with MLS or Swept-Sine signals applied to architectural acoustics: an in-depth analysis of the two methods and some case studies of measurements inside theaters. *Energy Procedia* **2015**, *78*, 1611–1616. [CrossRef]
34. Antoniadou, S.; Papadakis, N.M.; Stavroulakis, G.E. Measuring Acoustic Parameters with ESS and MLS: Effect of Artificially Varying Background Noises. In Proceedings of the Euronoise 2018, Heraclion, Crete, Greece, 27–31 May 2018.
35. Kleiner, M. *Acoustics and Audio Technology*; J. Ross Publishing: Richmond, VA, USA, 2011; p. 7.
36. Hak, C.; Wenmaekers, R.H.; Hak, J.P.; van Luxemburg, L. The source directivity of a dodecahedron sound source determined by stepwise rotation. In Proceedings of the Forum Acusticum, Aalborg, Denmark, 27 June–1 July 2011.
37. Quested, C.; Moorhouse, A.; Piper, B.; Hu, B. An analytical model for a dodecahedron loudspeaker applied to the design of omni-directional loudspeaker arrays. *Appl. Acoust.* **2014**, *85*, 161–171. [CrossRef]

38. Martellotta, F. Optimizing stepwise rotation of dodecahedron sound source to improve the accuracy of room acoustic measures. *J. Acoust. Soc. Am.* **2013**, *134*, 2037–2048. [CrossRef] [PubMed]
39. Behler, G.K.; Müller, S. *Technique for the Derivation of Wide Band Room Impulse Response*; Tecni Acustica: Madrid, Spain, 2000.
40. Vorländer, M. *Auralization: Fundamentals of Acoustics, Modelling, Simulation, Algorithms and Acoustic Virtual Reality*; Springer Science & Business Media: Berlin/Heidelberg, Germany, 2007; p. 24.
41. Borwick, J. *Loudspeaker and Headphone Handbook*; CRC Press: Boca Raton, FL, USA, 2012; p. 14.
42. Dick, D.A.; Neal, M.T.; Tadros, C.S.; Vigeant, M.C. Evaluation of a three-way omnidirectional sound source for room impulse response measurements. *J. Acoust. Soc. Am.* **2015**, *137*, 2394. [CrossRef]
43. Papadakis, N.M.; Stavroulakis, G.E. Low Cost Omnidirectional Sound Source Utilizing a Common Directional Loudspeaker for Impulse Response Measurements. *Appl. Sci.* **2018**, *8*, 1703. [CrossRef]
44. Bo, E.; Shtrepi, L.; Pelegrín Garcia, D.; Barbato, G.; Aletta, F.; Astolfi, A. The Accuracy of Predicted Acoustical Parameters in Ancient Open-Air Theatres: A Case Study in Syracusae. *Appl. Sci.* **2018**, *8*, 1393. [CrossRef]
45. Berardi, U.; Iannace, G.; Maffei, L. Virtual reconstruction of the historical acoustics of the Odeon of Pompeii. *J. Cult. Herit.* **2016**, *19*, 555–566. [CrossRef]
46. Martellotta, F.; Cirillo, E.; Carbonari, A.; Ricciardi, P. Guidelines for acoustical measurements in churches. *Appl. Acoust.* **2009**, *70*, 378–388. [CrossRef]
47. Manohare, M.; Dongre, A.; Wahurwagh, A. Acoustic characterization of the Buddhist temple of Deekshabhoomi in Nagpur, India. *Build. Acoust.* **2017**, *24*, 193–215. [CrossRef]
48. Trematerra, A.; Iannace, G. The acoustics of the catacombs of San Callisto in Rome. In Proceedings of the Meetings on Acoustics, Montreal, QC, Canada, 2–7 June 2013; 166ASA. p. 015001.
49. Lombardi, I.; Trematerra, A. The acoustics of the catacombs of Vigna Cassia in Syracuse. In Proceedings of the Meetings on Acoustics, Montreal, QC, Canada, 2–7 June 2013; 173EAA. p. 015009.
50. Fazenda, B.; Drumm, I. Recreating the sound of Stonehenge. *Acta Acust. United Acust.* **2013**, *99*, 110–117. [CrossRef]
51. Postma, B.; Katz, B. Acoustics of Notre-Dame Cathedral de Paris. *Intl. Cong. Acoust.* **2016**, *269*, 1–10.
52. Pentcheva, B.V. Hagia Sophia and multisensory aesthetics. *Gesta* **2011**, *50*, 93–111. [CrossRef]
53. Stevens, F.; Murphy, D. Spatial impulse response measurement in an urban environment. In Proceedings of the Audio Engineering Society Conference: 55th International Conference: Spatial Audio, Helsinki, Finland, 27–29 August 2014.
54. Van Renterghem, T.; Botteldooren, D. In-situ measurements of sound propagating over extensive green roofs. *Build. Environ.* **2011**, *46*, 729–738. [CrossRef]
55. Iannace, G.; Berardi, U.; Giordano, G. Acoustic Characteristics of Four Subway Stations in Naples, Italy. In Proceedings of the 44th International Congress and Exposition on Noise Control Engineering, San Francisco, CA, USA, 9–12 August 2018.
56. Iannace, G.; Trematerra, A. The acoustics of the caves. *Appl. Acoust.* **2014**, *86*, 42–46. [CrossRef]
57. Iannace, G.; Ianniello, C.; Ianniello, E. Acoustic measurements in underground rooms of Castelcivita Caves (Italy). In Proceedings of the Euronoise, Prague, Czech Republic, 10–13 June 2012.
58. Almahdi, R.; Diharjo, K.; Hidayat, R.L.G.; Suharty, N.S. In situ test: acoustic performance of eco-absorber panel based albizia wood and sugar palm fiber on meeting room in UNS Inn Hotel. In Proceedings of the IOP Conference Series: Materials Science and Engineering, Iasi, Romania, 16–17 May 2019; p. 012033.
59. Papadopoulos, A.; Don, C. A study of barrier attenuation by using acoustic impulses. *J. Acoust. Soc. Am.* **1991**, *90*, 1011–1018. [CrossRef]
60. Puglisi, G.E.; Prato, A.; Sacco, T.; Astolfi, A. Influence of classroom acoustics on the reading speed: A case study on Italian second-graders. *J. Acoust. Soc. Am.* **2018**, *144*, 144–149. [CrossRef] [PubMed]
61. ISO. *ISO 6926:2016. Acoustics—Acoustics—Requirements for the Performance and Calibration of Reference Sound Sources Used for the Determination of Sound Power Levels*; ISO: Geneva, Switzerland, 2016.
62. ISO. *ISO 3741:2010. Acoustics—Acoustics—Determination of Sound Power Levels and Sound Energy Levels of Noise Sources Using Sound Pressure—Precision Methods for Reverberation Test Rooms*; ISO: Geneva, Switzerland, 2010.
63. ISO. *ISO 3743-1:2010. Acoustics—Acoustics—Determination of Sound Power Levels and Sound Energy Levels of Noise Sources Using Sound Pressure—Engineering Methods for Small Movable Sources in Reverberant Fields—Part 1: Comparison Method for a Hard-Walled Test Room*; ISO: Geneva, Switzerland, 2010.

64. ISO. *ISO 3747:2010. Acoustics—Determination of Sound Power Levels and Sound Energy Levels of Noise Sources Using Sound Pressure—Engineering/Survey Methods For Use in Situ in a Reverberant Environment*; ISO: Geneva, Switzerland, 2010.
65. Commission, I.I.E. *IEC 60268-16:2011. Sound System Equipment-Part 16: Objective Rating of Speech Intelligibility by Speech Transmission Index*; IEC: Geneva, Switzerland, 2011.
66. Vorländer, M. *Auralization: Fundamentals of Acoustics, Modelling, Simulation, Algorithms and Acoustic Virtual Reality*; Springer Science & Business Media: Berlin/Heidelberg, Germany, 2007; p. 137.
67. Pätynen, J.; Katz, B.F.; Lokki, T. Investigations on the balloon as an impulse source. *J. Acoust. Soc. Am.* **2011**, *129*, 27–33. [CrossRef] [PubMed]
68. Horvat, M.; Jambrosic, K.; Domitrovic, H. A comparison of impulse-like sources to be used in reverberation time measurements. *J. Acoust. Soc. Am.* **2008**, *123*, 3501. [CrossRef]
69. Deihl, D.T.; Carlson, F.R., Jr. "N Waves" from Bursting Balloons. *Am. J. Phys.* **1968**, *36*, 441–444. [CrossRef]
70. Abel, J.S.; Bryan, N.J.; Huang, P.P.; Kolar, M.; Pentcheva, B.V. Estimating room impulse responses from recorded balloon pops. In Proceedings of the Audio Engineering Society Convention 129, San Francisco, CA, USA, 4–7 November 2010.
71. Griesinger, D. Beyond MLS-Occupied hall measurement with FFT techniques. In Proceedings of the Audio Engineering Society Convention 101, Los Angeles, CA, USA, 8–11 November 1996.
72. Cheenne, D.J.; Ardila, M.; Lee, C.G.; Bridgewater, B. A qualitative and quantitative analysis of impulse responses from balloon bursts. *J. Acoust. Soc. Am.* **2008**, *123*, 3909. [CrossRef]
73. Hargather, M.J.; Settles, G.S.; Madalis, M.J. Schlieren imaging of loud sounds and weak shock waves in air near the limit of visibility. *Shock Waves* **2010**, *20*, 9–17. [CrossRef]
74. Vernon, J.A.; Gee, K.L.; Macedone, J.H. Acoustical characterization of exploding hydrogen-oxygen balloons. *J. Acoust. Soc. Am.* **2012**, *131*, 243–249. [CrossRef] [PubMed]
75. Țopa, M.D.; Toma, N.; Kirei, B.S.; Homana, I.; Neag, M.; De Mey, G. Comparison of different experimental methods for the assessment of the room's acoustics. *Acoust. Phys.* **2011**, *57*, 199–207. [CrossRef]
76. Jambrosic, K.; Horvat, M.; Domitrovic, H. Reverberation time measuring methods. *J. Acoust. Soc. Am.* **2008**, *123*, 3617. [CrossRef]
77. Fausti, P.; Farina, A. Acoustic measurements in opera houses: comparison between different techniques and equipment. *J. Sound Vib.* **2000**, *232*, 213–230. [CrossRef]
78. Rusiana, A.A.; Aves, J.M.C.; Hofileña, K.C. Validation of Balloon Burst Method in Measurement of Reverberation Time in a Classroom. *Int. J. Innov. Sci. Res* **2015**, *17*, 131–135.
79. Freytag, J.C.; Begault, D.R.; Peltier, C.A. The acoustics of gunfire. In Proceedings of the INTER-NOISE and NOISE-CON Congress and Conference Proceedings, Honolulu, HI, USA, 3–6 December 2006; pp. 1165–1174.
80. Beck, S.D.; Nakasone, H.; Marr, K.W. Variations in recorded acoustic gunshot waveforms generated by small firearms. *J. Acoust. Soc. Am.* **2011**, *129*, 1748–1759. [CrossRef] [PubMed]
81. Maher, R.; Routh, T. Wideband audio recordings of gunshots: waveforms and repeatability. In Proceedings of the Audio Engineering Society Convention 141, Los Angeles, CA, USA, 29 September–1 October 2016.
82. Maher, R.; Routh, T. Advancing forensic analysis of gunshot acoustics. In Proceedings of the Audio Engineering Society Convention 139, New York, NY, USA, 29 October–1 November 2015.
83. Page, E.A.; Sharkey, B. SECURES: system for reporting gunshots in urban environments. In Proceedings of the Public Safety/Law Enforcement Technology, Orlando, FL, USA, 19–20 April 1995; pp. 160–173.
84. Maher, R.C.; Routh, T. Gunshot acoustics: pistol vs. revolver. In Proceedings of the Audio Engineering Society Conference: 2017 AES International Conference on Audio Forensics, Arlington, VA, USA, 15–17 June 2017.
85. Din, E. *ISO 17201-1: 2005–11: Akustik–Geräusche von Schießplätzen–Teil 1: Bestimmung des Mündungsknalls durch Messung (ISO 17201-1: 2005); Deutsche Fassung EN ISO 17201-1*; Beuth: Berlin, Germany, 2005.
86. Szłapa, P.; Boroń, M.; Zachara, J.; Marczak, W. A Comparison of Handgun Shots, Balloon Bursts, and a Compressor Nozzle Hiss as Sound Sources for Reverberation Time Assessment. *Arch. Acoust.* **2016**, *41*, 683–690. [CrossRef]
87. Pääkkönen, R.; Parri, A.; Tiili, J. Low-frequency noise emission of finnish large-calibre weapons. *J. Low Freq. NoiseVib. Act. Control* **2001**, *20*, 85–92. [CrossRef]

88. Settles, G.S.; Grumstrup, T.; Miller, J.; Hargather, M.; Dodson, L.; Gatto, J. Full-scale high-speed "Edgerton" retroreflective shadowgraphy of explosions and gunshots. In Proceedings of the Fifth Pacific Symposium on Flow Visualisation and Image Processing, Daydream Island, Australia, 26–29 Spetember 2005.
89. Lamothe, M.R.; Bradley, J. Acoustical characteristics of guns as impulse sources. *Can. Acoust.* **1985**, *13*, 16–24.
90. Deželak, F.; Čurović, L.; Čudina, M. Determination of the sound energy level of a gunshot and its applications in room acoustics. *Appl. Acoust.* **2016**, *105*, 99–109. [CrossRef]
91. Jambrosic, K.; Horvat, M.; Bogut, M. Comparison of impulse sources used in reverberation time measurements. In Proceedings of the ELMAR, 2009, International Symposium, Zadar, Croatia, 28–30 September 2009; pp. 205–208.
92. Bradley, J. Auditorium acoustics measures from pistol shots. *J. Acoust. Soc. Am.* **1986**, *80*, 199–205. [CrossRef]
93. Arana, M.; Vela, A.; San Martin, L. Calculating the impulse response in rooms using pseudo-impulsive acoustic sources. *Acta Acust. United Acust.* **2003**, *89*, 377–380.
94. Flamme, G.A.; Liebe, K.; Wong, A. Estimates of the auditory risk from outdoor impulse noise I: Firecrackers. *Noise Health* **2009**, *11*, 223. [CrossRef] [PubMed]
95. Bo, E.; Kostara-Konstantinou, E.; Lepore, F.; Shtrepi, L.; Puglisi, G.E.; Astolfi, A.; Barkas, N.; Mangano, B.; Mangano, F. Acoustic Characterization of the Ancient Thea-Tre of Tyndaris: Evaluation and Proposals For Its Reuse. In Proceedings of the 23rd International Congress on Sound and Vibrations. Athens: International Institute of Acoustics and Vibration, Athens, Grece, 10–14 July 2016.
96. Sharma, O.; Mohanan, V.; Singh, M. Characterisation of sound pressure levels produced by crackers. *Appl. Acoust.* **1999**, *58*, 443–449. [CrossRef]
97. Čudina, M.; Prezelj, J. Noise due to firecracker explosions. Proceedings of the Institution of Mechanical Engineers, Part C. *J. Mech. Eng. Sci.* **2005**, *219*, 523–537. [CrossRef]
98. Tandon, N. Firecrackers noise. *Noise Vib. Worldw.* **2003**, *34*, 9–12. [CrossRef]
99. Fletcher, N.H. Shock waves and the sound of a hand-clap–a simple model. *Acoust. Aust.* **2013**, *41*, 165–168.
100. Repp, B.H. The sound of two hands clapping: An exploratory study. *J. Acoust. Soc. Am.* **1987**, *81*, 1100–1109. [CrossRef]
101. Peltola, L.; Erkut, C.; Cook, P.R.; Valimaki, V. Synthesis of hand clapping sounds. *IEEE Trans. Audio Speech Lang. Process.* **2007**, *15*, 1021–1029. [CrossRef]
102. Rizzi, L.; Ghelfi, G.; Campanini, S.; Rosati, A. Rapid Room Acoustics Parameters Measurements with Smartphones. In Proceedings of the ISCV22, Florence, Italy, 12–16 July 2019.
103. Seetharaman, P.; Tarzia, S.P. The hand clap as an impulse source for measuring room acoustics. In Proceedings of the Audio Engineering Society Convention 132, Budapest, Hungary, 27–29 April 2012.
104. Huang, K. Impulse response of the Bayreuth Festspielhaus. *arXiv* **2017**, arXiv:1703.07080.
105. Polack, J.-D.; Christensen, L.S.; Juhl, P.M. An innovative design for omnidirectional sound sources. *Acta Acta Acust. United Acust.* **2001**, *87*, 505–512.
106. Cobo, P.; Ortiz, S.; Ibarra, D.; de la Colina, C. Point source equalised by inverse filtering for measuring ground impedance. *Appl. Acoust.* **2013**, *74*, 561–565. [CrossRef]
107. Ibarra, D.; Ledesma, R.; Lopez, E. Design and Construction of an Omnidirectional Sound Source with Inverse Filtering Approach for Optimization. *HardwareX* **2018**, *4*, e00033. [CrossRef]
108. Ramakrishnan, R.; Salikuddin, M.; Ahuja, K. Generation of desired signals from acoustic drivers. *J. Sound Vib.* **1982**, *85*, 39–51. [CrossRef]
109. Jing, X.; Fung, K.-Y. Generation of desired sound impulses. *J. Sound Vib.* **2006**, *297*, 616–626. [CrossRef]
110. Salikuddin, M.; Brown, W.; Ramakrishnan, R.; Tanna, H. *Refinement and Application of Acoustic Impulse Technique to Study Nozzle Transmission Characteristics*; NASA: Washington, DC, USA, 1983.
111. ANSI/ASA. *American National Standard Method for Determining the Acoustic Impedance of Ground Surfaces*; Acoustical Society of America: New York, NY, USA, 2010.
112. Kruse, R. Application of the two-microphone method for in-situ ground impedance measurements. *Acta Acust. United Acust.* **2007**, *93*, 837–842.
113. Ortiz, S.; Kolbrek, B.; Cobo, P.; González, L.M.; Colina, C.d.l. Point source loudspeaker design: advances on the inverse horn approach. *J. Audio Eng. Soc.* **2014**, *62*, 345–354. [CrossRef]
114. Kjaer, B.A. Lightweight Omnidirectional Loudspeaker-OmniSource Loudspeaker. Available online: https://www.bksv.com/en/products/transducers/acoustic/sound-sources/omni-source-loudspeaker-4295 (accessed on 8 July 2019).

115. ANSI, S. 18: Template method for ground impedance. In *American National Standard*; Acoustical Society of America: Ney York, NY, USA, 1999.
116. Sumarac-Pavlovic, D.; Mijic, M.; Kurtovic, H. A simple impulse sound source for measurements in room acoustics. *Appl. Acoust.* **2008**, *69*, 378–383. [CrossRef]
117. Smeatham, D.; Wheeler, P. On the performance of hearing protectors in impulsive noise. *Appl. Acoust.* **1998**, *54*, 165–181. [CrossRef]
118. Mijic, M.; Sumarac-Pavlovic, D. Acoustical characteristics of old wooden churches in Serbia. *J. Acoust. Soc. Am.* **2000**, *108*, 2648. [CrossRef]
119. Don, C.; Cramond, A.; McLeod, I.; Swenson, G. Shotshell primer impulse sources. *Appl. Acoust.* **1994**, *42*, 85–93. [CrossRef]
120. Cramond, A.; Don, C. Effects of moisture content on soil impedance. *J. Acoust. Soc. Am.* **1987**, *82*, 293–301. [CrossRef]
121. Papadakis, N.M.; Serras, A.; Stavroulakis, G.E. Mimicking the Sound Field of a Dodecahedral Loudspeaker by a Common Directional Loudspeaker for Reverberation Time Measurements. In Proceedings of the Euronoise 2018 Heraclion, Crete, Greece, 27–31 May 2018.
122. Sayin, U.; Artís, P.; Guasch, O. Realization of an omnidirectional source of sound using parametric loudspeakers. *J. Acoust. Soc. Am.* **2013**, *134*, 1899–1907. [CrossRef] [PubMed]
123. Westervelt, P.J. Parametric acoustic array. *J. Acoust. Soc. Am.* **1963**, *35*, 535–537. [CrossRef]
124. Pompei, F.J. The use of airborne ultrasonics for generating audible sound beams. In Proceedings of the Audio Engineering Society Convention 105, San Francisco, CA, USA, 26–29 September 1998.
125. Kruse, R.; Häußler, A.; van de Par, S. An omnidirectional loudspeaker based on a ring-radiator. *Appl. Acoust.* **2013**, *74*, 1374–1377. [CrossRef]
126. Kruse, R.; Häußler, A.; van de Par, S. A new omni-directional source based on a ring radiator. In Proceedings of the DAGA 2012, Darmstadt, Germany, 19–22 March 2012; pp. 715–716.
127. Małecki, P.; Wiciak, J.; Nowak, D. Acoustics of Orthodox Churches in Poland. *Arch. Acoust.* **2017**, *42*, 579–590. [CrossRef]
128. Soeta, Y.; Ito, K.; Shimokura, R.; Sato, S.-i.; Ohsawa, T.; Ando, Y. Effects of sound source location and direction on acoustic parameters in Japanese churches. *J. Acoust. Soc. Am.* **2012**, *131*, 1206–1220. [CrossRef] [PubMed]
129. Brezina, P. Measurement of intelligibility and clarity of the speech in romanesque churches. *J. Cult. Herit.* **2015**, *16*, 386–390. [CrossRef]
130. Beranek, L.L.; Mellow, T.J. *Acoustics: Sound Fields and Transducers*; Academic Press: Cambridge, MA, USA, 2012; pp. 241–242.
131. Beranek, L.L.; Mellow, T.J. *Acoustics: Sound Fields and Transducers*; Academic Press: Cambridge, MA, USA, 2012; p. 273.
132. Đorđević, Z.; Novković, D.; Andrić, U. Archaeoacoustic Examination of Lazarica Church. *Acoustics* **2019**, *1*, 423–438. [CrossRef]
133. Jambrosic, K.; Horvat, M.; Domitrovic, H. The influence of excitation type on reverberation type measurements. In Proceedings of the 3rd Congress of the Alps Adria Acoustics Association, Graz, Austria, 27–28 September 2007.
134. James, A. *Results of the Npl Study into Comparative Room Acoustic Measurement Techniques, Part 1 Reverberation Time in Large Rooms*; Institute of Acoustics: Norwich, UK, 2003.
135. Wallace, D.; Harvie-Clark, J. *Reverberation Times in School Halls: Measurement Oddities and Modelling*; Forum Acusticum: Krakow, Poland, 2014.
136. Qin, Q.; Attenborough, K. Characteristics and application of laser-generated acoustic shock waves in air. *Appl. Acoust.* **2004**, *65*, 325–340. [CrossRef]
137. Oksanen, M.; Hietanen, J. Photoacoustic breakdown sound source in air. *Ultrasonics* **1994**, *32*, 327–331. [CrossRef]
138. Gómez Bolaños, J.; Delikaris-Manias, S.; Pulkki, V.; Eskelinen, J.; Hæggström, E. Laser-induced acoustic point source for accurate impulse response measurements within the audible bandwidth. *J. Acoust. Soc. Am.* **2014**, *135*, EL298–EL303. [CrossRef] [PubMed]
139. Hosoya, N.; Nagata, M.; Kajiwara, I. Acoustic testing in a very small space based on a point sound source generated by laser-induced breakdown: stabilization of plasma formation. *J. Sound Vib.* **2013**, *332*, 4572–4583. [CrossRef]

140. Gómez Bolaños, J.; Pulkki, V.; Karppinen, P.; Hæggström, E. An optoacoustic point source for acoustic scale model measurements. *J. Acoust. Soc. Am.* **2013**, *133*, 221–227. [CrossRef] [PubMed]
141. Gómez-Bolaños, J.; Delikaris-Manias, S.; Pulkki, V.; Eskelinen, J.; Hæggström, E.; Jeong, C.-H. Benefits and applications of laser-induced sparks in real scale model measurements. *J. Acoust. Soc. Am.* **2015**, *138*, 175–180. [CrossRef] [PubMed]
142. Wyber, R. The design of a spark discharge acoustic impulse generator. *IEEE Trans. Acoust. SpeechSignal Process.* **1975**, *23*, 157–162. [CrossRef]
143. Wright, W.M. Propagation in air of N waves produced by sparks. *J. Acoust. Soc. Am.* **1983**, *73*, 1948–1955. [CrossRef]
144. Zel'Dovich, Y.B.; Raizer, Y.P. *Physics of Shock Waves and High-Temperature Hydrodynamic Phenomena*; Courier Corporation: North Chelmsford, MA, USA, 2012.
145. Ayrault, C.; Béquin, P.; Baudin, S. Characteristics of a spark discharge as an adjustable acoustic source for scale model measurements. In Proceedings of the Acoustics 2012, Nintes, France, 23–27 April 2012.
146. Rindel, J.H. Modelling in auditorium acoustics. From ripple tank and scale models to computer simulations. *Revista de Acústica* **2002**, *33*, 31–35.
147. Picaut, J.; Simon, L. A scale model experiment for the study of sound propagation in urban areas. *Appl. Acoust.* **2001**, *62*, 327–340. [CrossRef]
148. Kunigelis, V.; Senulis, M. Acoustic investigation of air-filled porous materials. *Acta Acust. United Acust.* **2002**, *88*, 14–18.
149. Salikuddin, M.; Dean, P.; Plumblee, H., Jr.; Ahuja, K. An impulse test technique with application to acoustic measurements. *J. Sound Vib.* **1980**, *70*, 487–501. [CrossRef]
150. Cabot, R. Impulse response testing of acoustic spaces. In Proceedings of the ICASSP'78. IEEE International Conference on Acoustics, Speech, and Signal Processing, Tulsa, OK, USA, 10–12 April 1978; pp. 820–823.
151. Cabot, R.C.; McDaniel, O.K.; Breed, H.E. Acoustic Impulse Generation by High Energy Spark Dischargers. In Proceedings of the Audio Engineering Society Convention 58, New York, NY, USA, 28 February–3 March 1978.
152. Hidaka, Y.; Yano, H.; Tachibana, H. Scale model experiment on room acoustics by hybrid simulation technique. *J. Acoust. Soc. Jpn.* **1989**, *10*, 111–117. [CrossRef]
153. Shibayama, H.; Fukunaga, K.; Kido, K.i. Directional characteristics of pulse sound source with spark discharge. *J. Acoust. Soc. Jpn.* **1985**, *6*, 73–77. [CrossRef]
154. Latham, H.G. The signal-to-noise ratio for speech intelligibility—An auditorium acoustics design index. *Appl. Acoust.* **1979**, *12*, 253–320. [CrossRef]

© 2019 by the authors. Licensee MDPI, Basel, Switzerland. This article is an open access article distributed under the terms and conditions of the Creative Commons Attribution (CC BY) license (http://creativecommons.org/licenses/by/4.0/).

Article

Measurements of Acoustical Parameters in the Ancient Open-Air Theatre of Tyndaris (Sicily, Italy) [†]

Arianna Astolfi [1,*], Elena Bo [1], Francesco Aletta [2] and Louena Shtrepi [1]

[1] Department of Energy, Politecnico di Torino, 10129 Turin, Italy; elena.bo@polito.it (E.B.); louena.shtrepi@polito.it (L.S.)
[2] UCL Institute for Environmental Design and Engineering, The Bartlett, University College London (UCL), London WC1H 0NN, UK; f.aletta@ucl.ac.uk
* Correspondence: arianna.astolfi@polito.it; Tel.: +39-011-090-4496
[†] Part of this work was presented at the ICSV 2016 Conference (Athens, Greece) and the International Conference on Metrology for Archaeology 2015 (Benevento, Italy).

Received: 14 July 2020; Accepted: 13 August 2020; Published: 15 August 2020

Abstract: The emerging field of archaeoacoustics is attracting increasing research attention from scholars of different disciplines: the investigation of the acoustic features of ancient open-air theatres is possibly one of its main themes. In this paper, the outcomes of a measurement campaign of acoustical parameters in accordance with ISO 3382-1 in the ancient theatre of Tyndaris (Sicily) are presented and compared with datasets from other sites. Two sound sources were used (firecrackers and dodecahedron) and their differences were analysed. A very good reproducibility has been shown between the two measurement chains, with differences on average of 0.01 s for reverberation time T_{20}, and less than 0.3 dB for Clarity C_{50} and C_{80} and for sound strength. In general, results show that the reverberation time and strength of sound values are relatively low when compared with other theatres because of the lack of the original architectural element of the scaenae frons. When combining this effect with the obvious condition of an unroofed space, issues emerge in terms of applicability of the protocols recommended in the ISO standard. This raises the question of whether different room acoustics parameters should be used to characterise open-air ancient theatres.

Keywords: open-air ancient theatres; acoustical parameters; ISO 3382-1; firecrackers

1. Introduction

In recent years, there has been a growing interest in the acoustical characterization of ancient open-air theatres, which led to the development of several research projects (see, e.g., in [1–4]) and a prolific production of scientific literature on the topic (see, e.g., in [5–14]). This reflects a general trend observable in the broader field of archaeoacoustics [15]. There is indeed a renewed attention to the "acoustics of past" [16–18], which is driven by both a pure scholarly interest and a more practical need to adapt historical facilities to contemporary use [9,19–21] and historical relevance, which could help to better understand the design and evolution of other performance spaces [22,23]. However, performing acoustical measurements poses a number of challenges because room acoustics parameters and standards are conceived for enclosed spaces and might not be straightforwardly applicable in outdoor contexts [24,25]. Above all, the ISO 3382-1:2009 standard [26] deals with "performance spaces" but mostly refers to indoor environments (e.g., auditoria, concert halls, etc.). The topic of acoustical characterisation has already been examined in detail for indoor spaces through statistical analysis, in order to investigate the reproducibility of the measures, the accuracy of the parameter calculation [27], the influence of source–receiver position displacement, the measurement chains of different systems [28–30] and the sensitivity to materials characteristics variation [31–33]. The aim of this work was to compare two measurement techniques for ancient theatres, using the archaeological site of Tyndaris as case study [34,35]. Particularly, traditional room acoustics measurements have been carried out according to the reference standard ISO 3382-1:2009, with both a dodecahedron source and with firecrackers. The behaviours of the reverberation time, early-to-late energy and sound energy parameters were assessed. The results of the measurements were compared with the findings of other measurement campaigns available in literature and considerations about the applicability of ISO 3382-1 for open-air ancient theatres are presented.

2. Case Study

The ancient theatre of Tyndaris was chosen as case study for this measurement campaign that was carried out in September 2015 by the Applied Acoustics Research Group of the Department of Energy of the Politecnico di Torino. The theatre has Greek origins and it was later changed into an arena by the Romans (for this reason Latin terminology has been used throughout this work). Its cavea, made of yellowish local sandstone, has a diameter (d) of 76 m and only fragments of the monumental scaenae frons still exist (Figure 1). At the time of the measurements the cavea was in part covered by wooden benches and a wooden platform covered the orchestra floor. To the best of the authors' knowledge, no previous room acoustics measurements have been carried out directly in the theatre of Tyndaris.

(a)

(b)

Figure 1. Present conditions of the ancient theatre of Tyndaris: view from the scaenae (**a**) and from the cavea (**b**).

3. In Situ Measurements

Acoustical measurements were performed in the theatre in unoccupied conditions, with omnidirectional sound sources and receivers, as per the protocol of the ISO 3382-1:2009 [26]. Further guidance on practical aspects related to the applicability of the ISO standard in the context of ancient open-air theatres was retrieved from other works in literature [1,25].

Nine receivers were positioned on three radial axes of the cavea (Figure 2), 1.2 m off the ground (ear height). An omnidirectional microphone (Schoeps CMC 5-U) was used to record the impulse responses (IRs). For most of the measurement positions, 2–3 repetitions were performed to subsequently evaluate the repeatability of the results. Two source positions were considered: S1 was shifted horizontally by 1 m from the centre of the orchestra, and S2 was located behind S1, closer to the ancient scaenae frons position. The distance between S1 to S2 was 6.6 m.

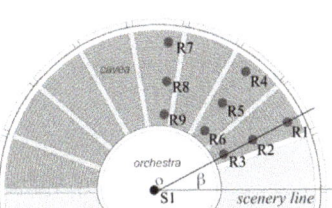

Figure 2. Measurement set-up: S1 and S2 indicate the source positions. R1–R9 indicate the receiver positions. O is the centre of the orchestra (at a distance of 1 m from S1, to the left, on the same horizontal axis) and β is the angle from the scenery line and the direction that joins the source and the receiver.

Table 1 shows the distances between the sources and the receivers. Two types of sources were used to measure the IRs as it was considered acoustically and metrologically relevant [36]: firecrackers ("Raudo Manna New Ma1b" and "Perfetto C00015 Raudo New") and a dodecahedral source (Bruel & Kjaer Omnipower sound source 4296). In the former case, the IRs were measured directly by recording the impulse produced by the firecracker blast, while in the latter case, the IRs were obtained after deconvolution of the recorded exponential sine sweep signal, which was 2.73 s long [26]. The firecracker measurements were only carried out with the source in position S1, apart from one measurement replication that was conducted at receiver R6 with the source in position S2. The IRs from the dodecahedral source were obtained for both source positions S1 and S2. Firecrackers maximize the signal-to-noise ratio (SNR) which is a considerable advantage in outdoor measurements, but they are also more affected by random effects (e.g., atmospheric conditions) [25,36].

Table 1. Source to receiver distances and angle β from the scenery line and the direction that joins the source and the receivers, related to the measurement campaign in September 2015 at the theatre of Tyndaris.

Receiver	S1-R (m)	S1-β (°)	S2-R (m)	S2-β (°)
R1	32.4	25	35.7	35
R2	24.0	25	27.4	37
R3	17.0	25	20.7	42
R4	32.1	52	37.5	58
R5	23.7	51	29.1	59
R6	16.7	50	22.2	61
R7	31.6	83	38.1	85
R8	23.2	82	29.7	84
R9	16.2	82	22.7	84

The Background Noise Level (BNL) was measured as an A-weighted equivalent sound level over a period of 10 min ($L_{Aeq, 10min}$), in between measurement sessions and did not overcome 34 dB(A). The air temperature and relative humidity were monitored during the whole measurement campaign, using a thermometer/hygrometer, Testo 608-H1 (Croydon South, VIC, Australia). The wind speed was measured by means of an anemometer, Testo 450-V1 (Croydon South, VIC, Australia). Table 2 shows a summary of the environmental conditions during the measurement campaign.

Table 2. Environmental conditions during the two measurement campaigns in September 2015 at the theatre of Tyndaris.

	5th September 2015	6th September 2015
Type of source	Dodecahedron	Firecrackers
Positions of source	2	1
Number of receivers	9	9
Measuring sessions	21:00–0:30	15:00–17:30
Temperature, t, (°C)	26.2–26.5 °C	26.9–28.8 °C
Relative Humidity, RH, (%)	77.4–79.4 %	45.0–69.9 %
Wind Speed, (m/s)	0.25–0.30 m/s	1.30–1.70 m/s

The dodecahedral source, powered by an amplifier (Lab.gruppen LAB-300), was connected to a laptop through a soundcard (Tascam US-144). The sound source was positioned at a height of 1.5 m off the ground, and a custom-made tripod was used to hold the firecrackers in a fixed position. Figure 3 shows the measurement chains of the two source types and a picture of the tripod customized for the firecrackers and the dodecahedron used during the measurements. Two kinds of acquisition software were used for the two measurement chains: Aurora for Audacity 2.4.1 and Dirac version 5.

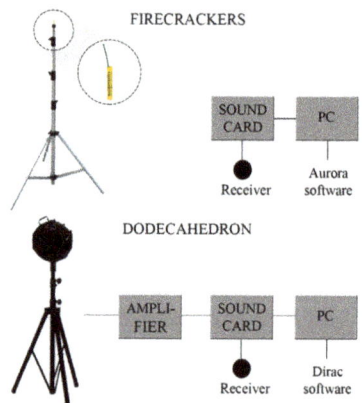

Figure 3. Measurement chains for the two source types—firecrackers (**top**) and dodecahedron (**bottom**).

The following parameters, which are commonly used for the acoustical characterization of open-air theatres [7], were computed from the IRs measured at each receiver position; Reverberation Time, T_{20} (s), which is s a measure of perceived reverberance; Clarity or early-to-late sound index C_{80} (dB), which is usually applied for clarity in music [37], and early-to-late sound index C_{50} (dB), which is usually applied for clarity for speech [38,39]; and Sound Strength, G (dB), which represents a measure of perceived level. A detailed description of the acoustical parameters is reported in the ISO 3382-1 [26] and in [25].

Figure 4 shows a typical IR measured with the firecracker source where, after the direct sound, the main recognisable reflection comes from the orchestra floor, that is, between 2.4 and 2.6 ms after the direct sound for the first row, between 2.8 and 3.0 ms for the second row and between 3.1 and 3.2 ms for the third row. Minor scattered reflections from the cavea steps are distinguishable in the latter part.

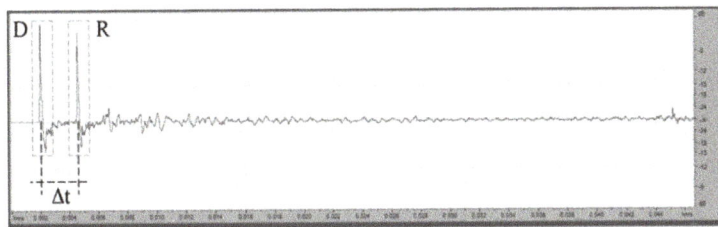

Figure 4. Measured IR in the theatre of Tyndaris for the S1–R6 measurement path and the firecracker source. Δt is the time interval between the direct sound (D) and the first reflection (R) from the orchestra floor [34].

The measurements dataset consists of the octave-band values, from 125 Hz to 8 kHz, of the acoustic parameters obtained from the IRs measured with the Dirac software, version 5, for T_{20}, C_{50} and C_{80}, in the case of both the dodecahedral and the firecrackers sources. The G estimation was performed through Dirac software and Aurora software for Audacity 2.4.1 for the dodecahedral and firecracker sources, respectively. Furthermore, a full compatibility has been checked and verified for the T_{20}, C_{50} and C_{80} parameters between the two software packages.

Sound Strength, G, is a measure that quantifies the amplification due to the space boundaries relative to a 10 m free-field reference [26]. A calibration is needed in order to obtain the reference measure. This can consist in the extrapolation of the free-field sound pressure level at 10 m from the source starting from measurements made at a short distance from the source, in situ, according to the basic spherical spreading law while accounting for source directivity through rotation of the source.

The Dirac software suggests an in situ free-field G calibration consisting of impulse response measurements with the microphone relatively close to the source, at several equally distanced microphone positions around the source [40]. When the measured impulse responses are loaded and a suitable time window, referring to the direct sound and the floor reflection gap, is entered, the system performs the calibration. In this case, three IRs, measured at a distance of 2.5 m from the source, at different angles all around, in steps of 120° (in order to average its directivity), were used selecting a time window of 6 ms, which allowed accurate results to be obtained, starting from the 250 Hz octave band.

Aurora software was used for the G measurement and the analyses of the IRs obtained with the firecrackers signals. According to the procedure suggested on the plugin website [41], the anechoic segment (direct sound) of any IR can be used for calibration, in a similar way to that in [42]. It is recommended to keep a length of the IR of at least 1 s and to silence the signal just after the end of the direct sound. In this way, the time spread caused by the octave filtering will not result in energy being pushed outside the time window, even at low frequencies, and the correct value of the signal level can be computed. A calibration file was obtained from each analysed IR and was used to calculate the G value for that measurement path, adding the exact source-to-receiver distance. The IR under analysis should be of the same length of the IR used for calibration (at least 1 s). Further details are given in [25].

According to the work in [43], in situ G calibration should be avoided because of the uncertainty of individual octave band values, which is much larger than the Just Noticeable Difference (JND) of 1 dB. Actually, repeatability for this measure can bring to differences up to 2 dB in closed theatres [42]. Nevertheless, the difficulties in performing accurate laboratory calibrations, due to the distant location of the theatre and the likely unsteady outdoor environmental conditions, determined the choice of conducting an in situ calibration.

4. Results of the Measurements

The measurement results are reported in Table 3, expressed as T_{20}, C_{50}, C_{80} and G acoustical parameters, with the dodecahedral source in positions S1 (receivers R1 to R9) and S2 (R1–R8), while

Table 4 reports those obtained with firecrackers in positions S1 (R1–R9) and S2 (R6). All the values are the averages of at least two repetitions (apart from some receivers with S2 with dodecahedron) in the same receiver position and of the central 500 Hz and 1 kHz octave band frequencies, as indicated in ISO 3382-1. Moreover, the spatial average is indicated for each row and as an overall value of the theatre. The Impulse Response-to-Noise Ratio (INR), which is a parameter that establishes the reliability of the acoustical measurements [40], was between 42 dB and 73 dB over all the measurements. According to ISO 3382-1, the source level should be at least 35 dB above the background noise level in the corresponding frequency band for the case of T_{20}. With good measurements most practical INR values range from 35 to 60 dB, thus all the measurements considered in this study, for the octave bands from 125 Hz to 8 kHz, comply with this rule.

The results shown in Tables 3 and 4 were obtained from one to five repetitions and give a general overview of the situation, but they are not robust enough to draw conclusions on the uncertainty of the results, due to the high variability of the standard deviations. It is worth highlighting that, in acoustics, parameters are indicators for the perceptual evaluation of an acoustic signal, namely, the average capability of a "conventional" listener to detect sound variations. An important factor that correlates the subjective field to objective measures is the JND, that is, the smallest perceivable change in a given acoustical parameter, which was defined in ISO 3382-1 for central frequencies (500 Hz and 1 kHz). Table 5 shows the JNDs of the acoustical parameters considered in this study [26,44]. Within this framework, a tendency of a uniform pattern of the standard deviations of the spatial means of the parameters can be seen, with values that are closer to the JND values for T_{20} and G, and slightly higher for C_{50} and C_{80}. In the case of G, as the distance from the source affects the parameter, spatial standard deviations are quite uniform across the different rows, as expected. As general observation, it can be seen that the standard deviations reveal much higher values than the JNDs with the firecracker than with the dodecahedron source.

Table 3. Mean values of the T_{20}, C_{80} and G acoustical parameters measured with a dodecahedral (D) source at the theatre of Tyndaris with the source in the $S1_D$ and $S2_D$ positions. The data refer to the averages of the 500 Hz and 1 kHz octave bands and to the repetitions for the same receiver position. The rows and overall spatial means are also reported. The standard deviations of the spatial means outside the JND range are shown in bold.

Acoustical Parameter	$S1_D$ Receivers											Overall
	First Row				Central Row				Last Row			
	R3	R6	R9	Sp. Mean	R2	R5	R8	Sp. Mean	R1	R4	R7	Sp. Mean	Sp. Mean
N. of repetitions	2	4	4		2	2	5		2	2	3		
T_{20} (s)	0.50	0.59	0.52	0.54	0.56	0.60	0.61	0.59	0.57	0.59	0.60	0.59	0.57
St. Dev. T_{20}	0.03	0.05	0.05	**0.05**	0.01	0.01	0.06	0.03	0.05	0.02	0.05	0.02	0.04
C_{50} (dB)	9.7	13.3	13.4	12.1	10.8	12.8	14.3	12.3	11.0	12.2	13.1	12.1	12.3
St. Dev. C_{50}	1.1	2.4	2.5	**2.1**	1.2	1.4	3.0	**1.8**	0.2	0.2	2.4	**1.1**	**1.5**
C_{80} (dB)	17.2	17.2	19.7	18.0	16.7	16.9	18.1	17.2	17.2	17.0	16.8	17.0	17.4
St. Dev. C_{80}	0.8	1.2	0.7	**1.5**	1.2	1.4	1.5	0.8	1.2	0.8	0.6	0.2	**1.0**
G (dB)	0.4	−2.2	−1.0	−0.9	−4.9	−4.1	−4.2	−4.3	−7.8	−6.8	−7.1	−7.2	−4.2
St. Dev. G	0.8	1.3	0.6	**1.3**	0.7	1.0	0.9	0.4	0.8	0.8	0.7	0.5	**2.8**

Acoustical Parameter	$S2_D$ Receivers											Overall
	First Row				Central Row				Last Row			
	R3	R6	R9	Sp. Mean	R2	R5	R8	Sp. Mean	R1	R4	R7	Sp. Mean	Sp. Mean
N. of repetitions	1	1			1	1	1		1	2	3		
T_{20} (s)	0.53	0.51		0.52	0.46	0.60	0.43	0.50	0.46	0.54	0.49	0.50	0.50
St. Dev. T_{20}	0.00	0.03		0.01	0.07	0.01	0.01	**0.09**	0.05	0.03	0.02	0.04	**0.06**
C_{50} (dB)	11.2	12.4		11.8	11.2	13.5	16.4	13.7	10.1	13.2	15.0	12.8	12.9
St. Dev. C_{50}	1.0	2.4		0.8	0.9	1.6	1.5	**2.6**	2.6	1.7	1.9	**2.5**	**2.1**
C_{80} (dB)	16.3	18.5		17.4	18.1	17.8	19.2	18.4	17.1	17.4	18.4	17.6	17.8
St. Dev. C_{80}	1.3	1.4		**1.6**	1.5	1.4	1.9	0.8	2.2	1.6	1.2	0.7	0.9
G (dB)	−2.6	−3.0		−2.8	−4.9	−6.2	−5.5	−5.6	−7.4	−7.6	−7.6	−7.5	−5.6
St. Dev. G	0.8	1.2		0.3	1.1	1.2	1.0	0.6	1.9	1.7	1.4	0.1	**2.0**

Table 4. Mean values of the T_{20}, C_{80} and G acoustical parameters measured with firecrackers (F) at the theatre of Tyndaris with the source in the $S1_F$ and $S2_F$ positions. The data refer to the averages of the 500 Hz and 1 kHz octave bands and to the repetitions for the same receiver position. The rows and overall spatial means are also reported. The standard deviations of the spatial means outside the JND range are shown in bold.

$S1_F$

Acoustical Parameter	First Row				Central Row				Last Row				Overall
	R3	R6	R9	Sp. Mean	R2	R5	R8	Sp. Mean	R1	R4	R7	Sp. Mean	Sp. Mean
N. of repetitions	3	2	2		2	2	2		2	2	2		
T_{20} (s)	0.48	0.68	0.55	0.57	0.51	0.63	0.61	0.58	0.54	0.56	0.63	0.57	0.58
St. Dev. T_{20}	0.10	0.07	0.07	**0.10**	0.10	0.06	0.05	**0.06**	0.04	0.06	0.02	**0.05**	**0.06**
C_{50} (dB)	8.2	14.0	12.4	11.5	9.7	12.9	14.1	12.3	10.8	10.9	14.7	12.1	12.0
St. Dev. C_{50}	2.6	1.2	1.2	**3.0**	2.5	0.6	0.8	**2.3**	0.9	0.5	0.8	**2.2**	**2.2**
C_{80} (dB)	15.8	16.9	18.1	17.0	16.6	17.1	17.5	17.1	18.2	16.5	18.0	17.6	17.2
St. Dev. C_{80}	1.5	1.3	1.3	**1.2**	1.0	0.4	1.7	0.5	1.1	1.0	1.2	0.9	0.8
G (dB)	0.0	−1.7	−1.2	−1.0	−4.6	−3.9	−4.2	−4.2	−6.8	−6.7	−6.3	−6.6	−3.9
St. Dev. G	0.4	0.4	0.1	0.8	0.7	0.9	0.3	0.4	0.5	0.6	1.0	0.3	**2.5**

$S2_F$

Acoustical Parameter	Receiver
	First Row
	R6
N. of repetitions	2
T_{20} (s)	0.52
	0.52
St. Dev. T_{20}	0.05
C_{50} (dB)	11.7
St. Dev. C_{50}	1.4
C_{80} (dB)	17.8
	17.8
	1.6
St. Dev. C_{80}	0.4
G (dB)	−1.9
	−5.2
	0.8
St. Dev. G	0.09

Table 5. Just Noticeable Difference (JND) of the T_{20}, C_{50}, C80 and G acoustical parameters according to ISO 3382-1 (Annex A) [26] and the work in [44].

Acoustical Parameter	JND
T_{20} (s)	5% ≈ 0.03
C_{50} (dB)	1.1
C_{80} (dB)	1
G (dB)	1

4.1. Reverberation Time

Figure 5 shows the averaged values of T_{20} of all the receivers, considering both source typologies and positions: dodecahedral source in positions S1 and S2, and firecrackers in position S1. The whole theatre mean value, calculated on 0.5–1 kHz octave band frequency range, considering the results from all the sources, is 0.57 s. Standard deviations are approximately 0.04–0.05 for frequencies from 250 Hz to 8 kHz. Only the firecracker source shows a higher standard deviation at 125 Hz. In general, measurements with both sources in position S1 give similar results for all octave band frequencies, showing higher values with respect to measurements with source S2 in the furthest position.

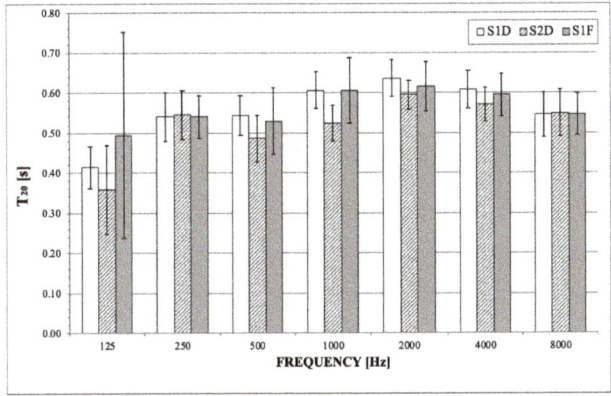

Figure 5. T_{20} measurement results (average values) for all the receivers and sources, expressed as a function of frequency. Standard deviations are reported on the error bars.

4.2. Early-to-Late Energy Parameters

Figures 6 and 7 show, respectively, the averaged values of C_{50} and C_{80} of all the receivers as a function of frequency. The whole theatre mean value, calculated on 0.5–1 kHz octave band frequency range, is 12.6 dB in the case of C_{50}, while it is 17.5 dB in the case of C_{80}. Due to the open-air conditions not allowing a proper reverberant tail (that is, late energy), it is not possible to assume that the results are in the typical ISO 3382-1 ranges. The standard deviation of the C_{50} is ~2.5 dB for all octave band frequencies, while for C_{80} is lower, ~1 dB. Compatible values are shown for all the sources with slightly lower values for S1$_F$, i.e., with firecrackers.

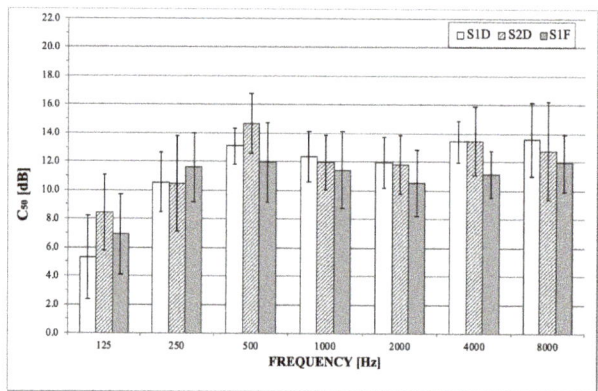

Figure 6. C_{50} measurement results (average values), for all the receivers and sources, expressed as a function of frequency. Standard deviations are reported on the error bars.

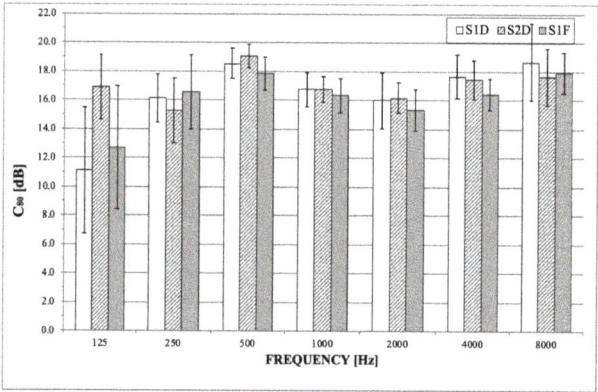

Figure 7. C_{80} measurement results (average values), for all the receivers and sources, expressed as a function of frequency. Standard deviations are reported on the error bars.

4.3. Sound Strength

In Figure 8, the averaged values of Sound Strength G for 0.5–1 kHz octave band frequency range, G_{mid}, are shown for each receiver as a function of source-to-receiver distance (d), expressed on a logarithmic scale (the free field linear regression is also reported as a reference). In Table 6, the values of the parameters A_i and B_i of the equation $G_{mid} = A_i + B_i \log_{10}(d)$ are reported for each source position. In all the considered cases, the decay of the G_{mid} with distance seems to follow quite well the behaviour around a source in free field. The gap between the curves with respect to the free field trend is approximately 3–4 dB depending on source typology and position. It is possible to conclude that in the theatre the sound strength is low, i.e., in the range from −1 to −8 dB. Moreover, the source typology seems not to influence the G_{mid} results, with differences between $S1_D$ and $S1_F$ on average values of the first and the second rows of 0.1 dB, and less than 0.6 dB for the third row, and overall across the theatre of 0.3 dB. This is very promising, as variations of G_{mid} results between 1 and 2 dB are reported in literature for closed theatres related to different calibration procedures and measurement chains [42,45].

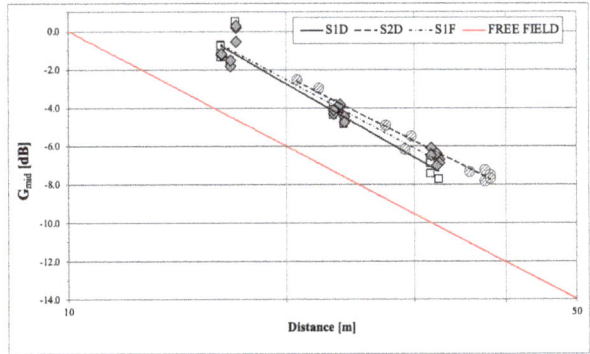

Figure 8. G_{mid} averaged 500–1 kHz measurement results for all the receivers and sources, expressed as a function of source-to-receiver distance.

Table 6. Linear regression parameters for the Sound Strength decays in Figure 8 ($G_{mid} = A_i + B_i \log_{10} d$), and the related R^2 coefficient.

Source	A_i (dB)	B_i (dB)	R^2
Free field	20	−20	1
S1D	25.6	−21.8	0.95
S2D	22.7	−19.2	0.98
S1F	23.9	−20.3	0.94

5. Discussion

5.1. General Results and Comparison with Other Ancient Theatres

A first analysis is conducted on the collected IRs structure. The sound field in ancient theatres is generally considered more similar to a free field than a diffuse one. As previously mentioned, there is a limited energetic contribution after the direct sound. This characteristic has consequences both on the extrapolation of the acoustical parameters and their analysis. In fact, all the parameters were calculated from the IRs through their integration in reverse time, strongly affected by the first reflections. Thus, before analysing the whole structure of the IRs, some conclusions are drawn focusing on the Early Decay Curve (EDC), calculated on the squared impulse response.

Figure 9 shows a measured EDC of a typical IR from open-air theatres. The analysis was made with Dirac version 5. As an example, receiver position R6 with dodecahedral source in S1 was chosen as typical EDC. On the x-axis, the arrival time of the direct sound, after the so-called "flight time" starting with the emission of the signal from the source, is indicated. The corresponding IR (S1$_D$-R$_6$) is shown in the following Figure 10.

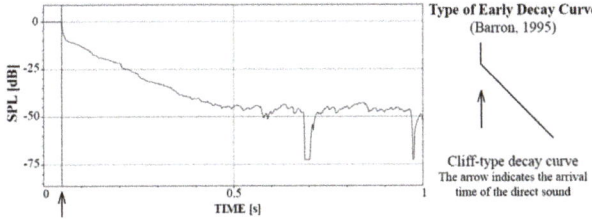

Figure 9. Measured EDC at 1 kHz for the S1–R6 measurement path with dodecahedral source.

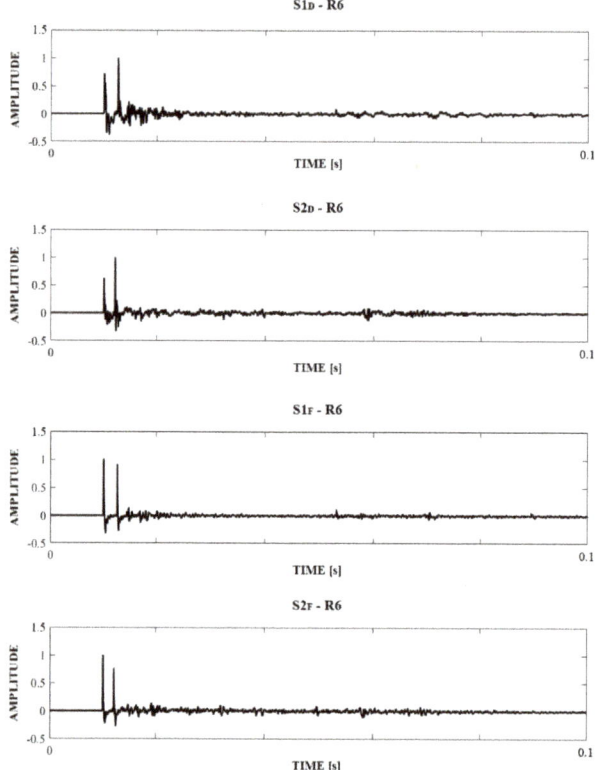

Figure 10. Measured IRs in TYN for R6 position, all the source positions (S1 and S2) and typologies (D and F) considered.

Considering the classification made by Barron [46], this type of EDC is described as "Cliff-type" decay. In this case, either the direct sound or the early reflections are very strong and the direct sound itself could turn out to be not significant with measurements made beyond 10 m from the source, as in the case of this example. Consequently, Barron reports that this type of EDC is usually leading to Early Decay Time (EDT) shorter than reverberation time such as T_{20} or T_{30}, due to the difference between the slopes considered in the decay. EDT is determined from a best fit straight line to the first 10 dB of sound decays obtained the reverse-time integration of the squared impulse response curve, while for T_{20} the evaluation range of the slope is from −5 dB to −25 dB.

In an open-air theatre, when the sound field substantially differs from a diffuse field, reverberation time is not in principle applicable, but despite this, it is largely used for the acoustic qualification of ancient theatres and for their comparison [2,3,10,21,47]. The main contributions to reverberation are few energetic reflections and the scattered sound energy coming from the steps in the cavea. Generally, reverberation time T_{20} or T_{30} show a limited variability in the theatre and for this reason they are used for comparison with other similar theatres. On the contrary, EDT shows a higher variability in the theatre as it is obtained from the very first part of the decay curve and its value is mostly affected by the arrival time and by the energy of the reflections after the direct sound. It is thus not used to obtain an overall qualification of the theatres that allows comparisons between different architectural typologies [7]. In this study, EDT resulted lower than reverberation time and is highly instable, thus in order to avoid the reader making misleading conclusions, it was decided not to show its values.

Figure 10 shows the fine structure of the typical IRs measured in the theatre with the two source typologies, i.e., the dodecahedron and the firecrackers, obtained for S1–R6 and S2–R6 paths with MATLAB version 2015b.

The IRs clearly show that the main contribution to the direct sound is coming from the first reflection from the orchestra, that is between 2.4 and 2.6 ms after the direct sound. Then, the first reflection is followed by minor scattered reflections from the cavea steps, distinguishable in the latter part. The absence of a scenic building is responsible for the absence of the third main reflection.

In Figure 10, the IRs obtained with the dodecahedral source show the direct sound with a lower level than the first reflection. This could happen in the case of acoustical focusing, that is, as already mentioned, an effect quite common in ancient theatres [2]. The effect is shown along the R4-R5-R6 and R7-R8-R9 lines for both S1 and S2 only for the dodecahedral source, and it is not observed along the R1-R2-R3 line. It could be due to the lack of omnidirectionality of the dodecahedral source at the highest frequencies [26,36] that determines selected focusing effects due to the loudspeakers position. Considering firecrackers, it does not exhibit directional characteristics or a very good repeatability [36,48]. Moreover, IR with the dodecahedral source results to be less sharp than those obtained with firecrackers. This indicates the firecracker source as more suitable than the dodecahedron to obtain defined IRs, at least to investigate the fine structure of the reflection pattern.

Figure 11 reports the frequency trend of the direct sounds extrapolated from the firecracker and dodecahedral sources and compared (measurement path: S1 and R6). The firecrackers direct sound is quite stable and flat until almost 10 kHz. However, the firecracker is more subject to problem of repeatability, as shown by the higher level of standard deviations obtained during measurements.

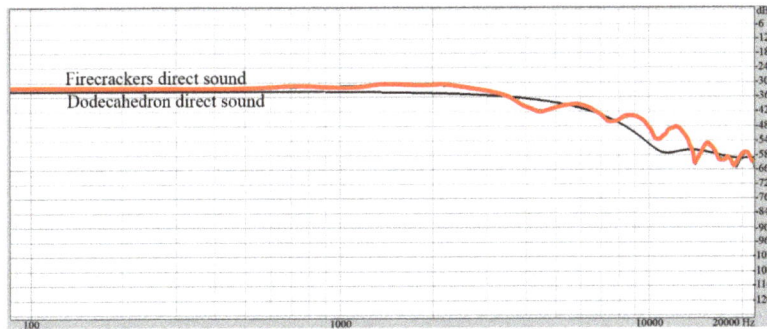

Figure 11. Frequency trend of the direct sound, from firecracker measurements (red line) and dodecahedral source (black line).

A further comparison with other case studies measurements is proposed. The cases considered are the in situ measurements realized on six theatres: Syracusae, Aspendos, Jerash, Taormina, Delphi and Segesta [6,25]. Regarding the in situ measurements, results for Syracusae (Italy, V cent. BCE; Greek-Roman theatre) in this study come from a measurement campaign conducted by the Acoustics Group at Politecnico di Torino in 2015 [25]. Aspendos (Turkey, I cent. CE; Roman theatre) and Jerash (Jordan, I cent. CE; Roman theatre) were the object of measurements during the aforementioned ERATO project. In particular, Aspendos is considered one of the best conserved ancient theatres in the world, as it still preserves architectural elements such as a complete scaenae frons and porticus; Jerash does not have any more a porticus, but it has a partially conserved scaenae frons. Instead, Taormina (Italy, 265-215 BCE, then modified in arena II-III cent. CE; Greek-Roman theatre), Delphi (Greece, II-I cent. BCE, Greek theatre) and Segesta (Italy, IV-I cent. BCE; Greek theatre) were object of another measurement campaign, performed by the University of Ferrara in collaboration with Kobe University [5]. Taormina has a partially conserved scaenae frons, while Delphi and Segesta do not have any stage building. This further comparison allows understanding if the performed measurements in Tyndaris are comparable to those realized in previous experiences.

The results are presented in terms of RT$_{mid}$, that is the average value between reverberation time in the octave bands 0.5, 1 and 2 kHz, together with the architectural characteristics of the aforementioned theatres as reported in [7], in the following Table 7. It is evident that the complete absence of scaenae frons corresponds to a drastic reduction of the reverberation inside the theatre. Furthermore, it is important to underline that in Tyndaris the theatre cavea is covered by grass and topsoil for some of the area.

Table 7. Main characteristics of the theatre investigated in this work (in bold) compared with other ancient theatres.

Theatre	Cavea Diameter (m)	Seating Capacity	Cavea Slope	Scaenae frons	Scaenae frons Height (m)	Porticus	RT$_{mid}$ (s)
Tyndaris	**76**	**7000**	**27°**	**No**	**-**	**No**	**0.57**
Syracusae	105	* 10,000	20–25°	No	-	No	0.78
Aspendos	98	15,000	33°	Yes	26.0	Yes	1.68
Jerash	63	* 3000	43°	Yes (part.)	8.5	No	1.19
Taormina	109	* 5500	39°	Yes (part.)	20.0	No	1.16
Delphi	50	5000	28°	No	-	No	0.50
Segesta	63	4000	26°	No	-	No	0.45

* Current conditions.

The G$_{mid}$ regression lines of the measurements performed in this research are compared to those presented in [7,25] in Figure 12. Considering the in situ measurements, it is evident that to a better state of conservation (i.e., Aspendos and Jerash) corresponds a higher acoustic performance. In particular, for Tyndairs results are in line with those obtained from theatres without preserved scenic building (i.e., Segesta and Delphi), as in all these cases it is possible to count only on the orchestra as acoustic mirror.

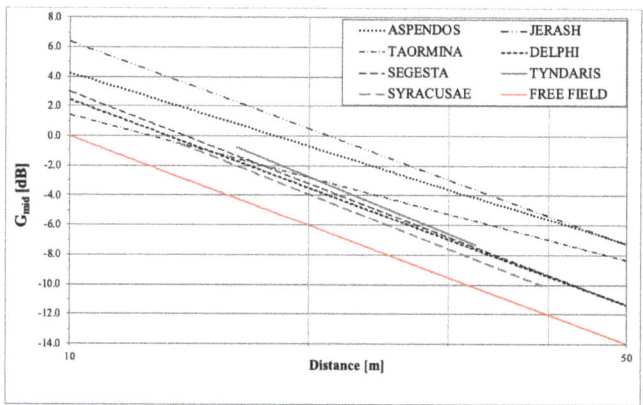

Figure 12. G$_{mid}$ averaged 0.5–1 kHz measurement results, for all the receivers and sources. Comparison with measurement in real theatres [7,25].

5.2. Applicability of the ISO 3382-1

Performing acoustic measurements at archaeological sites is a challenging task and researchers often face a number of practical issues that may require some deviations from standardized protocols [18]. In particular, ISO 3382-1 is meant for roofed performance spaces, thus its straightforward applicability to open-air ancient theatres is problematic at least. Table 8 reports some key recommendations mentioned in different sections of ISO 3382-1 and a brief comment on whether (if applicable and to what extent) those were implemented in the current study. A similar analysis was conducted in [25] and the emerging issues are comparable.

Table 8. ISO 3382-1 recommendations and their applications in the measurement campaign in Tyndaris.

ISO 3382-1 Section	Recommendation	Implemented	Notes
4. Measurement conditions	Temperature and Relative Humidity: these quantities should be measured with an accuracy of ±1 °C and 5%, respectively.	Yes	In the case of inter-measurement temperature change, the recommended deviation that allows for accurate measurements of room acoustic parameters is 2 °C [30].
	Equipment: omnidirectional sources and receivers. Maximum deviations of directivity for an omnidirectional source are indicated.	Yes	The deviation of directivity of the used sound source respected the maximum values indicated by the reference standards.
	Number of source positions: minimum 2, located where the natural sound source would take position. Height of sources: 1.5 m.	Yes	Besides taking measurements in two source positions, also measurements with two sound source types were performed.
	Number of microphone positions: Microphone positions should be at positions representative of positions where listeners would normally be located. For reverberation time measurements, it is important that the measurement positions sample the entire space; for the room acoustic parameters, they should also be selected to provide information on possible systematic variations with position in the room. Height of the receivers: 1.2 m.	Yes	
5. Measurement procedures	Integrated Impulse Response method: any source is allowed provided that its spectrum is broad enough to cover from 125 Hz to 4 kHz. The peak sound pressure level has to ensure a decay curve starting at least 35 dB above the BNL.	Yes	In some receiving positions, the 125 Hz frequency band did not guarantee the required 35 dB over the BNL, with the firecrackers.
6. Decay curves	Regression analysis: a least-squares fit line shall be computed for the decay curve.	No	The open-air condition is characterised by a cliff-decay curve linked to a few strong reflections, but this case is not considered by the standard.
7. Uncertainty	If the impulse response is not exactly repeatable, results should be the average of several repeated measurements at the same position.	No	The standard does not indicate a suggested number of repetitions, or a methodology to define it. Since the priority was to keep stable the boundary conditions (Temperature, air Velocity, Relative Humidity), it was not possible to repeat many measurements in each position.
A4. Positions	A minimum of between 6 and 10 microphone positions should be used, depending on the size of the hall. Above 2000 seats, at least 10 positions are suggested.	Yes	Open-air theatres easily reach more than 2000 seats, but at farther positions the measurements may have easily unreliable results.

The application of ISO recommendations to the open-air case study is questionable in the evaluation of the correct decay curves, of the measurement uncertainty (using Integrated Impulse Response method) and of the repeatability of the IRs. Thus, it seems fair to assume that the ISO 3382-1 and its parameters, although referred to performance environments, are not completely applicable to open-air spaces. A specific standard for the acoustical measurements in open-air conditions should probably be taken into consideration and added as a further part to the ISO 3382 series.

6. Conclusions

This work presents the results of an acoustical measurements campaign in the ancient open-air theatre of Tyndaris. Measurements based on ISO 3382-1 were conducted in situ in unoccupied conditions using omnidirectional sound sources and receivers. The impulse responses (IRs) were measured directly using a firecracker as the impulse source, and a dodecahedral source, which generated a sine sweep. The acoustical parameters described in the ISO 3382-1 standard, that is, Reverberation Time (T_{20}), Clarity (C_{50}) and (C_{80}) and Sound Strength (G), were obtained from the IRs measured at each receiver position. The following main results are highlighted.

- For Reverberation Time (T_{20}), the measurements with the firecrackers and the dodecahedron sources returned similar results. When looking at the average value across all source types and positions in the mid-frequencies range, T_{20} was 0.57 s; this value is relatively lower if compared to similar open-air ancient theatres, which is a common trait for those sites where the scaenae frons is no longer in place.
- For the Early-to-Late Energy parameters, the average theatre values in the mid-frequencies ranges are: C_{50} = 12.6 dB and C_{80} = 17.5 dB; yet, such figures should be interpreted with caution as due to operational constraints (open-air conditions) it was not possible to accurately assess the results in alignment with the ISO 3382-1 guidance.
- For the parameter Strength of Sound G, the values for the theatre in the mid-frequency range are typically low, going from −1 to −8 dB, depending on sound source type and position.

A very good reproducibility has been shown between the measurements obtained with the two different measurement chains, with differences on average spatial values for the whole theatre that are equal to 0.01 s for T_{20}, and less than 0.3 dB for C_{50}, C_{80} and G.

Future research should explore additional parameters that could be more suitable for the characterization of the ancient theatre, and unroofed historical spaces more broadly [49]. At other sites, the context should also be taken into account, considering the possible influence of the state of conservation of the architectural elements and materials.

Author Contributions: E.B. and A.A. conceived, designed the data collection campaigns and collected data on site; E.B. performed data analysis together with L.S. and A.A.; A.A., E.B. and F.A. drafted and curated the first version of the manuscript. All the authors revised the paper. All authors have read and agreed to the published version of the manuscript.

Funding: This research was funded through a Ph.D. scholarship awarded to the first author by the Politecnico di Torino (Turin, Italy).

Acknowledgments: The authors are grateful to Fabrizio Bronuzzi, Rocco Costantino and Maurizio Bressan from the Acoustics Laboratory of Politecnico di Torino for their technical contribution to this project. The authors are grateful to Nicola Prodi and Andrea Farnetani from the Univeristy of Ferrara, Angelo Farina from the University of Parma and Monika Rychtarikova from KU Leuven for their suggestions and advice.

Conflicts of Interest: The authors declare no conflict of interest.

References

1. Rindel, J.H. *ERATO*; Final Report; Technical University of Denmark: Copenaghen, Denmark, 2006.
2. Rindel, J.H. Roman theatres and revival of their acoustics in the ERATO project. *Acta Acust. United Acust.* **2013**, *99*, 21–29. [CrossRef]
3. Hak, C.; Hoekstra, N.; Nicolai, B.; Wenmaekers, W. Project ancient acoustics part 1 of 4: A method for accurate impulse response measurements in large open air theatres. In Proceedings of the 23rd International Congress of Sound Vibrations (ICSV23), Athens, Greece, 10–14 July 2016.
4. Wenmaerkers, R.H.; Nicolai, B.; Hoekstra, N.; Hak, C.C. Project ancient acoustics part 4 of 4: Stage acoustics measured in the odeon of Herodes Atticus and in the theatre of Argos. In Proceedings of the 23rd Internatinational Congress on Sound Vibrations (ICSV23), Athens, Greece, 10–14 July 2016.
5. Sato, S.; Sakai, H.; Prodi, N. Acoustical measurements in ancient Greek and Roman theatres. In Proceedings of the 3rd International Conference of Forum Acusticum, Sevilla, Spain, 16–20 September 2002.

6. Farnetani, A. Investigation on the Acoustics of Ancient Theatres by Mean of Modern Technologies. Doctoral Dissertation, Università degli Studi di Ferrara, Ferrara, Italy, 2005.
7. Farnetani, A.; Prodi, N.; Pompoli, R. On the acoustics of ancient Greek and Roman theatres. *J. Acoust. Soc. Am.* **2008**, *124*, 1557–1567. [CrossRef] [PubMed]
8. Iannace, G.; Trematerra, A.; Masullo, M. The large theatre of Pompeii: Acoustic evolution. *Build. Acoust.* **2013**, *20*, 215–227. [CrossRef]
9. Barkas, N. Contemporary sound environment around ancient Greek theatre: Current operation problems. In Proceedings of the Conference SMED "Echo days", Athens, Greece, 29 September–3 October 2013.
10. Psarras, S.; Hatziantoniou, P.; Kountouras, M.; Tatlas, N.A.; Mourjopolous, J.N.; Skarlatos, D. Measurements and analysis of the Epidaurus theatre acoustics. *Acta Acust. United Acust.* **2013**, *99*, 30–39. [CrossRef]
11. Iannace, G.; Trematerra, A. The rediscovery of Benevento Roman theatre acoustics. *J. Cult. Herit.* **2014**, *15*, 698–703. [CrossRef]
12. Girón, S.; Álvarez-Corbacho, Á.; Zamarreño, T. Exploring the Acoustics of Ancient Open-Air Theatres. *Arch. Acoust.* **2020**, *45*, 181–208.
13. Barkas, N. The Contribution of the Stage Design to the Acoustics of Ancient Greek Theatres. *Acoustics* **2019**, *1*, 337–353. [CrossRef]
14. Galindo, M.; Girón, S.; Cebrián, R. Acoustics of performance buildings in Hispania: The Roman theatre and amphitheatre of Segobriga, Spain. *Appl. Acoust.* **2020**, *166*, 107373. [CrossRef]
15. Scarre, C.; Lawson, G. *Archaeoacoustics*; Mc Donald Institute for Archaeological Research: Cambridge, UK, 2006.
16. Till, R. Sound Archaeology: A Study of the Acoustics of Three World Heritage Sites, Spanish Prehistoric Painted Caves, Stonehenge, and Paphos Theatre. *Acoustics* **2019**, *1*, 661–692. [CrossRef]
17. Witt, D.E.; Primeau, K.E. Performance Space, Political Theater, and Audibility in Downtown Chaco. *Acoustics* **2019**, *1*, 78–91. [CrossRef]
18. Aletta, F.; Kang, J. Historical Acoustics: Relationships between People and Sound over Time. *Acoustics* **2020**, *2*, 128–130. [CrossRef]
19. Bo, E.; Astolfi, A.; Pellegrino, A.; Pelegrin-Garcia, D.; Puglisi, G.E.; Shtrepi, L.; Ryctarikova, M. The modern use of ancient theatres related to acoustic and lighting requirements: Stage design guidelines for the Greek theatre of Syracuse. *Energy Build.* **2014**, *95*, 106–115. [CrossRef]
20. Bo, E.; Bergoglio, M.; Astolfi, A.; Pellegrino, A. Between the archaeological site and the contemporary stage: An example of acoustic and lighting retrofit with multifunctional purpose in the ancient theatre of Syracuse. *Energy Procedia* **2015**, *78*, 913–918. [CrossRef]
21. Berardi, U.; Iannace, G. The acoustic of Roman theatres in Southern Italy and some reflections for their modern uses. *Appl. Acoust.* **2020**, *170*, 107530. [CrossRef]
22. D'Orazio, D.; De Cesaris, S.; Morandi, F.; Garai, M. The aesthetics of the Bayreuth Festspielhaus explained by means of acoustic measurements and simulations. *J. Cult. Herit.* **2018**, *34*, 151–158. [CrossRef]
23. D'Orazio, D. Italian-Style Opera Houses: A Historical Review. *Appl. Sci.* **2020**, *10*, 4613. [CrossRef]
24. Mo, F.; Wang, J. The conventional RT is not applicable for testing the acoustical quality of unroofed theatres. *Build. Acoust.* **2013**, *20*, 81–86. [CrossRef]
25. Bo, E.; Shtrepi, L.; Pelegrín García, D.; Barbato, G.; Aletta, F.; Astolfi, A. The Accuracy of Predicted Acoustical Parameters in Ancient Open-Air Theatres: A Case Study in Syracusae. *Appl. Sci.* **2018**, *8*, 1393. [CrossRef]
26. International Organization for Standardization. *ISO 3382-1:2009 Acoustics—Measurement of Room Acoustic Parameters—Part. 1: Performance Spaces*; ISO: Geneva, Switzerland, 2009.
27. Katz, B.F. International Round Robin on Room Acoustical Impulse Response Analysis Software 2004. *Acoust. Res. Lett. Online* **2004**, *5*, 158–164. [CrossRef]
28. Pelorson, X.; Vian, J.-P.; Polack, J.-D. On the variability or Room Acoustical Parameters: Reproducibility and Statistical Validity. *Appl. Acoust.* **1992**, *37*, 175–198. [CrossRef]
29. Malecki, P.; Zastawnik, M.; Wiciak, J.; Kamisinski, T. The influence of the measurement chain on the impulse response of a reverberation room and its application listening test. *Acta Phys. Pol. A* **2011**, *119*, 1027–1030. [CrossRef]
30. Guski, M. Influences of External Error Sources on Measurements of Room Acoustic Parameters. Doctoral Dissertation, RWTH Aachen University, Aachen, Germany, 2015.

31. Shtrepi, L.; Astolfi, A.; Pelzer, S.; Vitale, R.; Rychtáriková, M. Objective and perceptual assessment of the scattered sound field in a simulated concert hall. *J. Acoust. Soc. Am.* **2015**, *138*, 1485–1497. [CrossRef] [PubMed]
32. Shtrepi, L.; Astolfi, A.; D'Antonio, G.; Guski, M. Objective and perceptual evaluation of distance-dependent scattered sound effects in a small variable-acoustics hall. *J. Acoust. Soc. Am.* **2016**, *140*, 3651–3662. [CrossRef] [PubMed]
33. Shtrepi, L.; Di Blasio, S.; Asltolfi, A. Listeners Sensitivity to Different Locations of Diffusive Surfaces in Performance Spaces: The Case of a Shoebox Concert Hall. *Appl. Sci.* **2020**, *10*, 4370. [CrossRef]
34. Bo, E.; Kostara-Konstantinou, E.; Lepore, F.; Puglisi, G.E.; Shtrepi, L.; Barkas, N.; Astolfi, A. Acoustic characterization of the ancient theatre of Tyndaris: Evaluation and proposals for its reuse. In Proceedings of the 23rd International Congress of Sound Vibrations (ICSV23), Athens, Greece, 10–14 July 2016.
35. Bo, E.; Shtrepi, L.; Pelegrín-García, D.; Barbato, G.; Astolfi, A. Acoustical measurements in ancient theatres: Uncertainty in prediction models. In Proceedings of the 1st International Conference on Metrology for Archaeology, Benevento, Italy, 22–23 October 2015.
36. Papadakis, N.M.; Stavroulakis, G.E. Review of Acoustic Sources Alternatives to a Dodecahedron Speaker. *Appl. Sci.* **2019**, *9*, 3705. [CrossRef]
37. Barron, M. *Auditorium Acoustics and Architectural Design*; Spon Press: London, UK, 1993.
38. Marshall, L.G. Speech intelligibility prediction from calculated C50 values. *J. Acoust. Soc. Am.* **1995**, *98*, 2845–2847. [CrossRef]
39. Bradley, J.S. Review of objective room acoustics measures and future needs. *Appl. Acoust.* **2011**, *72*, 713–720. [CrossRef]
40. Bruel Kjaer. *Dirac Room Acoustics Software*; Type 7841, User Manual: Version 5.0; Bruel Kjaer: Nærum, Denmark, 2010.
41. Farina, A. aurora-plugins.com. 2013. Available online: http://pcfarina.eng.unipr.it/Aurora_XP/index.htm (accessed on 7 April 2016).
42. Katz, B.F. In situ calibration of the sound strength parameter G. *J. Acoust. Soc. Am.* **2015**, *138*, EL167–EL173. [CrossRef]
43. Wenmaekers, R.H.; Hak, C.C. The sound power as a reference for Sound Strength (G), Speech Level (L), and Support (ST): Uncertainty of laboratory and in-situ calibration. *Acta Acust. United Acust.* **2015**, *101*, 892–907. [CrossRef]
44. Bradley, J.S.; Reich, R.; Norcross, S. A just noticeable difference in C50 for speech. *Appl. Acoust.* **1999**, *58*, 99–108. [CrossRef]
45. Hak, C.C.; Wenmaekers, R.H.; Hak, J.P.; van Luxemburg, L.C.; Gade, A.C. Sound Strength Calibration Methods. In Proceedings of the ICA 2010 Conference, Sydney, Australia, 23–27 August 2010.
46. Barron, M.B. Interpretation of Early Decay Time in concert auditoria. *Acustica* **1995**, *81*, 320–331.
47. Chourmouziadou, K.; Kang, J. Acoustic evolution of ancient Greek and Roman theatres. *Appl. Acoust.* **2008**, *69*, 514–529. [CrossRef]
48. Arana, M.; Vela, A.; San Martin, L. Calculating the impulse response in rooms using pseudo-impulsive. *Acta Acust. United Acust.* **2003**, *89*, 377–380.
49. Bo, E.; Shtrepi, L.; Aletta, F.; Puglisi, G.E.; Astolfi, A. Geometrical Acoustic Simulation of Open-air Ancient Theatres: Investigation on the Appropriate Objective Parameters for Improved Accuracy. In Proceedings of the 16th IBPSA Conference, Rome, Italy, 2–4 September 2019; pp. 18–25. [CrossRef]

© 2020 by the authors. Licensee MDPI, Basel, Switzerland. This article is an open access article distributed under the terms and conditions of the Creative Commons Attribution (CC BY) license (http://creativecommons.org/licenses/by/4.0/).

Article

Evolutionary Analysis of the Acoustics of the Baroque Church of San Luis de los Franceses (Seville)

Enedina Alberdi *, Miguel Galindo and Ángel L. León-Rodríguez

Instituto Universitario de Arquitectura y Ciencias de la Construcción (IUACC), Escuela Técnica Superior de Arquitectura, Universidad de Sevilla. Av. Reina Mercedes 2, 41012 Seville, Spain; mgalindo@us.es (M.G.); leonr@us.es (Á.L.L.-R.)
* Correspondence: ealberdi@us.es

Abstract: In the 16th century the Society of Jesus built a large number of churches following the Tridentine model of a Latin cross and a single nave. However, the shift towards this model did not entail the abandonment of the central floor plan, especially in the 17th century. The acoustics of these spaces can present phenomena linked to focalizations which increase the sound pressure level. The church of San Luis de los Franceses, built by the Jesuits for their novitiate in Seville (Spain), is an example of a Baroque church with a central floor plan. Although the church has hosted different congregations since its inauguration it is currently desacralized and used for theatres and concerts. The acoustics of this church were studied by the authors through in situ measurements and virtual models. The main objective was to analyse the evolution and perception of its sound field from the 18th to 21st centuries, considering the different audience distributions and sound sources and the modifications in furniture and coatings. Analysis of the evolution of its sound field shows that the characteristics have remained stable, with a notable influence of the dome on the results for the different configurations studied.

Citation: Alberdi, E.; Galindo, M.; León-Rodríguez, Á.L. Evolutionary Analysis of the Acoustics of the Baroque Church of San Luis de los Franceses (Seville). *Appl. Sci.* **2021**, *11*, 1402. https://doi.org/10.3390/app11041402

Academic Editor: Nikolaos M. Papadakis
Received: 3 January 2021
Accepted: 29 January 2021
Published: 4 February 2021

Publisher's Note: MDPI stays neutral with regard to jurisdictional claims in published maps and institutional affiliations.

Copyright: © 2021 by the authors. Licensee MDPI, Basel, Switzerland. This article is an open access article distributed under the terms and conditions of the Creative Commons Attribution (CC BY) license (https://creativecommons.org/licenses/by/4.0/).

Keywords: worship space acoustics; acoustics simulation; acoustic heritage

1. Introduction

After the Council of Trent (1545–1563), in his book *Instructiones Fabricae et Supellectilis Ecclesiasticae* [1] Cardinal Carlos Borromeo recorded his "*Instructions for ecclesiastical construction and decoration*", identifying the Latin cross plan with a single nave as the most suitable for churches. This was considered a symbol of Christianity, and was favoured over central floor plans that were more characteristic of pagan temples at the time. The importance of acoustics in churches inevitably leads to the debate on the best way to cover them, as Sendra and Navarro [2] have analysed based on the documentation of four churches (three Jesuit churches and one Franciscan one), identifying the model of single nave churches with wooden roofs as the best option.

The main Jesuit church, Il Gesú (1568–1584) by Giacomo Barozzi da Vignola, follows the Tridentine model and the austere spirit of the Society of Jesus, where the main functions of preaching and administration of the sacraments benefit from a model with a single nave where the parishioners gather. This guarantees an adequate visual connection with the presbytery, while the members of the religious community and schoolchildren of the Society were housed in tribunes which allowed them to follow the services independently from the people. Although this initial model of a Latin cross plan and single nave was reproduced by the Jesuits in numerous churches, circular, Greek cross or elliptical plans were also used to a lesser extent.

The rapid expansion of the Jesuit Order, following its foundation in 1534, allowed the establishment, on 7 January 1554, of the province of Andalusia, with the kingdoms of Jaén, Córdoba, Granada and Seville, as well as the region of Fregenal de la Sierra, south of Badajoz and the Canary Islands [3].

The modo nostro followed by the Jesuits required them to send the designs or plans of the buildings built in each province to Rome for review. This required a degree of coordination and standardization in terms of building typologies. In Andalusia, the churches built in the 16th century mainly follow the model of a single nave, transept, vaults and hemispherical dome [4]. Notable examples of this model are the church of Santa Catalina in Córdoba (1564–1589), the Anunciación in Seville (1565–1579), and the Encarnación in Marchena (Seville) (1584–1593). As a result, of the centralizing tendencies of the Jesuits, throughout the 17th century churches with central typologies were introduced, especially for those cases which required churches with a smaller capacity.

In Andalusia the first example of this model was the church of San Hermenegildo in Seville (1614–1620), with an elliptical plan. This was followed by the church of the College of San Sebastian in Malaga (1626–1630), with a circular plan, churches with the Greek cross floor plan, as seen in the churches of the College of San Torcuato de Guadix (Granada) (1626) and that of the Novitiate of San Luis de los Franceses in Seville (1699–1731). This study focuses on the Baroque church of San Luis de los Franceses, considered a jewel both among the churches of the city of Seville and those built by the Order in the province of Andalusia (Figure 1).

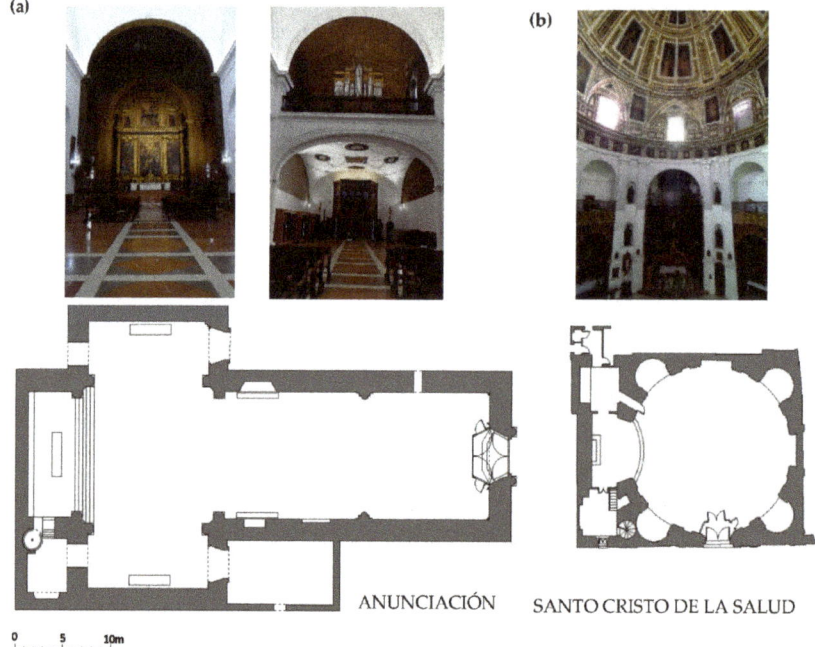

Figure 1. Jesuit church. (**a**) Single nave. Church of the Anunciación (Seville). (**b**) Central plan. Church of Santo Cristo de la Salud (Málaga).

Archaeoacoustics, as a method of analysis of historical heritage, makes it possible to study the complex relationship established between architecture and acoustics, introducing as an aspect of analysis the characteristics of the sound field of the analysed spaces, as well as the relationship between people and sound [5,6]. Ecclesial spaces from different historical periods have been analysed from this point of view: Byzantine [7]; Romanesque, such as the Cathedral of Santiago de Compostela [8] or the Abbey of Cluny [9]; Gothic, such as the Spanish cathedrals [10]; and Baroque, such as the Church of Santa María Magdalena [11], as well as unique spaces such as the Mosque Cathedral of Córdoba [12].

The conditions of the sound field in Baroque central spaces have been studied by Cirillo et al. [13] for the churches of St. Luca e Martina (Rome), St. Agnese in Agone (Rome), St. Lorenzo (Turin) and the Basilica of Superga (Turin), where in situ measurements were carried out based on the statement that in all cases *"the small dimension of the church (in plan) allow short source–receiver distance with relatively high C_{80} values"*. Furthermore, Carvalho analysed the church of Dos Clérigos in Porto [14], presenting the main monaural parameters derived from impulse responses together with the RASTI index, conducting subjective studies in relation to the evaluation of intelligibility and live music performances.

Spaces with large domes can give rise to unexpected acoustic phenomena [15–17] and focalizations that can increase the sound pressure level or cause colorations in the sound or echoes [18,19]. These aspects have been studied by Alberdi et al. in the church of San Luis de los Franceses [20], where the Bayesian analysis performed showed the presence of double slopes in the energy decay curves, for the different frequencies, especially when the source is located under the dome or near a lateral altar. The double slope phenomenon could be associated with an uneven distribution of sound energy due to acoustic coupling between different sub-volumes, as confirmed by directional intensity maps. The receivers under the dome receive the early reflections mainly from the side walls, while reflections that are more delayed in time do so from the hemispherical surfaces delimiting the volume of the central dome.

The main aim of this research is the evolutionary analysis of the acoustic conditions of the church of San Luis de los Franceses from its inauguration in 1731 to the present day. The church currently displays the same formal configuration inside, but it is deconsecrated and used to stage cultural events such as the Bienal de Flamenco. Over time, the positions of sources and receivers have undergone variations, and this work analyses the impact on the properties of the sound field, in light of the influence that the large central space covered with the large dome could have on the results obtained.

2. The Church of San Luis de los Franceses

2.1. Origins and Description

The church of San Luis de los Franceses is located on calle San Luis, in the northern part of the historic centre of Seville (Spain). The church is part of a group of buildings, corresponding to the novitiate of San Luis and belonging to the Society of Jesus (Figure 2).

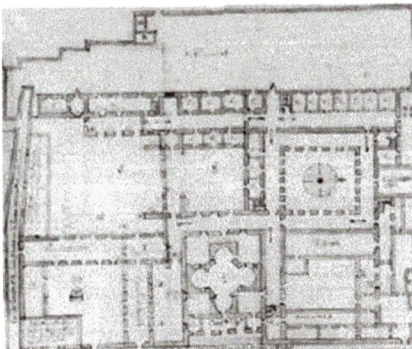

Figure 2. Plan of the Royal Convent of San Diego in Seville (1784), formerly known as the Novitiate of San Luis. Archivo Histórico Nacional. Consejos n° 1423.

The Company was established in Seville in 1554. The foundation of the novitiate is thought to date back to 1609, when it was established in some houses near the church of Santa Marina, where there was a small church whose presbytery collapsed in 1695. As a result, work began on the church of San Luis (1699–1731), under the supervision of the architect Leonardo de Figueroa [21,22].

The plan of the church of San Luis follows the model of the Greek cross within a square (Figure 3) so that four semi-circular apses are found at the sides of the cross. The need to separate the interior of the church from calle San Luis led to the incorporation of an atrium before the enclosure, thus preventing direct communication through the apse next to calle San Luis [23].

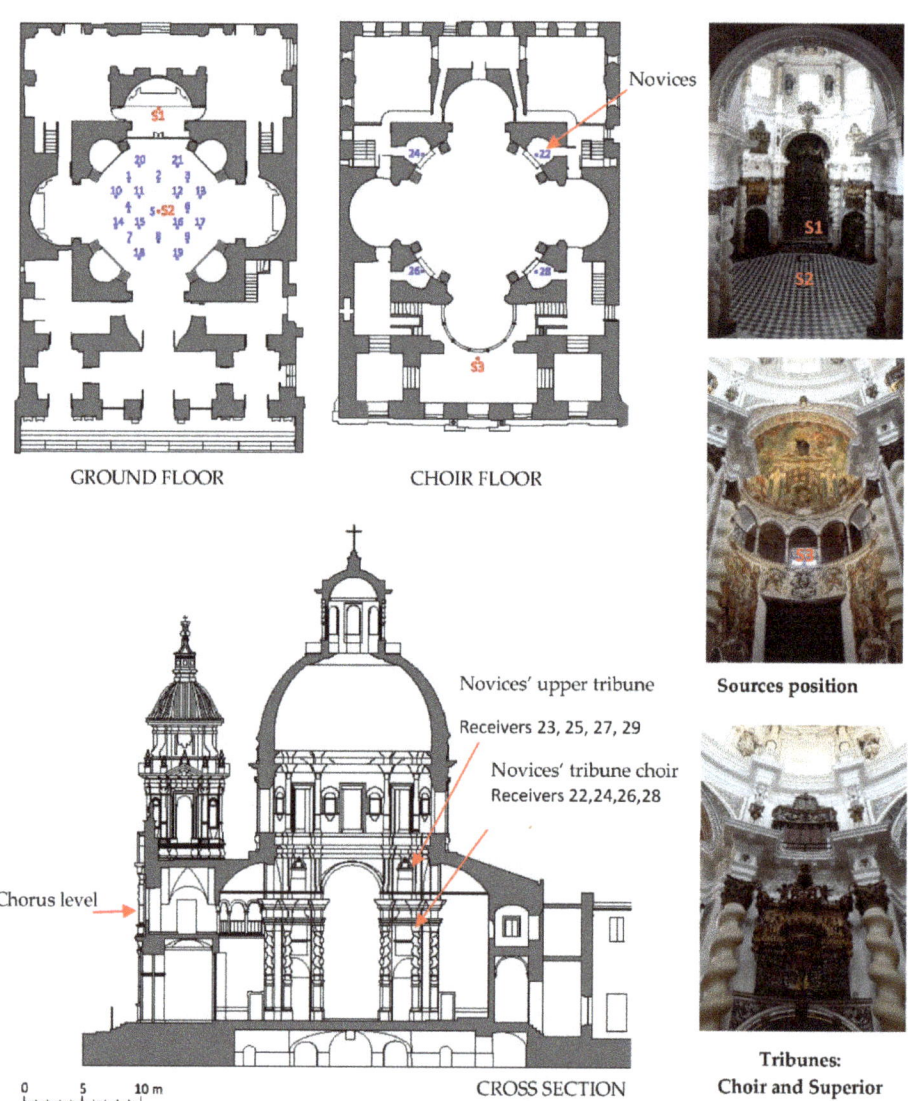

Figure 3. Church of San Luis de los Franceses. Ground floor and choir. Cross section.

This space provides enough space on the upper floor for the location of the choir, accessed from the novitiate enclosure on its upper floors. Another aspect to note in the importance of this previous space which provides the interior space, when closed, with better acoustic insulation from the bustle of calle San Luis. The main geometric features of the church are shown in Table 1.

Table 1. Main geometric characteristics.

Inner volume	4804 m^3
Ground floor surface	231.80 m^2
Choir floor surface	58.40 m^2
Main axis length	22.35 m
Transverse axis length	22.10 m
Inner diameter dome	12.84 m
Higher height under lantern	34.95 m

In the layout of the plan, the dome takes on great importance, following the model inspired by Father Pozzo's treatise on perspective [24].

The intervention by Figueroa is clearly seen in the execution of the dome, both in the layout of the drum, with large windows on a cylinder resting on pillars avoiding the use of pendentives, and in the materials used, red brick and ceramic tiles which give the cupola an appearance characteristic of Figueroa's work. The dome rests on four large pillars that connect to small altars on the ground floor. In the same position on the upper floors there are tribunes which provide a visual and acoustic connection between the novitiate and the main space under the dome.

2.2. Evolutionary Stages

From its inauguration in 1731 to the present day, the church and its novitiate complex have welcomed different communities and collectives who have used it for different purposes, placing the audience areas, furniture, coatings and sound sources in different positions.

During the 18th and 19th centuries, the church was used by the Jesuits—intermittently, due to their expulsion from Spain in 1767—before the church was ceded to the Franciscan brothers of San Diego in 1784. In the 18th century, the parishioners were not seated on benches or chairs, so for the study hypothesis, the public was considered standing occupying the space under the dome [23]. In addition, the necessary separation between novices and parishioners suggests that the novices occupied the tribunes linking the novitiate with the church, allowing the ceremony to continue with no contact between the two.

During the First Republic (1873), the building continued to be used as a hospice. This meant that during the Spanish Civil War in the 20th century the building was spared and in an acceptable state of conservation, although it remained in disuse and empty from 1968 to 1976, when it opened to worship. It can thus be said that worship was maintained in the church during the 20th century, albeit intermittently, with benches added on the ground floor for the parishioners, following the canons that mark the current liturgy.

Since 1984, major intervention projects have been carried out for the full conservation of the monument. The complex, with its deconsecrated church, served as the headquarters for the Andalusian Centre for Performing Arts from 1992 to 2010, and was again restored to its use as a scenic place in 2016, when it became the headquarters of the Bienal de Flamenco. When used as a scenic space for the Bienal de Flamenco, the public sits on wooden chairs, with the sound source located on a small stage on the main altar, while for theatrical performances the sound source is placed under the dome (Figure 4).

Figure 4. (**a**) Representation of the play Don Juan Tenorio. Classical Theatre Company of Seville (2005). http://www.clasicodesevilla.com/Don-Juan-Tenorio. (**b**) María Terremoto. Bienal de Flamenco, Seville (2016). http://www.labienal.com/galeria/.

3. Methodology

The characteristics of the sound field in the church of San Luis were studied through in situ measurements of the acoustic conditions of the empty church in its current state and the creation of three-dimensional models reproducing the configurations of sources and audience areas of audience between the 18th and 21st centuries, allowing analysis of the characteristics of the sound field. The CATT-Acoustic software [25] developed by the company CATT (*Computer Aided Theatre Technique*) from Gothenburg (Sweden) was used for this analysis, with two different calculation engines, CATT-*Acoustic v.9.1b* and *CATT TUCT v2.0b*.

For each of the models, the interior of the church has been reproduced using *SketchUp v.14* software, which allows the geometry to be exported to the *CATT* program. The models made simplify the interior space, eliminating any elements which, due to their smaller size, are not decisive when it comes to obtaining adequate results. However, the acoustic properties of the materials of the different surfaces (absorption and scattering) are more decisive for obtaining adequate results [26]. The environmental conditions (humidity and temperature), as well as the background noise obtained from in situ measurements, were considered for the simulations with *TUCT algorithm* 2 (Table 2).

Table 2. Calculation conditions.

Background Noise (dB)	125	250	500	1000	2000	4000
	43	43	49	52	48	41
Environmental parameters	Temperature			18.70 °C		
	Humidity			73.00%		
Calculation conditions	Calculation algorithm			2		
	Number of rays			100,000		
	Echogram/impulse response			4000 ms		
	Air density			1.20 kg/m³		
	Air absorption			activated		
	Diffraction			activated		
Number of planes	Hypothesis 0 (H0)			764		
	Hypothesis 1 (H1)			797		
	Hypothesis 2 (H2)			797		
	Hypothesis 3 (H3)			774		
	Hypothesis 4 (H4)			774		
	Hypothesis 5 (H5)			791		
	Hypothesis 6 (H6)			817		

3.1. Acoustic Parameters

For the evaluation of the sound field of San Luis de los Franceses, objective acoustic parameters are obtained from in situ measurement following standard ISO 3382-1 [27].

These parameters are associated with a subjective characteristic that allows the listener to assess the acoustics of the room. After validation of the church model in its current state, the characteristics of the sound field in the different evolutionary models were assessed. The reverberation of the inner space is quantified with the T_{30} parameter, while the perceived reverberation is judged from the Early Decay Time EDT, more closely linked to the subjective reverberation time. The sound strength parameter G is used to assess the subjective sound level; the perceived musical clarity of sound, C_{80}; definition, D_{50}; the apparent source width from early lateral energy fraction, J_{LF}; and listener envelope from the early inter-aural cross correlation coefficient $IACC_E$. The results for all these parameters are obtained spatially averaged by frequency, between 125 and 4000 Hz. The Speech Transmission Index parameter, STI, using a standardized scale to rate the intelligibility of speech based on a standardized scale, is also analysed [28].

3.2. In Situ Measurement

Measurements in the church were carried out in situ without an audience to characterize its behaviour [27]. The measurement chain was used to obtain impulse responses (IR) from which all the acoustic parameters that define the sound field can be obtained. At each point where a receiver was located, the IR was obtained from sweeps of sinusoidal wave signals, with frequency increasing exponentially with time [28]. Both the frequency range and the duration of each sweep were adjusted to suit the environmental conditions in order to have impulse responses of adequate quality, so that the signal-to-noise ratio exceeded 45 dB in all the octave bands analysed, 125 to 4000 Hz.

WinMLS2004 software with a Roland Edirol UA 101 sound card was used to generate the signal, recordIRs, and analyse results. The signal generated by the laptop was fed to a Behringer Eurolive B1800DProas amplifier connected to an AVM D012 01 dB omnidirectional source. IRs were recorded using an Audio-Technica AT4050 multi-pattern microphone with omnidirectional configuration connected to the Soundfield polarization source (SMP 200). A Head Acoustic HMS III torso simulator pointed towards the sound source was used, together with the OPUS 01dB signal conditioner, to obtain the cross-correlation coefficients. The background noise spectrum is measured with a Brüel & Kjær B&K 4165 microphone connected to a Svantek SVAN 958 noise analyser.

In the acoustic measurement of the church, the sources were placed in four positions. Positions S1 (main altar), S2 (dome) and S3 (choir) corresponded to the positions described in Figure 3, used in the study of the different hypotheses. Source S4, on a side altar, was considered only as an in situ measurement. The positions for the receivers located in the audience area coincided with those in Figure 3, numbered from 1 to 9. The sources and receivers were placed at heights of 1.50 m and 1.20 m, respectively.

With respect to the interior material used (Table 3), the plaster of the walls and dome accounts for more than half of the surfaces, while the marble of the flooring and decorative elements also has a great impact. The wooden altarpieces also occupy a large surface area, with other materials such as glass or ceramic flooring found, albeit to a lesser extent, on the upper floor.

Table 3. Materials, location, surface area, and percentage.

Materials	Location	Surface Area (m^2)	Surface Area. (%)
Plaster	Walls and dome	1527.80	58.62
Marble	Ground and choir floor. Solomonic and choir columns.	577.34	23.15
Wooden altarpieces	Altars	283.70	10.90

Table 3. *Cont.*

Materials	Location	Surface Area (m²)	Surface Area. (%)
Glass	Dome and choir windows	82.80	3.20
Hole	Communication gap with the Novitiate	43.70	1.70
Ceramic flooring	Choir and tribunes floor.	38.26	1.50
Organ	Choir organ	23.50	0.93

Table 4 shows the results for the acoustic parameters of the ISO 3382-1 standard, obtained for the position of the source on the main altar (S1), the source position considered for the validation of the computer model.

Table 4. Acoustic parameter values in frequency octave band and single number frequency averaging [27].

	125 Hz	250 Hz	500 Hz	1 kHz	2 kHz	4 kHz	Single Number
T_{30} (s)	3.59	3.72	3.72	3.39	2.81	2.16	3.55
EDT (s)	3.030	3.09	3.03	2.75	2.20	1.71	2.89
G (dB)	15.07	15.63	14.28	10.97	10.37	9.42	12.62
C_{80} (dB)	−0.42	−1.18	−1.80	−0.61	0.41	1.52	−1.20
D_{50} (-)	0.38	0.32	0.27	0.35	0.40	0.45	0.31
J_{LF} (-)	0.13	0.27	0.40	0.35	0.31	0.28	0.28
$IACC_E$ (-)	0.96	0.84	0.58	0.57	0.40	0.34	0.51

Based on the measurement results evaluated by Alberdi et al. [20], it can be stated that "the analysis showed negligible differences in the reverberation time for all sources, as also confirmed by conventional criteria proposed by standards to evaluate the curvature of decays. Nevertheless, for EDT values, the differences in early energy growth gave rise to a different behaviour as a function of the position of the source clearly appeared. Bayesian analysis showed that several double slopes appeared in decay curves, spanning different frequencies, particularly when the source was under the dome or close to a lateral altar. The most affected octave band frequencies were 2 kHz and 4 kHz. Although not particularly evident, the double slope phenomenon could be associated to an uneven distribution of sound energy due to acoustic coupling between different sub-volumes". The influence of the large volume of the dome in relation to the perception of the sound field for the different relationships between the position of the source and the receivers is evaluated using the computer models created for the different hypotheses.

3.3. Model Validation

Once the data of the model of the current state with the empty church had been entered in CATT-Acoustic, it was validated to adjust it to the results obtained in the in situ measurement, so that its acoustic behaviour was similar to the current one (Figure 5).

To achieve this objective, the absorption and scattering coefficients of the materials, especially those which present greater uncertainty, must be modified. This adjustment was applied to the reverberation time (T_{30}) so that the values obtained in the calculation were similar to those measured in situ for the different octave bands. As a validation criterion [29], it was estimated that the coincidence was adequate if they differed less than the perceptible threshold Just Noticeable Difference (JND), that is, 1 JND for T_{30} (less than 5% of the values measured for each octave band). The rest of the parameters were considered adequate below 2 JND, according to consensus [30]. In contrast, as indicated by Martellotta [31] in his study for the value of JND in C_{80} and Ts for three churches, the value of 1 JND for clarity and sharpness in widely reverberant spaces can be modified in relation to the regulations in place, considering 1.5 dB for C_{80}. In the case studied, although a faithful adjustment of the reverberation times was achieved, the validation criterion could not be met in almost any of the parameters. EDT values, corresponding to the early decay time, are more closely related to the subjective perception of reverberation, given its dependence on the energy associated with the early reflections obtained in the in situ measurement. Major variations in EDT were observed based on the position of the sound source and in its behaviour in relation to the source–receiver distance.

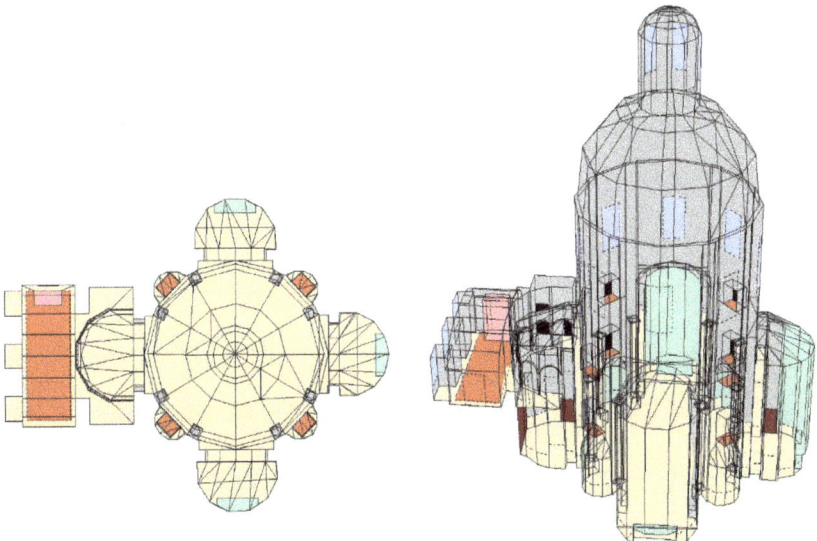

Figure 5. San Luis de los Franceses church. Colours correspond to different materials.

This highlights the presence of double slopes in the energy decay curves, showing that the dome behaves as a coupled space. For this reason, EDT was chosen as the adjustment parameter, allowing improved behaviour of most of the acoustic parameters which determine the sound characteristics of the church. The model was validated with the source located at the main altar (S1) and the receivers in the positions used in the in situ measurement. Table 5 shows the absorption and scattering coefficients of the different materials found inside the room used for the validation of the model.

The composition of the plaster walls, which account for more than 50% of the total interior surface, is not known exactly. Therefore, in the interactive process of fitting the model, the initial absorption coefficient values were modified to ensure that the space simulated presents the same acoustic behaviour as the real room.

Another aspect to consider in the model is that the room is not independent from the novitiate building, as the church and novitiate are connected by spaces with no door. These connections occur at the levels of the choir floor and the upper tribune, in the spaces arranged for the visual connection between both parts of the building, considered in the modelling as flat virtual surfaces with an absorption coefficient close to 100%.

Three options were considered for the scattering coefficients. In general, a default value of 10% was allowed in all octave bands for all materials, except for those with an irregular real surface, which were simulated using flat surfaces, as is the case of altarpieces. In this case, variable dispersion coefficients of between 30 and 80% were considered for the octave bands. For the spaces connecting with the rest of the novitiate units, a surface was modelled with an absorption coefficient close to 100% and a dispersion of 1%.

Table 5. Finishes: materials, references, surface (%), and absorption (up) and scattering (down) coefficients. * Material used for the adjustment.

Material	Area (%)	125 Hz	250 Hz	500 Hz	1 kHz	2 kHz	4 kHz
Plaster *	58.60	0.12	0.12	0.11	0.13	0.15	0.17
		0.10	0.10	0.10	0.10	0.10	0.10
Marble [30]	23.10	0.10	0.10	0.10	0.20	0.20	0.20
		0.10	0.10	0.10	0.10	0.10	0.10
Altarpieces [29]	10.90	0.12	0.12	0.15	0.15	0.18	0.18
		0.30	0.40	0.50	0.60	0.70	0.80
Wooden door [30]	0.80	0.14	0.10	0.06	0.08	0.10	0.10
		0.10	0.10	0.10	0.10	0.10	0.10
Glass [32]	3.2	0.04	0.04	0.03	0.03	0.02	0.02
		0.10	0.10	0.10	0.10	0.10	0.10
Opening	1.70	0.99	0.99	0.99	0.99	0.99	0.99
		0.10	0.10	0.10	0.10	0.10	0.10
Ceramic flooring [30]	0.80	0.02	0.02	0.03	0.03	0.04	0.05
		0.10	0.10	0.10	0.10	0.10	0.10
Organ [33]	0.9	0.12	0.14	0.16	0.16	0.16	0.16
		0.30	0.40	0.50	0.60	0.70	0.80

Following the comparison of the results of the in situ measurement and the computer-simulated model, it can be concluded that the model adopted for the church of San Luis de los Franceses adequately reproduces the acoustic behaviour of the room in its current state.

Figure 6 shows the JND differentials of the different parameters, obtained as the difference between the mean values of the parameters at the different frequencies, from the modelling and the in situ measurement, and divided by the JND value obtained in the measurement from the indications given in of UNE-EN ISO 3382-1 standard.

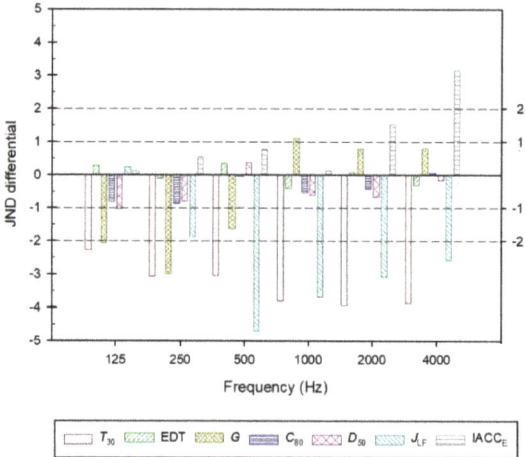

Figure 6. JND differentials spatially averaged versus frequency octave band for the acoustic parameters evaluated: difference between the simulated and measured results divided by JND of the measured value.

In general, the results are between −1.00 and +1.00 for the different frequencies. The value of early lateral energy fraction (J_{LF}) is that with the lowest adjustment as this parameter is difficult to simulate. However, the $IACC_E$ value is adequate, except for in the 4000 Hz band. The JND differential for reverberation time is higher than the rest of the parameters, obtaining results between −2.00 and −4.00 JND, possibly due to the arrangement of coupled spaces.

3.4. Analysis of the Evolution of Acoustic Conditions

After validating the model of the empty church in its current state, the different audience and source positions from the 18th to 21st centuries were studied. Models were drawn up to simulate these changes and any variations in coverings and furniture, allowing the analysis of the characteristics of the sound field in each of the stages studied.

4. Evolutionary Models

The necessary modifications are carried out on the initial model to adapt it to the models representing the different evolutionary stages in the history of San Luis, from its establishment in the 18th century to the present day (Figure 7). Six specific hypotheses were established (Table 6) to represent the historical moment (18th-19th century, 20th century, 21st century), the position of the source (S1, S2, S3), and the configuration of the audience (standing, wooden benches, wooden chairs).

Figure 7. Ground floor. Hypothesis audience distribution 1 to 6.

Table 6. Evolutionary hypotheses (* liturgy).

	H1 *	H2 *	H3 *	H4 *	H5 (Theatre)	H6 (Flamenco B.)
18–19th c.	X	X				
20th c.			X	X		
21st c.					X	X
S1 (main altar)	X		X			X
S2 (dome)					X	
S3 (chair)		X		X		
Standing audience	X	X				
Wooden benches audience			X	X		
Wooden seats audience					X	X
Novices in tribune	X	X				
Stage in main altar						X

Table 7 defines the acoustic properties (absorption and scattering) of the modified elements with respect to the initial model, specifically, the planes that simulate the audiences in the different hypotheses, as well as the wooden stage introduced in Hypothesis 6 (H6) and on which S1 is placed 1.50 m above the stage.

Table 7. Modified materials, references, and absorption (up) and scattering (down) coefficients.

Material	125 Hz	250 Hz	500 Hz	1 kHz	2 kHz	4 kHz
Standing audience	0.16	0.29	0.55	0.80	0.92	0.90
(1p/m^2) [30]	0.10	0.10	0.10	0.10	0.10	0.10
Wooden occupied	0.23	0.37	0.83	0.99	0.98	0.98
bench [34]	0.30	0.40	0.50	0.60	0.70	0.80
Wooden occupied chair	0.24	0.40	0.78	0.98	0.96	0.87
(2chairs/m^2) [30]	0.30	0.40	0.50	0.60	0.70	0.80
Stage: wooden	0.18	0.12	0.10	0.09	0.08	0.07
platform [30]	0.10	0.10	0.10	0.10	0.10	0.10

With respect to audience receivers, those numbered in Figure 2 and within the audience area of each hypothesis are considered. As for the eight positions representing the novices located in tribunes (choir level and upper tribune) of Hypotheses 1 and 2, these correspond to the receivers numbered from 22 to 29, as indicated in Figure 8.

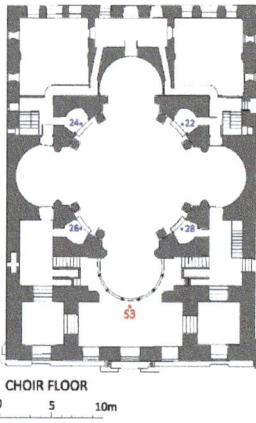
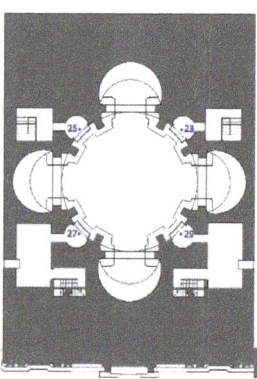

CHOIR FLOOR   UPPER TRIBUNE FLOOR

Figure 8. Choir and upper tribune floor. Receivers located in stands for novices.

5. Results and Discussion

The acoustic behaviour of the church of San Luis de los Franceses was analysed, taking into account the different uses, occupations and sound sources throughout history. Six acoustic simulation models were generated: H1 and H2, corresponding to religious use and with the audience standing in the 18th and 19th centuries; H3 and H4 with the audience seated on wooden benches in the 20th century; H5 and H6 with the church desacralized for the 21st century; H5 for use as a theatre; and H6 for concerts within the Bienal de Flamenco. In H1, H3 and H6 the sound source is located in the main altar, in H2 and H4 in the choir and in H5 in the centre of the audience under the dome.

To evaluate the sound sensation of the listeners in each model, together with the reverberation time (T_{30}), the acoustic parameters related to the different subjective sensations are presented: Perceived reverberance (EDT), Subjective level of sound (G), Perceived clarity of sound (C_{80} for music and D_{50} for speech), Apparent source width (J_{LF}), and Listener envelopment ($IACC_E$).

5.1. Global Analysis

Figure 9 represents the spatially averaged values versus frequency in octave bands and the standard deviations of these parameters. Reverberation, evaluated based on the values of T_{30} and EDT, even with the presence of the public, in all models over the centuries is excessive, both for religious music and for speech. Only the values obtained for the 4000 Hz octave band fall within the optimum range [35]. The averaged values and their errors for T_{30} and EDT are very similar in each octave band, both in time and between both parameters.

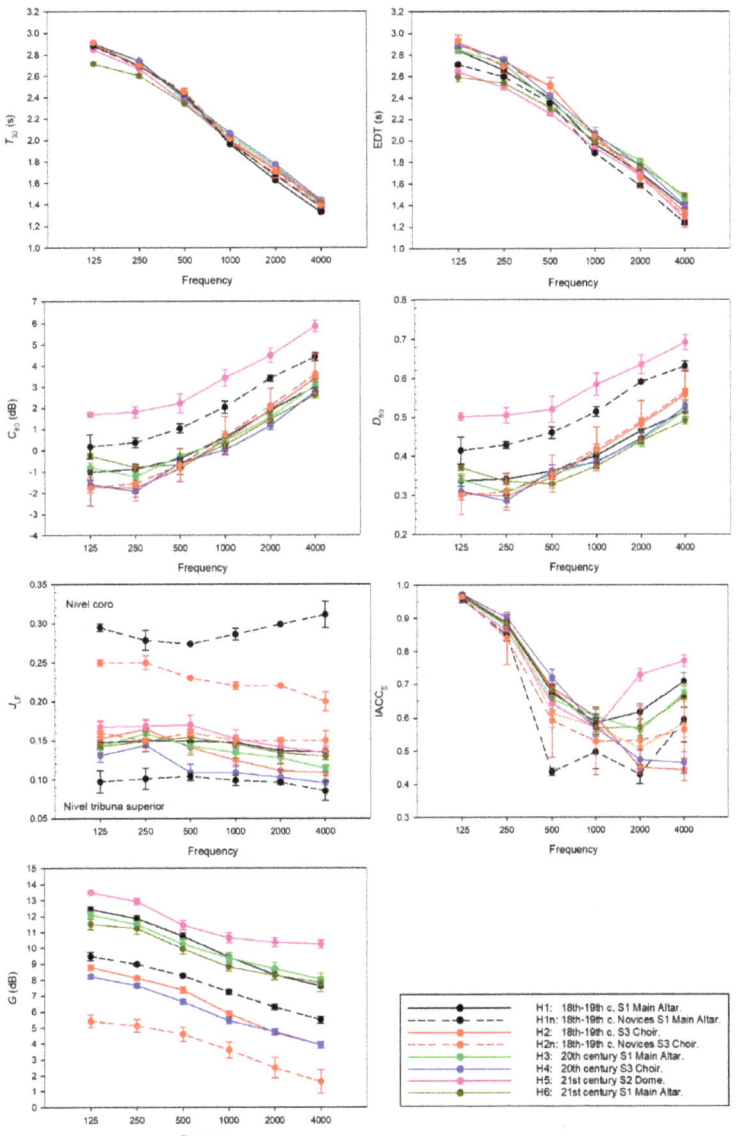

Figure 9. T_{30}, EDT, G, C_{80}, D_{50}, J_{LF} and $IACC_E$ spatially averaged versus frequency octave band and error bars (standard error) for the different simulation hypotheses.

C_{80} and D_{50} follow the same behaviours and trends. The 21st-century setting for theatre performances displays the best performance. The improvement experienced by the novices in the 18th and 19th centuries when the source is placed on the main altar is especially noteworthy. In the rest of the models, the behaviour is very similar, with significant differences only at low frequencies (125 Hz).

In addition to the previous models, the figure showing the sensation of the width of the source (J_{LF}) incorporates the analysis of the novices in the tribunes at the height of the choir and in the upper floor due to their different behaviour. This is not evident in the rest of the parameters. The differences between the two tribunes are revealed in the first temporary models (H1 and H2), with the source located in the main altar or choir, and with the best subjective sensations being obtained in the upper tribune due to the permanence of the sound late reflections in the dome lantern. Interestingly, when the source is placed in the main altar, the best source width is achieved in the upper tribune and the worst in the lower ones.

For the audience located on the ground floor, the best sensation appears with the source located under the dome, followed by the models that locate the source in the main altar, and the worst results are obtained when the source is located in the choir. In this last case, the subjective sensation worsens in the 20th century compared to the previous ones.

The enveloping sensation of the listener, evaluated in the octave bands of interest (500–2000 Hz) is the most suitable for novices in the 18th-19th centuries, followed by the models where the source is placed in the choir. Next, are the models in which, throughout history, the sound source has been located in the main altar and, lastly, due to the location of the source, the 21st century model for theatrical performances under the dome.

Finally, the G values are high enough in all models to create an adequate subjective sound level. The highest values correspond to the 21st-century model, with the source located under the dome. When it is located on the main altar, the sound level remains similar in its temporal evolution to a choir position, although there is a significant decrease compared to the main altar. The same happens to the novices in stands with respect to the faithful on the ground floor.

For the purpose of a global qualification, Table 8 presents the results of the simulations, spectrally and spatially averaged for each of the acoustic parameters, sources and hypotheses. In addition, the receivers located on the ground floor and those located in novices' tribunes in the 18th century are analysed separately, since the results show significant differences in some parameters. The lowest and highest values of the acoustic parameter analysed have been highlighted in red and blue, for the audience on the ground floor and in the stands. The largest and smallest of the audience areas are shaded in the same colours.

Table 8. Unique number values for sources S1 (main altar), S2 (dome) and S3 (choir) in the different hypotheses.

SOURCE	MODEL	Ground Floor Audience Receivers							
		T_{30m} (ms)	EDT_m (ms)	G_m (dB)	C_{80m} (dB)	D_{50m}	STI	J_{LFm}	$IACC_{Em}$
S1 MAIN ALTAR	H1	2.19	2.18	10.10	0.12	0.38	0.51	0.15	0.63
	H3	2.21	2.21	9.82	0.11	0.37	0.51	0.14	0.61
	H6	2.17	2.15	9.38	−0.20	0.35	0.50	0.15	0.61
S2 DOME	H5	2.18	2.09	11.06	2.83	0.55	0.56	0.16	0.65
S3 CHOIR	H2	2.21	2.28	6.63	−0.16	0.38	0.52	0.15	0.58
	H4	2.23	2.25	6.05	−0.26	0.37	0.51	0.12	0.59
SOURCE	MODEL	RECEIVERS NOVICES' TRIBUNES							
		T_{30m} (ms)	EDT_m (ms)	G_m (dB)	C_{80m} (dB)	D_{50m}	STI	J_{LFm}	$IACC_{Em}$
S1 MAIN ALTAR	H1	2.12	2.21	7.75	1.54	0.49	0.57	0.20	0.46
S3 CHOIR	H2	2.21	2.27	4.11	0.02	0.39	0.52	0.20	0.55

5.2. Evolutionary Analysis

The values for reverberation time, calculated using the parameter T_{30m}, are very similar throughout history for the different uses. In general, the variation between the different models in reverberation times is within a range of 4.00%, regardless of the position of the source. This was foreseeable, since there are no significant changes in volume or in the absorption of the coatings of the church in the models (Figure 10). In addition, if these average values are compared with the average values of the optimal reverberation times, the church shows a degree of reverberance since in all cases they remain above the recommended limit, 1.59 s for musical use and 1.19 s for speech [35].

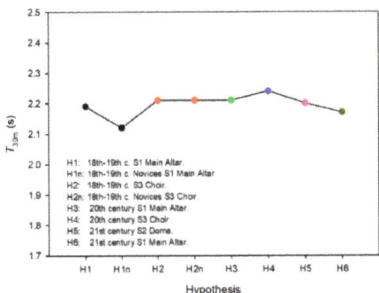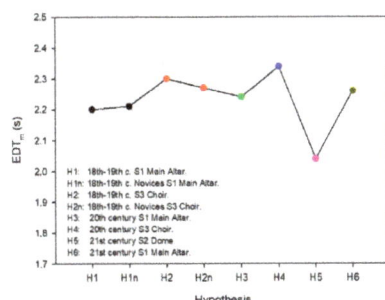

Figure 10. Acoustic parameter evolution T_{30} and EDT. Values averaged to a single number.

EDT results support those obtained for T_{30m}. The lowest reverberation occurs in the 21st-century layout when the source is located under the dome that exceeds 1 JND compared to the hypotheses of the 18th to 20th centuries.

The degree of amplification produced by the audience room located on the ground floor or in the stands is lower when the source is located in the choir when compared to its location on the main altar or under the dome, exceeding 2.5 JND. In addition, the subjective sound level of the novices is lower than that of the audience on the ground floor.

Figure 11 represents the distribution of G, in the ground floor audience area, at a frequency of 1000 Hz, for the different hypotheses. No major differences are observed in the behaviour of the room between hypotheses H1–H2 and H3–H4, which correspond to the 18th–19th centuries and the 20th century, respectively. However, the introduction of wooden benches for the liturgy in the 20th century leads to a decrease of 1 JND in the perception of the sound level for the audience located in the benches when the sound source is located in the main altar. This last hypothesis, H3, is very similar to that obtained in the 21st century for the Bienal de Flamenco (H6). In H5, when the sound source is placed under the dome, the amplification of the room increases in the audience area on the ground floor due to the decrease in distances to receivers, and a very similar spatial distribution is achieved for the four audience distributions.

The sound clarity perceived, both musical, evaluated using C_{80m}, and speech, through D_{50m}, presents values that remain below 1 JND over time for the different sound sources and audiences. However, two exceptions must be noted. The clarity of the novices during the 18th and 19th centuries is clearly superior to that of the audience on the ground floor when the source is located in the main altar. In addition, the arrangement of the audience in the 21st century, with the source under the dome, obtains the best results of musical clarity.

If evaluating the intelligibility of the church using the STI (Speech Transmission Index) parameter, the results confirm the results of C_{80m} and D_{50m}, allowing the results to be qualified within the acceptable range (0.45–0.60).

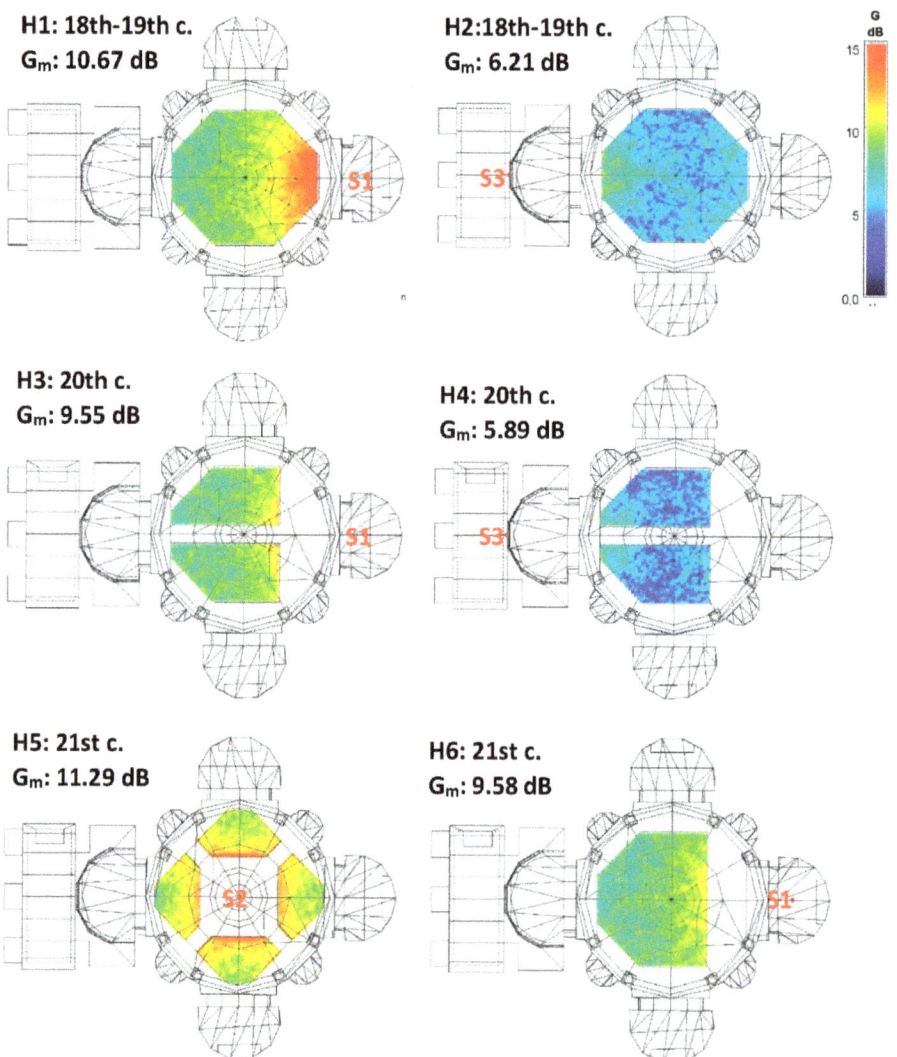

Figure 11. Sound force mapping 1 kHz octave band and for the different hypotheses. S1 (main altar), S2 (dome) and, S3 (choir).

When evaluating the apparent sensation of the source, the J_{LFm} values obtained for the ground floor audience for all models are low (<0.20), with variations between positions and hypotheses below 1 JND. However, for the novices' positions analysed in the 18th and 19th centuries, the values are the highest, so that the differences with respect to the public on the ground floor are 1 JND.

The analysis is concluded with an evaluation of the enveloping sensation of the listener using the parameter $IACC_{Em}$. For the receivers of the ground floor the sensation of spatiality has remained constant over time, with a range of values less than 1 JND and showing the best results when the source is located in the choir. There is a notable improvement for the novices with respect to the audience on the ground floor, especially when the source is located in the main altar.

A global evaluation of the results shows that, for the audience on the ground floor, when the source is located under the dome there is an improvement in all subjective sensations, while similar results to the rest of hypotheses are obtained for the enveloping sensation. The results for receivers located on the ground floor and the source in the main altar (S1) are worse that those with the source located under the dome (S2). The best results correspond to the H1 model, which simulates the Catholic liturgy in the 18th-19th centuries with the parishioners standing. Notably, the worst musical clarity results are found in the current arrangement for the Bienal de Flamenco (H6). When the source is located in the choir (S3), the acoustic conditions are the same as when the source is located in the main altar (S1), with a worsening in reverberation, sound level and musical clarity. The acoustic conditions of the listeners on the ground floor are very similar, although with a slight worsening in the 20th century compared to the two preceding centuries.

Furthermore, the novices' tribunes, during the first two centuries of the study, present better subjective acoustic sensations when the sound source is placed in the main altar. Compared to the rest of the hypotheses, there is an improvement in perceived clarity and a better spatial sensation in the tribunes.

5.3. Strengths and Limitations

The ISO 3382-1 standard is applicable in performance spaces. The deconsecrated church of San Luis de los Franceses is currently being used to stage theatrical performances, concerts and cultural events and therefore the application of this standard would be relevant. However, it must be emphasized that this standard was designed for concert halls and auditoriums whose acoustic behaviour differs greatly from that of churches, where absorption distribution is not homogeneous and the presence of chapels, domes and aisles results in a distribution of energy far from the diffuse field. In this case, depending on the position of the source and the receiver, as well as the cupola or choir which act as coupled spaces, the decrease in energy can cease to be linear, giving rise to two or more slopes. Caution must therefore be exercised when considering reverberation time. For the same reason, early energy is highly variable and intrinsic to the location of the source and receiver, so that EDT loses the homogeneity of values obtained in concert halls and auditoriums. The main limitation of the application of the standard consists of the JND associated with each acoustic parameter established in theatres. The Martellota [31] criteria for C_{80} and Ts in this type of space have been taken into account for the analysis carried out. Finally, the typical values collected in the standard for each acoustic parameter do not have to coincide with those obtained in churches.

Moreover, acoustic simulations are also subject to limitations. The main limitation is connected with the wave nature of the sound "rays" not being taken into account. In this case, the Schroeder frequency of each evolutionary period is around 45 Hz and therefore statistical acoustics would be valid from 250 Hz. Although all the simulations estimated the effect of diffraction and scattering, they were not able to capture the real wave phenomena. The level of detail used was estimated following the recommendations of Vorländer [30]. However, the treatment of the dome as a set of flat surfaces hinders the adequate reproduction of its focal behaviour. In all hypotheses, the final number of rays assumed was considered so that the results obtained in different simulations of the same hypothesis do not differ by more than 2%. All the absorption and scattering coefficients were taken from the specialized literature. Only retouching below 1% was assumed in plaster to adjust the measured and simulated results.

6. Conclusions

The church of San Luis de los Franceses has maintained its spatial configuration in terms of volume and interior cladding since the 18th century, except for the variations resulting from modifications in the furniture for the audience. In the 18th–19th centuries, the public remained standing, whereas in the 20th and 21st centuries, they went on to sit on wooden benches and chairs, respectively. In the case of the performances for the Bienal de Flamenco, a small wooden stage is also set up at the main altar. The plaster covering the walls and domes of the church accounts for over 50% of the surfaces in all models and presents slight variations which remain within 2% with respect to the total surfaces for all models between the 18th and 21st centuries. The rest of the materials, such as altarpieces, marble or glass, show variations below 1% between models.

Given the continuity in volume and coatings since its inauguration in 1731 until the 21st century, the characteristics of the sound field of the church of San Luis have remained stable and the variations detected are a consequence of the different relationships between the position of the source and the receivers, introduced by different users in different historical periods.

The evolutionary analysis of the acoustic conditions of the church of San Luis de los Franceses makes it possible to state that the model currently used for theatrical performances (20th c.), with the audience distributed around the sound source located under the dome, improves the sound clarity, the perceived source width and the sound strength in the audience located on the ground floor. However, the worst results are obtained when evaluating the listener envelopment.

During the 18th and 19th centuries, the sound clarity and the spatial sensation improved for the novices who attended religious acts in front of the standing audience on the ground floor when the sound source is in the main altar.

Author Contributions: Conceptualization, E.A., M.G. and Á.L.L.-R.; methodology, E.A., M.G. and Á.L.L.-R.; software, E.A., M.G. and Á.L.L.-R.; validation, E.A., M.G. and Á.L.L.-R.; formal analysis, E.A., M.G. and Á.L.L.-R.; investigation, E.A., M.G. and Á.L.L.-R.; resources, E.A., M.G. and Á.L.L.-R.; data curation, E.A., M.G. and Á.L.L.-R.; writing—original draft preparation, E.A., M.G. and Á.L.L.-R.; writing—review and editing, E.A., M.G. and Á.L.L.-R.; visualization, E.A., M.G. and Á.L.L.-R.; supervision, E.A., M.G. and Á.L.L.-R.; project administration, E.A., M.G. and Á.L.L.-R.; funding acquisition, E.A., M.G. and Á.L.L.-R. All authors have read and agreed to the published version of the manuscript.

Funding: This research received no external funding.

Institutional Review Board Statement: Not applicable.

Informed Consent Statement: Not applicable.

Data Availability Statement: Not applicable.

Acknowledgments: The authors wish to thank the Provincial Council of Seville for facilitating access to the church to carry out acoustic measurements and Fernando Mendoza Castells, architect in charge of the restoration of the church, for the graphic documentation provided.

Conflicts of Interest: The authors declare no conflict of interest.

References

1. Borromei, C. *Instructionum Fabricae et Supellectilis Ecclesiasticae*; Da Ponte, Pacífico: Milán, Italia, 1557.
2. Sendra, J.J.; Navarro, J. *La Evolución de las Condiciones Acústicas en las Iglesias del Paleocristiano al Tardobarroco*; Universidad de Sevilla, Institutо Universitario de Ciencias de la Construcción: Sevilla, Spain, 1997; pp. 52–62.
3. Fernando García Gutierrez, S.J. Introducción: Coordenadas Histórico-Geográficas de la Provincia Bética de la Compañía de Jesús (Soto Artuñedo, W.). In *El arte de la Compañía de Jesús en Andalucía (1554–2004)*; Publicaciones Obra Social y Cultural; CajaSur: Córdoba, Spain, 2004; pp. 18–25.
4. Rodriguez, G.; de Ceballos, A. *Bartolomé de Bustamante y los Orígenes de la Arquitectura Jesuítica en España*; Institutum Historicum S.I.: Roma, Italy, 1967; p. 332.
5. Aletta, F.; Kang, J. Historical Acoustic: Relationships between People and Sound over Time. *Acoustics* **2020**, *2*, 9. [CrossRef]
6. Scarre, C.; Lawson, G. *Archaeoacoustics*; MacDonald Institute for Archaeological Research, University of Cambridge: Cambridge, UK, 2006.
7. Đorđević, Z.; Novković, D.; Andrić, U. Archaeoacoustic. Examination of Lazarica Church. *Acoustic* **2019**, *1*, 24. [CrossRef]
8. Suárez, R.; Alonso, A.; Sendra, J.J. Intangible cultural heritage: The sound of the Romanesque cathedral of Santiago de Compostela. *J. Cult. Herit.* **2015**, *16*, 239–243. [CrossRef]
9. Suárez, R.; Alonso, A.; Sendra, J.J. Archeoacoustics of intangible cultural heritage: The sound of the Maior Ecclesia of Cluny. *J. Cult. Herit.* **2016**, *19*, 567–572. [CrossRef]
10. Alonso, A.; Suárez, R.; Sendra, J.J. The Acoustic of the Choir in Spanish Cathedrals. *Acoustic* **2019**, *1*, 4. [CrossRef]

11. Alberdi, E.; Galindo, M.; León-Rodríguez, A.L. Acoustic behaviour of polychoirs in the Baroque church of Santa María Magdalena, Seville. *Appl. Acoust.* **2021**, *175*. [CrossRef]
12. Suárez, R.; Sendra, J.J.; Navarro, J.; León, A.L. The sound of the Cathedral-Mosque of Cordoba. *J. Cult. Herit.* **2005**, *6*, 307–312. [CrossRef]
13. Cirillo, E.; Martellotta, F. *Worship, Acoustic, and Architecture*; Multi-Science publishing Co. Ltd.: Brentwood, UK, 2006; pp. 147–171.
14. Carvalho, A. Acoustical Measures in Churches Portos's Clérigos Church. A comprehensive example. In Proceedings of the 7th ICSV, Garmisch-Partenkirchen, Germany, 4–7 July 2000; Volume III, pp. 1645–1652.
15. Tzekalkis, E.G. Reverberation time of the Rotunda of Thesalonikki. *Acoust. Soc. Am.* **1975**, *57*, 1207–1209. [CrossRef]
16. Su-Guhl, Z.; Yilmazer, S. The Acoustical Characteristics of the Kocate Mosque in Ankara, Turkey. *Archit. Sci. Rev.* **1964**, *51*, 21–30. [CrossRef]
17. Elicio, L.; Martellotta, F. Acoustics as a cultural heritage: The case of Orthodox churches and of the "Russian church" in Bari. *J. Cult. Herit.* **2015**, *16*, 912–917. [CrossRef]
18. Moreno, A.; Zaragoza, J.; Alcantarilla, F. Generation and suppression of flutter echoes in spherical domes. *J. Acoust. Soc. Jnp.* **1981**, *E2*, 197–202. [CrossRef]
19. Vercammen, M.L.S. Sound Concentration Caused by Curved Surfaces. Ph.D. Thesis, Technische Universiteit Eindhoven, Eindhoven, The Netherlands, 2012. [CrossRef]
20. Alberdi, E.; Martellotta, F.; Galindo, M.; León, A.L. Dome sound effects in San Luis. *Appl. Acoust.* **2019**, *156*, 56–65. [CrossRef]
21. De la Banda y Vargas, A. *La IGLESIA de San Luis de los Franceses*; Exma; Diputación Provincial de Sevilla: Sevilla, Spain, 1977; pp. 19–37.
22. Bonet Correa, A. *Andalucía Barroca: Arquitectura y Urbanismo*; Barcelona Ediciones Polígrafa: Barcelona, Spain, 1978; pp. 88–91.
23. Castilla, M. Influencia del humanismo en la arquitectura de los Jesuitas: Iglesia de San Luis de los Franceses de Sevilla. *Liño 23 Rev. Anu. Hist. Arte* **2017**, 21–29. [CrossRef]
24. Pozzo, A. *Perspective in Architecture and Painting*; (1642) Dover: New York, NY, USA, 1989.
25. Dalenbäck, B.-I.L. *CATT-Acoustic v9 Powered by TUCT*; User's Manual; CATT: Gothenburg, Sweden, 2011.
26. Rindel, J.H.; Shiokawa, H.; Christensen, C.L.; Gade, A.C. Comparisons between computer simulations of room acoustical parameters and those measured in concert halls. In Proceedings of the Joint Meeting of the Acoustical Society of America and the European Acoustics Association, Berlín, Germany, 14–19 March 1999; p. AAa3. [CrossRef]
27. ISO 3382-1:2009(E). *Acoustics-Measurement of Room Acoustic Parameters-Part 1: Performance Spaces*; International Organisation for Standardisation: Geneva, Switzerland, 2009.
28. ICE 60268-16:2011. Sound System Equipment-Part 16: Objetive Rating of Speech Intelligibility by Speech Transmission Index. Available online: https://webstore.iec.ch/publication/1214#additionalinfo (accessed on 1 December 2020).
29. Galindo, M.; Zamarreño, T.; Girón, S. Acoustic simulations of mudejar-gothic churches. *J. Acoust. Soc. Am.* **2009**, *126*, 1207–1218. [CrossRef] [PubMed]
30. Vorländer, M. *Auralization, Fundamentals of Acoustics, Modelling, Simulation, Algorithms and Acoustic Virtual Reality*; Springer: Berlin/Heidelberg, Germany, 2008; pp. 303–310.
31. Martellotta, F. The just noticeable difference of center time and clarity index in large reverberant spaces. *J. Acoust. Soc. Am.* **2010**, *128*, 654–663. [CrossRef] [PubMed]
32. Arau, H. *ABC de la Acústica Arquitectónica.*; CEAC: Barcelona, Spain, 1999; p. 264.
33. Carmona, C.; Zamarreño, T.; Girón, S.; Galindo, M. Acústica virtual de la iglesia de San Lorenzo de Sevilla. *Rev. Acústica* **2009**, *40*, 7–12.
34. Martellotta, F.; D'alba, M.; Della Crociata, S. Laboratory measurement of sound absorption of occupied pews and standing audiences. *Appl. Acoust.* **2011**, *72*, 341–349. [CrossRef]
35. Beranek, L. *Acoustics*; American Institute of Physics, Acoustical Society of America: New York, NY, USA, 1954.

Article

Historically Based Room Acoustic Analysis and Auralization of a Church in the 1470s

Hanna Autio [1,*], Mathias Barbagallo [1,2,*], Carolina Ask [3], Delphine Bard Hagberg [1], Eva Lindqvist Sandgren [4] and Karin Strinnholm Lagergren [5,6]

1. Division of Engineering Acoustics, Lund University, Box 118, 221 00 Lund, Sweden; Delphine.Bard@construction.lth.se
2. Brekke & Strand Akustik AB, Box 122, 201 21 Malmö, Sweden
3. Freelance Consultant, 222 20 Lund, Sweden; caro.ask@gmail.com
4. Department of Art History, Uppsala University, Box 630, 751 26 Uppsala, Sweden; eva.sandgren@konstvet.uu.se
5. Department of Music and Art, Linnæus University, 351 95 Växjö, Sweden; karin.strinnholm.lagergren@lnu.se
6. Alamire Foundation, Leuven University, B-3001 Leuven, Belgium
* Correspondence: hanna.autio@construction.lth.se (H.A.); mathias.barbagallo@construction.lth.se (M.B.)

Citation: Autio, H.; Barbagallo, M.; Ask, C.; Bard Hagberg, D.; Lindqvist Sandgren, E.; Strinnholm Lagergren, K. Historically Based Room Acoustic Analysis and Auralization of a Church in the 1470s. *Appl. Sci.* **2021**, *11*, 1586. https://doi.org/10.3390/app11041586

Academic Editor: Nikolaos M. Papadakis

Received: 28 December 2020
Accepted: 5 February 2021
Published: 10 February 2021

Publisher's Note: MDPI stays neutral with regard to jurisdictional claims in published maps and institutional affiliations.

Copyright: © 2021 by the authors. Licensee MDPI, Basel, Switzerland. This article is an open access article distributed under the terms and conditions of the Creative Commons Attribution (CC BY) license (https://creativecommons.org/licenses/by/4.0/).

Abstract: Worship space acoustics have been established as an important part of a nation's cultural heritage and area of acoustic research, but more research is needed regarding the region of northern Europe. This paper describes the historical acoustics of an important abbey church in Sweden in the 1470s. A digital historical reconstruction is developed. Liturgical material specific to this location is recorded and auralized within the digital reconstruction, and a room acoustic analysis is performed. The analysis is guided by liturgical practices in the church and the monastic order connected to it. It is found that the historical sound field in the church is characterized by the existence of two distinct acoustical subspaces within it, each corresponding to a location dedicated to the daily services of the monastical congregations. The subspaces show significantly better acoustic conditions for liturgical activities compared to the nave, which is very reverberant under the conditions of daily services. Acoustic transmission from the two subspaces is limited, indicating that the monastic congregations were visually and acoustically separated from the visitors in the nave and each other.

Keywords: archaeo-acoustics; worship space acoustics; acoustic subspaces; auralization

1. Introduction

The acoustics of worship spaces are a significant element of a nation's cultural heritage. The concept of cultural heritage is described by UNESCO as

"those sites, objects and intangible things that have cultural, historical, aesthetic, archaeological, scientific, ethnological or anthropological value to groups and individuals" [1].

Since UNESCO's adoption of the *Convention for the Safeguarding of the Intangible Cultural Heritage* [2] in 2003, the acoustics and acoustical experiences of churches have been established as an important area of research. The interaction between ritualistic and cultural expressions in churches and their acoustics have been the topic of several research projects [3–10]. The comprehensive review article on church acoustics by Girón et al. [11] discusses the efforts of several research teams who have acoustically characterized a large number of churches across the world and, specifically, Europe.

Most of this research has focused on churches located in countries around the Mediterranean sea, resulting in less scientific literature regarding churches in northern Europe. While Polish researchers have made interesting analyses on some churches around the Baltic Sea [12,13] and there is research on some Russian churches [14], there is very little, if any, acoustic research on religious buildings in Scandinavia and, specifically, Sweden. The research presented in this paper sheds new light on intangible cultural heritage in this part of the world by presenting a room acoustic analysis on a digital reconstruction of Vadstena abbey church in Sweden. This work is part of a larger research project [15].

Such *archaeoacoustical* [16] projects, where acoustic simulations in digital reconstructions of historical spaces are performed, have already been undertaken [17–21]. The results of these projects may be combined with visual models to produce Virtual Reality experiences, which has been done by several teams [22–24] and is the goal of this project. Efforts such as these require a tight collaboration between acousticians, 3D artists, historians, and musicologists to tackle the intrinsic multidisciplinary nature of the research. The large amount of heritage objects not yet investigated with such techniques, the relative novelty of the underlying technologies, and the challenges posed by such collaborations motivates further research projects such as the one reported on in this paper.

The combination of archaeoacoustic modeling and visual models often requires some form of *auralization*. Auralization can be defined as "... the process of rendering audible, by physical or mathematical modeling, the soundfield of a source in a space, in such a way as to simulate the binaural listening experience at a given position in the modeled space" [25]. Application and implementation of auralization is in itself a large research area [26–28]. Although auralized audio samples are presented in this paper, the main focus is placed on the development of the 3D model, and on the room acoustic analysis. A foundational course in auralization can be found in [29].

Liturgical practices are characterized by significant auditory elements, such as prayers, chants, or preaching. Understanding of these practices and how they function within a church benefits from room acoustic analyses of churches. Such analyses are based on room acoustic parameters typically computed from impulse responses, which can be estimated using acoustic simulation software. For example, the perception of chant is strongly related to early reverberation [4], and the intelligibility of song and speech can be related to clarity parameters. As within any volume, room acoustic parameters in a church may have a weaker or stronger spatial dependence.

Spatial variations of room acoustic parameters within churches depend partly on varying materials and partly on varying geometry. Churches are often large, complex buildings with vaulted ceilings and spaces such as choirs, transepts, chapels, or apses. This may lead to the formation of acoustic subspaces [30] that are characterized by room acoustic parameters significantly deviating from the rest of the space. For instance, such subspaces may result in varying perceived reverberation as indicated by EDT within the apse, or degradation of clarity far from the chancel.

In some cases, the structural separation of such subspaces serves a liturgical purpose. For example, smaller chapels may be dedicated for funeral procedures, or the choir for the hourly services of a monastic congregation. Distinct liturgical purposes may coincide with distinct liturgical activities, such as the monastical service being characterized by the chanting of the congregation. As the liturgical activities benefit from proper acoustics, the acoustics in a given subspace must be evaluated with respect to the activities at that location. Pedrero et al. [7] presents such research regarding the cathedral in Toledo, which indicates that the acoustical properties in the subspaces supported the activities performed there. Similar results regarding the choir have been seen in several research projects [31–33].

The interaction between religious rites and acoustics is not yet fully understood, especially when spatial variations within churches are considered. It has been established that the acoustic requirements of worship spaces differ from conventional acoustic guidelines [7,34], but there is not yet a strong consensus on what these requirements are. As of yet, the international standard on measurements of room acoustic parameters, ISO 3382, does not contain guidelines for acoustic measurements of worship spaces. Although there exist established methods for such measurements [35], the lack of an official standard may lead to variations in the measurement methods which in turn cause issues when comparing the results of such measurements. In addition, the lack of guidelines regarding acoustic requirements may cause issues when acousticians are tasked with constructing new worship spaces, or improve existing spaces. The analysis in this paper, where room acoustic parameters within the abbey are combined with a discussion of the religious purposes of various subspaces, may help shed light on these questions.

The goal of this paper is twofold: First, describing the process of constructing a historically accurate archeo-acoustical digital model of an abbey church in Sweden. Second, presenting the analysis of its room acoustic properties. To start, some background information on the abbey is briefly presented. Then, certain aspects of the historically based digital reconstruction are discussed. This description shows that the conclusions of the acoustical analysis rely on sound historical research, and act as inspiration for future, similar projects. Subsequently, recordings of material for auralization are briefly presented. The process of acoustic simulation is described, and the simulated acoustic field is analyzed in detail both globally and locally within acoustic subspaces.

2. Background on the Church

The church targeted for reconstruction in this paper is a Gothic abbey church located in the south of Sweden, built in the 14th and 15th century. It played an important religious and cultural role in the Nordic countries in the middle ages. After the Swedish reformation (1527), the monastery was eventually dissolved and the abbey church fell into disrepair and neglect for almost two centuries. Although some artifacts remain and most of the abbeys extensive library has been preserved, the interior space of the church has been significantly altered due to several renovations [36].

The church itself is oriented west–east, with the chancel in the west. Its nave is divided into three aisles of five bays each, every bay measuring about 11×11 m^2. In the west, the central aisle is extended by an apse of approximately the same size as the bays. The church walls and pillars are built of limestone, and the ceiling vaults are of plastered brick. In the present day, the interior walls are of naked stone, but their rough surface indicate that they were originally plastered. It is also known that plaster was removed during a renovation in the 19th century [36]. A photograph of the modern day church can be seen in Figure 1, and a 3D model of the interior walls and vaulted ceiling is shown in Figure 2.

Figure 1. Present day church. View from east to west.

Figure 2. 3D scan showing only the church's original 15th century parts. View to the north, with the chancel and the location of the monks' choir to the west, on the left.

The church and its abbey was the mother abbey of a religious order of both monks and nuns. The monks and nuns lived in separate enclosures, but shared the same church. Both congregations had a specific location within the church where their services were celebrated. The order was characterized by the premiership of women, and is the source of the only known monastical office exclusive for women [37].

The information about the order guides the acoustic analysis to focus on the dedicated spaces for the monastic congregations, especially the nuns'. These two areas are digitally reconstructed, as described in Section 3.1. The monks' choir and the gallery may both be interpreted as choirs. As the choir often has an important role in the acoustics of a church [32,33], a deep acoustic analysis is motivated both by religious and acoustic considerations. However, only traces of the structures remain in the church today, and a historically based digital reconstruction is the only practical option to evaluate the historical sound field of the space.

Below, the processes of digital reconstruction, recording of liturgical elements, and acoustic simulations of the church are presented. The foundations of the historical model are described thoroughly, as one of the goals of the paper is to inspire future researchers in similar projects, and to validate the acoustic results based on historical information.

3. Digital and Historical Reconstruction

The digital reconstruction aims at a time period around year 1470, as this coincides with a period in the abbey's history for which the historical source material is rich. The acoustics of daily religious practices are examined by recording and auralizing material from an ordinary Friday sext. This condition was also the target of room acoustic analysis, characterizing the acoustic properties of the space under normal conditions.

In the following sections, the process of creating the elements for auralization and room acoustic simulation are described. First, the construction of the digital model is discussed. The reconstructions of the spaces for monks and nuns in particular are presented. The recording of material for the auralization is discussed, and finally a few comments are made regarding the adaptation of the high-detail visual model for acoustic simulation software.

3.1. Model Creation

A wide range of historical sources were used in the construction of the digital model. Due to the importance of the church in question, there are many sources directly related to it. This includes drawings, documents, plans, maps, traces in the church room, and historical objects such as sculptures and textiles. In addition, there is a range of earlier research [36,38–42]. This material could be complemented by more general historical information regarding practices for the given time period and geographical location. The collected information was translated into concrete 3D suggestions of what lost and altered parts of the church interior may have looked like.

One of the first steps in the creation of a digital model was the laser-scanning of the complete interior space of the church, aimed at obtaining a high-resolution 3D model to be used as reference. This model was then further processed and refined using the graphical modeling software Blender. First, elements of the scan dating later than the 15th century, as based on historical information, were removed from the digital model. The resulting model of the 15th century shell (Figure 2) was a starting point for the digital reconstruction.

The 15th century shell was extended by 3D suggestions of lost and altered parts of the interior, based on information in historical sources. Different suggestions were then evaluated to find which were more plausible. This process was iterative, and involved primarily the art historian and 3D modeler within the project. Experts on medieval construction, theology, and archaeology were consulted when appropriate.

The evaluation and appraisal of the different 3D suggestions benefited from working in a spatial environment. Formulating the constructions in a 3D space made it clear that some suggestions were incompatible with the church room itself, historical accounts, or other constructions. For example, in some cases several constructions would need to occupy the same space, or they might lack necessary physical support structures. Further refinement could be achieved when liturgical practices were considered. Line of sight to certain spots important for the liturgy, processional walkways, and easy access to key locations was necessary for the religious functionality of the space, and could thus be used to dismiss less appropriate solutions.

The next sections will focus on key areas of the church. First, the furnishings in the nave are discussed briefly. Second, the reconstruction of the gallery and the choir are described and motivated.

3.1.1. The Nave Area

The historical nave of the church was filled with over 60 altars [39] and a multitude of richly decorated grave chapels, giving it a very different impression compared to today (Figure 1). Examples of high-detail reconstructions of chapels and altars are shown in Figure 3. These types of structures are expected to have had a significant impact on the acoustics of the medieval nave. The large number of decorated and complex surfaces lead to increased scattering of the acoustic field, which may have contributed to increased diffusivity. On the other hand, the presence of these structures may have lead to the formation of acoustical subspaces which would imply an increased acoustic heterogeneity.

An additional aspect affecting the acoustic field in the nave was the prevalence of textiles. Textiles played an important part of the furnishing in a late medieval church, and were used as curtains creating small enclosures around most altars (see Figure 3), covers for altars not in use, decorations on altars and walls, and carpets.

The historical reconstruction is expected to reduce reverberation in the nave compared to today, both due to textiles and wood structures such as chapels. The large empty stone volume of the nave today offers very little absorption area, and the reverberation time is long [43]. Wooden grave chapels are expected to increase the amount of absorption across all frequencies, most significantly for the mid-high range. In addition, the large amount of textiles will increase absorption for high frequencies. Together, these effects will lead to a decrease in reverberation across the full frequency range. Such results have been found in other projects [5,44].

Figure 3. 3D model of three reconstructed chapels (left foreground, middle, right background). In the far left corner is the side of an altar with textile curtains, and a platform on its front covered by a carpet. View from the northeastern side of the nuns' gallery (compare with Figure 4) towards southwest.

3.1.2. The Nuns' Gallery

The construction of an elevated platform for the nuns in the middle of the church is prescribed in the building instructions of the abbey [45]. The only visible remnant of this platform today is a niche near the ceiling vaults that was once its entrance (Figure 4). Research has several suggestions on the specific form and placement of the nuns' gallery [36,38], but there is so far no consensus on its specific location, size, or configuration.

The suggested configuration of the gallery in this project is based on historical documents and physical traces in the church room, and shown in Figure 4. It is surrounded by high panels, and its size was estimated to $10 \times 18\,\mathrm{m}^2$. This is large enough to accommodate the congregation of nuns and the furnishings required for religious purposes [45,46]. In addition, this size allows for the display of a series of paintings along the interior western side of the gallery, which are recorded in [47]. The high panels surrounding the gallery ensure that the nuns could not be seen by anyone else in the church, although see-through lattices allow them visual access to key altars, as required by religious documents [48].

The location of the gallery was determined by using information from a geophysical investigation of the church floor. This investigation revealed anomalies which were interpreted as foundations of support pillars for the gallery. The final model of the gallery, shown in Figure 4, satisfies requirements regarding function and size, is structurally sound and matches the physical traces well.

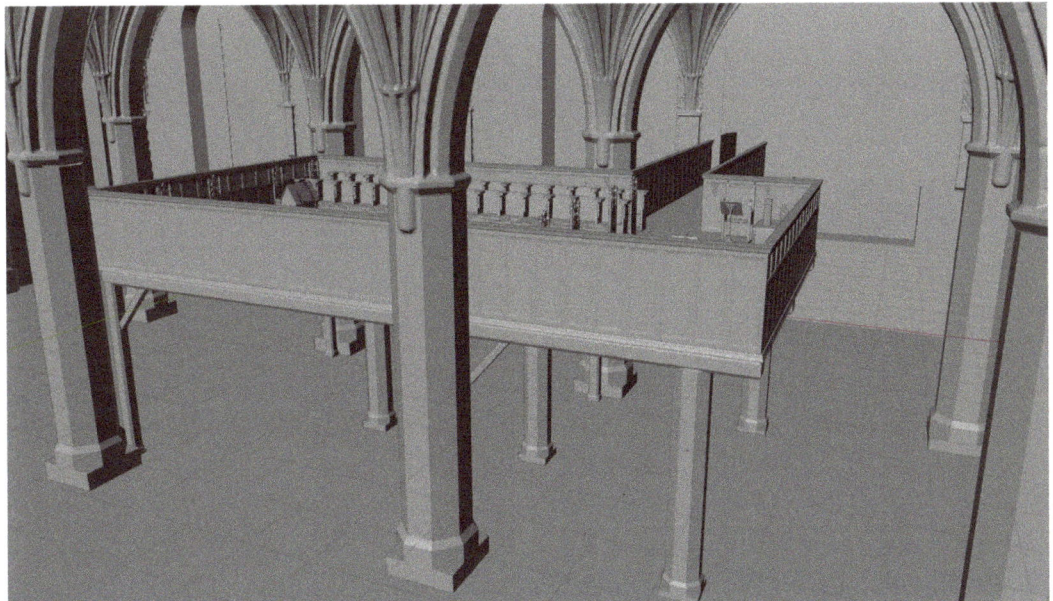

Figure 4. 3D model of the nuns' gallery (other nave elements mentioned in the text are not visualized here), showing the entrance niche in the northern wall, leading through a crossing to the gallery. The panels on the shorter sides have small windows for visual access. The gallery is supported from beneath by four brick pillars. View from south to north.

This interpretation of the gallery is located close to the vaulted ceiling and is separated from the rest of the nave by its floor and panel walls. As such, it likely acts as an acoustical subspace for the congregation of nuns, as is often found in the choir [7,31,32]. As the congregation of nuns gathered there, it is also plausible that the space was characterized by increased absorption by the presence of nuns and their clothing, leading to decreased reverberation and increased clarity. In addition, the gallery's central position close to the vaulted ceiling may have had a significant effect on the transmission of sounds from the gallery.

3.1.3. The Monks' Choir

The monks were located in the choir behind the chancel, in the apse in the west of the church. (left side, Figure 2). As the ground level in this area is about 1 m below the nave, a wooden platform has been suggested to bring the monks to an elevation more liturgically favorable [36]. Possible beam holes and records of the removal of wooden elements from this location support this theory, and it is adopted in this reconstruction.

The shape and size of the wooden platform were determined based on physical traces in the church. Beam holes, book niches, confessionals, and a door opening prescribed the solution shown in Figure 5. Choir stalls were based on duplicates of two choir stalls which remain today.

Similar to the nuns' gallery, this space served the purpose of housing a monastical congregation. Some differences with the nuns' gallery are noted. First, the distance between monks and the vaulted ceiling was significantly larger than that between nuns and the ceiling. Second, the monks' choir was in a recessed position nestled in the apse behind the chancel. It is thus enclosed by close, hard walls on three of four sides, which may affect the sound field within the choir to the point where it differed significantly from that in the gallery and the nave. The acoustic characteristics of this space, especially as compared to the gallery, are further investigated in Section 4.1.

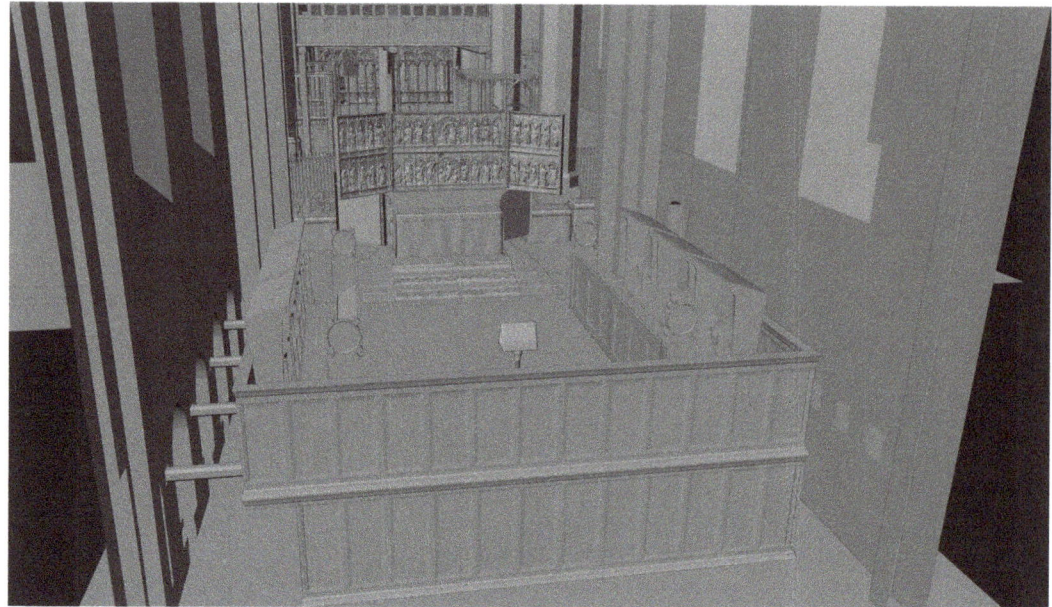

Figure 5. 3D model of the monks' choir with beams fitted in the wall niches to the left, supporting the platform. The back part of the stalls are modeled based on preserved stalls used by the monks. View from the west towards the east. The main altar is visible east of the platform.

3.2. Recording Liturgy

This section describes the overall process of acquiring anechoic recordings of appropriate liturgical material for the auralization task. The full process of selecting material, recording material, and choosing performers will not be fully described here. Only the final choice in material will be presented, and the recordings themselves will be briefly discussed.

The liturgical practices of the order located in this abbey are quite well documented. The daily services consisted of monophonic Gregorian chant, from the respective divine offices of the monks and nuns. The nuns followed their unique liturgical office, as mentioned in Section 2, while the monks observed the liturgy of the diocese where the abbey was located [37]. This information made it possible to recreate a plausible Friday sext, from which elements were chosen for recording. The selected material was chosen to include the important liturgical elements of the service: short responsories, antiphons, prayers, and psalms.

The number of participants in the Friday sext was estimated to about 10 for each congregation. There were to be 60 nuns and 25 monks in the monastery, and only 13 of the monks were ordained and participated in all services. All members of the congregations had responsibilities which may excuse them from services and in particular they might be absent from the small hours (prime, terce, sext, and none). It was concluded that eight male voices and twelve female voices would suffice for the recordings.

3.2.1. Recordings

The recordings were made in the anechoic chamber at Engineering acoustics, LTH, Lund University. Four male and three female singers, familiar with the style of music and the material, were recruited for the recordings. They were recorded using four close mics model Milab VM-44, hanging from the walls.

In order to reach a plausible number of voices in the recordings, the singers were recorded multiple times. During each recording, all participants belonging to the same group were performing together in the anechoic chamber. One of them was wearing headphones, playing back any earlier recordings and fed with generic live 6 s reverb effect; this time length was chosen as the singers deemed it more helpful to achieve good results as compared to no, or 3 s, reverb. The others followed this individual. The variations caused by this method are thought to be consistent with the assumption that these individuals were performing a daily task, albeit as part of a service. Recording and mixing were done in Cubase 11.0.

Singing in a anechoic chamber may pose challenges for singers both due to lack of support and response from the room, and due to the physical influence such environment may have on humans. Such circumstances have an impact on the performance itself, and may result in a presentation that is quite different from how the same material would be presented in a more traditional space, such as a church. The addition of artificial reverb, the choice of singers, and great care during the recording process were used to combat these effects, but it is not possible to guarantee that the presentation is not affected by the discrepancy between the recording conditions and the church. Such problems have been encountered in other research projects, and have not yet been solved [22,49].

A musicologist familiar with the material was present for the full duration of the recordings, ensuring that the performance was as accurate as possible. As the act of performing in an anechoic chamber can be very challenging, the singers took frequent breaks. During these breaks, the performers, the producer, and the musicologist listened together to the recording, both dry and with added reverb, and determined whether it was of an appropriate tempo and quality.

In addition to elements of the sext, some additional sounds were recorded. These were background sounds including prayers, said by female and male voices, the sound of historically representative clothing, sounds from a rosary, sounds from walking with leather shoes against a stone surface, and coughing.

3.3. Room Acoustic Simulations

The digital model constructed in Section 3.1 was exported from Blender to Google Sketch-up and therein adapted for simulations in ODEON 16.0 [50]. Simulations were performed both to characterize the historical sound field in the reconstructed abbey and with the goal of producing auralizations of the sext itself. This section describes the process of adapting the digital model for simulation, as well as the choices made for simulation and auralization.

During the simulation, air absorption was tuned to conditions of 18°C and 50% relative humidity.

3.3.1. Adaptation of the Digital Model

The high-detail model used for visual presentation was transformed into a digital model for acoustic simulation, primarily by simplification of various surfaces. The structure of the ceiling vaults was simplified significantly, and detailed models of sculptures and altar decorations were replaced by simple geometric forms. This improves the performance of the acoustic simulations with regards to accuracy and to calculation speed. Examples of simplifications are shown in Figures 6 and 7. The simplified model is exported from Google Sketch-up using ODEON's exporting tool and imported to ODEON. The exported model counts 5600 surfaces and takes around 1.5 s to export to ODEON format.

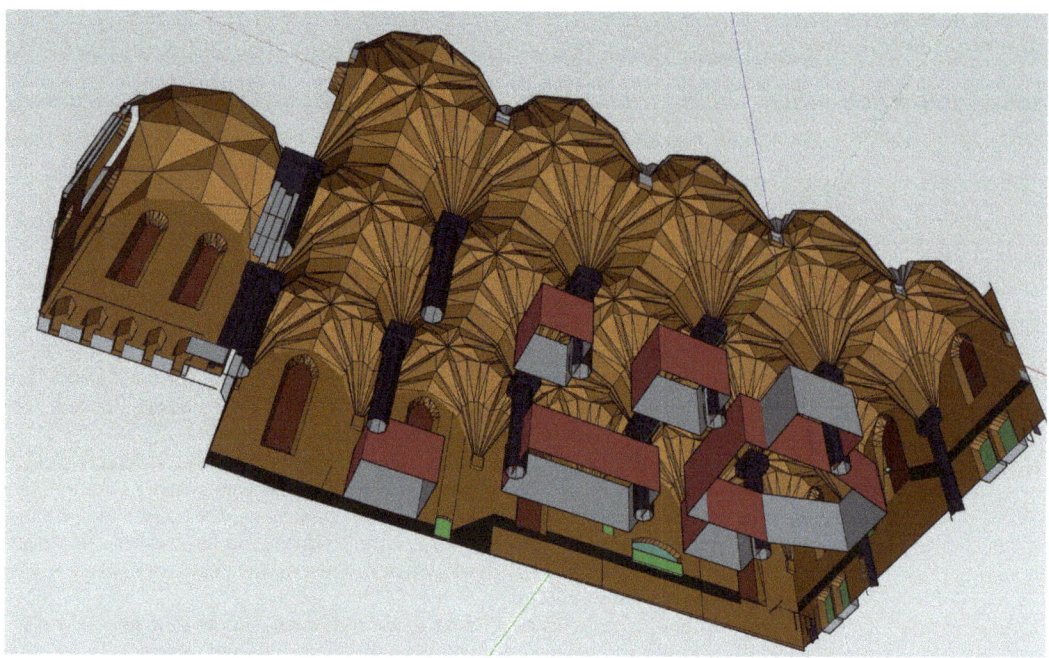

Figure 6. Simplification of vaults and chapels in the acoustic model, view from below with floor layer being hidden. The elaborate wooden walls around the altars and graves (compare to Figure 3) are rendered with boxes which are assigned a 45% transparency in ODEON. Screenshot from Google Sketch-up.

Figure 7. Simplification of statues from the visual model to the acoustic model. Screenshot from Google Sketch-up.

3.3.2. Acoustic Properties of Materials in the Church and Calibration of the Model

The acoustic properties of the various materials in the church needed to be estimated. As an initial step, table values for absorption and scattering from several different sources were used. The full table of material parameters is given in Appendix A.

Whereas there are many table values for absorption coefficients, scattering coefficients are more difficult to measure and may vary depending on the software used. In ODEON, scattering coefficients for 707 Hz are provided by the user, which are then extrapolated to the full frequency range using a built-in algorithm. For the simulations in this project, table values were primarily used. When these were unavailable, mid-frequency scattering was estimated from the characteristic depth of the structure in question, as discussed in [51].

To improve the quality of the absorption coefficient estimates, calibration by comparison to modern reverberation time measurements (measured by the integrated impulse response method [52] and a B&K type 2270 analyzer) was performed. As the modern interior of the church differs so significantly from the historical configuration (see, e.g., Figures 1 and 3–5), only a few materials can be calibrated: the vaulted ceilings, the glass windows, and the stone floor. These are assumed to have the same material properties as during the 1470s, whereas the walls which were historically covered by plaster are now bare and can not be calibrated.

The calibration is performed using ODEON's built-in genetic algorithm optimization tool, in a digital model corresponding to the modern church. This model comprises the exterior shell, modern wooden pews, and the main altar. This corresponds to the conditions under which the modern measurements were performed. The average error of the simulated reverberation time in the calibration model compared to measurements is within ±1.4 JND (see in Figure 8 for detailed comparison for each octave band and position). The calibrated material parameters were then used in the historical digital model.

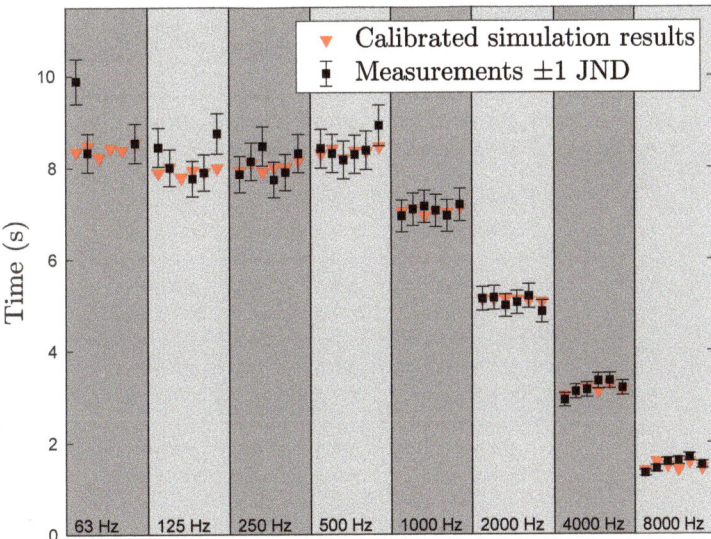

Figure 8. Comparison between simulated and measured T_{20} values in the modern church. The red triangles show the results of the calibrated simulation, and the black boxes correspond to measurements together with frequency-dependent 1 JND error bars. The results for six different listener–receiver combinations are shown across center octave-band frequencies from 63 to 8000 Hz. For low frequencies, some values could not be extracted from the measurements and only the simulated results are shown. The average error in JND is highest in the 63 Hz and 8000 Hz octave bands with 1.2 and 1.4, respectively; the remaining octave bands have values between 0.2 and 0.7.

3.3.3. Listener and Source Positions

Sound source and sound receiver positions were chosen according to the goals of the simulations. The source positions for room acoustic simulations are shown in Figure 9. The nuns' gallery was identified as a location of interest, and source P1 and a receiver are located there. Similarly, both a source (P2) and a receiver are in the monks' choir. In addition, one sound source was located close to the high altar (point P3). Additional receiver positions are distributed on the floor of the nave. In the auralization, twelve nuns and eight monks modeled as point sources are placed in their respective choir stalls in the gallery and choir.

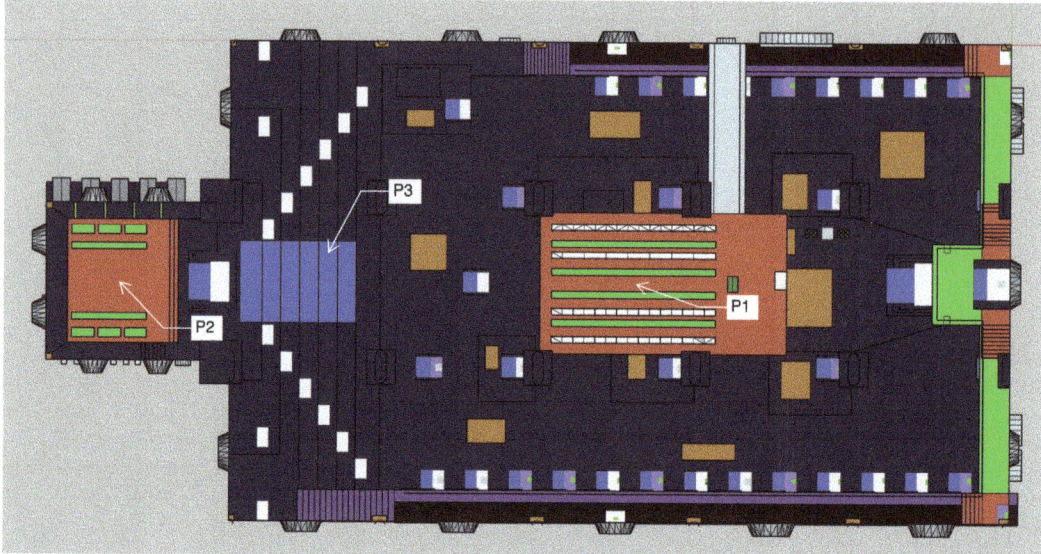

Figure 9. Top-down view of the digital reconstruction of the abbey. North is up in the figure. The monks' choir is the red area to the left, and the nuns' gallery is the red area in the center. Three source positions—P1, P2, and P3—are marked.

4. Room Acoustic Analysis

The sound field in the reconstructed church was evaluated from room acoustic simulations. Three questions were specifically examined. First, three different visitor conditions were evaluated. The three conditions corresponded to an empty church, an ordinary sext (about 30 individuals), and a more festive event (about 140 individuals). The sext condition corresponds to the situation targeted in the auralization, with twelve nuns in the gallery, eight monks in the choir, and about ten other visitors in the central-eastern part of the nave. In the festive condition, denoted "full", there were 70 nuns in the gallery, 13 monks in the choir, and about 70 visitors in the nave.

Second, the sound fields within the choir and gallery are examined. These locations are important for the religious practices of the order, and acoustic analysis of these spaces may provide new insights. Third, the sound field in the nave is examined.

The auralizations are provided as supplementary material, and are not further analyzed in this text.

Four room acoustical parameters are presented, as defined in ISO 3382-1:2009 [52]. The reverberation time T_{20} is presented, due to its traditional importance in acoustics. It is primarily useful as a tool for comparison to other spaces, as the reverberation time is a commonly measured acoustic characteristic. The early decay time (EDT) is also presented. It describes the rate of sound energy decay in the first parts of the impulse response, and is closely related to perceived reverberation and the presentation quality of Gregorian chant.

According to guidelines proposed by Martellotta et al. [4], EDT values in the range 2.1 s to 4.2 s are appropriate for churches. In addition, C_{80} is presented as an indicator of the clarity and intelligibility of chant. As the liturgical practices of this region do not contain significant spoken elements, C_{50} (speech clarity) is not presented separately. For concert halls, C_{80} values of above 0 dB are usually considered "good", but no such guidelines have been defined for worship spaces. Finally, the sound strength (G) is presented. This value shows the total sound pressure level (SPL) at the receiver, as compared to what would be perceived from the sound source in free field at a distance of 10 m. G is positive in enclosed spaces.

A summary of the results is shown in Table 1. This table gives values for T_{20}, C_{80}, EDT, and G, averaged according to ISO 3382-1:2009 [52]. In addition, the results are divided into categories based on the location of sources and receivers. If the source and the receiver are both within the gallery (point P1 in Figure 9), the data is denoted "Gallery". If both source and receiver are in the choir (point P2 in Figure 9), the data is denoted "Choir". Finally, if the sound has travelled through the nave, the data is categorized as "Nave". This includes when either the source or the receiver is located in the nave, but also the conditions where one is in the choir and one is in the gallery.

Table 1. Room acoustic parameters in various configurations of the reconstructed abbey. The values have been averaged according to ISO 3382-1:2009 [52].

	Gallery			Choir			Nave		
	Empty	Sext	Full	Empty	Sext	Full	Empty	Sext	Full
T_{20} (s)	5.17	4.90	4.00	4.06	3.89	3.26	5.37	5.19	4.36
EDT (s)	1.21	1.10	0.62	2.41	2.33	2.21	5.47	5.29	4.55
C_{80} (dB)	8	8	10	1	1	1	−10	−10	−10
G (dB)	16	16	15	15	15	15	6	6	4

One result seen in Table 1 is that the number of visitors and the visitor condition have an impact on all room acoustic parameters presented. There are no or almost no differences between the empty and the sext condition, but the full condition leads to significant decreases of EDT and T_{20}. This is an expected result of the increased amount of absorption when the number of visitors increase. Increased absorption should also lead to an increased C_{80} and a decrease in G. These patterns can be seen, but are weak. The only differences larger than 1 JND are seen for C_{80} in the gallery and G in the nave.

The results in Table 1 show a clear difference between the sound field within the gallery and within the choir, compared to the other configurations. All of the room acoustic parameters presented above are significantly different for these two subspaces as compared to the rest of the church. The subspaces are analyzed separately in Section 4.1. The results for the rest of the church is presented in Section 4.2.

4.1. Acoustics within the Choir and Gallery

The acoustic simulation results within the gallery (source and receiver both near point P1 in Figure 9, reflecting the experiences of the nuns), and within the choir (source and receiver near point P2, reflecting the experience of the monks) are presented in Figure 10. In these graphs, room acoustical parameters T_{20}, EDT, C_{80}, and G are shown in octave band resolution. The gallery data are shown together with lines denoting ±1 JND, such that when the choir data falls within these lines there is no perceivable difference between the two locations.

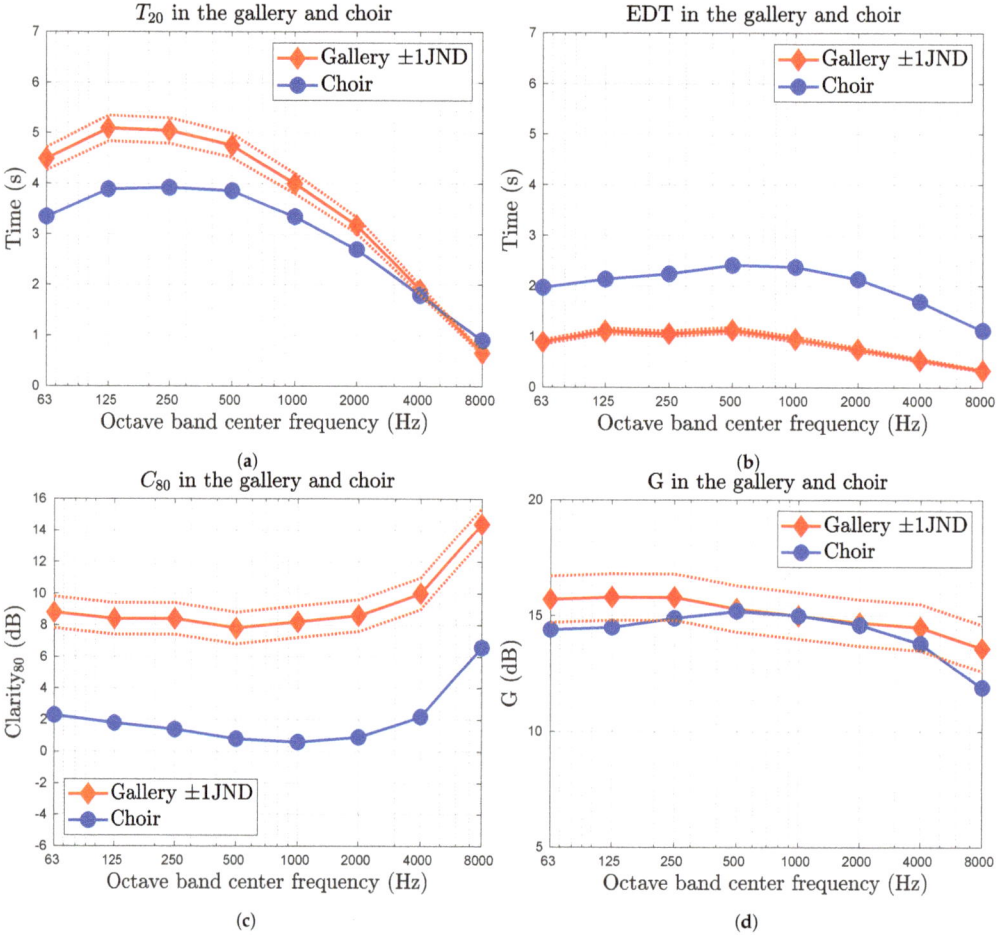

Figure 10. Graphs showing (**a**) T_{20}, (**b**) early decay time (EDT), (**c**) C_{80}, and (**d**) G in octave band resolution in the reconstructed abbey choir and gallery. The results for the gallery (red) are shown with dotted lines corresponding to ±1 JND. When the results for the choir (blue) falls within these lines, there is no perceivable difference between the two data sets.

The reverberation time is significantly lower in the choir compared to the gallery. A review of the full results in Table 1 shows that the results in the gallery and the results in the rest of the church only differ by ±1-2 JND, whereas the discrepancy with the choir is larger.

The second reverberation parameter, EDT, shows a different pattern. The EDT is much smaller than T_{20} for both positions, and the EDT in the gallery is significantly lower than that in the choir. The large discrepancy between the EDT and the T_{20} may indicate that the decay curves in these spaces follow a multi-slope decay pattern, which would be consistent with these spaces acting as acoustically distinct subspaces. The energy decay curves are more closely investigated in Figure 11.

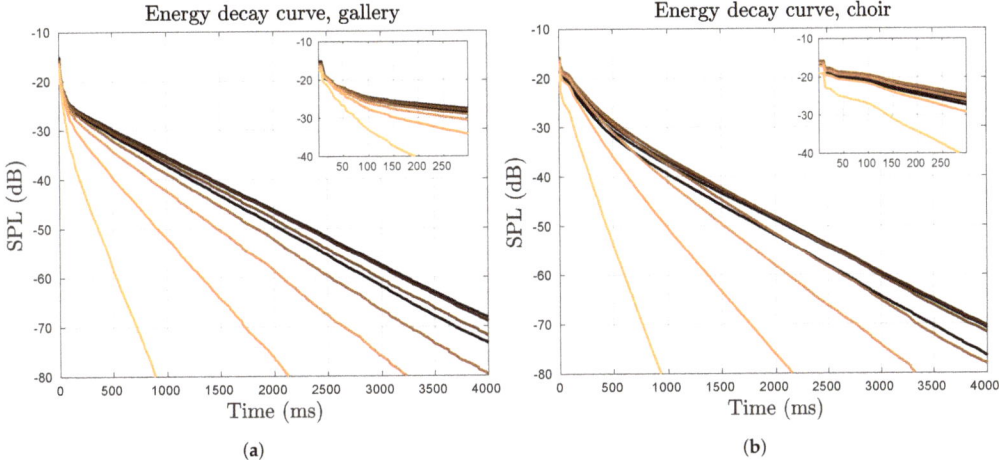

Figure 11. Energy decay curves in the (**a**) nuns' gallery and (**b**) monks' choir in the reconstructed abbey. The small insets show the early parts of the decay. Each line corresponds to the results in one octave band.

Reviewing the energy decay curves in Figure 11 shows a distinctly curved decay pattern in both gallery and choir. As expected from the results in Figure 10, the acoustic energy in the gallery decreases rapidly in the early parts of the decay, and then flattens out to a much slower decay rate. This is consistent with the formation of an acoustic subspace within the gallery, coupled to the larger, more reverberant space in the nave. As the acoustic energy within the subspace decreases, the reverberance of the nave becomes more dominant. The transition seems to occur after about 100 ms. The energy decay curve in the choir seems more complex, and is likely influenced by the three close, hard walls on three of four sides as well as the more distant vaulted ceiling. It is possible that these decay curves are formed by more than two distinct decay patterns. It can, however, be seen that the decay process after about 750 ms is more smooth. This transition time is later than in the gallery.

Returning to the results in Figure 10, it is seen that C_{80} is also significantly different between the gallery and the choir across all octave bands. Clarity is much better in the gallery compared to the choir, which is in turn much better than what is found in the rest of the church (Table 1). The reason for these differences can be explained by the physical configuration and particularly the effects of the ceiling vaults. A time series illustrating a raytracing model of the energy distribution over time is shown in Figure 12.

On the left hand side in Figure 12, the energy distribution over time for a sound source located in the choir is shown. The energy emitted from the source travels upward as time progresses, to be reflected from the ceiling vaults. By 80 ms, the third figure from the top, the reflected wavefront can be seen in the space above the choir. Consequently, it has not reached the receiver by 80 ms, and thus has a detrimental effect on C_{80}.

On the right hand side in Figure 12, the corresponding time series for a source and receiver in the gallery is shown. As the gallery is closer to the ceiling, the strong reflected wavefront from the ceiling has already reached the receiver in the gallery by 80 ms, thus improving C_{80}.

The G within the gallery and choir can be seen in Figure 10d. As shown in the graph, there are no or very small differences between the gallery and the choir. The results in both spaces are, however, significantly better than outside these spaces. This can be explained by strong reflections, from the ceiling in the gallery and by the walls in the choir. The positions of these surfaces ensure that a significant amount of sound energy is reflected back to the space, thus increasing the total G.

Figure 12. Images illustrating the dispersion of acoustic energy, as approximated by an acoustic particle model, over time. Subfigure (**a**) shows the progression when the source is located in the choir, and subfigure (**b**) shows when the source is located in the gallery. From top to bottom, the pictures show snapshots at 20 ms, 60 ms, 80 ms, and 120 ms.

Finally, some brief comments are made regarding sound transmission between the gallery and choir, corresponding to the acoustic perception of monks from the nuns' position and vice versa. It is found that acoustic transmission between the two locations is very similar to the transmission between choir and nave. As those results are presented in Section 4.2, no further comments are made here.

4.2. Acoustics in the Nave

Within the nave, the acoustic simulations aim first at characterizing the acoustics of the space itself and, second, at evaluating any difference between various source locations. As such, the data in this section are presented separately, according to the source position. The three source positions are in the gallery (point P1 in Figure 9, nuns' position), in the choir (point P2, monks' position), or in the nave, by the high altar (point P3). T_{20}, EDT, and G results are shown in Figure 13. As the results in this section are in general spatially averaged, the standard deviations are also shown as an estimate of the spatial variations between listener positions.

The average reverberation time in the reconstructed abbey is shown in octave band resolution Figure 13a. Results for the three different source positions are shown, and are very similar; also shown in the graph is a gray area delimiting values that are within 1 JND of the global average. All lines fall entirely within this area, indicating that there are no significant differences in the reverberation time in the nave depending on the location of the sound source; also shown in Figure 13d are the standard deviations of T_{20} for each of the source positions. These are shown together with the global JND for reverberation time. All standard deviations fall below the JND line, indicating that the variations of the reverberation time are on average imperceptible. This indicates that the late acoustic response within the nave was, in general, diffuse and not characterized by significant spatial variations.

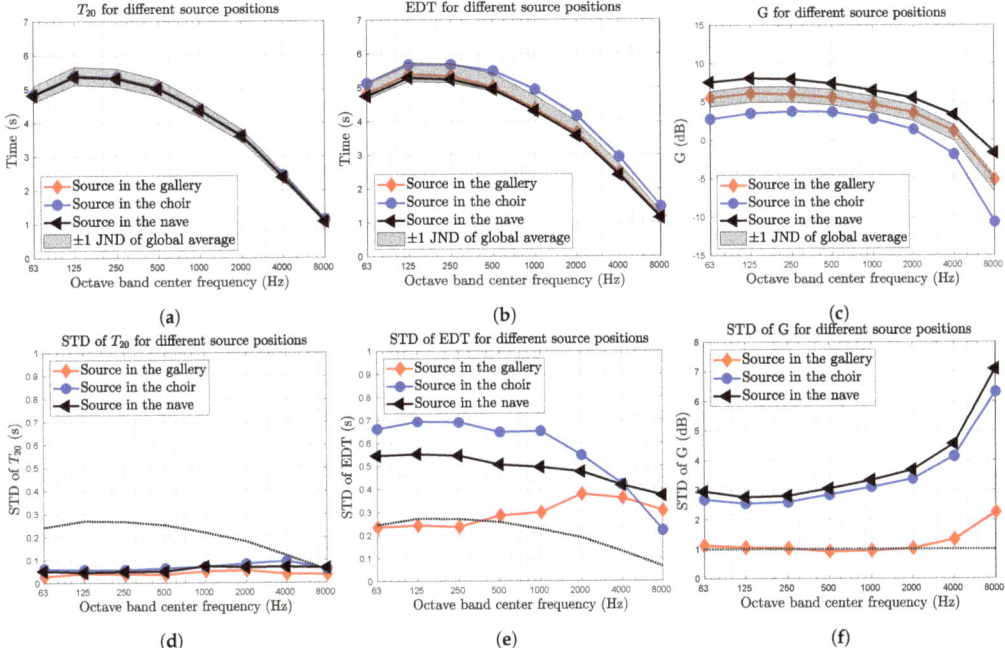

Figure 13. Overview of room acoustic parameters in the reconstructed church. Results are shown for three different source positions in octave band resolution. (**a**) The average T_{20} and (**d**) its standard deviation. (**b**) The EDT and (**e**) its standard deviation. (**c**) The G and (**f**) its standard deviation. All standard deviations are shown compared to the global average JND of that parameter (dotted line). The gray areas in graphs (**a**–**c**) delimit values within 1 JND of the global average.

In general, the reverberation time is long, but comparable to similar spaces. T_{20} is significantly decreased compared to modern day measurements, as expected by the discussion in Section 3.1.1. In Table 2, the mid-frequency reverberation time (T_m) is shown together with measurement results from the modern day church (from in [43]) and some other Gothic churches around the Baltic Sea (from in [11]). These results allow for a basic comparison between the reconstructed abbey and similar spaces.

Table 2. Mid-frequency reverberation times for some Gothic churches in Northern Europe.

Church	Volume (m^3)	T_m (s)
Swedish church, current configuration	29,000	7.79
Swedish church, historical configuration	29,000	5.15
Church of our Lady, Krakow, Poland	9500	6.5
Church of St Thomas, Lipsk, Germany	18,000	4.05
Marien Church, Lübeck, Germany	100,000	5.50

Spatially averaged EDT values for each source position are shown in Figure 13b. Again, the gray area delimits values within 1 JND of the global average. The variations for different source positions are greater than those seen for T_{20}, but only the results for the choir are significant compared to the JND. This difference is, however, small. Comparing the global EDT values to the guidelines proposed by Martellotta et al [34] shows that the average EDT exceeds the recommended values for the reconstruction.

In Figure 13e, the standard deviation of the EDT is shown for each source position together with the global JND. The standard deviation for a source in the gallery is smaller than the others, and falls below 1 JND for low frequencies. The typical variations for the sources in the nave and in the choir exceed 1 JND. This shows that the perceived reverberance of sources in those two locations varies significantly depending on the listener's position. It also indicates that the small difference in average EDT between sources in the choir and elsewhere (seen in Figure 13e) may be too small to be relevant, compared to the variations between receiver positions.

G in the nave is shown in Figure 13e, in octave band resolution for each of the three source positions under consideration. The gray area delimits values within 1 JND of the global average. In this case, there are significant differences between the average results for the three positions. Sound emitted from the source by the high altar is on average heard louder, and sound from sources in the choir are on average heard more quietly. Part of the explanation for this fact may be that there are more receivers with a direct line-of-sight to the sound source in the nave than for the sources in the gallery and the choir. The contribution of the direct sound has a significant impact on the overall sound pressure level, and thus on G.

The standard deviations of G is also shown, in Figure 13f. Again, these are shown together with the corresponding JND. This graph shows that for sources located in the choir or by the high altar, the spatial variations of G are of a very similar magnitude, and significantly larger than when the source is located in the gallery. This can partially be explained by the position of the gallery in the middle of the church. G is affected by the distance between the source and the receiver, and the gallery's position ensures that the distance between source and receiver varies minimally.

C_{80} simulation results are shown in octave band resolution in Figure 14. As C_{80} is not expected to be uniform in a space, all measurement points are shown, together with a line indicating the spatial average. The average C_{80} for sources in the choir and gallery fall in the region of -15 dB to -5 dB, significantly below what would be characterized as "good" for a concert hall. Clarity for sources located by the high altar is better, at about -5 dB to about 2 dB.

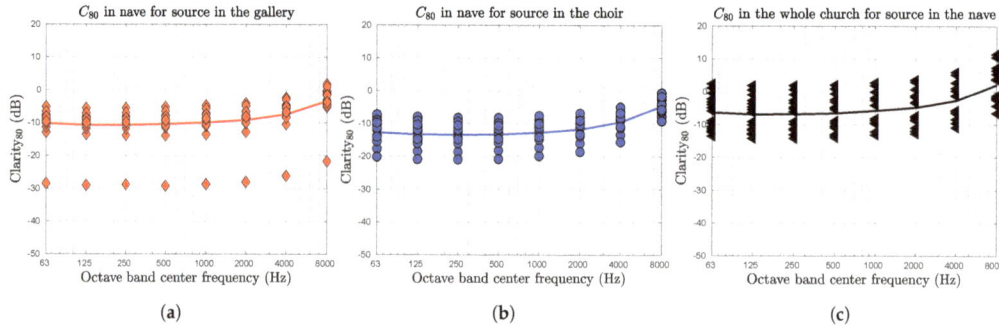

Figure 14. C_{80} in octave bands simulated for the recreated church, shown for 13 different receiver positions. The spatial mean (line) is also shown. (**a**) C_{80} when the sound source is located in the nuns' gallery. One position produces significantly lower clarity. (**b**) C_{80} when the sound source is located in the choir. (**c**) C_{80} when the sound source is located by the high altar in the nave.

Although C_{80} is not expected to be uniform in a space, the results in Figure 14 show that most values fall in clusters around the average value. However, there is one outlier in Figure 14a, when the source is located in the gallery. This estimate comes from the receiver located underneath the gallery, below point P1 in Figure 9. It is separated from the sound source by the physical structure of the gallery itself, and there is no direct line of sight. A review of the acoustic simulation reveals that the first sound energy reaching this location from the source in the gallery is a second-order reflection, showing that sound energy reaching this location from the gallery consists solely of reflections of order two or higher. This significantly decreases the cohesion and energy level in early parts of the impulse response, leading to a very poor clarity of sources in the gallery as perceived by listeners beneath it.

The acoustic field in the nave is further analyzed to determine whether the spatial variations for sources in the choir and by the high altar, seen in Figures 13e,f and 14, are caused by acoustic subspaces. It is found that the spatial variations can be explained well by the distance between source and receiver. Linear regression models for mid-frequency G and C_{80} to source–receiver distance are shown in Figure 15, and the R^2-values are presented in the caption. Except for sources in the gallery, more than 70% of the typical spatial variations in G and C_{80} can be explained by the source–receiver distance.

In Figure 15, the results from within the gallery, within the choir, and the outlier found in Figure 14 are marked, and not included in the linear regression models. They deviate significantly from the pattern defined by the regression line, indicating that the assumption that these locations are governed by the acoustic properties of a certain subspace is accurate. However, no other measurement points show a similar deviation from the prediction. Thus, there are no indications that there are distinct acoustical subspaces in the reconstructed church except those already identified.

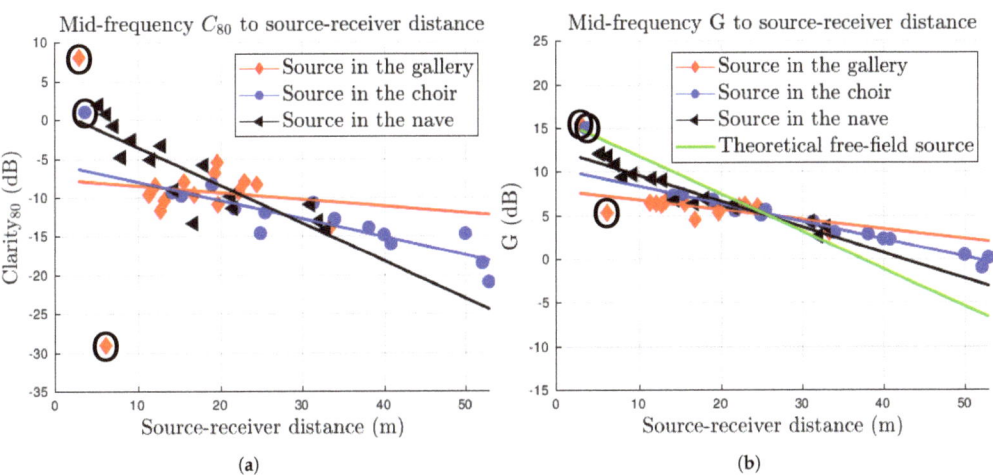

Figure 15. (a) Mid-frequency C_{80} and (b) mid-frequency G as they vary with source–receiver distance; also shown are linear regression models estimating the influence of source–receiver distance for each source locations. R^2 values for the models in subfigure (**a**): Source in the gallery: ($R^2 \approx 0.06$), choir: ($R^2 \approx 0.78$), and nave ($R^2 \approx 0.75$). In subfigure (**b**), the corresponding values are gallery: $R^2 \approx 0.48$, choir: $R^2 \approx 0.98$, and nave: $R^2 \approx 0.92$. In subfigure (**b**), the regression line for a theoretical omnidirectional source in free field is also shown. Highlighted data points have been excluded from the regression lines. These correspond to configurations entirely within the gallery or choir and the outlier case when the source is in the gallery and the receiver below it.

Another pattern emerges in the evaluation of source–distance dependence, showing that the influence of distance varies depending on the location of the sound source. As shown in Figure 15a, C_{80} decreases as source–receiver distance increases and the reverberant field becomes more dominant. The linear regression quantifies the influence of distance for each source position. The effects of distance is most strongly seen when the sound source is located in the nave, by the high altar. Moving the sound source to the choir decreases this effect slightly. When the sound source is in the gallery, the effect is almost entirely gone, and clarity is not significantly affected by increasing the distance to a sound source in the gallery. This analysis is confirmed by a review of the R^2-values for the respective linear regression models. These indicate that only about 6% of the variations of C_{80} in the nave for sources in the gallery can be explained by the distance between receiver and source. The values are significantly higher for sources in the choir and nave.

In Figure 15b, the distance dependence of G is examined and compared to the SPL decay of a free-field source. To facilitate comparison, the free-field condition is approximated with a linear regression model and is not normalized to the source's SPL at a distance of 10 m. As would be expected, the G decreases with increased source–receiver distance across all data series. However, the slope of the decays within the church are softer than that for the free field condition. This is due to the sustained reflections of the enclosed space. The G decay is fastest for the free field condition, followed by the source by the high altar. For sources in the choir, the effect is smaller and for sources in the gallery smaller still.

These results indicate that the acoustic perception of sources by the high altar (P3 in Figure 9) strongly depends on the listener's position, which implies that the sound field created by such a source is less homogeneous. This may be explained by the lack of nearby reflecting structures around that source position, as compared to the choir and gallery. Accordingly, there are no strong early reflections supporting the transmission of sound from this location. Thus, acoustic energy is dispersed in all directions, whereas the vaulted ceiling above the gallery and the walls of the apse reinforce early reflection and direct the the spread of acoustic energy from the gallery and the choir into the nave.

There is also a significant difference between the choir and the gallery in distance dependence, which can be understood further by again turning to the raytracing model shown in Figure 12. From these graphs, it appears that, after about 100 ms, the acoustic energy is distributed much more evenly in the nave when the source is located in the gallery, rather than the choir. This behavior can be explained by two factors. First, the gallery is located in a much more advantageous location in the middle of the church. This leads to a much more even distribution of sound energy initially. Second, the gallery is larger and interacts with multiple ceiling vaults, while the acoustic energy from the choir only is reflected by the vaults of one bay. The reflections from multiple ceiling vaults leads to a much greater scattering of the acoustic field, and a greater diffusion overall. These two factors together give a more even distribution of acoustic energy when the source is located in the gallery.

5. Discussion

The room acoustic analysis of the reconstructed abbey reveals the presence of two distinct acoustical subspaces, which coincides with locations of significant liturgical importance. The nuns' gallery and the monks' choir are characterized by shorter EDT, greater G, and improved C_{80} compared to the rest of the church. These better acoustic conditions facilitate the auditory elements of liturgical practices, which are a fundamental part of the monastical congregations' daily tasks. Examples of such subspaces have been found in many other worship spaces [31–33] and are sometimes referred to as a "church within a church", indicating their role as an exclusive environment for the initiated. The presence of two such locations, rather than one, is an expected consequence of the presence of two separate monastic enclosures within the same church.

Despite their similarity as acoustic subspaces with improved conditions for liturgical activities, there are distinct differences between the gallery and the choir. The proximity of the gallery to the vaulted ceiling both improves the clarity within it and results in a rather homogenous sound field in the nave when a sound source is located in the gallery. Within the choir, as compared to within the gallery, the reverberation time is more significantly different to that in the nave. This could imply a weaker coupling between the acoustic subspace in the choir and the nave, than between the gallery and the nave. This could be a reason for the poor acoustic transmission from the choir to the nave.

Such differences may reflect a religious intent, irrespective of whether there was an acoustical intent in the design. The position of the gallery reflects a religious intent to premier the nuns within this abbey, and the improved acoustic transmission from this location supports this intent. Its location makes the gallery both acoustically and visually characteristic for the whole church, and establishes the nuns' position as central within the monastic order. The reflections from the ceiling vaults cause the nuns' chants to be perceived as coming "from above". This could reflect an acoustic intent of making the nuns sound more heavenly.

The spatial variations in the church can be heard in the auralizations (as presented in the supplementary material) of elements from a Friday sext. When both source and listener is in the gallery, or both in the choir, intelligibility is acceptable and the acoustics support chants, prayers, and responsories sufficiently to seem plausible. Reverberation from the nave can be heard, but is not strong enough to dominate the sound field. The services as perceived outside the respective monastical subspace give a very different impression. Individual syllables can not be distinguished, the locations of the sound sources are difficult to determine and the reverberance dominates perceived sound. This is the case for the perception of monks from within the gallery, nuns from within the choir and both from within the nave. Monks and nuns sound distant, yet omnipresent. Liturgical interactions between monks, nuns or people in the nave were thus likely not possible during daily services.

Further analysis of the auralizations themselves, with listening tests, could lead to additional insights regarding the experiences of historical visitors to this place. For example, the introduction of HRTFs is needed in order to evaluate the perceptual impact of the nuns' elevated position, to further evaluate the theory of their voices sounding more heavenly. Such research could show more clearly the perceptual differences between the experiences of the monastical congregations and the pilgrims. Furthermore, it may be possible to use the results of the simulation as a tool for VR performances, where singers experience the "live" simulated acoustics of the historical space as they sing. Such setups may minimize the effects of recording in an anechoic chamber.

The reverberation time within the nave is long but comparable to other churches around the Baltic sea. Although no comparison could be made for other Scandinavian or Nordic churches, comparisons to Gothic churches in other countries around the Baltic Sea (Germany and Poland) indicate some similarities in the acoustic cultural heritage. However, as the sample size is so small, further research is needed before any conclusions can be drawn.

The significant reverberation within the nave, and especially the long EDT, indicates that the space in the sext configuration is not suitable for the Gregorian chants usually performed there. This conclusion is supported by listening to the auralizations. However, in the more festive condition examined, with an increased number of visitors and members of the congregations, the EDT is decreased sufficiently within the nave to be within Martellotta et al.'s [34] guidelines for Gregorian chant. While there are no indications that this was intentional, it implies that during events aimed at a more general public, the acoustics of the church supported such events. Accordingly, the acoustics could be considered somewhat self-regulatory; when there are few listeners in the nave, acoustics are sufficient only for the important monastical congregations, and when the number of visitors in the nave increases, the larger number of visitors experience an acceptable acoustic field.

However, caution should be applied when making quantitative comparisons of archaeoacoustical simulations to guidelines or measurements, such as the comparisons made above. The rate of uncertainty in the model may be significant, both due to uncertainties in the material parameters and in the geometric modeling techniques [53,54]. The calibration procedure employed in this paper reduces these uncertainties, but can not remove them entirely.

Despite these words of caution, it should be noted that the major conclusions of this work concern the acoustic subspaces in the gallery and nave. These effects are primarily caused by the geometrical configuration of the space, which has been established based on the thorough historical research presented in Section 3.1. As such, the qualitative conclusions regarding improved acoustic conditions for the monastical congregations holds, despite the uncertainties presented regarding the absolute values of calculated parameters.

6. Conclusions

The room acoustic analysis in this paper shows that acoustical subspaces were formed in locations of religious and liturgical importance. These acoustical subspaces offered improved acoustics for the monastical congregations for which the church was built. Of the two congregations, the nuns were more central, and this is reflected in the design and acoustics of the church.

Although some of the results of the acoustic analysis of the gallery may have been possible with a less accurate digital model, the thorough acoustical analysis has benefited from the substantial historic research underlying it and finds its validation from it. The precise determination of the gallery's size, elevation, position, and form has been shown to have consequences for the sound field within the gallery and within the nave. This shows that archeo-acoustical modeling benefits from tight collaboration between acousticians, historians, and 3D-artists.

Supplementary Materials: The following are available online at https://www.mdpi.com/2076-3417/11/4/1586/s1.

Author Contributions: Conceptualization, H.A., M.B., C.A., D.B.H., E.L.S. and K.S.L.; methodology, M.B., C.A., E.L.S. and K.S.L.; formal analysis, H.A. and M.B.; investigation, M.B., C.A., E.L.S. and K.S.L.; writing—original draft preparation, H.A. and C.A.; writing—review and editing, H.A. M.B., C.A. and D.B.H.; visualization, H.A. and C.A.; supervision, D.B.H. and E.L.S.; funding acquisition, D.B.H., E.L.S. and K.S.L. All authors have read and agreed to the published version of the manuscript.

Funding: This research was funded by the Swedish Research Council grant number 2016-01784.

Institutional Review Board Statement: Not applicable.

Informed Consent Statement: Not applicable.

Data Availability Statement: The data presented in this study are available on request from the corresponding authors.

Acknowledgments: The authors express gratitude to: Brekke & Strand Akustik AB for providing ODEON 16.0 and Cubase 11.0 licences; Marcin Brycki from Brekke & Strand Akustik AB for leading the recording sessions and performing the mixing; Milab Microphones AB for providing microphones for recording; the singers involved in the recordings; Nikolaos-Georgios Vardaxis and Erling Nilsson from Engineering Acoustics LTH for valuable inputs; Tim Näsling from Brekke & Strand Akustik AB for useful tips and discussions during room acoustic modelling; Stefan Lindgren at LU Humanities Lab for the 3D scan of the church; Mattias Hallgren at Traditionsbärarna for input regarding historical construction.

Conflicts of Interest: The authors declare no conflict of interest. The funders had no role in the design of the study; in the collection, analyses, or interpretation of data; in the writing of the manuscript; or in the decision to publish the results.

Appendix A. Material Parameters for Acoustic Simulation

Table A1. Material coefficients used for the room acoustic simulations and auralization of the reconstructed abbey.

Material	Absorption Factor								Mid-Frequency Scattering
	63 Hz	125 Hz	250 Hz	500 Hz	1000 Hz	2000 Hz	4000 Hz	8000 Hz	
Thin silk textile, freely suspended [1]	0.05	0.05	0.06	0.39	0.63	0.70	0.73	0.73	0.01
Wool textile [1]	0.07	0.07	0.31	0.49	0.75	0.70	0.60	0.60	0.01
Thick wool (carpet) [1]	0.02	0.02	0.06	0.14	0.37	0.60	0.65	0.65	0.01
Heavy velvet [1]	0.03	0.03	0.03	0.15	0.4	0.50	0.50	0.50	0.01
Thick linen against stone [1]	0.01	0.01	0.02	0.05	0.15	0.30	0.40	0.40	0.01
Wooden construction, not painted [2]	0.09	0.09	0.09	0.08	0.08	0.10	0.07	0.07	0.60
Wooden construction, painted [2]	0.11	0.11	0.11	0.10	0.10	0.10	0.07	0.07	0.40
Wooden decoration, painted [2,3]	0.12	0.12	0.12	0.15	0.15	0.18	0.18	0.19	0.99
Hollow wooden structure, painted [1]	0.40	0.40	0.30	0.20	0.17	0.15	0.10	0.10	0.60
Plastered brick [6]	0.102	0.029	0.144	0.097	0.007	0.016	0.008	0.003	0.20
Limestone [6]	0.028	0.074	0.005	0.034	0.09	0.028	0.084	0.009	0.003
Plastered limestone [4]	0.07	0.06	0.05	0.05	0.04	0.04	0.04	0.04	0.001
Ceiling vaults [5]	0.12	0.09	0.09	0.05	0.04	0.03	0.03	0.03	0.30
Leaded glass windows [6]	0.254	0.259	0.24	0.016	0.101	0.039	0.495	0.003	0.14
Iron lattice [5]	0.01	0.01	0.01	0.02	0.06	0.03	0.03	0.03	0.001
People [1]	0.62	0.62	0.72	0.80	0.83	0.84	0.85	0.85	

[1] From an ODEON standard material [50]; [2] Suarez et al. [55]; [3] Alonso et al. [56]; [4] Postma et al. [51]; [5] Own data; [6] Determined by genetic algorithm optimization, see Section 3.3.2.

References

1. UNESCO. *Concept of Digital Heritage*; UNESCO: London, UK, 2019.
2. UNESCO. *Convention for the Safeguarding of the Intangible Cultural Heritage*; UNESCO: London, UK, 2003.
3. Lubman, D.; Kiser, B.H. The History of Western Civilization Told through the Acoustics of its Worship Spaces. In Proceedings of the 17th International Congress on Acoustics, Rome, Italy, 2–7 September 2001.
4. Martellotta, F. Optimal Reverberation Conditions in Churches. In Proceedings of the International Congress on Acoustics 2007, Madrid, Spain, 2–7 September 2007; Sociedad Espanola de Acustica: Madrid, Spain, 2007.
5. Howard, D. Recordings of Music Written for St. Mark's. In *Word, Image, and Song: Essays on Early Modern Italy*; University of Rochester Press: Rochester, NY, USA, 2013; Volume 1, pp. 89–100.
6. Navarro, J.; Sendra, J.J.; Muñoz, S. The Western Latin church as a place for music and preaching: An acoustic assessment. *Appl. Acoust.* **2009**, *70*, 781–789. [CrossRef]
7. Pedrero, A.; Ruiz, R.; Díaz-Chyla, A.; Díaz, C. Acoustical study of Toledo Cathedral according to its liturgical uses. *Appl. Acoust.* **2014**, *85*, 23–33. [CrossRef]
8. Aletta, F.; Kang, J. Historical Acoustics: Relationships between People and Sound over Time. *Acoustics* **2020**, *2*, 128–130. [CrossRef]
9. Đorđević, Z.; Novković, D.; Andrić, U. Archaeoacoustic Examination of Lazarica Church. *Acoustics* **2019**, *1*, 423–438. doi:10.3390/acoustics1020024. [CrossRef]
10. Elicio, L.; Martellotta, F. Acoustics as a cultural heritage: The case of Orthodox churches and of the "Russian church" in Bari. *J. Cult. Herit.* **2015**, *16*, 912–917. [CrossRef]
11. Girón, S.; Álvarez-Morales, L.; Zamarreño, T. Church acoustics: A state-of-the-art review after several decades of research. *J. Sound Vib.* **2017**, *411*, 378–408. [CrossRef]
12. Kosała, K. Calculation Models for Acoustic Analysis of St. Elizabeth of Hungary Church in Jaworzno Szczakowa. *Arch. Acoust.* **2016**, *41*, 485–498. [CrossRef]
13. Niemas, M.; Sadowski, J.; Engel, Z. Acoustic Issues of Sacral Structures. *Arch. Acoust.* **1998**, *23*, 87–104.
14. Kanev, N. Resonant Vessels in Russian Churches and Their Study in a Concert Hall. *Acoustics* **2020**, *2*, 399–415. doi:10.3390/acoustics2020023. [CrossRef]
15. Swedish Research Council. The Multisensory World of Vadstena Abbey in the Late Middle Ages. 2016. Available online: https://www.swecris.se/betasearch/details/project/201601784VR?lang=en (accessed on 11 January 2021).
16. Scarre, C.; Lawson, G. *Archaeoacoustics*; McDonald Institute for Archaeological Research: Cambridge, UK, 2006.
17. Boren, B.; Longair, M. A Method for Acoustic Modeling of Past Soundscapes. In Proceedings of the Acoustics of Ancient Theatres Conference, Patras, Greece, 18–21 September 2011; European Acoustics Association and the Hellenic Institute of Acoustics: Patras, Greece, 2011.
18. Berardi, U.; Iannace, G.; Maffei, L. Virtual reconstruction of the historical acoustics of the Odeon of Pompeii. *J. Cult. Herit.* **2016**. [CrossRef]

19. Martellotta, F.; Álvarez-Morales, L. Virtual acoustic reconstruction of the church of Gesú in Rome: A comparison between different design options. In Proceedings of the Forum Acusticum 2014, Krakow, Poland, 7–12 September 2014; The Polish Acoustical Society: Krakow, Poland, 2014. [CrossRef]
20. Alonso, A.; Sendra, J.J.; Suárez, R.; Zamarreño, T. Acoustic evaluation of the cathedral of Seville as a concert hall and proposals for improving the acoustic quality perceived by listeners. *J. Build. Perform. Simul.* **2014**, *7*, 360–378. [CrossRef]
21. Sender, M.; Planells, A.; Perelló, R.; Segura, J.; Giménez, A. Virtual acoustic reconstruction of a lost church: application to an Order of Saint Jerome monastery in Alzira, Spain. *J. Build. Perform. Simul.* **2018**, *11*, 369–390. [CrossRef]
22. Selfridge, R.; Cook, J.; McAlpine, K.; Newton, M. Creating Historic Spaces in Virtual Reality using off-the-shelf Audio Plugins. In Proceedings of the 2019 AES International Conference on Immersive and Interactive Audio, York, UK, 27–29 March 2019; Audio Engineering Society: York, UK, 2019.
23. Postma, B.N.J.; Katz, B.F.G. Perceptive and objective evaluation of calibrated room acoustic simulation auralizations. *J. Acoust. Soc. Am.* **2016**, *140*, 4326–4337. [CrossRef]
24. Katz, B.F.G.; Poirier-quinot, D.; Postma, B.N.J. Virtual reconstructions of the Théâtre de l'Athénée for archeoacoustic study. In Proceedings of the 23rd International Congress on Acoustics integrating 4th EAA Euroregio 2019, Aachen, Germany, 9–13 September 2019; Deutsche Gesellschaft für Akustik: Aachen, Germany, 2019; pp. 303–310.
25. Kleiner, M.; Dalenbäck, B.I.; Svensson, P. Auralization—An Overview. *J. Audio Eng. Soc.* **1993**, *41*, 861–875.
26. Mehra, R.; Rungta, A.; Golas, A.; Lin, M.; Manocha, D. WAVE: Interactive Wave-based Sound Propagation for Virtual Environments. *IEEE Trans. Vis. Comput. Graph.* **2015**, *21*, 434–442. [CrossRef] [PubMed]
27. Pelzer, S.; Aspöck, L.; Schröder, D.; Vorländer, M. Interactive Real-Time Simulation and Auralization for Modifiable Rooms. *Build. Acoust.* **2014**, *21*. [CrossRef]
28. Poirier-Quinot, D.; Katz, B.; Noisternig, M. EVERTims: Open source framework for real-time auralization in VR. In Proceedings of the AM '17, London, UK, 23–26 August 2017; ACM: New York, NY, USA; London, UK, 2017.
29. Vorländer, M. *Auralization. Fundamentals of Acoustics, Modelling, Simulation, Algorithms and Acoustic Virtual Reality*, 1st ed.; Springer: Berlin, Germany, 2008; pp. 1–335. [CrossRef]
30. Anderson, J.S.; Bratos-Anderson, M. Acoustic coupling effects in St Paul's Cathedral, London. *J. Sound Vib.* **2000**, *236*, 209–225. [CrossRef]
31. Álvarez-Morales, L.; Lopez, M.; Alvarez-Corbacho, A.; Bustamante, P. Mapping the acoustics of Ripon cathedral. In Proceedings of the 23rd International Congress on Acoustics, Aachen, Germany, 9–13 September 2019; Deutsche Gesellschaft für Akustik: Aachen, Germany, 2019.
32. Boren, B.; Longair, M.; Orlowski, R. Acoustic Simulation of Renaissance Venetian Churches. *Acoust. Pract.* **2013**, *1*, 17–28. [CrossRef]
33. Alonso, A.; Suárez, R.; Sendra, J. The Acoustics of the Choir in Spanish Cathedrals. *Acoustics* **2018**, *1*, 35–46. [CrossRef]
34. Martellotta, F. Subjective study of preferred listening conditions in Italian Catholic churches. *J. Sound Vib.* **2008**, *317*, 378–399. [CrossRef]
35. Martellotta, F. Identifying acoustical coupling by measurements and prediction-models for St. Peter's Basilica in Rome. *J. Acoust. Soc. Am.* **2009**, *126*, 1175–1186. [CrossRef] [PubMed]
36. Andersson, I.; Ljungstedt, S.; Malm, G. *Vadstena Klosterkyrka. 1, Kyrkobyggnaden*; Riksantikvarieämbetet och Kungl. Vitterhets-, Historie-, och Antikvitetsakademien: Borås, Sweden, 1991.
37. Strinnholm Lagergren, K. The invitatory antiphons in Cantus sororum: a unique repertoire in a world of standard chant. *Plainsong Mediev. Music.* **2018**, *27*, 121–142. [CrossRef]
38. Berthelson, B. *Studier i Birgittinerordens Byggnadsskick 1, Anläggningsplanen och dess Tillämpning*; Kungl. Vitterhets-, historie- och antikvitetsakademien: Stockholm, Sweden, 1947.
39. Andersson, A. *Vadstena Klosterkyrka II. Inredning*; Riksantikvarieämbetet och Kungl Vitterhets-, historie- och antikvitetsakademien: Borås, Sweden, 1983.
40. Sigurdson, J.M.; Zachrisson, S. *Aplagårdar och Klosterliljor: 800 år kring Vadstena Klosters Historia*; Artos: Skellefteå, Sweden, 2010.
41. Nyberg, T.S. *Birgittinische Klostergründungen des Mittelalters*; Gleerup: Lund, Sweden, 1965.
42. Bennett, R. *Vadstena klosterkyrka III. Gravminnen*; Riksantikvarieämbetet och Kungl. Vitterhets-, Historie-, och Antikvitetsakademien: Borås, Sweden, 1985.
43. Autio, H.; Bard, D. A statistical method for parameter estimation from Schroeder decay curves. In Proceedings of the INTER-NOISE 2018, The 47th International Congress and Exposition on Noise Control Engineering, Chicago, IL, USA, 26–29 August 2018; INCE-USA: Chicago, IL, USA, 2018.
44. Alonso, A.; Sendra, J.J.; Suárez, R. Sound Space Reconstruction in the Cathedral of Seville for major feasts celebrated around the main chancel. In Proceedings of the Forum Acusticum 2014, Krakow, Poland, 7–12 September 2014; The Polish Acoustical Society: Krakow, Poland, 2014.
45. Sancta Birgitta. *Reuelaciones Extrauagantes*; Almqvist & Wiksell: Stockholm, Sweden, 1956.
46. Klemming, G.E. (Ed.); Lucidarium. In *Heliga Birgittas Uppenbarelser*; P. A. Nordstedt & Söner: Stockholm, Sweden, 1884; Volume 5.
47. Ugglas, C.R.A. En svensk korsväg med målade stationsbilder, del 1. *Fornvännen* **1938**, *33*, 230–272.
48. Sancta Birgitta. *Opera Minora, Vol 1: Regula Saluatorie*; Almqvist & Wiksell: Stockholm, Sweden, 1975.

49. Boren, B.; Abraham, D.; Naressi, R.; Grzyb, E.; Lane, B.; Merceruio, D. Acoustic Simulation of Bach's Performing Forces in the Thomaskirche. In Proceedings of the 1st EAA Spatial Audio Signal Processing Symposium, Paris, France, 6–7 September 2019; Sorbonne Université: Paris, France, 2019.
50. ODEON Room Acoustics Software. 2020. Available online: https://www.geonoise.com/odeon-room-acoustics-software/ (accessed on 26 December 2020).
51. Postma, B.N.J.; Katz, B.F.G. Creation and calibration method of acoustical models for historic virtual reality auralizations. *Virtual Real.* **2015**, *19*, 161–180. [CrossRef]
52. ISO 3382-1:2009. *Acoustics—Measurement of Room Acoustic Parameters—Part 1: Performance Spaces*; Technical Report; ISO: Geneva, Switzerland, 2009.
53. Pilch, A. Optimization-based method for the calibration of geometrical acoustic models. *Appl. Acoust.* **2020**, *170*. [CrossRef]
54. Vorländer, M. Computer simulations in room acoustics: Concepts and uncertainties. *J. Acoust. Soc. Am.* **2013**, *133*, 1203–1213. [CrossRef]
55. Suárez, R.; Alonso, A.; Sendra, J.J. Archaeoacoustics of intangible cultural heritage: The sound of the Maior Ecclesia of Cluny. *J. Cult. Herit.* **2016**, *19*, 567–572. [CrossRef]
56. Alonso, A.; Suárez, R.; Sendra, J.J. Virtual reconstruction of indoor acoustics in cathedrals: The case of the Cathedral of Granada. *Build. Simul.* **2017**, *10*, 431–446. [CrossRef]

Article

Potential of Room Acoustic Solver with Plane-Wave Enriched Finite Element Method

Takeshi Okuzono [1],*, M Shadi Mohamed [2] and Kimihiro Sakagami [1]

[1] Environmental Acoustics Laboratory, Department of Architecture, Graduate School of Engineering, Kobe University, Kobe 657-8501, Japan; saka@kobe-u.ac.jp
[2] School of Energy, Geoscience, Infrastructure and Society, Heriot–Watt University, Edinburgh EH14 4AS, UK; M.S.Mohamed@hw.ac.uk
* Correspondence: okuzono@port.kobe-u.ac.jp; Tel.: +81-78-803-6577

Received: 27 February 2020; Accepted: 11 March 2020; Published: 13 March 2020

Abstract: Predicting room acoustics using wave-based numerical methods has attracted great attention in recent years. Nevertheless, wave-based predictions are generally computationally expensive for room acoustics simulations because of the large dimensions of architectural spaces, the wide audible frequency ranges, the complex boundary conditions, and inherent error properties of numerical methods. Therefore, development of an efficient wave-based room acoustic solver with smaller computational resources is extremely important for practical applications. This paper describes a preliminary study aimed at that development. We discuss the potential of the Partition of Unity Finite Element Method (PUFEM) as a room acoustic solver through the examination with 2D real-scale room acoustic problems. Low-order finite elements enriched by plane waves propagating in various directions are used herein. We examine the PUFEM performance against a standard FEM via two-room acoustic problems in a single room and a coupled room, respectively, including frequency-dependent complex impedance boundaries of Helmholtz resonator type sound absorbers and porous sound absorbers. Results demonstrated that the PUFEM can predict wideband frequency responses accurately under a single coarse mesh with much fewer degrees of freedom than the standard FEM. The reduction reaches $\mathcal{O}(10^{-2})$ at least, suggesting great potential of PUFEM for use as an efficient room acoustic solver.

Keywords: frequency domain; PUFEM; room acoustics; wave-based method

---

1. Introduction

1.1. Background

Acoustic simulation methods are necessary tools for predicting impulse responses or frequency responses of room spaces in architectural acoustics design. These quantities are necessary to evaluate room acoustics with acoustical parameters such as reverberation times, and the clarity of speech or music. These can also be used for the visualization and auralization of sound fields. Wave-based numerical methods, which solve a wave equation or a Helmholtz equation numerically, are physically reliable simulation methods with the capability of capturing wave phenomena such as interference and diffraction, and also of modeling boundary effects accurately by sound diffusers and sound absorbers. The finite element method (FEM) [1–5], boundary element method (BEM) [6], and finite difference time domain (FDTD) [7–9] method exemplify the often-used numerical methods for room acoustic simulations. Although they entail a huge computational effort for acoustic simulations especially at kilohertz frequencies in a real-sized room, their application to room acoustics prediction is increasing gradually by virtue of the progress of computer technology and the continuous development of efficient methods [9–24]. In addition, some recent studies [16,18,22,25] use extended-reaction boundary conditions to address both the frequency dependent and incident-angle dependent absorption characteristics of sound absorbers accurately, whereas many studies use the simplest local-reaction boundary conditions, which simplify the incident-angle dependence of surface impedance. Nevertheless, wave-based predictions are still time-consuming. Therefore, the development of more efficient methods or optimizing the performance has a marked impact in room acoustics field. This report presents a preliminary study to this end particularly using FEM in the frequency domain solving the Helmholtz equation with a few degrees of freedom (DOF).

The huge computational effort necessary for performing reliable acoustic simulations in real-sized rooms using FEM stems from the large dimension of spaces, the broad frequency range of interest, the complicated boundary conditions, and an inherent error property of FEM. The volumes of architectural spaces such as offices, lecture rooms, and concert halls range from the order of 10 m^3 to 10,000 m^3. The human audible frequency is 20 Hz to 20 kHz. In addition, the sound-absorption characteristics of boundary materials, which depend on both frequency and incident-angle of sound, should be modeled accurately. However, FEM is well known to have inherent spatial discretization error called dispersion error, which is an error evaluated in sound speed or in wavelength. To maintain the error within an acceptable level, the discretization of spaces, i.e., mesh generation, must be performed with consideration of a rule of thumb, e.g., for linear elements spatial discretization of 10 elements per wavelength at least. This discretization rule imposes the use of a large FE models with many DOF for acoustic simulation of a real-sized room, making the solution of the problem prohibitively expensive. For instance, in earlier works [4,5] conducted with high-order finite elements [26], the acoustics in a multi-purpose hall of 37,000 m^3 were simulated using an FEM model of 2,630,435 DOF at 125 Hz. The acoustics in a small hall with complicated diffusers were analyzed using an FEM mesh with 8,926,001 DOF. A recent study [27] conducting simulations at kilohertz frequencies used FEM models with the order of ten million DOF for both acoustic simulations of a simply shaped concert hall and the reverberation absorption coefficient measurement though the use of a dispersion-reduced scheme [28]. Furthermore, when using extended-reaction boundary conditions, the required elements increase because of the discretization of materials. Therefore, the development of room acoustic FEM solvers able to perform reliable simulations with an FE model having much fewer DOF is one direction for enhancing the applicability of FEM to room acoustic problems.

1.2. Partition of Unity Finite Element Method for Acoustic Problems

As a numerical method with such potential, acoustic numerical methods based on the Partition of Unity FEM [29] (PUFEM) have been formulated and examined in some studies [30–40]. Two papers [36,37] present demonstrations of PUFEM on 2D car interior analyses at high frequencies including porous absorbing materials modeled respectively using an equivalent fluid model [41–44] and poroelastic material model [45]. A very recent study [35] examined a 2D plane wave-scattering problem by which PUFEM can significantly reduce the DOF of the FEM model at high frequencies. The study also includes demonstration of partition of unity isogeometric analysis of 2D car interior sound field analysis at 20 kHz. The ability of acoustic PUFEM or PU-based method derives from enriching the approximation of sound fields by incorporating the general solution of Helmholtz equation into finite element approximation. For example, when using plane waves as the general solution, the sound pressure at a nodal point is expressed by the superposition of plane waves propagating various directions. It incorporates into local finite element approximation via the partition of the unity property. Additionally, sound fields are approximated up to high frequencies using q-refinement, which is a refinement technique by which plane waves propagating in various directions are added at nodal points of a fixed mesh gradually with increasing frequency. The most notable feature of acoustic PUFEM is that it obviates re-meshing according to frequencies, which is necessary for conventional FEM. Actually, PUFEM analysis can use elements of many times greater length than the wavelength of analyzed frequency, whereas conventional linear FEM must use an element size satisfying less than one-tenth of the wavelength of the analyzed frequency, as described above. Some closely related approaches with PUFEM exist. A review article presents additional details [46].

The PUFEM feature is apparently favorable for room acoustic simulations. In addition, the plane wave enrichment is applicable to any FEM mesh. That characteristic is useful to improve the existing FEM code performance. However, acoustic PUFEM is still developing. Various aspects remain to be studied for application to practical room acoustic simulations. In most earlier works [35–37], the potential of PUFEM for practical applications has been presented on 2D acoustic analysis in a small car having an area of less than 3 m^2, without quantitative comparison with reference solutions. Additionally, those studies examine only pure tone analyses at high frequencies. Therefore, from the perspective of room acoustic applications, it remains unclear how PUFEM will perform robustly and accurately for problems of calculating wideband frequency responses in larger interior sound fields such as architectural spaces. This is an extremely important aspect because room acoustic simulations specifically address large room models and require wideband frequency components with fine frequency resolution especially for calculating room impulse responses. For reference, such an evaluation using frequency-domain methods can be found in earlier reports of the literature [16,23,24]. This study is the first attempt at revealing the PUFEM performance for wideband frequency response analysis in room acoustic problems with quantitative evaluation in accuracy. The architectural spaces treated here have 13–20 times larger dimensions than earlier studies.

1.3. Purpose of This Study

This study was conducted to discuss the potential of the plane-wave-enriched FEM as a room acoustic solver via performance examination on 2D real-scale room acoustic problems through numerical experiments. This is a preliminary study toward constructing an efficient 3D room acoustic solver. Plane-wave-enriching low-order FEs are used herein to discretize spaces. Local-reaction impedance boundary modeling is used to address the absorption characteristics of sound absorbers such as Helmholtz resonators and porous absorbers. As the main result, we can report whether or not PUFEM can predict wideband frequency responses robustly and accurately from low frequencies to high frequencies with a fine frequency resolution and with a single coarse mesh having a small amount of DOF. Our study also includes a performance comparison against conventional linear FEM having second-order accuracy with respect to dispersion errors. The performance is measured in terms of both the prediction accuracy of sound pressure level, and in terms of the reduction effect of DOF. We use two problems including realistic boundary impedance of sound absorbers commonly used in the room acoustics field. Because the two problems have no analytical solutions, reference solutions are calculated using a fourth-order accurate FEM with fine meshes. The remainder of this study is organized as follows. For reader convenience, PUFEM approximation is briefly introduced in Section 2 where also important aspects related to choosing the numerical parameters are discussed. Section 3 presents examination of the applicability of PUFEM via both the sound field analysis in a single room and a more complex coupled room composed of four rooms. Section 4 concludes this study.

2. Brief Preliminaries for Room Acoustic Simulations Using PUFEM

2.1. Interior Sound Field Analysis

We consider a sound propagation problem in an interior sound field Ω with boundary Γ composed of three boundary conditions: a rigid boundary Γ_0, a vibration boundary Γ_V, and an impedance boundary Γ_Z, as shown in Figure 1. This problem is described in terms of the sound pressure p using the following Helmholtz equation as

$$\nabla^2 p + k^2 p = 0, \quad \text{in } \Omega \tag{1}$$

where k represents the wavenumber. The three boundary conditions are given as

$$\frac{\partial p}{\partial n} = \begin{cases} 0 & \text{on } \Gamma_0 \\ -j\omega\rho_0 v & \text{on } \Gamma_v \\ -jk\frac{1}{z_n}p & \text{on } \Gamma_z \end{cases} \qquad (2)$$

where ω represents the angular frequency, ρ_0 expresses the air density, v signifies the vibration velocity, z_n stands for the normalized impedance ratio, and $-j$ denotes the imaginary unit ($j^2 = -1$). With the arbitrary weight function ϕ, the weak form of Helmholtz equation for finite element discretization is given as

$$\int_\Omega (-\nabla\phi\nabla p + k^2 \phi p)\, d\Omega + \int_\Gamma \phi \frac{\partial p}{\partial n}\, d\Gamma = 0. \qquad (3)$$

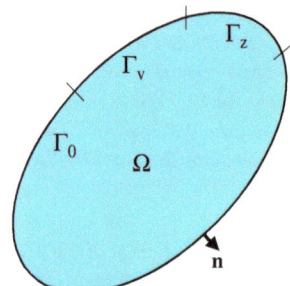

Figure 1. Interior sound field Ω where Γ_0, Γ_v and Γ_z respectively represent a rigid boundary, a vibration boundary, and an impedance boundary. In addition, **n** is the outward normal.

2.2. Plane-Wave Enriched Finite Elements in 2D Analysis

In FEM, sound pressure at an arbitrary point (x, y) within an element Ω_e is approximated by product of shape functions $N_i(\xi, \eta)$ and nodal values of sound pressure p_i

$$p(x,y) = \sum_{i=1}^n N_i(\xi, \eta) p_i, \qquad (4)$$

where ξ and η are the local coordinate system. In plane-wave enriched finite elements, the plane wave, which is the general solution of Helmholtz equation, is incorporated into shape functions via the partition of unity property [33,34]. Nodal values of sound pressure p_i are approximated with the superposition of plane waves propagating in various directions as

$$p_i = \sum_{l=1}^q A_i^l e^{jk(x\cos\theta_l + y\sin\theta_l)}, \qquad (5)$$

where q stands for the number of plane waves, θ_l denotes the angles of plane waves in polar coordinate systems, and A_i^l expresses the amplitude of plane wave propagating in a direction θ_l. As presented in this report, θ_l are set as evenly distributed angles $\theta_l = 2\pi l/q$. Substituting Equation (5) into Equation (4) engenders a plane wave enriched approximation of sound pressure at an arbitrary point within an element as

$$p(x,y) = \sum_{i=1}^n \sum_{l=1}^q N_i(\xi, \eta) e^{jk(x\cos\theta_l + y\sin\theta_l)} A_i^l. \qquad (6)$$

As presented in Equation (6), the plane wave enrichment is applicable to any finite element with a different shape function. By defining a new shape function P as the product of the shape function and the plane wave with unit amplitude, Equation (6) can be expressed as shown below:

$$p(x,y) = \sum_{i=1}^{n} \sum_{l=1}^{q} P_{(i-1)q+l} A_i^l \tag{7}$$

The equation above is used for PUFEM discretization of the weak form of Equation (3).

2.3. Semi-Discretized Matrix Equation

We apply PUFEM discretization [33,34] with a Galerkin approach to the integral equation of Equation (3); then, consideration of the three boundary conditions in Equation (2) engenders the following semi-discretized matrix equation as

$$\sum_{e}^{n_e} \left[\int_{\Omega_e} \left(\nabla \boldsymbol{P}^{\mathrm{T}} \nabla \boldsymbol{P} - k^2 \boldsymbol{P}^{\mathrm{T}} \boldsymbol{P} \right) d\Omega + j\frac{k}{z_n} \int_{\Gamma_{e,z}} \boldsymbol{P}^{\mathrm{T}} \boldsymbol{P} d\Gamma \right] \boldsymbol{A} = \sum_{e}^{n_e} \left[-j\omega\rho_0 v \int_{\Gamma_{e,v}} \boldsymbol{P}^{\mathrm{T}} d\Gamma \right], \tag{8}$$

where $\boldsymbol{P}$ is the shape function vector having components of the new shape function P, $\boldsymbol{A}$ is the nodal amplitude vector, and n_e represents the number of plane-wave enriched elements. For illustration in this report, we use plane-wave-enriched linear quadrilateral elements for spatial discretization. A high-order Gauss–Legendre rule is used for evaluating the domain integral and the boundary integral. Finally, the complex sound pressure in the domain Ω can be computed via Equations (6) or (7) with the plane wave amplitude A_i^l obtained as the solution of the linear system of equations presented above.

2.4. Numerical Setup of PUFEM

To perform the efficient PUFEM analysis, it is important to apply a proper setup in some numerical parameters. First, it is necessary to use a proper number of Gauss points n_g in the evaluations of domain integral and boundary integral according to the frequency to be analyzed. However, an established rule that can perform well in wide frequency ranges from low to high frequency remains insufficient. For this study, we applied the following rule obtained from a preliminary numerical experiment on plane wave propagation problem in a duct:

$$n_g = \begin{cases} \text{int}(10 n_w + 1) & (n_w \geq 1) \\ 10 & \text{otherwise} \end{cases} \tag{9}$$

Here, n_w represents the maximum element length $h_{\max}$ relative to wavelength λ defined as $n_w = h_{\max}/\lambda$, which represents the number of the wavelength included in each element. For high-frequency analyses, the well-used rule exists: around ten integration points per wavelength contained within each element. We applied the rule to frequencies $n_w \geq 1$ as in Equation (9). That is, we defined the frequencies satisfying $n_w \geq 1$ as high frequencies. Other recent proposed integration rules [39] may become a better alternative. However, the analytical integration scheme is limited to elements with straight edges and would be difficult to implement if the edges are curved. Since we are interested in real-world applications, we decided to use the numerical integration scheme as it is a more general approach and can be applied without alteration to any type or shape of elements. Furthermore, a proper setup for a way of adding how many plane waves at each nodal point along with frequency is important, but it remains insufficient as an established setup for wideband frequency analysis from low frequencies to high frequencies. We applied the following equation referred from earlier work reported in the literature [36]:

$$q = \text{round}[kh_{\max} + C(kh_{\max})^{\frac{1}{3}}], \tag{10}$$

In Equation (10), $h_{\max}$ is the maximum element length in all elements. The constant C adjusts the resulting accuracy. However, no way exists to set an appropriate value of C in advance. An alternative mode of Equation (10) has been proposed for high frequencies. It has a discretization level of around 2.5 degrees of freedom per wavelength. This paper selected the use of Equation (10).

3. Numerical Experiments

We examine the PUFEM performance using plane-wave enriched quadrilateral elements through two numerical examples on the calculation of sound fields in a single room and in a coupled room, respectively, including local reaction impedance boundaries of a resonant absorber and a porous absorber. The two room models were created from reference to an existing office plan of the authors in Kobe University. The accuracy of PUFEM is shown in comparison with the standard FEM using linear quadrilateral elements, which has second-order accuracy with respect to the dispersion error. In addition, the PUFEM efficiency is measured by the reduction effect of DOF for achieving the equivalent level of accuracy. In both rooms, sound fields generated by acoustic emission from a loudspeaker placing in the room were calculated at 20 Hz to 2.5 kHz with a 1 Hz interval. Because the two numerical examples have no analytical solutions, we used reference solutions calculated using a fourth-order accurate FEM with a dispersion reduction technique called modified integration rules [47]. The performance of fourth-order accurate FEM for 3D room acoustic simulations was described in earlier reports of the literature [16,18,28,48]. For this paper, a 2D version is used in the frequency domain [49]. The speed of sound and the air density were set, respectively, as 340 m/s and 1.205 kg/m^3.

3.1. Measurement of Accuracy

The accuracy of numerical solutions was evaluated in terms of frequency responses of the sound pressure level (SPL). We define the following RMS error $L_{\text{rms}}(f)$ with respect to the spatial distribution of SPL as

$$L_{\text{rms}}(f) = \sqrt{\frac{1}{N_{\text{point}}} \sum_{i=1}^{N_{\text{point}}} [L_{\text{fem}}(f,i) - L_{\text{ref}}(f,i)]^2}, \tag{11}$$

where N_{point} signifies the number of receivers, $L_{\text{fem}}(f,i)$ stands for the SPLs in a receiver i at frequency f calculated using the PUFEM and standard FEM, and $L_{\text{ref}}(f,i)$ denotes the SPL calculated using the reference solution. Furthermore, we performed 1/3 octave band averaging to the RMS error to capture the error behavior easily:

$$\overline{L}_{\text{rms}}(f_c) = \frac{1}{N_f} \sum_{f=f_l}^{f_u} L_{\text{rms}}(f). \tag{12}$$

Therein, $\overline{L}_{\text{rms}}(f_c)$ represents the RMS error at 1/3 octave band center frequency f_c, N_f denotes the number of frequencies included within 1/3 octave band, and f_l, f_u respectively denote the lower and upper limit frequencies.

3.2. Sound Propagation in a Single Room

3.2.1. Problem Description and Numerical Setup

Figure 2 portrays a single room with area S_a of 39.92 m^2, including a small rectangular area S assuming a loudspeaker where Γ_V is treated as the vibration boundary. The room's boundaries comprise a weakly absorbing impedance boundary $\Gamma_{z,1}$ and an impedance boundary $\Gamma_{z,m}$ of honeycomb-backed microperforated panel (MPP) absorber with Helmholtz resonator type sound-absorption characteristics. The weakly absorbing surface $\Gamma_{z,1}$ has real valued impedance corresponding to normal incidence sound absorption coefficient $\alpha_0 = 0.05$. For the MPP sound absorber, the sound absorption characteristics are frequency dependent in both α_0 and z_n, as shown in Figure 2. A theoretical impedance model [24,50] considering a limp MPP was used to calculate the z_n. The geometrical parameters of MPP are 0.5 mm in hole diameter, 1 mm in plate thickness, 0.75% in perforation ratio, and 1.13 kg/m² in surface density. The backing honeycomb core thickness is 0.015 m. With those specifications, the MPP absorber shows peak sound absorption at around 1 kHz. We applied $v = 1$ m/s for the vibrating surface, assuming a speaker cone.

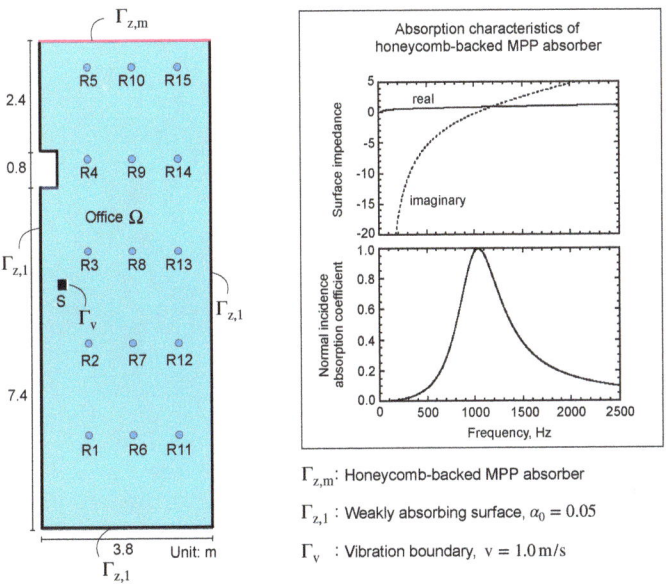

Figure 2. Single-room model including a vibration boundary, and impedance boundaries of weakly absorbing surfaces and honeycomb-backed microperforated panel absorbers: S and R respectively represent sound source and receivers.

Figure 3a,b show two PUFEM meshes, Mesh 1 and Mesh 2, with different spatial resolutions. Both meshes consist of elements larger than the wavelength of upper-limit frequency. Mesh 1 is a uniform mesh discretized with square elements of 0.2 m having 1.47 times larger size than the wavelength at 2.5 kHz. Mesh 2 is a non-uniform mesh discretized by rectangular elements of 0.2–0.4 m having 2.97 times larger size at maximum. The total numbers of elements N_{ele} and nodes N_{node} are 922 and 994, respectively, for Mesh 1. They are 267 and 307, respectively, for Mesh 2. The constant C in Equation (10) was set as 5–14 with one interval. For the reference solution, two fine meshes with different spatial resolution were used for different frequency ranges. We used a mesh discretized with 0.01 m square elements with 400,720 DOF for analyses up to 1.5 kHz. Its spatial resolution is 22 elements per wavelength at 1.5 kHz. At higher frequencies, we used a mesh with 1,599,840 DOF discretized by 0.005 m square elements. Its spatial resolution is 27 elements per wavelength. Standard FEM analysis used the same meshes as those used for a reference solution.

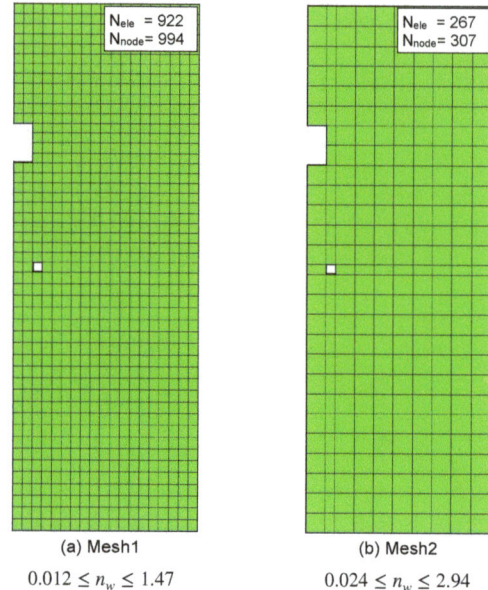

Figure 3. PUFEM mesh with two spatial resolutions: (**a**) Mesh 1 and (**b**) Mesh 2. The range of n_w is also shown for reference.

3.2.2. Results and Discussion

As an example of comparison of frequency responses, Figure 4a,b respectively show comparisons of frequency response calculated using PUFEM and standard FEM at R1 with reference solution (Ref): (a) Ref versus PUFEM (C = 13, Mesh2), and (b) Ref versus standard FEM (0.01 m mesh). This figure visually presents the effectiveness of PUFEM. The standard FEM using 0.01 m mesh with 400,720 DOF shows a marked difference from the reference solution at higher frequencies than 1 kHz because of the inherent large dispersion error, i.e., sound speed increases at higher frequencies, and the error magnitude becomes higher for larger domains. The PUFEM result shows much better agreement with the reference solution at entire frequencies despite the use of coarse mesh with 267 elements. Regarding the PUFEM analysis, Figure 5a,b show changes in the DOF for the entire frequency range when using Mesh 2 with C = 13. The DOF in PUFEM is defined as the product of plane wave numbers q for enrichment and nodes N_{node} i.e., DOF = $q \times N_{node}$. The DOF of PUFEM analysis changes from 2149 to 16,271 at frequencies of 20 Hz to 2.5 kHz, with a change in the plane wave numbers of 7–53. The PUFEM analysis has only 16,271 DOF at 2.5 kHz, which is 1/25 smaller than the DOF of standard FEM.

Figure 6a,b show RMS errors of both the standard FEM (0.01 m mesh) and the PUFEM with different values of C for Mesh 1 and Mesh 2. We only present results with C = 5, 7, 9, 11, and 13 for ease of illustration of the error behavior. In the PUFEM results, the RMS errors are reduced overall with larger C. The larger value of C is necessary to reduce the error at higher frequencies. This is true for Mesh 1 and Mesh 2. In comparison with standard FEM results, the PUFEM with $C \geq 6$ offers more accurate results at frequencies higher than 315 Hz in both Mesh 1 and Mesh 2. It is a very interesting capability of plane wave enrichment because the standard FEM with non-enriched elements uses very fine mesh at 315 Hz, where the spatial resolution is 108 elements per wavelength. Regarding the RMS error at frequencies below 315 Hz, PUFEM shows error of less than 1 dB when using $C \geq 7$ for Mesh1 and $C \geq 6$ for Mesh 2. It is an acceptable error magnitude for practical applications. Furthermore, standard FEM shows RMS error of 8.1 dB at 2 kHz although the used mesh has 400,720 DOF. By contrast, PUFEM shows a significantly low error value of 0.76 dB in both Mesh 1 and Mesh 2 when using C = 13. Additionally, we performed the standard FEM analysis with

0.005 m mesh having 1,599,840 DOF, which has spatial resolution of 34 elements per wavelength at 2 kHz. However, the result still shows an RMS error of 7.4 dB at 2 kHz because, for large scale sound field analysis at high frequencies, the standard FEM results include large dispersion error as described previously. That result also demonstrates that the standard FEM requires a finer mesh to obtain the equivalent level of accuracy as the PUFEM results at 2 kHz. Based on those results, the PUFEM with Mesh 2 can probably perform more accurate analysis with at least 1/100 fewer DOF than the standard FEM does.

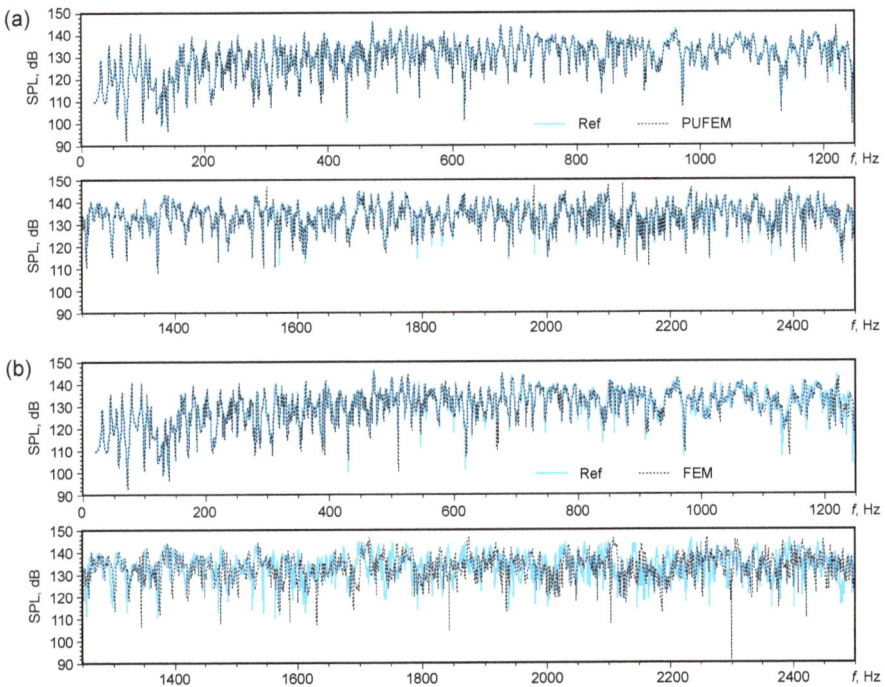

Figure 4. Comparison of frequency responses at R1: (**a**) reference solution (Ref) vs. PUFEM ($C = 13$, Mesh 2), and (**b**) reference solution vs. standard FEM (0.01 m mesh).

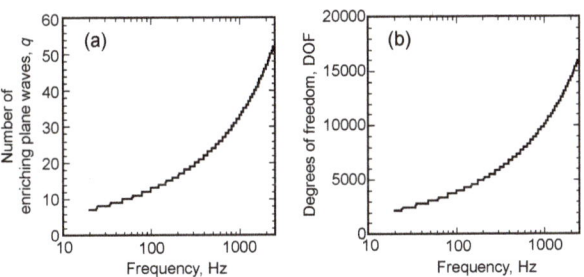

Figure 5. Changes in (**a**) the enriched plane waves number q and (**b**) DOF when using Mesh 2 with $C = 13$.

Furthermore, we present the convergence behavior of a solution against DOF and examine whether a coarse mesh or a fine mesh is effective. Figure 7a,d show relations between DOF and RMS error in PUFEM analysis using Mesh 1 and Mesh 2 at four higher frequency bands of 1 kHz, 1.25 kHz, 1.6 kHz, and 2 kHz. They show that the coarse mesh, Mesh 2, achieves a practically acceptable error magnitude of less than 1 dB with fewer DOF than the finer mesh. This result suggests that the use of coarse mesh enriched by many plane waves is more effective than that of fine mesh enriched with a few plane waves from the aspect of computational cost. Additionally, we can observe an important aspect in the Mesh 2 results at 1 kHz and 1.25 kHz. There exists a proper number of plane waves for enrichment, i.e., an increase of DOF does not always produce more accurate results. Similar results were obtained from an earlier study [33]. The earlier report described this as attributable to the ill-conditioning of the resulting linear system, showing an increase of the condition number when continuously increasing the attached plane wave numbers. In addition, the present paper showed the mesh size effect on the resulting accuracy in limited conditions. Therefore, comprehensive investigations on this topic will be shown in future works with well-organized numerical experiments.

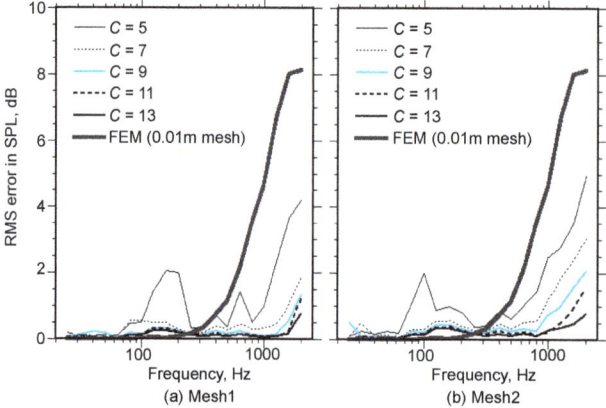

Figure 6. Comparison of RMS error for (**a**) Mesh 1 and (**b**) Mesh 2. An FEM result obtained using 0.01 m mesh is also shown for comparison.

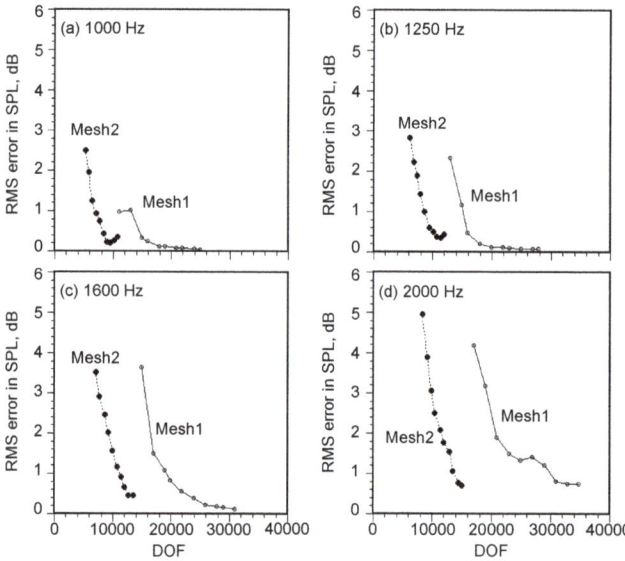

Figure 7. Relations between DOF and RMS error in PUFEM analysis at four higher frequency bands: (**a**) 1 kHz, (**b**) 1.25 kHz, (**c**) 1.6 kHz, and (**d**) 2 kHz. The DOF in the x-axis is the value at a center frequency within each band.

3.3. Sound Propagation Problem in a Coupled Room

3.3.1. Problem Description and Numerical Setup

In the second example, a real-world acoustic application is considered again where the model is based on an existing office plan in Kobe University and the boundary conditions are based on actual sound absorber installed in the offices. The installed absorber is a porous type absorber, which is different from the previous example. Additionally, as a further demonstration of the effectiveness of PUFEM, we present computed SPL distributions inside rooms with a fine spatial resolution, comparing with the reference solution and the standard FEM. Figure 8 presents a coupled room ($S_a = 60.48$ m^2) composed of four rooms where the largest room is the same as that used in the first numerical example and where the other three are soundproof rooms with highly absorbing boundaries. The room boundaries comprise reflecting impedance boundary $\Gamma_{z,1}$ and highly absorbing impedance boundary $\Gamma_{z,p}$ of a rigid-backed porous sound absorber. We gave a real valued surface impedance corresponding to $\alpha_0 = 0.05$ to the reflecting boundary $\Gamma_{z,1}$. An equivalent fluid model [51] with Miki's empirical equation [52] was used to calculate the surface impedance of the porous absorber. Regarding the porous material, glass wool of 32 kg/m^3 with 100 mm thickness was assumed. The flow resistivity was set as 13,900 Pa s/m^2. The absorption characteristics of z_n and α_0 are presented in Figure 8. The sound absorber has high absorption coefficient greater than 0.9 at frequencies higher than 500 Hz. Therefore, in this problem, sound fields have high SPL difference among the largest sized room and the three soundproof rooms. We applied $v = 1$ m/s for the vibration boundary Γ_v. We placed 18 sound receivers in the room.

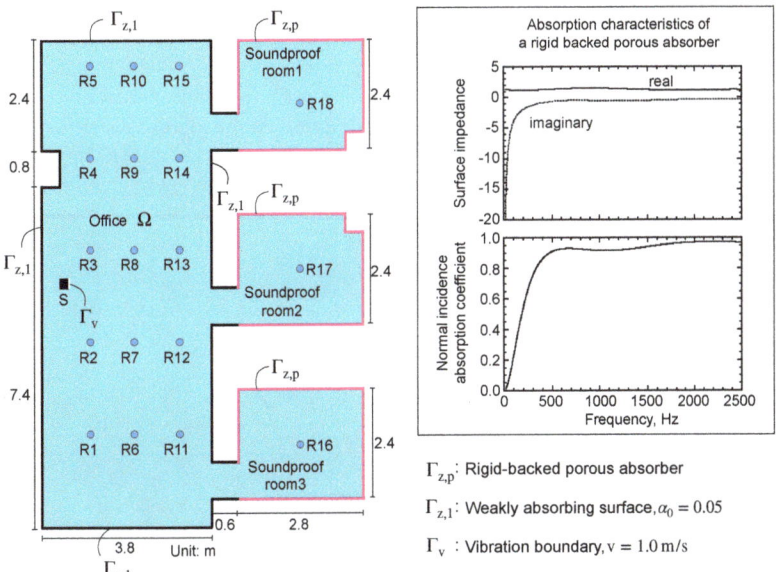

Figure 8. Coupled room model including a vibration boundary, with impedance boundaries of a reflecting surface and a rigid-backed porous absorber: S, R respectively represent the sound source and receivers.

Figure 9 shows a PUFEM mesh discretized using rectangular elements of various lengths in the range of 0.2–0.4 m. The mesh includes three times larger elements than wavelength at the upper-limit frequency. The constant C for plane wave enrichment varies from 5 to 14 with an interval of one. In addition, N_{ele} and N_{node} respectively denote 413 and 493. However, similarly to the earlier numerical example, we used two FE meshes respectively discretized with 0.01 m square elements and 0.005 m square elements to calculate the reference solutions. The 0.01 m mesh was used to calculate frequency responses at 20 Hz to 1500 Hz. For higher frequencies, we used 0.005 m mesh. The DOF of the meshes are, respectively, 612,000 and 2,448,000.

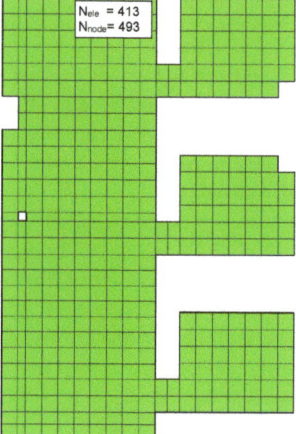

Figure 9. PUFEM mesh for coupled room analysis: The n_w range is the same as Mesh 2 in Figure 3.

3.3.2. Results and Discussion

Figure 10 presents a comparison of RMS errors among results with different values of C. The figure also includes standard FEM results obtained using 0.01 m mesh having 612,000 DOF. The RMS error of PUFEM is reduced at higher frequencies with larger values of C. The PUFEM with $C \geq 6$ were more accurate than the standard FEM results at above 500 Hz band. At the highest 2 kHz band, PUFEM using $C = 13$ shows a smaller RMS value of 1.4 dB than the 7.5 dB in the standard FEM. The magnitude is 1/5.4 of the standard FEM. Here, the change in DOF of PUFEM is the same as that shown in Figure 5; in addition, the DOFs change from 3451 to 26,129 at 20 Hz to 2.5 kHz. The results demonstrate clearly that the PUFEM can also perform well for more complex coupled fields with a small amount of DOF.

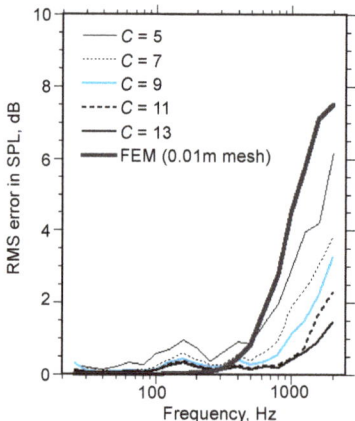

Figure 10. Comparison of RMS errors among results with different values of C: The standard FEM results obtained using 0.01 m mesh are also shown.

Furthermore, Figure 11 depicts SPL distributions at 2 kHz and 2.5 kHz among Ref, PUFEM, and standard FEM, where Ref and the standard FEM used 0.005 m mesh having 2,448,000 DOF. The PUFEM used C = 12 at 2 kHz and C = 13 at 2.5 kHz. The DOFs are, respectively, 21,692 and 26,129. In addition, PUFEM results were shown at the same 2,448,000 nodal points as in Ref and standard FEM. RMS error L_{rms} was calculated with all nodal points. The L_{rms} values for PUFEM and standard FEM are also included in Figure 11. The SPL distributions at 2 kHz and 2.5 kHz of PUFEM agree very well with those of Ref, exhibiting smaller values of L_{rms} than the standard FEM results. For PUFEM results, the value is 1.3 dB at 2 kHz and 1.5 dB at 2.5 kHz. The standard FEM shows large errors of 5.6 dB at 2 kHz and 7.1 dB at 2.5 kHz. We can observe in the standard FEM results by which SPL distributions differ from those of Ref in both frequencies. For example, at 2.5 kHz, SPL values in the four rooms show lower values than Ref. Clearer dips in SPL distributions are apparent in the three soundproof rooms. These results suggest that the DOF reduction in PUFEM reaches at least 1/100 of the standard FEM even for large problems including practical boundary conditions.

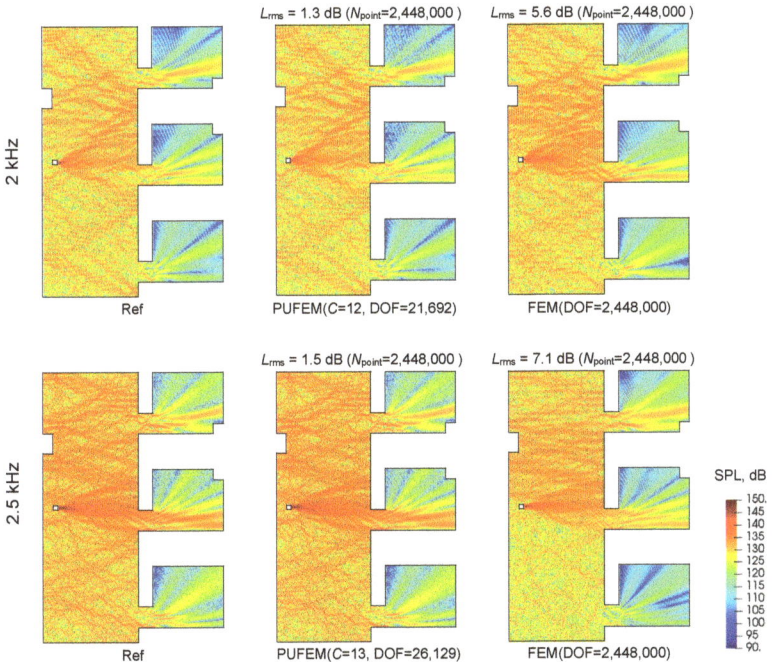

Figure 11. SPL distributions at 2 kHz (**upper**) and 2.5 kHz (**lower**) among Ref (**left**), PUFEM (**center**), and standard FEM (**right**).

However, we must state an important aspect in the use of PUFEM at this stage. In Figure 11, we present results of PUFEM with C = 12 instead of C = 13 at 2 kHz because the result with C = 13 showed noisy SPL distributions, as shown in Figure 12. In the present figure, it is apparent that the solution did not converge to the reference solution in the three soundproof rooms because of the ill-conditioning of the linear system, as described previously. From this result, it can be inferred that there exists a proper number of plane waves for enrichment to obtain reliable results. Finally, in Figure 13, we present comparisons of SPL distributions between Ref and PUFEM with C = 13 at three lower frequencies of 125 Hz, 250 Hz, and 500 Hz, where modal behavior is enhanced because of low diffuseness. The figure also includes RMS error values calculated at 615,020 nodal points. The agreement of SPL distributions between PUFEM and Ref is excellent with RMS errors less than

0.25 dB at all frequencies. The PUFEM results used $C = 13$, but lower values of C can be used at the low frequencies shown in Figure 10.

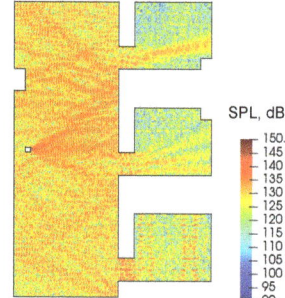

Figure 12. SPL distributions at 2 kHz calculated using PUFEM with $C = 13$.

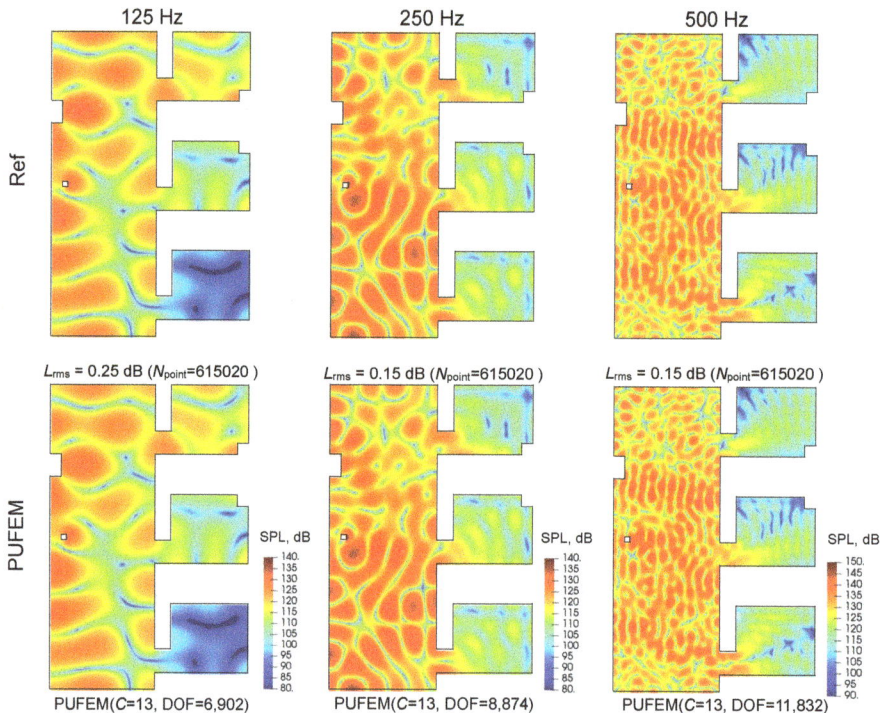

Figure 13. Comparisons of SPL distributions between Ref (**upper**) and PUFEM (**lower**) at 125 Hz (**left**), 250 Hz (**center**) and 500 Hz (**right**).

4. Conclusions

This report described a study of a room acoustics solver based on plane-wave-enriched FEM using low-order quadrilateral elements and local-reaction impedance boundaries. We discussed the potential of plane-wave-enriched FEM as a room acoustic solver to solve the issues on wave-based room acoustic simulations via performance examination on 2D real-scale room acoustic problems with realistic boundary conditions in comparison with the standard second-order accurate FEM. In particular, we examined its accuracy for multi-frequency analysis including fine frequency resolution quantitatively by comparison with reference solutions calculated using a fourth-order accurate FEM. The tested frequency range is 20 Hz to 2.5 kHz with 1 Hz interval. We used two numerical room models of a single room and a coupled room, each with 13–20 times larger interior sound fields than those used in earlier studies, and also including frequency-dependent impedance boundary conditions of sound absorbers commonly used in room acoustic simulations i.e., Helmholtz resonators and porous absorbers. The numerical results clearly revealed advantages of using plane wave enriched finite elements against the standard finite elements in terms of the resulting accuracy, required DOF to obtain the same accurate results, and also the ease of mesh generation. More specifically, the two numerical examples demonstrated that PUFEM can predict a broadband frequency response at low to high frequencies accurately using a single coarse mesh with much less DOF than the standard FEM. The reduction in DOF reaches at least 1/100 of the standard FEM. In the single-room problem, PUFEM produced acceptably accurate results at all frequencies with only 16,271 degrees of freedom, whereas the standard FEM showed unacceptable results because of the inherent larger dispersion error property despite the use of mesh with 1,599,840 DOF. Similarly, in the coupled room problem, the PUFEM results obtained using coarse mesh with 26,129 DOF showed much lower error values than those with 2,448,000 DOF of the standard FEM. Results demonstrate that PUFEM using plane wave enrichment has noteworthy potential for increasing the applicability of wave-based room acoustic simulations. However, numerical results also indicate that an appropriate number of plane waves exists for enrichment. When inappropriate numbers of plane waves are added to nodal points, the PUFEM produces unstable results because of the resulting ill-conditioned linear systems. Therefore, we expect to develop the mode of adding plane waves properly in future studies so that multi-frequency analysis can perform efficiently and robustly. Furthermore, we showed the mesh size effect on PUFEM results in limited conditions. Therefore, our future works will include showing the effect in detail. Application of PUFEM solver to 3D room acoustic simulations will be presented in future reports.

Author Contributions: Conceptualization, T.O.; Investigation, T.O.; Methodology, T.O. and M.S.M.; Project administration, K.S.; Validation, K.S.; Visualization, T.O.; Writing—original draft, T.O.; Writing—review and editing, T.O., M.S.M., and K.S. All authors have read and agreed to the published version of the manuscript.

Funding: This work was supported by JSPS KAKENHI Grant No. 17K14771. The second author is partially supported by JSPS Invitational Fellowship No. L19554.

Conflicts of Interest: The authors declare that they have no conflict of interest related to this report or the study it describes.

References

1. Easwaran, V.; Craggs, A. On further validation and use of the finite element method to room acoustics. *J. Sound Vib.* **1995**, *187*, 195–212. [CrossRef]
2. Easwaran, V.; Craggs, A. Transient response of lightly damped rooms: A finite element approach. *J. Acoust. Soc. Am.* **1996**, *99*, 108–113. [CrossRef]
3. Otsuru, T.; Tomiku, R.; Toyomasu, M.; Takahashi, Y. Finite element sound field analysis of rooms in built environment. In Proceedings of the Eighth International Congress on Acoustics, Hong Kong, China, 2–6 July 2001.
4. Otsuru, T.; Okamoto, N.; Okuzono, T.; Sueyoshi, T. Applications of large-scale finite element sound field analysis onto room acoustics. In Proceedings of the 19th International Congress on Acoustics, Madrid, Spain, 2–7 September 2007.
5. Okamoto, N.; Tomiku, R.; Otsuru, T.; Yasuda, Y. Numerical analysis of large-scale sound fields using iterative methods part II: Application of Krylov subspace methods to finite element analysis. *J. Comput. Acoust.* **2007**, *15*, 473–493. [CrossRef]
6. Yasuda, Y.; Sakamoto, S.; Kosaka, Y.; Sakuma, T.; Okamoto, N.; Oshima, T. Numerical analysis of large-scale sound fields using iterative methods part I: Application of Krylov subspace methods to boundary element analysis. *J. Comput. Acoust.* **2007**, *15*, 449–471. [CrossRef]
7. Botteldooren, D. Finite-difference time-domain simulation of low-frequency room acoustic problems. *J. Acoust. Soc. Am.* **1995**, *98*, 3302–3308. [CrossRef]

8. LoVetri, J.; Mardare, D.; Soulodre, G. Modeling of the seat dip effect using the finite-difference time-domain method. *J. Acoust. Soc. Am.* **1996**, *100*, 2204–2212. [CrossRef]
9. Sakamoto, S.; Nagatomo, H.; Ushiyama, A.; Tachibana, H. Calculation of impulse responses and acoustic parameters in a hall by the finite-difference time-domain method. *Acoust. Sci. Technol.* **2008**, *29*, 256–265. [CrossRef]
10. Kowalczyk, K.; Walstijn, M. Formulation of locally reacting surfaces in FDTD/K-DWM modelling of acoustic spaces. *Acta Acust. United Acta* **2008**, *94*, 891–906. [CrossRef]
11. Okuzono, T.; Otusru, T.; Tomiku, R.; Okamoto, N. Fundamental accuracy of time domain finite element method for sound-field analysis of rooms. *Appl. Acoust.* **2010**, *71*, 940–946. [CrossRef]
12. Mehra, R.; Raghuvanshi, N.; Savioja, L.; Lin, M.C.; Manocha, D. An efficient GPU-based time domain solver for the acoustic wave equation. *Appl. Acoust.* **2012**, *73*, 83–94. [CrossRef]
13. Simonaho, S.P.; Lähivaara, T.; Huttunen, T. Modeling of acoustic wave propagation in time-domain using the discontinuous Galerkin method—A comparison with measurements. *Appl. Acoust.* **2012**, *73*, 173–183. [CrossRef]
14. Okuzono, T.; Otsuru, T.; Tomiku, R.; Okamoto, N. A finite element method using dispersion reduced spline elements for room acoustics simulation. *Appl. Acoust.* **2014**, *79*, 1–8. [CrossRef]
15. Hornikx, M.; Krijnen, T.; Harten, L. The open source pseudospectral time-domain method for acoustic propagation. *Comput. Phys. Commun.* **2016**, *203*, 298–308. [CrossRef]
16. Okuzono, T.; Sakagami, K. A frequency domain finite element solver for acoustic simulations of 3D rooms with microperforated panel absorbers. *Appl. Acoust.* **2018**, *129*, 1–12. [CrossRef]
17. Yoshida, T.; Okuzono, T.; Sakagami, K. Numerically stable explicit time-domain finite element method for room acoustics simulation using an equivalent impedance model. *Noise Control Eng. J.* **2018**, *66*, 176–188. [CrossRef]
18. Okuzono, T.; Shimizu, N.; Sakagami, K. Predicting absorption characteristics of single-leaf permeable membrane absorbers using finite element method in a time domain. *Appl. Acoust.* **2019**, *151*, 172–182. [CrossRef]
19. Rabisse, K.; Ducourneau, J.; Faiz, A.; Trompette, N. Numerical modelling of sound propagation in rooms bounded by walls with rectangular irregularities and frequency-dependent impedance. *J. Sound Vib.* **2019**, *440*, 291–314. [CrossRef]
20. Wang, H.; Sihar, I.; Pagán Muñoz, R.; Hornikx, M. Room acoustics modelling in the time-domain with the nodal discontinuous Galerkin method. *J. Acoust. Soc. Am.* **2019**, *145*, 2650–2663. [CrossRef]
21. Pind, F.; Engsig-Karup, A.; Jeong, C.H.; Hesthaven, J.S.; Mejling, M.S; Strømann-Andersen, J. Time domain room acoustic simulations using the spectral element method. *J. Acoust. Soc. Am.* **2019**, *145*, 3299–3310. [CrossRef]
22. Toyoda, M.; Eto, D. Prediction of microperforated panel absorbers using the finite-difference time-domain method. *Wave Motion* **2019**, *86*, 110–124. [CrossRef]
23. Yasuda, Y.; Saito, K.; Sekine, H. Effects of the convergence tolerance of iterative methods used in the boundary element method on the calculation results of sound fields in rooms. *Appl. Acoust.* **2020**, *157*, 106997. [CrossRef]
24. Hoshi, K.; Hanyu, T.; Okuzono, T.; Sakagami, K.; Yairi, M.; Harada, S.; Takahashi, S.; Ueda, Y. Implementation experiment of a honeycomb-backed MPP sound absorber in a meeting room. *Appl. Acoust.* **2020**, *157*, 107000. [CrossRef]
25. Yasuda, Y.; Ueno, S.; Kadota, M.; Sekine, H. Applicability of locally reacting boundary conditions to porous material layer backed by rigid wall: Wave-based numerical study in non-diffuse sound field with unevenly distributed sound absorbing surfaces. *Appl. Acoust.* **2016**, *113*, 45–57. [CrossRef]
26. Otsuru, T.; Tomiku, R. Basic characteristics and accuracy of acoustic element using spline function in finite element sound field analysis. *Acoust. Sci. Technol.* **2000**, *21*, 87–95. [CrossRef]
27. Okuzono, T.; Sakagami, K.; Otsuru, T. Dispersion-reduced time domain FEM for room acoustics simulation. In Proceedings of the 23rd International Congress on Acoustics, Aachen, Germany, 9–13 September 2019.
28. Okuzono, T.; Otsuru, T.; Tomiku, R.; Okamoto, N. Application of modified integration rule to time-domain finite-element acoustic simulation of rooms. *J. Acoust. Soc. Am.* **2012**, *132*, 804–813. [CrossRef] [PubMed]
29. Melenk, J.M.; Babuška, I. Partition of unity finite element method: Basic theory and applications. *Comput. Methods Appl. Mech. Eng.* **1996**, *139*, 289–314. [CrossRef]
30. Laghrouche, O.; Bettess, P.; Astley, J. Modelling of short wave diffraction problems using approximating systems of plane waves. *Int. J. Numer. Meth. Eng.* **2002**, *54*, 1501–1533. [CrossRef]
31. Laghrouche, O.; Bettess, P.; Perrey-Debain, E.; Trevelyan, J. Wave interpolation finite elements for Helmholtz problems with jumps in the wave speed. *Comput. Methods Appl. Mech. Eng.* **2005**, *194*, 367–381. [CrossRef]

32. Laghrouche, O.; Mohamed, M.S. Locally enriched finite elements for the Helmholtz equation in two dimensions. *Comput. Struct.* **2010**, *88*, 1469–1473. [CrossRef]
33. Mohamed, M.S.; Laghrouche, O. Some numerical aspects of the PUFEM for efficient solution of 2D Helmholtz problems. *Comput. Struct.* **2010**, *88*, 1484–1491. [CrossRef]
34. Mohamed, M.S. Numerical Aspects of the PUFEM for Efficient Solution of Helmholtz Problems. Ph.D. Thesis, Heriot–Watt University, Edinburgh, UK, 2010.
35. Diwan, G.C.; Mohamed, M.S. Pollution studies for high order isogeometric analysis and finite element for acoustic problems. *Comput. Methods Appl. Mech. Eng.* **2019**, *350*, 701–718. [CrossRef]
36. Chazot, J.D.; Nennig, B.; Perrey-Debain, E. Performances of the Partition of Unity Finite Element Method for the analysis of two-dimensional interior sound fields with absorbing materials. *J. Sound Vib.* **2013**, *332*, 1918–1929. [CrossRef]
37. Chazot, J.D.; Perrey-Debain, E. The partition of unity finite element method for the simulation of waves in air and poroelastic media. *J. Acoust. Soc. Am.* **2014**, *135*, 724–733. [CrossRef] [PubMed]
38. Christodoulou, K.; Laghrouche, O.; Mohamed, M.S.; Trevelyan, J. High-order finite elements for the solution of Helmholtz problems. *Comput. Struct.* **2017**, *191*, 129–139. [CrossRef]
39. Banerjee, S.; Sukumar, N. Exact integration scheme for planewave-enriched partition of unity finite element method to solve the Helmholtz problem. *Comput. Methods Appl. Mech. Eng.* **2017**, *317*, 619–648. [CrossRef]
40. Dinachandra, M.; Raju, S. Plane wave enriched Partition of Unity Isogeometric Analysis (PUIGA) for 2D-Helmholtz problems. *Comput. Methods Appl. Mech. Eng.* **2018**, *335*, 380–402. [CrossRef]
41. Craggs, A. A finite element model for rigid porous absorbing materials. *J. Sound Vib.* **1978**, *61*, 101–111. [CrossRef]
42. Craggs, A. Coupling of finite element acoustic absorption models. *J. Sound Vib.* **1979**, *66*, 605–613. [CrossRef]
43. Easwaran, V.; Munjal, M.L. Finite element analysis of wedges used in anechoic chambers. *J. Sound Vib.* **1993**, *160*, 333–350. [CrossRef]
44. Allard, J.F.; Atalla, N. Sound propagation in porous materials having a rigid frame. In *Propagation of Sound in Porous Media: Modeling Sound Absorbing Materials*, 2nd ed.; John Wiley & Sons, Ltd.: Chichester, UK, 2009; pp. 73–109.
45. Allard, J.F.; Atalla, N. Finite element modeling of poroelastic materials. In *Propagation of Sound in Porous Media: Modeling Sound Absorbing Materials*, 2nd ed.; John Wiley & Sons, Ltd.: Chichester, UK, 2009; pp. 309–349.
46. Hiptmair, R.; Moiola, A.; Perugia, I. Survey of Trefftz methods for the Helmholtz equation. In *Building Bridges: Connections and Challenges in Modern Approaches to Numerical Partial Differential Equations*; Springer: Cham, Switzerland, 2016; pp. 237–278.
47. Guddati, M.N.; Yue, B. Modified integration rules for reducing dispersion error in finite element methods. *Comput. Methods Appl. Mech. Eng.* **2004**, *193*, 275–287. [CrossRef]
48. Otsuru, T.; Okuzono, T.; Tomiku, R.; Asniawaty, K.; Okamoto, N. Large-scale finite element sound field analysis of rooms using a practical boundary modeling technique. In Proceedings of the 19th International Congress on Sound and Vibration, Vilnius, Lithuania, 8–12 July 2012.
49. Okuzono, T.; Sakagami, K. A finite-element formulation for room acoustics simulation with microperforated panel sound absorbing structures: Verification with electro-acoustical equivalent circuit theory and wave theory. *Appl. Acoust.* **2015**, *95*, 20–26. [CrossRef]
50. Sakagami, K.; Morimoto, M.; Yairi, M. A note on the effect of vibration of a microperforated panel on its sound absorption characteristics. *Acoust. Sci. Technol.* **2005**, *26*, 204–207. [CrossRef]
51. Allard, J.F.; Atalla, N. Acoustic impedance at normal incidence of fluids. Substitution of a fluid layer for a porous layer. In *Propagation of Sound in Porous Media: Modeling Sound Absorbing Materials*, 2nd ed.; John Wiley & Sons, Ltd.: Chichester, UK, 2009; pp. 15–27.
52. Miki, Y. Acoustical properties of porous materials—Modifications of Delany–Bazley models. *J. Acoust. Soc. Jpn.* **1990**, *11*, 19–24. [CrossRef]

© 2020 by the authors. Licensee MDPI, Basel, Switzerland. This article is an open access article distributed under the terms and conditions of the Creative Commons Attribution (CC BY) license (http://creativecommons.org/licenses/by/4.0/).

Article

Time Domain Room Acoustic Solver with Fourth-Order Explicit FEM Using Modified Time Integration

Takumi Yoshida [1,2,*], Takeshi Okuzono [2] and Kimihiro Sakagami [2]

1. Technical Research Institute, Hazama Ando Corporation, Tsukuba 305-0822, Japan
2. Environmental Acoustics Laboratory, Department of Architecture, Graduate School of Engineering, Kobe University, Kobe 657-8501, Japan; okuzono@port.kobe-u.ac.jp (T.O.); saka@kobe-u.ac.jp (K.S.)
* Correspondence: yoshida.takumi@ad-hzm.co.jp; Tel.: +81-29-858-8811

Received: 16 April 2020; Accepted: 25 May 2020; Published: 28 May 2020

Abstract: This paper presents a proposal of a time domain room acoustic solver using novel fourth-order accurate explicit time domain finite element method (TD-FEM), with demonstration of its applicability for practical room acoustic problems. Although time domain wave acoustic methods have been extremely attractive in recent years as room acoustic design tools, a computationally efficient solver is demanded to reduce their overly large computational costs for practical applications. Earlier, the authors proposed an efficient room acoustic solver using explicit TD-FEM having fourth-order accuracy in both space and time using low-order discretization techniques. Nevertheless, this conventional method only achieves fourth-order accuracy in time when using only square or cubic elements. That achievement markedly impairs the benefits of FEM with geometrical flexibility. As described herein, that difficulty is solved by construction of a specially designed time-integration method for time discretization. The proposed method can use irregularly shaped elements while maintaining fourth-order accuracy in time without additional computational complexity compared to the conventional method. The dispersion and dissipation characteristics of the proposed method are examined respectively both theoretically and numerically. Moreover, the practicality of the method for solving room acoustic problems at kilohertz frequencies is presented via two numerical examples of acoustic simulations in a rectangular sound field including complex sound diffusers and in a complexly shaped concert hall.

Keywords: discretization error; explicit method; finite element method; high order scheme; room acoustic simulations; time domain

1. Introduction

1.1. Study Background

For room acoustic design, it is crucially important to predict accurate impulse responses to provide comfortable acoustic environments necessary for various rooms such as concert halls, offices, and classrooms. Using impulse responses, one can design room acoustics via visualization and auralization of sound fields as well as calculating room acoustic parameters such as reverberation times and the speech transmission index. Computer simulation methods are indispensable tools for room acoustic modeling because they can virtually simulate acoustics in architectural spaces. Furthermore, simulation can facilitate parametric studies more readily than scale model experiments for designing basic room shapes and interior finishes that is, selection of acoustical absorptive or reflective materials [1]. Geometrical acoustics simulation methods have flourished since their original proposition in the 1960s [2,3] to today, as a practical room acoustic design tool [4,5]. Geometrical acoustics methods can be undertaken with low computational costs through simplified approximation of wave phenomena by which wave propagation is modeled as a ray propagation. Because of such approximation, geometrical acoustics methods have less ability to accommodate or reflect wave phenomena at low and mid-range frequencies, although studies for alleviating these difficulties have been conducted, for example, the implementations of diffraction model [6] and scattering coefficients [7]. Nevertheless, recent progress in computer technology has increased the applicability of inherently accurate wave acoustic methods; those circumstances promote its use in practical applications [1]. Wave acoustics methods are classified into frequency domain methods and time domain methods. The latter, time domain methods, can calculate room impulse responses directly. They have attracted more intense interest for application as room acoustic solvers.

In general time domain wave acoustics methods, the wave equation or its first-order forms, that is, the continuity equation and Euler equation, are discretized in both space and time. The discretized equation is solved numerically with appropriate initial and boundary conditions. They are inherently accurate, including all wave phenomena naturally. However, wave acoustics methods demand considerable computational effort, entailing enormous memory requirements and long computational timed, especially for acoustic simulations of actual rooms at kilohertz frequencies because one must control the inherent discretization error, so-called dispersion error, to obtain reliable results. The dispersion error is frequency and directional angle dependent error with respect to sound speed caused by both spatial and temporal discretization. Fundamentally, a sufficiently finer spatial discretization than the wavelength of frequency to be analyzed and the controlling the phase error coming from time discretization must be achieved to reduce the numerical dispersion. However, that requirement engenders large-scale problems that entail many degrees of freedom (DOF) and time steps. Consequently, various efficient time domain methods have been developed to date for practical application of wave acoustic methods.

The finite difference time domain (FDTD) method [8–12] is the most popular and well-used method. The FDTD method models sound propagation in rooms by discretizing the partial differential equation straightforwardly using finite difference approximation in both space and time. The implementation is simple and attractive for computational efficiency, but it often suffers from both dispersion error and approximation error in room shapes coming from staircase approximation. To alleviate these shortcomings, higher-order accurate methods have been proposed based on compact difference method [11], modified equation method [12], and application of unstructured grid based on finite volume modeling [13]. Additionally, some studies using graphics processing units have been presented for solving large concert hall models [14,15].

Other time-domain room acoustic solvers are based on the finite element method framework. They are called the time domain finite element method (TD-FEM). Actually, TD-FEM can accommodate complex geometries naturally. Therefore, it has attracted considerable attention in recent years [16–22]. In TD-FEM, the weak form of a wave equation is spatially discretized using finite elements (FEs). Then the resulting semi-discretized equation is discretized temporally using a time integration method such as Newmark β method [23]. In general, the resulting time marching scheme becomes implicit as opposed to the explicit scheme in FDTD method. It is also adversely affected by dispersion error, but it offers attractive capabilities for room-shape modeling and offers stability in computations. Similar to the FDTD method, two efficient room acoustic TD-FEMs exist. They use low-dispersion FEs, a highly accurate time integration method, and Krylov subspace iterative solvers for large linear systems. In several reports of the relevant literature [17,18], an efficient room acoustic solver with a high-order spline TD-FEM has been used for acoustic simulations in existing rooms such as a multi-purpose hall and a reverberation room. A further improved method has been developed recently for high frequencies [20]. Another solver uses fourth-order accurate linear TD-FEM in both space and time [19]. Its applicability has been demonstrated via acoustic simulations at kilohertz frequencies of a simply shaped concert hall and of reverberation room absorption coefficient measurements [22]. Although the two solvers offer some promising potential for their ability to predict sound fields in actual-sized rooms accurately, they are still time-consuming methods because of their implicit formulation. Therefore, recent studies have included some attempts at developing a room acoustic solver using explicit TD-FEM.

1.2. Room Acoustic Solver Using Explicit TD-FEM and Contributions of the Present Paper

The room acoustic solver using explicit TD-FEM [24–26] is based on simultaneous first-order ordinary differential equations (ODEs) equivalent to a second-order ODE derived by application of the Galerkin method to a wave equation. It achieves fourth-order accuracy in both space and time. It also uses dispersion-reduced FEs as in the implicit TD-FEM described above. The fully explicit and stable scheme is achieved by introducing a matrix lumping into a dissipation term [25]. Earlier study [24] revealed that the explicit TD-FEM has better performance than the fourth-order accurate implicit TD-FEM in terms of computational times to achieve similar accurate results. In subsequent study [26], an extended reaction model for permeable membrane absorbers is implemented to increase its applicability. However, at this stage, the explicit method has a shortcoming by which fourth-order accuracy in time is maintained for the case using only square or cubic FEs. As a result, staircase approximation is introduced into room shape modeling. In general, room acoustic simulations address sound propagation in complex shaped rooms. When using staircase approximation in room shape modeling of complex shaped rooms, sufficiently small size of elements must be used to maintain the approximation error acceptable. That treatment increases computational costs significantly. Therefore, introducing the approximation reduces its attractiveness as a room acoustic solver. In addition, use of higher-order accurate time integration methods are essential to perform acoustic simulations at kilohertz frequencies efficiently.

The purpose of the present study is to overcome shortcomings inherent in designing a special time integration method suitable for explicit TD-FEM. We propose a novel room acoustic solver using fourth-order accurate explicit TD-FEM in both space and time. Then we demonstrate its practicality as a room acoustic design tool. The proposed explicit TD-FEM can fit room boundaries using irregularly shaped FEs. Moreover, it can eliminate use of the staircase approximation in room shape modeling while maintaining fourth-order accuracy in time without additional computational complexity compared to that achieved using the conventional method. The discretization error characteristics of the proposed method are investigated theoretically and numerically to elucidate its basic performance. Additionally, the effectiveness of the proposed method for practical applications is examined using numerical examples.

Notably efficient room acoustic solvers have been produced using the discontinuous Galerkin method [27] and the spectral element method [28]. To reduce dispersion error, they use higher-order elements and Runge–Kutta method, exhibiting its attractive capabilities for room acoustic modeling. As opposed to these higher-order methods, the proposed explicit TD-FEM in the present paper is an attempt to achieve higher order accuracy using low-order linear FEs. An important advantage is that the resulting sparse matrix has narrower bandwidth than those of higher-order methods. Moreover, our designed time-integration method requires less computational complexity than higher-order Runge–Kutta method. Although the present method has a slight dissipation error, the error magnitude is controllable by the time interval used.

The present paper is organized as follows. Section 2 introduces the theory of conventional explicit TD-FEM applied to room acoustic simulations, including detailed explanations of its shortcomings which must be overcome. Section 3 presents the theory of the novel explicit TD-FEM. To reduce both spatial and temporal discretization errors, frequency domain and time domain dispersion analyses are performed respectively. Also, a new time marching scheme is derived with a specially designed time integration method. The section also includes derivation of its stability condition. Section 4 presents both dispersion and dissipation characteristics of the new method both theoretically and numerically. Section 5 demonstrates the performance of the present method via two practical numerical examples, each with acoustic simulations in a rectangular sound field including complex sound diffusers and in a complexly shaped concert hall. Section 6 presents a summary of the contributions of this paper.

2. Room Acoustic Solver Using Conventional Explicit TD-FEM

For reader convenience, this section overviews the room acoustic solver with conventional explicit TD-FEM [24–26], which becomes a basis of constructing new solver.

2.1. Basic Equation and Its Discretization in Space and Time

We consider the following nonhomogeneous wave equation to simulate sound propagation in a closed sound field Ω_f with a boundary Γ.

$$\frac{\partial^2 p}{\partial t^2} - c_0^2 \nabla^2 p = \rho_0 c_0^2 \frac{\partial q}{\partial t}, \tag{1}$$

where p, c_0, ρ_0, q, and t respectively represent the sound pressure, the sound speed, the air density, the added fluid mass per unit volume, and the time. ∇^2 represents the Laplacian of p. Using Green's theorem, the weak form of the nonhomogeneous wave equation is expressed as

$$\int_{\Omega_f} \phi_f \frac{\partial^2 p}{\partial t^2} dV + c_0^2 \int_{\Omega_f} \nabla \phi_f \nabla p \, dV = c_0^2 \int_{\Gamma} \phi_f \frac{\partial p}{\partial n} dA + \rho_0 c_0^2 \int_{\Omega_f} \phi_f \frac{\partial q}{\partial t} dV. \tag{2}$$

Here, ϕ_f denotes the arbitrary weight function. Applying the Galerkin method and incorporating three boundary conditions (a rigid boundary, a vibrating boundary, and an impedance boundary) engender the semi-discretized matrix equation as shown below.

$$M\ddot{p} + c_0^2 K p + c_0 C \dot{p} = f, \tag{3}$$

with

$$M = \sum^{N_e} M_e = \sum^{N_e} (\int_{\Omega_e} N^T N dV),$$

$$K = \sum^{N_e} K_e = \sum^{N_e} (\int_{\Omega_e} \nabla N^T \nabla N dV), \tag{4}$$

$$C = \sum^{N_b} C_e = \sum^{N_b} (\frac{1}{z_n} \int_{\Gamma_b} N^T N dA).$$

Therein, M, K, and C respectively represent the global mass matrix, the global stiffness matrix, and the global dissipation matrix, each composed of their element matrix M_e, K_e and C_e. In the equations, N is the shape function. Here, p and f respectively denote the sound pressure vector and the external force vector. Also, N_e and N_b respectively stand for the number of volume elements Ω_e and the number of boundary elements Γ_b. Symbols $\cdot$ and $\cdot\cdot$ signify first-order and second-order time derivatives. Also, z_n is the surface impedance of the boundary. The discussion presented herein uses an equivalent impedance approach [25] to model the boundary absorption effect. To construct an explicit time marching scheme, we introduce a lumped mass matrix D lumped from M using the row-sum method [29,30] and a vector $v = \dot{p}$. Consequently, the second-order ODE of Equation (3) is transformed into the following simultaneous first-order ODEs.

$$D\dot{p} = Mv, \tag{5}$$

$$D\dot{v} = f - c_0^2 K p - c_0 C \dot{p}. \tag{6}$$

For the numerically stable computation, $\dot{p}$ in Equation (5) is discretized using the first-order accurate forward difference. Also, $\dot{v}$ in Equation (6) is discretized using the first-order accurate backward difference [25]. Consequently, the time marching scheme for the explicit TD-FEM is expressed as

$$p^n = p^{n-1} + \Delta t D^{-1} M v^{n-1}, \tag{7}$$

$$(D + \Delta t c_0 C) v^n = D v^{n-1} + \Delta t (f^n - c_0^2 K p^n). \tag{8}$$

Here, Δt and n respectively denote the time interval and the time step. Equation (8) includes an implicit expression, but it is calculable explicitly by the diagonalization in C with the row-sum method. Two sparse matrix-vector products (MVP) per time step are the main operation of the scheme.

2.2. Fourth-Order Accurate Dispersion Reduced Scheme

Conventional explicit TD-FEM uses dispersion-reducing four-node square or eight-node cubic FEs, respectively, for 2D or 3D analysis to achieve fourth-order accuracy in both space and time. The dispersion-reducing FEs are constructed using modified integration rules [31], which are two points Gauss–Legendre quadrature with modified integration points. In general, element matrices M_e and K_e of four-node quadrilateral FEs are constructed using Gauss–Legendre quadrature with two integration points in each direction as explained below.

$$K_e = \sum_{i=1}^{2} \sum_{j=1}^{2} \nabla N(\alpha_{k,i}, \alpha_{k,j})^T \nabla N(\alpha_{k,i}, \alpha_{k,j}) \det(J), \tag{9}$$

$$M_e = \sum_{i=1}^{2} \sum_{j=1}^{2} N(\alpha_{m,i}, \alpha_{m,j})^T N(\alpha_{m,i}, \alpha_{m,j}) \det(J). \tag{10}$$

Therein, α_k and α_m represent local coordinates of integration points for the computation of K_e and M_e. For this discussion, J represents the Jacobian matrix. For the two points rule, local coordinates of integration points are usually set as $\alpha_k = \alpha_m = \pm\sqrt{1/3}$. However, time-domain dispersion error analysis [31] revealed that the use of modified integration points produces fourth-order accurate scheme instead of a conventional second-order accurate scheme. The modified integration points are given as [31]

$$\alpha_k = \pm\sqrt{\frac{2}{3}}, \quad \alpha_m = \pm\sqrt{\frac{1}{3}(4 - \tau^2)}, \tag{11}$$

where τ represents the Courant number defined as $c_0 \Delta t / h$. Therein, h denotes the length of square FEs. The dispersion error analysis yields a useful expression to evaluate the dispersion error of the fourth-order accurate explicit scheme. The numerical sound speed c^h can be evaluated as

$$c^h \approx c_0 \left(1 - \frac{(kh)^4}{1440}((8 - 10\tau^2 + 2\tau^4) - (19 - 10\tau^2 + 5\tau^4)\cos^2\theta \sin^2\theta)\right), \tag{12}$$

where k denotes the wavenumber and θ represents the propagation direction of a plane wave in a polar coordinate system. This conventional dispersion-reduced explicit TD-FEM is computationally efficient by virtue of its use of linear elements and low-order finite difference approximation in time. However, the scheme can achieve fourth-order accuracy in both space and time for the case using square elements because the modified integration point in α_m requires the length of square elements h. This requirement poses an important hindrance to room acoustic modeling in that the staircase approximation is introduced into the room boundary model. In a later section, we present a means of solving this difficulty.

3. New Room Acoustic Solver Using Explicit TD-FEM with Modified Adams Method

This section presents a description of the novel time domain room acoustic solver using explicit FEM, which is suitable for acoustic simulations in complex shaped rooms. The proposed method uses four-node quadrilateral FEs with modified integration points for spatial discretization and a specially designed higher-order time integration method for time discretization. First, frequency domain dispersion error analysis is applied to derive modified integration points that can maintain fourth-order accuracy in space. Then, a fourth-order accurate time integration method suitable for the present room acoustic solver is designed using a time domain dispersion error analysis. Additionally, we present the stability condition of this new method.

3.1. Frequency Domain Dispersion Analysis

We consider the following frequency domain expressions of the semi-discretized matrix equations, Equations (5) and (6), assuming the time factor of $e^{i\omega t}$.

$$i\omega \mathbf{D}\mathbf{p} = \mathbf{M}\mathbf{v}, \tag{13}$$

$$i\omega \mathbf{D}\mathbf{v} = \mathbf{f} - c_0^2 \mathbf{K}\mathbf{p} - i\omega c_0 \mathbf{C}\mathbf{p}. \tag{14}$$

Therein, i and ω respectively represent the imaginary unit and the angular frequency. For dispersion error analysis, we assume an ideal condition, which is plane wave propagation in a free field. Figure 1 shows plane wave propagation in discretized free field using square FEs with element length of h. The plane wave in polar coordinate system is expressed as

$$p_{x,y} = e^{i(kx\cos\theta + ky\sin\theta)}. \tag{15}$$

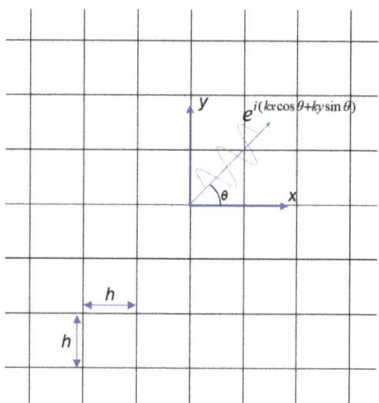

Figure 1. Two-dimensional plane wave propagation in a free field under a polar coordinate system discretized by square finite elements (FEs) with element size of h for dispersion analysis.

To realize the ideal condition in dispersion error analysis, we remove the source and the dissipation terms from Equations (13) and (14). Then reconstructing the two equations in terms of sound pressure p engenders the following expression as

$$(c_0^2 \mathbf{D}^{-1}\mathbf{M}\mathbf{D}^{-1}\mathbf{K} - \omega^2 \mathbf{I})\mathbf{p} = 0, \tag{16}$$

where $\mathbf{I}$ represents a unit matrix. The first term in Equation (16) can be rewritten as

$$\mathbf{D}^{-1}\mathbf{M}\mathbf{D}^{-1}\mathbf{K}\mathbf{p} = A p_{x,y}, \tag{17}$$

with

$$A = -\frac{1}{2h^2}(\alpha_m^2(C_x - 1) - C_x - 1)(\alpha_m^2(C_y - 1) - C_y - 1)(1 + \alpha_k^2(C_x - 1)(C_y - 1) - C_xC_y),$$
$$C_x = \cos(k^h h\cos\theta), \quad C_y = \cos(k^h h\sin\theta), \tag{18}$$

where k^h represents the numerical wavenumber. More detailed calculation procedures can be found in one report of the literature [31] for 2D case and in another [24] for the 3D case. The substitution of Equation (17) into Equation (16) leads to a dispersion relation as

$$c_0 = c^h k^h \sqrt{\frac{1}{A}}. \tag{19}$$

Taking the Taylor expansion with respect to k^h, the numerical sound speed is evaluated as

$$c^h \approx c_0(1 - \frac{(kh)^2}{24}((4 - 3\alpha_m^2) - 2(3\alpha_k^2 - 2)\cos^2\theta\sin^2\theta)). \tag{20}$$

The use of the following integration points clearly eliminates the second-order dispersion error term in Equation (20).

$$\alpha_k = \pm\sqrt{\frac{2}{3}}, \quad \alpha_m = \pm\sqrt{\frac{4}{3}} \tag{21}$$

With Equation (21), the fourth-order accuracy in space is achieved as shown below.

$$c^h \approx c_0(1 - \frac{(kh)^4}{1440}(8 - 19\cos^2\theta\sin^2\theta)) \tag{22}$$

3.2. Designing a Higher-Order Time Integration Method

3.2.1. Linear Multi-Step Method

A linear multi-step method is a numerical time integrator in which a time marching of function Y is expressed as

$$Y^n = \sum_{j=1}^{l}(a_j Y^{n-j} + b_j F^{n-j}\Delta t) + b_0 F^n \Delta t, \tag{23}$$

where F is the function representing time gradient of Y. Also, α_j and β_j represent j-th weight coefficients. In the linear multi-step method, high accuracy can be realized without an increase of the number of MVP per time step by storing past values of F. By contrast, single-step methods such as Runge–Kutta method cannot avoid an increase of the number of MVP per time step to achieve high accuracy. Adams methods are the most popular linear multi-step technique setting $a_1 = 1$ and $a_j = 0$ ($j \neq 1$) in Equation (23). Adams methods are classified into explicit Adams–Bashforth method with $b_0 = 0$ and implicit Adams–Moulton method with $b_0 \neq 0$. This paper uses Adams–Bashforth methods for time integration of p, thereby addressing $D^{-1}Mv$ as a time gradient of p as shown below.

$$p^n = p^{n-1} + \sum_{j=1}^{l}(b_j D^{-1}Mv^{n-j}\Delta t). \tag{24}$$

In Adams–Bashforth methods, the order of accuracy, which can be achieved, corresponds to the number of required time steps. To achieve the fourth-order accuracy, $l \geq 4$ is demanded. Furthermore, the following condition must be satisfied for stable computation.

$$\sum_{j=1}^{l} b_j = 1. \tag{25}$$

In subsequent sections, we use Equation (8) for the time marching of v and the modified integration points in Equation (21).

3.2.2. Conventional Fourth-Order Adams–Bashforth Method

In general, the weight coefficients of fourth-order Adams–Bashforth method are given as $l = 4$, $b_1 = 55/24$, $b_2 = -59/24$, $b_3 = 37/24$, and $b_4 = -9/24$ [32]. With the coefficients, Equation (24) is rewritten as

$$p^n = p^{n-1} + \frac{\Delta t}{24} D^{-1}(55Mv^{n-1} - 59Mv^{n-2} + 37Mv^{n-3} - 9Mv^{n-4}). \tag{26}$$

In this case, the time marching scheme of explicit TD-FEM comprises Equations (26) and (8). However, a time-domain dispersion error analysis reveals that the simple use of general weight coefficients do not keep the fourth-order accuracy in time as below.

$$c^h \approx c_0(1 - i\frac{\omega \Delta t}{4} + \frac{5(\omega \Delta t)^2}{96} - i\frac{(\omega \Delta t)^3}{128} - \frac{(kh)^4}{1440}(\frac{16143}{64}\tau^4 + 8 - 19\cos^2\theta \sin^2\theta)). \tag{27}$$

Equation (27) shows that the resulting time marching scheme has fourth-order accuracy in space and first-order accuracy in time. In addition, odd-order terms of dispersion error include an imaginary number. These results suggest that the resulting scheme has low accuracy and that it is highly dissipative. They also suggest the necessity of designing appropriate weight coefficients to increase the accuracy in time of explicit TD-FEM.

3.2.3. Modified Fourth-Order Adams–Bashforth Method

To achieve the fourth-order accuracy in the resulting time marching scheme, we design appropriate weight coefficients of Adams–Bashforth method using time domain dispersion error analysis. With $l = 4$, the linear multi-step form without the source and the dissipation terms for dispersion error analysis is expressed as

$$(p^{n+1} - 2p^n + p^{n-1}) + \Delta t^2 c_0^2 D^{-1} M D^{-1} K(b_1 p^n + b_2 p^{n-1} + b_3 p^{n-2} + b_4 p^{n-3}) = 0. \tag{28}$$

Here, the same two-dimensional free field used in frequency domain dispersion analysis (Section 3.1) is assumed, but the plane wave in time domain is defined as

$$p_{x,y}^n = e^{i(kx\cos\theta + ky\sin\theta - \omega n \Delta t)}. \tag{29}$$

Using Equation (29), Equation (28) is transformed into

$$2(\cos\omega\Delta t - 1)p_{x,y}^n + \Delta t^2 c_0^2 A(b_1 + b_2 e^{i\omega\Delta t} + b_3 e^{2i\omega\Delta t} + b_4 e^{3i\omega\Delta t})p_{x,y}^n = 0. \tag{30}$$

From Equation (30), the dispersion relation is represented as

$$c_0 = \sqrt{\frac{(1 - \cos c^h k^h \Delta t)}{A\Delta t^2 (b_1 + b_2 e^{ic^h k^h \Delta t} + b_3 e^{2ic^h k^h \Delta t} + b_4 e^{3ic^h k^h \Delta t})}}. \tag{31}$$

By taking Taylor expansion of k^h in Equation (31), c^h is evaluated as

$$c^h \approx c_0(1 + i\frac{(kc_0\Delta t)}{2}X_1 + \frac{(kc_0\Delta t)^2}{24}X_2 - i\frac{(kc_0\Delta t)^3}{48}X_3 + \mathcal{O}(k^4)), \tag{32}$$

with

$$\begin{aligned}
X_1 &= b_2 + 2b_3 + 3b_4, \\
X_2 &= (b_1 - 2b_2)^2 - 22b_1b_3 + 8b_2b_3 + 13b_3^2 - 52b_1b_4 - 4b_2b_4 + 32b_3b_4 + 28b_4^2, \\
X_3 &= b_1^2(5b_2 + 34b_3 + 111b_4) + b_2^2(-8b_1 + 2b_2 + 8b_3 + 22b_4) + b_3^2(-76b_1 + b_2 + 10b_3 + 35b_4) \\
&\quad + b_4^2(-264b_1 - 70b_2 + 40b_3 + 30b_4) - 30b_1b_2b_3 - 24b_2b_3b_4 - 250b_3b_4b_1 - 16b_4b_1b_2.
\end{aligned} \tag{33}$$

By eliminating the first-order to the third-order dispersion error terms in Equation (32) while satisfying Equation (25), we obtain the following modified weight coefficients as

$$b_1 = 14/12, \quad b_2 = -5/12, \quad b_3 = 4/12, \quad b_4 = -1/12. \tag{34}$$

With the coefficients in Equation (34), c^h is defined as

$$c^h \approx c_0(1 - \frac{(kh)^4}{1440}(57\tau^4 + 8 - 19\cos^2\theta\sin^2\theta) - i\frac{(\omega\Delta t)^5}{24}). \tag{35}$$

Equation (35) shows clearly that by using the designed Adams–Bashforth method with the modified weight coefficients of Equation (34), the resulting time marching scheme has fourth-order accuracy in both space and time. In addition, dispersion error analysis shows that the resulting scheme includes the fifth-order dissipation error term, but its magnitude is small; moreover, it is controllable in a simple manner. The effect of dissipation error is evaluated theoretically and numerically in Section 4, including the presentation of a simple control method. With the modified weight coefficients of Equation (34) the complete expression of time marching scheme of sound pressure in Equation (24) is

$$p^n = p^{n-1} + \frac{\Delta t}{12}D^{-1}(14Mv^{n-1} - 5Mv^{n-2} + 4Mv^{n-3} - Mv^{n-4}). \tag{36}$$

The proposed explicit TD-FEM using the modified Adams method comprises Equations (36) and (8); it also uses the modified integration points of Equation (21) in element matrices calculation.

3.3. Stability Analysis

The stability condition of proposed explicit TD-FEM using the modified Adams method is derived using Von Neumann's stability analysis [33]. Here, we assume the use of square elements. By introducing a time marching amplifier B, Equation (28) transforms into

$$[(B - 2 + \frac{1}{B}) + \frac{\Delta t^2 c_0^2 A}{12}(14 - 5\frac{1}{B} + 4\frac{1}{B^2} - \frac{1}{B^3})]p^n = 0. \tag{37}$$

For stable computation, we must fulfill $|B| \leq 1$ regarding the plane wave propagation in all directions at arbitrary frequencies. This result engenders $0 \leq A \leq 2$. Here, $|C_x| \leq 1$ and $|C_y| \leq 1$ are clear because of their definition. Moreover, for the case with square elements using integration points of Equation (21), A becomes a monotonically decreasing function with respect to C_x and C_y, taking a minimum value of 0 with $C_x = C_y = 1$. By substituting $C_x = C_y = -1$, the stability limit in time interval, Δt_{limit}, is assessed as

$$\Delta t_{\text{limit}} = \frac{0.459279h}{c_0}. \tag{38}$$

Derivation of the stability condition for the case with irregular shaped FEs is considerably difficult because A becomes a complicated function. It remains as a task for future work.

4. Discretization Error Characteristics

Dispersion and dissipation characteristics of explicit TD-FEM using modified Adams method were investigated both theoretically and numerically. As presented in Equation (35), the numerical sound speed of the proposed scheme becomes a complex number, including a dissipation effect as in the sound propagation in porous sound absorbing materials. Therefore, dispersion error and dissipation error can be evaluated separately by calculating the phase velocity c_p and an attenuation constant C_α from a propagation constant γ based on the analogy with sound absorber modeling. From Equation (35), γ, c_p and C_α can be expressed as

$$\gamma = i\omega/c^h = \frac{i\omega}{c_0(1 - \frac{(kh)^4}{1440}(57\tau^4 + 8 - 19\cos^2\theta\sin^2\theta) - i\frac{(\omega\Delta t)^5}{24})}$$

$$= i\omega \frac{1 - \frac{(kh)^4}{1440}(57\tau^4 + 8 - 19\cos^2\theta\sin^2\theta) + i\frac{(\omega\Delta t)^5}{24}}{c_0((1 - \frac{(kh)^4}{1440}(57\tau^4 + 8 - 19\cos^2\theta\sin^2\theta))^2 + \frac{(\omega\Delta t)^{10}}{576})} \quad (39)$$

$$= i\omega\frac{1}{c_p} - C_\alpha,$$

with

$$c_p = c_0\frac{(1 - \frac{(kh)^4}{1440}(57\tau^4 + 8 - 19\cos^2\theta\sin^2\theta))^2 + \frac{(\omega\Delta t)^{10}}{576}}{1 - \frac{(kh)^4}{1440}(57\tau^4 + 8 - 19\cos^2\theta\sin^2\theta)}, \quad (40)$$

$$C_\alpha = k\frac{\frac{(\omega\Delta t)^5}{24}}{1(-\frac{(kh)^4}{1440}(57\tau^4 + 8 - 19\cos^2\theta\sin^2\theta))^2 + \frac{(\omega\Delta t)^{10}}{576}}. \quad (41)$$

The unit of attenuation constant is expressed as Np/m: Np/m is transformed into dB/m with multiplication by 20/Ln10.

4.1. Theoretical Dispersion Error Characteristics

With the Equation (40), the dispersion error in proposed explicit TD-FEM is evaluated theoretically as the relative error from the exact sound speed c_0. The relative error $e_{\text{dispersion}}$ is defined as shown below.

$$e_{\text{dispersion}} = \left|\frac{c_0 - c_p}{c_0}\right| \times 100 \, [\%] \quad (42)$$

The dispersion error is well known to show anisotropic behavior in terms of the direction of sound wave propagation in multidimensional analyses. We first evaluate anisotropic characteristics in the dispersion error of the proposed method. We calculated the dispersion error using the parameters of $c_0 = 343.7$ m/s, $k = 45.7$, $h = 0.02$, and two time interval values $\Delta t = \Delta t_{\text{limit}}$ and $0.01\Delta t_{\text{limit}}$. Under these conditions, the spatial resolution, in terms of points per wavelength (PPW), corresponds to 6.87. Figure 2a presents the dispersion error as a function of sound propagation directions θ for the two time interval settings. The result indicates that dispersion errors are symmetric with diagonal direction ($\theta = \pi/4$) as a center. They take a maximum value at the axial directions ($\theta = 0, \pi/2$). Furthermore, the result shows the error values as maximal at the critical time interval. Actually, the result becomes lower with a smaller time interval. We present the convergence of the dispersion error with respect to the spatial resolution PPW in Figure 2b, where the dispersion error is an averaged value in terms of propagation directions. Here, the critical time interval value Δt_{limit} was used. The dispersion error can be found to decrease with fourth-order convergence in terms of spatial resolution. To maintain the error magnitude within 1% or 0.5%, the proposed explicit TD-FEM requires spatial resolution of 5.45 or 6.48 PPW.

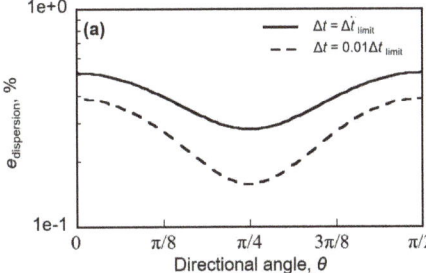 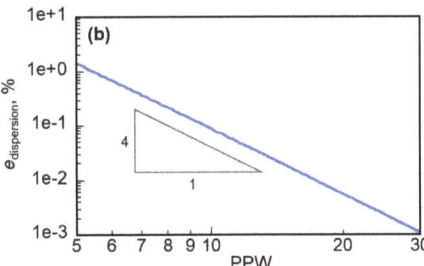

Figure 2. Dispersion characteristics of explicit time domain finite element method (TD-FEM) using modified Adams method: (**a**) anisotropic characteristics in the dispersion error in terms of sound propagation directions with two time interval values of $\Delta t = \Delta t_{\text{limit}}$ and $\Delta t = 0.01 \Delta t_{\text{limit}}$ and (**b**) convergence of dispersion error relative to spatial resolution points per wavelength (PPW).

4.2. Numerical Dispersion Error Characteristics

The numerical performance of explicit TD-FEM using modified Adams method is tested using a numerical simulation of circular wave propagation in a 2D free field. We calculated waveforms at two receiving points 2 m distant from the sound source in the axial and diagonal directions. The modulated Gaussian pulse [10,19] of the upper-limit frequency of 2.5 kHz was used as a source signal. The upper-limit frequency is a frequency with -3 dB gain in the frequency spectrum. The waveform and its frequency characteristics of the sound source signal are presented in Figure 3. We used three FE meshes (Mesh-1, -2 and -3) discretized with square elements of different sizes. Table 1 presents the element sizes and the number of PPW for the three meshes. Mesh-1 has lower spatial resolution than the well known rule of thumb for linear elements: 10 PPW [34]. Mesh-2 and Mesh-3 are, respectively, two and four times finer meshes. The critical time intervals were used with each mesh. The computed waveforms were compared with the reference solution calculated using fourth-order accurate implicit TDFEM [19,31] in both space and time with sufficiently fine mesh of Mesh-3. Figure 4 presents comparisons of the computed waveforms with the reference solution for the three meshes, where the upper and lower panels respectively show the results at the receiver in axial direction and in diagonal direction. The results demonstrate that the proposed method shows considerably good agreement with the reference solutions even when using the coarsest mesh. Slight wave fluctuation can be found in only the result of Mesh-1 at the axial direction, where the proposed method has the maximum error, as shown in Figure 2a. The theoretical results in Section 4.1 show that the results in Mesh-1 include approximately 0.5% of dispersion. Conventional TD-FEM using linear elements cannot achieve this level of accuracy with a mesh that does not fulfill the rule of thumb. The proposed method also uses low-order FEs, but it can produce much better results even in such a condition, as shown in this section.

Table 1. Details of three FE meshes (Mesh-1, -2 and -3) for convergence test: h and PPW respectively denote the length of square FE and the number of points per wavelength.

Mesh	h, m	PPW
Mesh-1	0.02	6.87
Mesh-2	0.01	13.75
Mesh-3	0.005	27.45

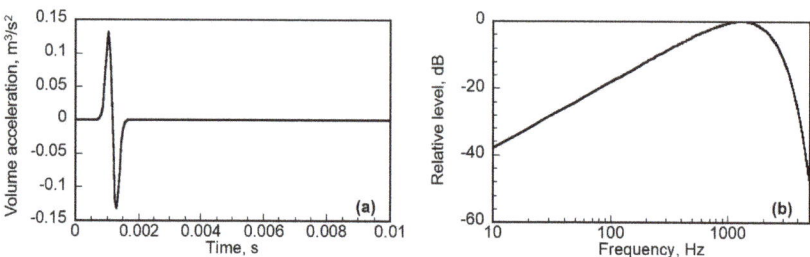

Figure 3. Modulated Gaussian pulse with upper-limit frequency of 2.5 kHz: (**a**) waveform and (**b**) frequency characteristics.

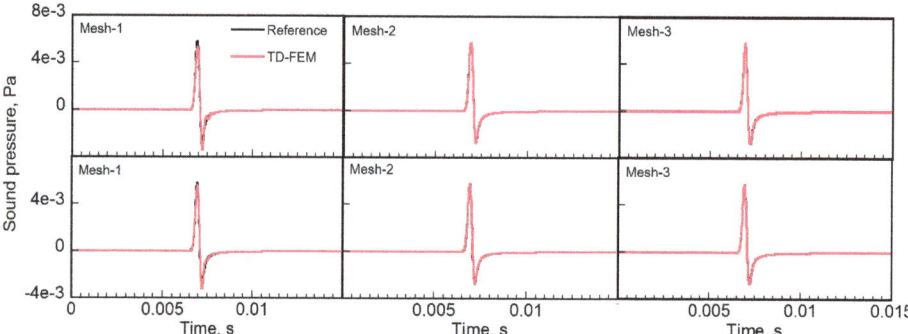

Figure 4. Waveforms at the receivers in the axial direction (upper row) and the diagonal direction (lower row) between the reference solution and the proposed method for three FE meshes (Mesh-1, -2 and -3) listed in Table 1.

4.3. Theoretical Dissipation Error Characteristics

The dissipation error in the proposed explicit TD-FEM is evaluated theoretically using Equation (41). As an example of the evaluation, Figure 5a shows whether or not the dissipation error has anisotropic behavior in terms of the direction of sound wave propagation. It presents results at 2 kHz and 3 kHz with the following settings of $c_0 = 340$ m/s, $h = 0.0125$ m, and $\Delta t = 1/59{,}224$ s. The results indicate that the dissipation is isotropic in terms of wave propagation direction; higher frequencies include larger energy dissipation. Then, Figure 5b shows the convergence of dissipation error with respect to the time interval. The error was calculated at 3 kHz for plane wave propagation at an axial direction with time length of 0.1 s. Two element sizes of 0.0125 m and 0.00625 m were tested. The results revealed that the dissipation error is dependent only on the time interval. Also, the use of smaller time intervals produces lower dissipation errors. These findings suggest that the proposed method can control the dissipation error easily, merely by adjustment of the time interval using Equation (41).

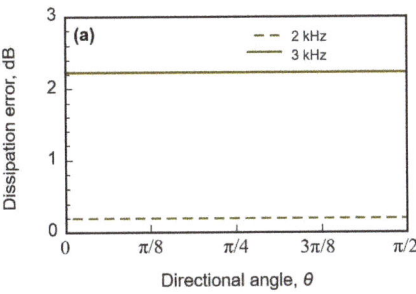 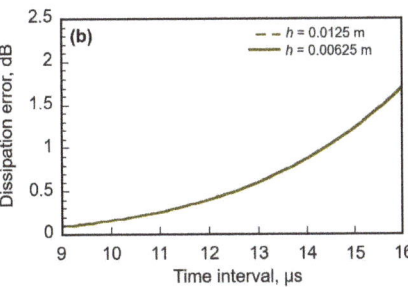

Figure 5. Theoretical dissipation error characteristics of the proposed method: (a) anisotropic characteristics in terms of sound propagation directions at 2 kHz and 3 kHz and (b) convergence of dissipation error relative to time resolution.

4.4. Numerical Dissipation Characteristics

To assess the dissipation error control method using Equation (41), we performed a numerical experiments, which is plane wave propagation in a long duct, as shown in Figure 6. The duct has 400 m length with 0.05 m width. We calculated the plane wave propagation up to 1.0 s with FE mesh of $h = 0.0125$ m and $\Delta t = 1/59{,}224$ s. The sound speed was set as 340 m/s. The plane wave incidence with the waveform of Gaussian pulse was considered at the tube inlet. Waveforms were calculated at receivers located at $x = 10$–330 m and were converted into frequency responses via discrete Fourier transformation. In this examination, to avoid numerical error occurring near the source, numerical dissipation error after x m propagation, $e_n(x)$, was evaluated as

$$e_n(x) = L_n(x) - L_n(1), \tag{43}$$

where $L_n(x)$ represents the numerical sound pressure level at point x m apart from the source. Then the numerically calculated error $e_n(x)$ was compared with the theoretically evaluated dissipation error by Equation (41) as shown below.

$$e_{\text{reference}}(x) = C_\alpha - C_\alpha x \tag{44}$$

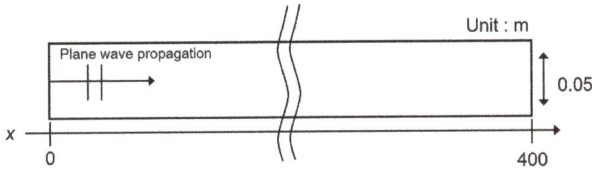

Figure 6. Long numerical duct model with 400 m length.

Figure 7a,b respectively present comparisons of dissipation error between $e_n(x)$ and $e_{\text{reference}}(x)$ and absolute errors between $e_n(x)$ and $e_{\text{reference}}(x)$. The results demonstrate that theoretically estimated values have good agreement with the numerical values. As in Figure 7b the discrepancy increases at larger propagation distances and at higher frequencies. This increase is attributable to the contribution of truncated higher order terms in c^h on theoretical analysis. However, the theoretical estimation gives acceptable accuracy with maximum absolute error below 0.5 dB. As the results above showed, one can assess a recommended time interval value easily using Equation (41) in advance.

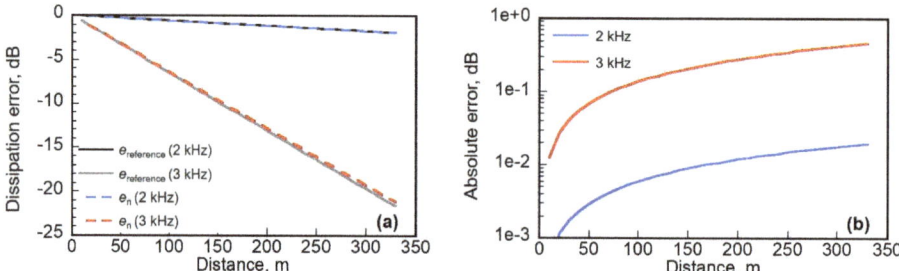

Figure 7. Theoretical and numerical dissipation errors: (**a**) $e_{\text{reference}}$ vs. e_n at 2 kHz and 3 kHz and (**b**) absolute errors from $e_{\text{reference}}$ at 2 kHz and 3 kHz.

5. Numerical Experiments with Practical Sound Fields

We demonstrate the performance of the proposed explicit TD-FEM using a modified Adams method via two room acoustic problems at kilohertz frequencies in a large rectangular room including complicated sound diffusers, and in a concert hall with two conditions. The two sound fields have no analytical solution. Therefore, we used a reference solution calculated using well-developed fourth-order accurate implicit TD-FEM [19] to assess the performance. In addition, the performance is shown in comparison to a standard second-order accurate implicit TD-FEM using a Newmark β method.

5.1. Standard Implicit TD-FEM and Dispersion Reducing Implicit TD-FEM

Standard implicit TD-FEM solves the second-order ODE of Equation (3) with a time integration method called constant averaged acceleration method: CAA. Actually, CAA is known to be an unconditionally stable Newmark method with parameter $\beta = 1/4$. With the space discretization using linear quadrilateral FEs, the resulting implicit time marching scheme has second-order accuracy in both space and time. The implicit time marching scheme is expressed as

$$(M + \frac{c_0^2 \Delta t^2}{4} K + \frac{c_0 \Delta t}{2} C) \ddot{p}^{n+1} = f^{n+1} - c_0 C P_1 - c_0^2 K P_2, \tag{45}$$

$$p^{n+1} = p^n + \Delta t \dot{p}^n + \frac{\Delta t^2}{4} \ddot{p}^n + \frac{\Delta t^2}{4} \ddot{p}^{n+1}, \tag{46}$$

$$\dot{p}^{n+1} = \dot{p}^n + \frac{\Delta t}{2} \ddot{p}^n + \frac{\Delta t^2}{2} \ddot{p}^{n+1}, \tag{47}$$

where

$$\begin{aligned} P_1 &= \dot{p}^n + \frac{\Delta t}{2} \ddot{p}^n, \\ P_2 &= p^n + \Delta t \dot{p}^n + \frac{\Delta t^2}{2} \ddot{p}^n. \end{aligned} \tag{48}$$

A Conjugate Gradient (CG) iterative solver is useful to solve a linear system of equations of Equation (45) easily with convergence tolerance of 10^{-4}. The simplest diagonal scaling is used as a preconditioning technique to facilitate the convergence of an iterative solver. The main operation of standard implicit TD-FEM at each time step comprises sparse MVPs at each iteration process of CG solver and an additional two MVPs. Fourth-order accurate implicit TD-FEM [19,31] for calculating the reference solution also solves the second-order ODE of Equation (3). However, it uses linear quadrilateral FEs with modified integration rules and a highly accurate Newmark method called Fox–Goodwin method. The main operation is the same as standard implicit TD-FEM, but the convergence of iterative solver becomes much better because of the additional effect of modified integration rules. The theoretically estimated dispersion error presented in earlier reports [19,31] shows that this dispersion-reducing implicit TD-FEM has a lower error magnitude than the proposed explicit TD-FEM.

5.2. Sound Propagation Problem in a Rectangular Room Including Acoustic Diffusers

Figure 8a shows the analyzed rectangular room of 10.2 m × 10.8 m surrounded by rigid boundaries with a source point and 27 receivers. The acoustic diffuser, which includes eight periods with a period of 1.2 m, is installed periodically in front of a wall with air spaces. Figure 8b presents details of diffusers. This room cannot be modeled with staircase approximation unless one uses very small rectangular elements because the diffusers comprise rigid reflectors of 0.25 m thickness inclined at various angles against the back wall. However, the proposed explicit TD-FEM can fit the inclined boundaries with fewer elements because of discretization using irregularly shaped FEs. We discretized this room using the irregular FEs with 0.017 m maximum edge length, which corresponds to 6.74 PPW at 3 kHz. The discretization result around the diffusers is shown in Figure 8c. The degrees of freedom (DOF) in this problem are 552,054. As a sound source signal, the Gaussian pulse with the upper-limit frequency of 3 kHz was used. The sound pressure waveforms were calculated at 27 receivers up to 0.1 s with the time interval of 1/120,000 s, which was determined from Equation (38) with the minimum edge length of used mesh. For the standard implicit TD-FEM, the same mesh and time interval were applied. Regarding the measurement accuracy, we use the relative error in sound pressure at each receiver between the reference solution and numerical solution as

$$e_{relative} = \sqrt{\frac{1}{N_{step}} \frac{\sum_{i=1}^{N_{step}} (p_{reference}(i) - p_{FEM}(i))^2}{\sum_{i=1}^{N_{step}} (p_{reference}(i))^2}}, \quad (49)$$

where N_{step} represents the total number of computed time steps. Also, $p_{reference}(i)$ and $p_{FEM}(i)$ respectively denote the reference solution and numerical sound pressure at i-th time step. Regarding the calculation of reference solution, sufficiently fine mesh of 15.3 PPW at 3 kHz was used.

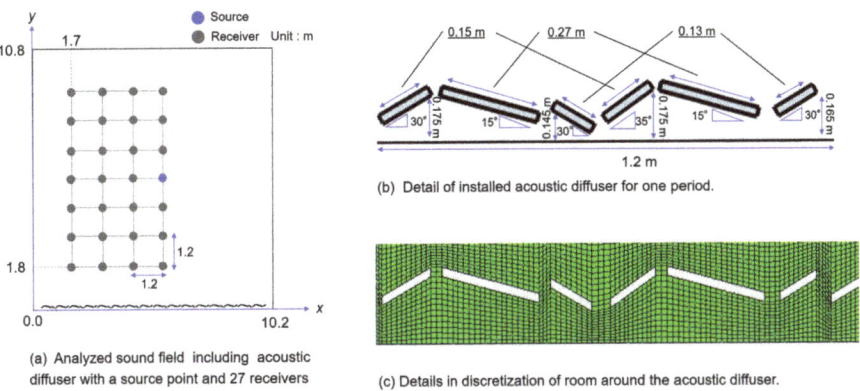

(a) Analyzed sound field including acoustic diffuser with a source point and 27 receivers located on a grid of 1.2 m × 1.2 m.

(b) Detail of installed acoustic diffuser for one period.

(c) Details in discretization of room around the acoustic diffuser.

Figure 8. (**a**) Analyzed sound field including acoustic diffuser with a source point and 27 receivers located on a grid of 1.2 m × 1.2 m. (**b**) Details of installed acoustic diffuser for one period. (**c**) Details of discretization of the room around the acoustic diffuser.

5.2.1. Results and Discussion

First, a comparison of sound propagation among the reference solution, the proposed explicit TD-FEM, and the standard implicit TD-FEM are portrayed in Figure 9, where the results at t = 16, 32, and 64 ms are presented. The sound propagation properties of proposed explicit TD-FEM agree very well with the reference solution. It shows an isotropic and less dispersive sound propagation.

In contrast, the standard implicit TD-FEM shows a dispersive and an anisotropic sound propagation, where the sound speed in axial directions is faster than those in oblique directions. This result in sound interference occurs at an earlier time, as in the result at 32 ms. Additionally, the wavefront becomes indistinct at later than 64 ms.

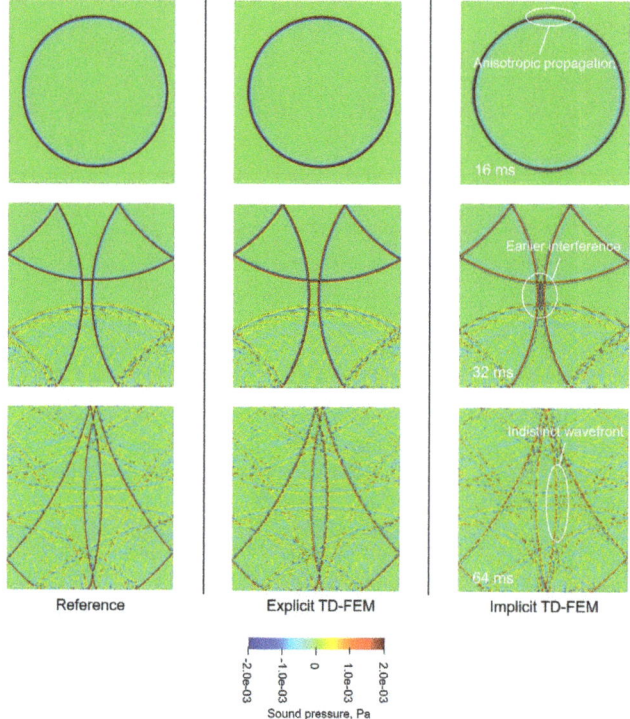

Figure 9. Comparison of sound propagations at t = 16, 32, and 64 ms among (a) Reference, (b) the proposed explicit TD-FEM and (c) the standard implicit TD-FEM.

Figure 10 presents comparisons of sound pressure waveforms at a receiver (x, y) = (5.3, 3.0) among the reference solution, the proposed explicit TD-FEM, and the standard implicit TD-FEM. The proposed explicit TD-FEM shows much better agreement with the reference solution, but it includes slight difference in amplitude. This agreement confirms that the proposed explicit TD-FEM has superior performance for dealing with discretization of sound fields including complex geometries. However, the standard implicit TD-FEM shows marked discrepancy from the reference solution because of the large dispersion error. As described, the effect appears as an increase of sound speed. The effects are visible even in the direct sound. Moreover, they accumulate with time. Additionally, one can observe that the standard implicit TD-FEM cannot capture reflected waves with high amplitude accurately at around t = 0.06 and 0.07 s. More quantitatively, regarding spatial averaged relative errors, the proposed explicit TD-FEM has one-third lower relative error of 0.316% compared to the error 1.031% in standard implicit TD-FEM. In addition, the proposed explicit TD-FEM has other benefits for computational time. The proposed method can compute 4.55 times faster with the total number of MVP of 24,000 in the proposed method and 109,109 in the standard method.

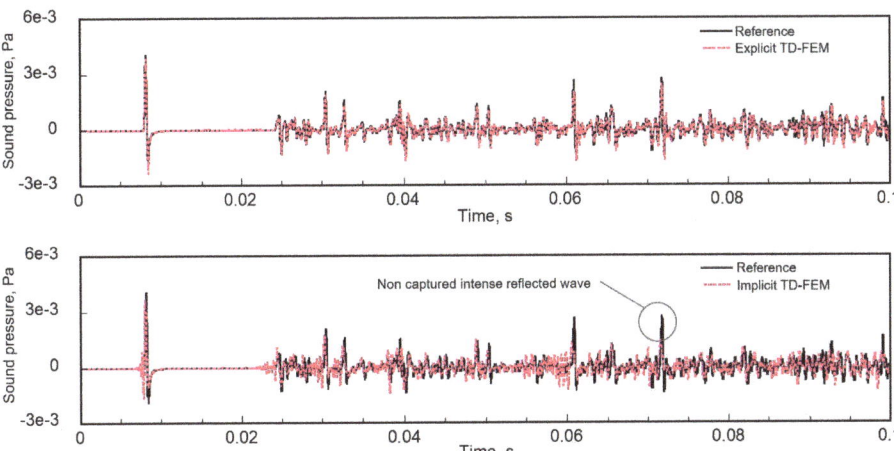

Figure 10. Comparisons of sound pressure waveforms at a receiver $(x, y) = (5.3, 3.0)$ among the reference solution, the proposed explicit TD-FEM, and the standard implicit TD-FEM.

5.3. Sound Propagation in a Concert Hall

A second numerical example demonstrates the performance in a more practical application. Figure 11 presents analyzed concert hall models of 296 m² surface area with two conditions. In the figure, Cond. 1 is a basic model; Cond. 2 is a model including an acoustic reflector above the stage and rib structures at a back wall in the stage for sound scattering. The reflector has 5.86 m length with 0.4 m width; the rib structure has a 0.4 m period length with 0.1 m depth and 0.2 m width. We used the simplest equivalent impedance model to address sound absorption effects at boundaries. For the seat and the back wall in the audience area, represented respectively as red and yellow lines in Figure 11, we gave $z_n = 3.87$ and $z_n = 7.14$ respectively. They correspond to Paris's statistical absorption coefficients, α_{Paris}, of 0.8 and 0.6, which are calculated as

$$\alpha_{\text{Paris}} = \frac{1}{z_n^2}\left(1 + z_n - \frac{1}{1+z_n} - 2\ln(1+z_n)\right). \tag{50}$$

Other boundaries including the reflector surfaces were set as $z_n = 71.519$ corresponding to $\alpha_{\text{Paris}} = 0.1$. As a sound source, we used an impulse response of fourth-order Butterworth type bandpass filter with 1/3 octave band width centered at 2 kHz. The source waveform and its frequency spectrum are portrayed in Figure 12. The band-limited room impulse responses were calculated up to 2 s at 15 receivers R1–R15, as listed in Table 2. The concert hall models were discretized using four-node quadrilateral FEs with maximum edge length of 0.02 m, which corresponded to 7.65 PPW at the upper-limit frequency of 2 kHz 1/3 octave band. The discretized FEM models have 806,478 DOF and 880,796 DOF, respectively, for Cond. 1 and Cond. 2. We set Δt as 1/98,000 s. As for the measure of accuracy, because one important evaluation in room acoustics is conducted using room acoustical parameters such as reverberation time, clarity, and strength, the absolute error in time integrated sound pressure level might become a useful measure. It is defined as shown below.

$$e(t) = \left|20\log\left(\frac{\int_0^t p_{\text{reference}}(\tau)d\tau}{\int_0^t p_{\text{TDFEM}}(\tau)d\tau}\right)\right| \text{ [dB]} \tag{51}$$

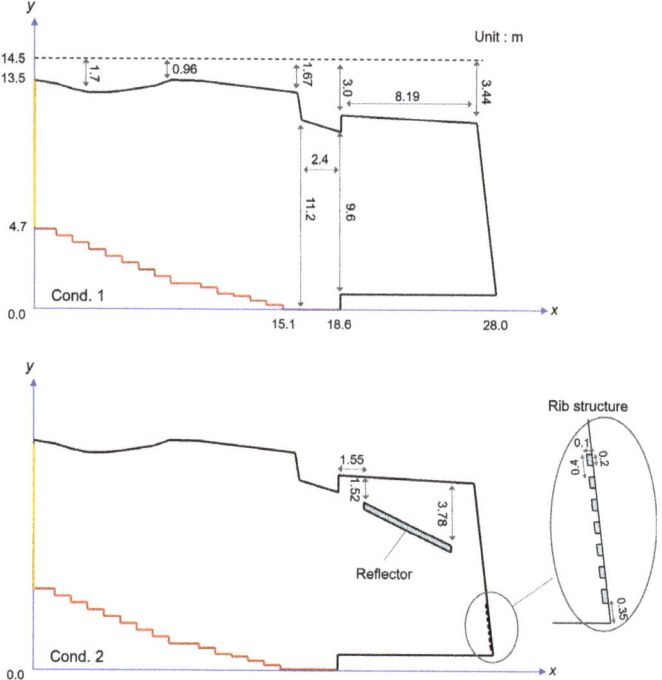

Figure 11. Concert hall models of two conditions (Cond. 1 and Cond. 2) where red and yellow lines respectively represent seat zone with $z_n = 3.87$ and back wall with $z_n = 7.14$.

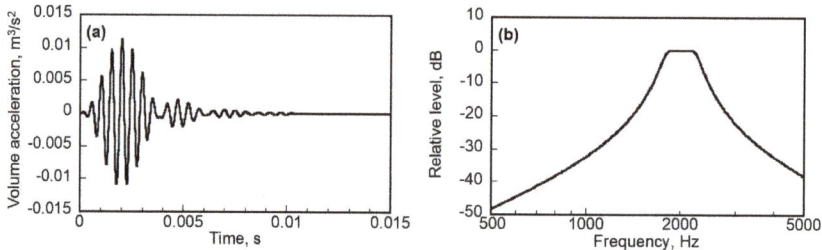

Figure 12. Source signal of fourth-order Butterworth type 1/3 octave bandpass filter with center frequency of 2 kHz: (**a**) impulse response and (**b**) frequency characteristic.

Table 2. List for 15 receiver positions (R1–R15) in concert hall model.

Receiver	(x, y), m	Receiver	(x, y), m
R1	(15.1, 1.25)	R9	(6.3, 3.5)
R2	(14.1, 1.5)	R10	(5.3, 3.9)
R3	(13.1, 1.7)	R11	(4.3, 4.3)
R4	(12.1, 2.0)	R12	(3.3, 4.7)
R5	(11.1, 2.1)	R13	(2.3, 5.1)
R6	(10.1, 2.5)	R14	(1.3, 6.3)
R7	(8.3, 2.7)	R15	(0.8, 6.7)
R8	(7.3, 3.1)		

In that equation, $p_{TDFEM}(t)$ and $p_{reference}(t)$ respectively denote the numerical solution and reference solution at time t. For computing the reference solution, the same FE meshes were used because the theoretically estimated maximum dispersion error of reference method with these meshes is lower than 0.1%.

5.3.1. Results and Discussion

Figure 13 portrays comparisons of direct sound waveforms at R1 $(x, y) = (15.1, 1.25)$ for the reference solution and the proposed explicit TD-FEM and the standard implicit TD-FEM for Cond. 1. Results for Cond. 2 were omitted because similar results were obtained. The proposed explicit TD-FEM shows good agreement with the reference solution, whereas the standard implicit method shows a marked discrepancy with an increased sound speed because of the larger dispersion property. Figure 14 presents a comparison of impulse responses up to 1.0 s for Cond. 1 and Cond. 2. The proposed explicit TD-FEM shows much better agreement with the reference solution in the entire time range for both conditions. However, the standard implicit TD-FEM shows poor approximation capability. In the early time region, it underestimates the amplitude of reflected waves, which is also observed for reflected waves at around 0.4 s for Cond. 1. Figure 15 presents a comparison of absolute error $e(t)$ between the proposed method and standard method for Cond. 1 and Cond. 2, where the error is the averaged value for all receivers. It is calculated at the time after arrival of the direct sound. The proposed explicit TD-FEM presents much lower errors for both conditions with magnitude below 0.2 dB. The standard implicit TD-FEM has error magnitude of approximately 0.4 dB and 0.6 dB for Cond. 1 and Cond. 2. In addition, the error level increases for more complicated Cond. 2. Regarding the advantage of using explicit TD-FEM, it can simulate sound propagation in concert halls, with approximately one-fifth less computational time than that of the implicit standard TD-FEM and one-third times the fourth-order accurate implicit TD-FEM used for reference calculation. These results suggest the high efficiency of the proposed explicit TD-FEM for acoustic simulations in complicated sound fields at kilohertz frequencies. For reference, total numbers of MVP required for the respective methods are 392,000 for the proposed explicit method, 1,995,506–2,053,979 for the standard implicit method, and 1,219,622–1,306,921 for the fourth-order implicit method.

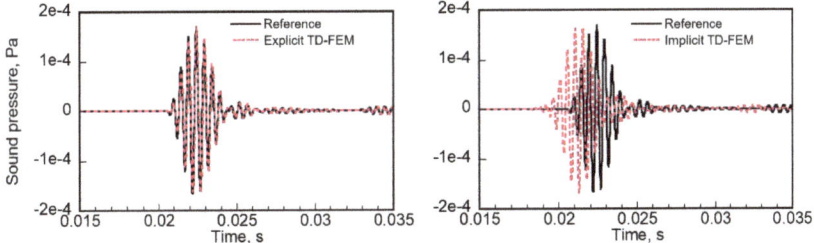

Figure 13. Comparisons of direct sound waveform at R1: the reference solution vs. the proposed explicit TD-FEM (Left) and the reference solution vs. the standard implicit TD-FEM (right).

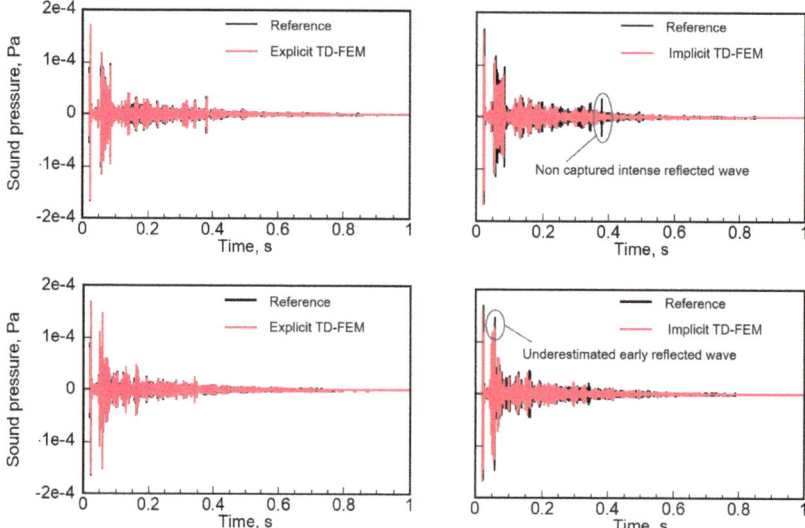

Figure 14. Band-limited impulse responses at R1 among the reference solution, the proposed explicit TD-FEM, and the standard implicit TD-FEM for Cond. 1 (upper row) and Cond. 2 (lower row).

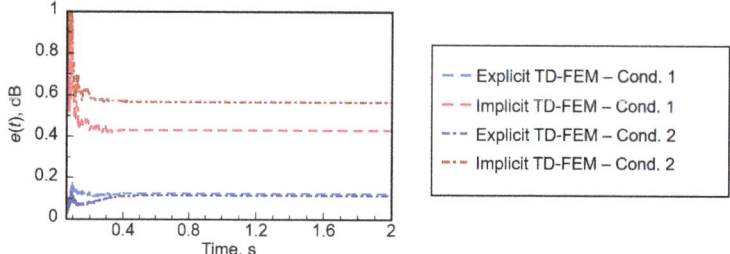

Figure 15. Comparison of absolute errors in time integrated sound pressure levels between the proposed explicit TD-FEM and the standard implicit TD-FEM for Cond. 1 and Cond. 2.

Figure 16 presents sound propagation in the concert hall for both conditions calculated using the proposed explicit TD-FEM. It shows again that the sound propagation is isotropic. Additionally, one can observe from the result of Cond. 2 that the effect of increased diffuseness of resulting sound fields can be gained by virtue of the sound scattering by the rib structures. It is also apparent from the effect of the reflector that strong reflection of sound from the reflector comes at an earlier time to the audience area than reflection waves from the ceiling and back wall of the stage. Furthermore, Figure 17 shows time-integrated SPL until 0.1 s for Cond. 1 and Cond. 2. By installing the reflector, it is readily apparent that early incident sound energies on the audience area increase and that sound energies around the back wall of the stage decrease. These results demonstrate the effectiveness of the proposed explicit TD-FEM clearly as a room acoustic design tool in practical applications.

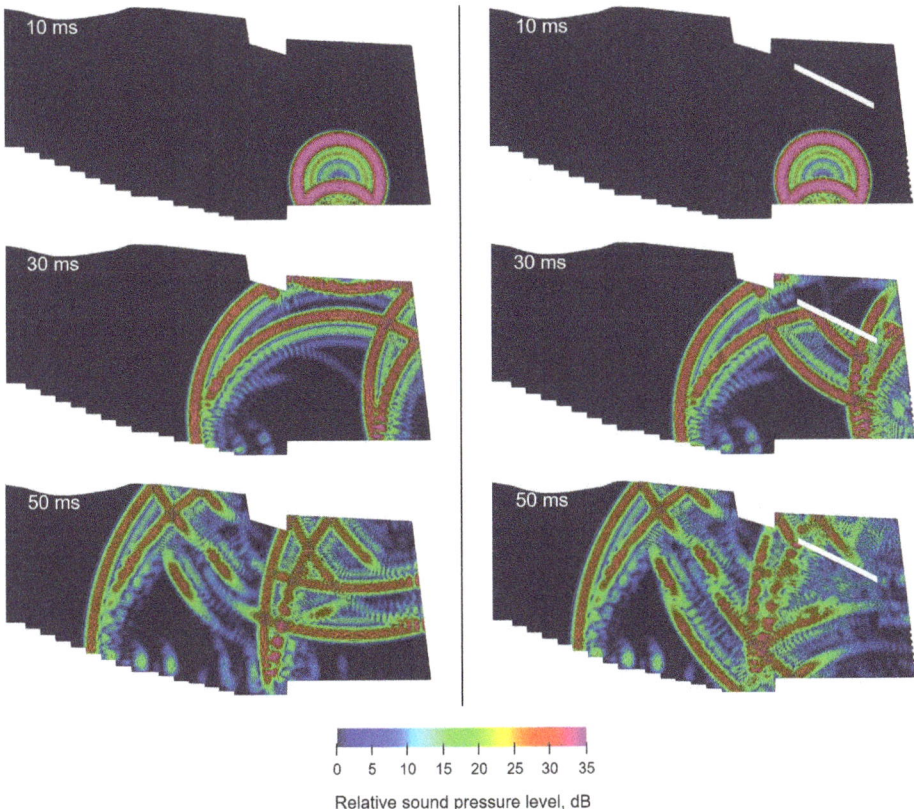

Figure 16. Sound propagation in a concert hall model for Cond. 1 (left column) and Cond. 2 (right column) at t = 10, 30 and 50 ms.

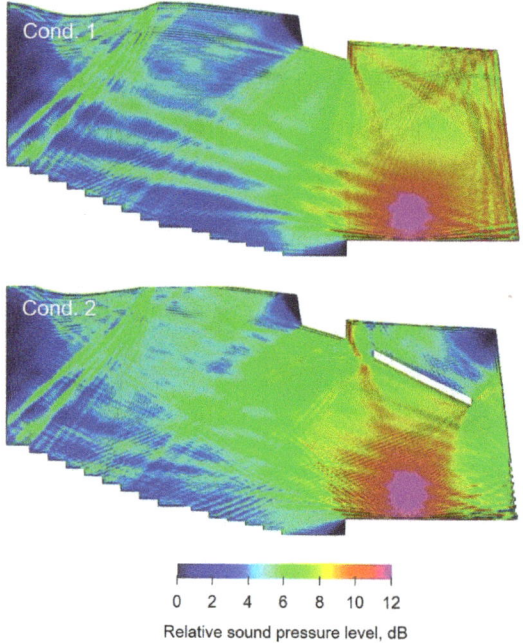

Figure 17. Comparison of time-integrated sound pressure level until 0.1 s for Cond. 1 (upper) and Cond. 2 (lower).

6. Conclusions

The study described in this report examined a proposed time domain room acoustic solver using an explicit TD-FEM and demonstrated its practicality as a room acoustic design tool. The present TD-FEM achieves fourth-order accuracy in both space and time using a dispersion-reduced low-order FEs and a specially designed linear multi-step time integration method. It completely overcomes the shortcomings of earlier presented methods fourth-order accurate explicit formulation [25] by which the staircase approximation is introduced into boundary modeling to maintain higher order accuracy in time discretization. In addition, as a noteworthy feature of the proposed method, it does not necessitate any additional computational cost in the matrix-vector product operations per time step. Therefore, the main operation complexity of the present method is the same as that of an earlier method. We conducted both dispersion and dissipation error analyses of the present method. The dispersion error analysis revealed that the maximum and minimum dispersion errors occur respectively at sound propagation in the axial direction and in an oblique direction $\theta = \pi/4$ in a polar coordinate system. In addition, the dispersion error decreases with a smaller time interval. The dispersion analysis also revealed that the present method includes a dissipation error that appears as a numerical complex sound speed in the resulting expression. However, we demonstrated that the dissipation error magnitude is dependent only on time interval values. It can simply reduce the control of time interval values using a proposed control method. Performance of the proposed explicit TD-FEM was examined against the standard implicit TD-FEM using sound propagation problems in a concert hall and in a rectangular room including acoustic diffusers. The results clearly demonstrated that the proposed method can predict complex sound fields at kilohertz frequencies accurately with a much lower requirement for computational resources, suggesting its promising potential for use as a time domain room acoustic solver. Although this paper only describes 2D analysis and its results, extension of the proposed method to 3D analysis is a trivial task. It will be a subject of future research.

Author Contributions: Conceptualization, T.Y. and T.O.; Investigation, T.Y.; Methodology, T.Y.; supervision, T.O.; Project administration, K.S.; Validation, T.Y.; Visualization, T.Y.; Writing—Original draft, T.Y and T.O.; Writing—Review & editing, T.Y., T.O. and K.S. All authors have read and agreed to the published version of the manuscript.

Funding: This research received no external funding.

Conflicts of Interest: The authors declare that they have no conflict of interest related to this report or the study it describes.

References

1. Sakuma, T.; Sakamoto, S.; Otsuru, T. Introduction. In *Computational Simulation in Architectural and Environmental Acoustics—Methods and Applications of Wave-Based Computation*; Springer: Tokyo, Japan, 2014; pp. 1–7.
2. Allred, J.C.; Newhouse, A. Application of the Monte Carlo method to architectural acoustics. *J. Acoust. Soc. Am.* **1958**, *30*, 1–3. [CrossRef]
3. Krokstad, A.; Strom, S.; Sørsdal, S. Calculating the acoustical room response by the use of a ray tracing technique. *J. Sound Vib.* **1968**, *8*, 118–125.
4. Vorländer, M. Simulation of sound in rooms. In *Auralization: Fundamentals of Acoustics, Modelling, Simulation, Algorithms and Acoustic Virtual Reality*; Springer Science & Business Media: Berlin/Heidelberg, Germany, 2007; pp.175–226.
5. Savioja, L.; Svensson, U.P. Overview of geometrical room acoustic modeling techniques. *J. Acoust. Soc. Am.* **2005**, *138*, 708–730. [CrossRef] [PubMed]
6. Rindel, J.H.; Nielsen, G.B.; Christensen, C.L. Diffraction around corners and over wide barriers in room acoustic simulations. In Proceedings of the 16th International Congress on Sound and Vibration, Kraków, Poland, 5–9 July 2009.
7. Rindel, J.H. Computer simulation techniques for acoustical design of rooms. *Acoust. Aust.* **1995**, *23*, 81–86.
8. Botteldooren, D. Finite-difference time-domain simulation of low-frequency room acoustic problems. *J. Acoust. Soc. Am.* **1995**, *98*, 3302–3308. [CrossRef]
9. LoVetri, J.; Mardare, D.; Soulodre, G. Modeling of the seat dip effect using the finite-difference time-domain method. *J. Acoust. Soc. Am.* **1996**, *100*, 2204–2212. [CrossRef]
10. Sakamoto, S. Phase-error analysis of high-order finite difference time-domain scheme and its influence on calculation results of impulse response in closed sound field. *Acoust. Sci. Technol.* **2007**, *28*, 295–309. [CrossRef]
11. Kowalczyk, K.; Van Walstijn, M. Room Acoustics Simulation Using 3-D Compact Explicit FDTD Schemes. *IEEE Trans. Audio Speech Lang. Process.* **2010**, *19*, 34–46. [CrossRef]
12. Hamilton, B.; Bilbao, S. FDTD Methods for 3-D Room Acoustics Simulation with High-Order Accuracy in Space and Time. *IEEE Trans. Audio Speech Lang. Process.* **2017**, *25*, 2112–2124. [CrossRef]
13. Bilbao, S. Modeling of Complex Geometries and Boundary Conditions in Finite Difference/Finite Volume Time Domain Room Acoustics Simulation. *IEEE Trans. Audio Speech Lang. Process.* **2013**, *21*, 1524–1533.
14. Hamilton, B.; Webb, C.J.; Fletcher, N.; Bilbao, S. Finite difference room acoustics simulation with general impedance boundaries and viscothermal losses in air: Parallel implementation on multiple GPUs. In Proceedings of the International Symposium on Musical and Room Acoustics ISMRA 2016, La Plata, Argentine, 11–13 September 2016.
15. Azad, H.; Siebein, G.W.; Ketabi, R. A Study of Diffusivity in Concert Halls Using Large Scale Acoustic Wave-Based Modeling and Simulation. In Proceedings of the 47th International Congress and Exposition on Noise Control Engineering, Chicago, IL, USA, 26–28 August 2018.
16. Craggs, A. The transient response of a coupled plate-acoustic system using plate and acoustic finite elements. *J. Sound Vib.* **1971**, *15*, 509–528. [CrossRef]
17. Otsuru, T.; Okamoto, N.; Okuzono, T.; Sueyoshi, T. Applications of large-scale finite element sound field analysis onto room acoustics. In Proceedings of the 19th International Congress on Acoustics, Madrid, Spain, 2–7 September 2007.
18. Okuzono, T.; Otsuru, T.; Tomiku, R.; Okamoto, N. Fundamental accuracy of time domain finite element method for sound-field analysis of rooms. *Appl. Acoust.* **2010**, *71*, 940–946. [CrossRef]

19. Okuzono, T.; Otsuru, T.; Tomiku, R.; Okamoto, N. Application of modified integration rule to time-domain finite-element acoustic simulation of rooms. *J. Acoust. Soc. Am.* **2012**, *132*, 804–813. [CrossRef]
20. Okuzono, T.; Otsuru, T.; Tomiku, R.; Okamoto, N. A finite-element method using dispersion reduced spline elements for room acoustics simulation. *Appl. Acoust.* **2014**, *79*, 1–8. [CrossRef]
21. Papadakis, N.M.; Stavroulakis, G.E. Effect of Mesh Size for Modeling Impulse Responses of Acoustic Spaces via Finite Element Method in the Time Domain. In Proceedings of the Euronoise 2018, Crete, Greece, 27–31 May 2018.
22. Okuzono, T.; Sakagami, K.; Osturu, T. Dispersion-reduced time domain FEM for room acoustics simulation. In Proceedings of the 23rd International Congress on Acoustics, Aachen, Germany, 9–13 September 2019.
23. Newmark, N.M. A method of computation for structural dynamics. *J. Eng. Mech. Div.* **1959**, *85*, 67–94.
24. Okuzono, T.; Yoshida, T.; Sakagami, K.; Otsuru, T. An explicit time-domain finite element method for room acoustics simulations: Comparison of the performance with implicit methods. *Appl. Acoust.* **2016**, *104*, 76–84. [CrossRef]
25. Yoshida, T.; Okuzono, T.; Sakagami, K. Numerically stable explicit time-domain finite element method for room acoustics simulation using an equivalent impedance model. *Noise Control Eng. J.* **2018**, *66*, 176–189. [CrossRef]
26. Yoshida, T.; Okuzono, T.; Sakagami, K. A three-dimensional time-domain finite element method based on first-order ordinary differential equations for treating permeable membrane absorbers. In Proceedings of the 25th International Congress on Sound and Vibration, Hiroshima, Japan, 8–12 July 2018.
27. Wang, H.; Sihar, I.; Muoz, R.P.; Hornikx, M. Room acoustics modeling in the time-domain with the nodal discontinuous Galerkin method. *J. Acoust. Soc. Am.* **2019**, *145*, 2650–2663. [CrossRef]
28. Pind, F.; Engsig-Karup, A.P.; Jeong, C.H.; Hesthaven, J.S.; Mejling, M.S.; Strømann-Anderson, J. Time domain room acoustic simulations using the spectral element method. *J. Acoust. Soc. Am.* **2019**, *145*, 3299–3310. [CrossRef]
29. Hughes, T.J.R. Formulation of parabolic, hyperbolic, and elliptic-eigenvalue problems. In *The Finite Element Method: Linear Static and Dynamic Finite Element Analysis*; Dover: New York, Ny, USA, 2000; pp. 436–446.
30. Zienkiewicz, O.C.; Taylor, R.L.; Zhu, J.Z. The time dimension: Semi-discretization of field and dynamic problems. In *The Finite Element Method: Its Basis and Fundamentals*, 7th ed.; Butterworth-Heinemann: Oxford, UK, 2013; pp. 382–386.
31. Yue, B.; Guddati, M.N. Dispersion-reducing finite elements for transient acoustics. *J. Acoust. Soc. Am.* **2005**, *118*, 2132–2141. [CrossRef]
32. Scherer, P.O.J. Equations of motion. In *Computational Physics: Simulation of Classical and Quantum Systems*, 3rd ed.; Springer Nature: Berlin/Heidelberg, Germany, 2017; pp. 306–308.
33. Von Neumann, J.; Richtmyer, R.D. A method for the numerical calculation of hydrodynamic shocks. *J. Appl. Phys.* **1950**, *21*, 232–237. [CrossRef]
34. Thompson, L.L. A review of finite-element methods for time-harmonic acoustics. *J. Acoust. Soc. Am.* **2006**, *119*, 1315–1330. [CrossRef]

 © 2020 by the authors. Licensee MDPI, Basel, Switzerland. This article is an open access article distributed under the terms and conditions of the Creative Commons Attribution (CC BY) license (http://creativecommons.org/licenses/by/4.0/).

Article

An Energy Model for the Calculation of Room Acoustic Parameters in Rectangular Rooms with Absorbent Ceilings

Erling Nilsson [1,*] and Emma Arvidsson [2]

1 Saint-Gobain Ecophon AB, Yttervägen 1, 265 75 Hyllinge, Sweden
2 Engineering Acoustics, Lund University, John Ericssons väg 1, 221 00 Lund, Sweden; emma.arvidsson@construction.lth.se
* Correspondence: erling.nilsson@ecophon.se

Abstract: The most common acoustical treatment of public rooms, such as schools, offices, and healthcare premises, is a suspended absorbent ceiling. The non-uniform distribution of the absorbent material, as well as the influence of sound-scattering objects such as furniture or other interior equipment, has to be taken into account when calculating room acoustic parameters. This requires additional information than what is already inherent in the statistical absorption coefficients and equivalent absorption areas provided by the reverberation chamber method ISO 354. Furthermore, the classical diffuse field assumption cannot be expected to be valid in these types of rooms. The non-isotropic sound field has to be considered. In this paper, a statistical energy analysis (SEA) model is derived. The sound field is subdivided into a grazing and non-grazing part where the grazing part refers to waves propagating almost parallel to the suspended ceiling. For estimation of all the inherent parameters in the model, the surface impedance of the suspended ceiling has to be known. A method for estimating the scattering and absorbing effects of furniture and objects is suggested in this paper. The room acoustical parameters reverberation time T_{20}, speech clarity C_{50}, and sound strength G were calculated with the model and compared with calculations according to the classical diffuse field model. Comparison with measurements were performed for a classroom configuration. With regard to all cases, the new model agrees better with measurements than the classical one.

Keywords: room acoustics; calculation models; absorption; scattering; airflow resistivity

1. Introduction

Many people spend most of their working hours in rooms such as offices, and education and healthcare premises. For the wellbeing of the people in those work places, the acoustical conditions are an important factor. The most common acoustical treatment in these type of public rooms is a suspended absorbent ceiling. The acoustical design is often aimed at reducing noise levels, improving speech intelligibility or, as in open-plan offices, preventing sound propagation. Due to the fact that most of the sound absorption located at the ceiling and other surfaces can be quite sound reflecting, the decay of sound energy and its relation to absorption is not properly explained by the classical assumption of a linear decay under diffuse field condition. These room types comprise a group of rooms where the diffuse field assumption is not valid and the sole use of reverberation time for characterization of the acoustical conditions is not sufficient.

The aim of this paper is to present a model for calculation of reverberation time T_{20}, speech clarity C_{50}, and sound strength G, as defined in ISO 3382-1 [1] and ISO 3382-2 [2]. The model was particularly designed for rooms with suspended absorbent ceilings. For public rooms, such as classrooms, offices, health-care premises, dining rooms, sport arenas, retail premises and similar kind of spaces, the typical acoustical treatment is a suspended absorbent ceiling. The model presented is based on a statistical energy analysis (SEA) approach used to describe the conditions at steady state and during the sound decay.

Rooms, as mentioned above, are places where large numbers of people spend most of their time during the day. It is obvious that the environment where we spend so many of our working hours should contribute to well-being and the ability to perform working tasks in the best possible way. The acoustical conditions are important in this respect. The purpose of the model presented in this paper is to obtain an estimation of room acoustic parameters for a relevant characterization of the acoustical conditions.

Schools are one of our largest work places. For learning and for the well-being of students and staff in educational premises, acoustic conditions play a central part. It has been recognized in several studies [3–6] that learning and the ability to remember and concentrate are affected by acoustic conditions as well are general well-being and the onset of stress-related symptoms. The effect of different signal-to-noise ratios on the ability to recall words shows that noisy surroundings in classrooms impair learning [7–9].

The effect of room acoustic improvement on the work situation in schools has been investigated in [10,11]. It has been shown that, with improved room acoustic conditions, the students social behavior becomes calmer and the teachers experience less physiological load (heart rate) as well as less fatigue. Poor acoustics in classrooms can result in high vocal loading of teachers, which presents a risk factor for voice disorders [12]. Keeping speakers' acoustics conditions in mind, measurement methods for the prediction of voice support and room gain in classrooms have been developed [13,14].

The high activity-based noise levels in preschools have been thoroughly investigated [15,16]. However, the long-term effects on children and staff are still a topic for investigations [17].

The sound environment in hospitals is diverse due to different activities that take place, the sound of medical equipment, and alarms and background noise. This can contribute to stress symptoms among staff as well as being a hinderance to patient recovery [18].

The acoustically challenging environments that open-plan spaces involve have received a great deal of attention in recent years [19]. Standards have been developed that present new measurement methods relevant for the typical scenarios occurring in open-plan offices as well as guidelines for creating good acoustic quality in these environments [20,21].

The knowhow relating to characterization of the acoustical conditions in public rooms has increased in recent years. Several investigations [22–25] have pointed out the necessity of addressing several acoustic parameters to achieve a relevant characterization of the acoustic environment. As has been shown, parameters relating to noise levels and speech intelligibility are an important complement to reverberation time. In [26], the speech clarity parameter U_{50}, i.e., C_{50}, including the effect of background noise, is used for designing good speech conditions in classrooms.

In [27,28], Barron presents a model for calculating clarity index and sound strength in rooms assuming linear sound decay. In [29], special effort was focused on explaining the non-diffusivity effect of the sound fields in public rooms with ceiling treatment and how these circumstances influence these parameters.

Since Sabine's [30] discovery and his classical formula, reverberation time has been the key parameter in room acoustics. In many standards and regulations, it is still the main parameter defining target values for good acoustics [31]. However, today, there are some new standards that have included measures, such as C_{50} and speech transmission index STI [32], as complements to reverberation time [33].

The idea of two rooms with approximately the same reverberation times being perceived as different is not a new finding and is mentioned in textbooks on acoustics [34,35] as well. This is especially the case in public rooms with ceiling treatment.

Many suggestions for improvement of the reverberation time formulas have been made. Several examples of such refinements are given in [34,35].

The influence of different corrections to Sabine's formula has been investigated by Joyce [36,37]. In support of Sabine's formula, Joyce shows that understated conditions of weak absorption and irregular reflections provides the correct answer.

In [38] Fitzroy presents an empirically derived formula for the reverberation time in rooms with non-uniform distribution of absorption. A modified version of Fitzroy's formula is presented by Neubauer [39]. The non-uniform distribution of absorption is also dealt with by the formula of Arau-Puchades [40]. The effect of location of absorbent material in a mock-up of a classroom and in a reverberation chamber has recently been studied by Cuchrero et al. [41].

In [42], Sakuma uses an image source method where the image sources are grouped as axial, tangential, and oblique groups corresponding to normal modes in wave acoustics. Scattering is taken into account by introducing the scattering coefficient. The non-linear decay in rooms with non-uniform distribution of absorption as well as the importance of scattering are apparent in the results.

In [43], Bistafa and Bradley compared experimental results with analytical and computer predictions of reverberation time in a simulated classroom. Their paper emphasizes the need to quantify the amount of scattering due to furniture and other objects in a room. The influence of scattering is also experimentally investigated by Prodi et al. [44].

A general problem in many reverberation time formulas is the use of a random absorption coefficients as input data. This is of course natural, as most manufacturers of absorbent products provide this data measured according to ISO 354 [45]. However, the non-isotropic properties in rooms with ceiling treatment differ from the almost diffuse conditions in reverberation chambers. In fact, even in reverberation chambers, the concept of a diffuse sound field is hard to achieve [46]. In [47,48], Nilsson presented a model particularly developed for rooms with suspended absorbent ceilings. The non-diffuse conditions were dealt with by introducing two sound fields related to grazing and non-grazing sound waves. The idea of subdividing the sound field into a grazing and non-grazing group were also adopted in [49].

To deal with the non-diffuse conditions in the model presented in this paper, an estimation of the surface impedance of the ceiling is used. The reason is to take into account the angle-dependent properties of the ceiling absorber. This is a major difference to the other energy models referred to above. Another difference to the referred models is the handling of the scattering effect of interior objects such as furniture. In rooms with absorbent ceiling treatments, the directional scattering effect of objects is important. A method for estimation of the directional scattering effect is suggested as an outcome of the model formulation.

When evaluating the reverberation T_{20} or T_{30} according to ISO 3382-2 [2], the dynamical ranges -5 to -25 dB and -5 to -35 dB are used, respectively. This means that the early reflections of the impulse response are neglected. Therefore, T_{20} and T_{30} are often referred to as late reverberation times. In a room with absorbent ceiling treatment, the late reverberation times are often related to energy travelling in the horizontal plane, comprising grazing waves in relation to the absorbent ceiling.

The importance of early reflections for design of auditoria was already observed by Lochner and Burger [50]. Chiara et al. [51] has investigated the subjective influence of early diffuse reflections on speech intelligibility and spatial perception. In [52], Bradley et al. show the importance of early reflections for speech intelligibility both for normal- and hearing-impaired listeners. These investigations show the benefits of using parameters incorporating the early reflections such as speech clarity.

The examples in the text above show that public rooms with acoustic ceiling treatment comprise a large and important group of rooms that deserve closer examination. This involves investigation into how different acoustical treatment affects the sound field and how this impact can be predicted in a more accurate way than by the classical diffuse field assumption. Further, elucidate the limitations related to only using reverberation time as a descriptor characterising the acoustics.

This paper presents a model that considers the special features of rooms with ceiling treatment and gives an estimation of several room acoustic parameters that are important for the subjective perception of the acoustics. The model takes into account the mounting height of ceiling absorbers and absorbent wall panels, as well as the scattering effect of furnishing, diffusers, or other objects. The purpose is to serve the user with a model that gives an estimation of room acoustic parameters that are reasonably consistent with measurements in rooms with ceiling treatment and thus, also to emphasize phenomena that influence the subjective perception of the acoustics.

2. General Description of the Model

A general discussion of the model is presented in this chapter. The model is based on a statistical energy analysis (SEA) approach [53,54]. The model addresses rectangular rooms with absorbent ceilings, i.e., rooms where the main contribution to the total absorption is related to the ceiling. A more precise requirement for this condition is given further on.

Important considerations are, firstly, that the surface impedance of the absorbent ceiling, including the air cavity behind the absorber, has to be known, and secondly, that the absorbing and scattering effects of furniture and other interior fittings have to be estimated. A method for measuring the scattering effect is proposed in Section 3.2.5. This method takes into account the directional scattering of objects due to the orientation towards the ceiling.

With the exception of the ceiling, other surfaces in the room are characterized by the statistical absorption coefficient. Further, added wall panels are defined by their statistical absorption coefficient, as measured according to ISO 354 [45].

The room acoustic parameters calculated are reverberation time T_{20} according to ISO 3382-2, speech clarity C_{50} in dB, and sound strength G in dB according to ISO 3382-1.

Speech clarity is defined as

$$C_{50} = 10\log\left(\frac{\int_0^{0.05} p^2(t)dt}{\int_0^{\infty} p^2(t)dt}\right) \quad (1)$$

where $p(t)$ is the impulse response at the measurement point.

Sound strength is defined as

$$G = 10\log\left(\frac{\int_0^{\infty} p^2(t)dt}{\int_0^{\infty} p_{10}^2(t)dt}\right) \quad (2)$$

where $p(t)$ is the impulse response at the measurement point and $p_{10}(t)$ is the impulse response measured at 10 m in a free field.

An omni-directional sound source is required for measurement of the acoustical parameters.

The model comprises the following steps:

Basic formulas are derived in Section 3.1 comprising

- Establish a general expression for the energy sound decay in a two-system SEA model.
- Express the total sound energy decay in the parameter sound strength G as defined in ISO 3382-1.
- From the expression for the total sound energy decay, derive an expression for the speech clarity C_{50} and the reverberation time T_{20}.

Estimation of the inherent parameters in the basic formulas are presented in Section 3.2 comprising

- Subdivide the total sound field into a grazing and non-grazing part where grazing refers to sound waves propagating almost parallel to the absorbent ceiling.
- Calculate the angle-dependent absorption coefficient, Section 3.2.1.
- Estimate the number of modes in the grazing subsystem as well as a representative absorption coefficient, Section 3.2.2.
- Estimate the number of modes in the non-grazing subsystem as well as a representative absorption coefficient, Section 3.2.3. Two approaches for estimation of the number of non-grazing waves were used: one empirical and one theoretical.

- Based on a 2-dim and 3-dim reverberation formula, estimate the reverberation times T_g and T_{ng} corresponding to the grazing and non-grazing subsystem, respectively. See Section 3.2.4.
- By knowing T_g and T_{ng} and the number of modes in each subsystem, the energy ratio C for the grazing and non-grazing sound fields in the formula for sound strength G can be calculated.

As an effect of the subdivision of the total sound field into a grazing and non-grazing part, the scattered and absorbed sound, due to objects such as furniture in the room, can be interpreted as a coupling loss factor between the two subsystems, see Figure 1. The coupling loss factor is reformulated as an equivalent scattering absorption area, denoted as A_{sc}. A corresponding measurement method of A_{sc} is suggested. See further Section 3.2.5.

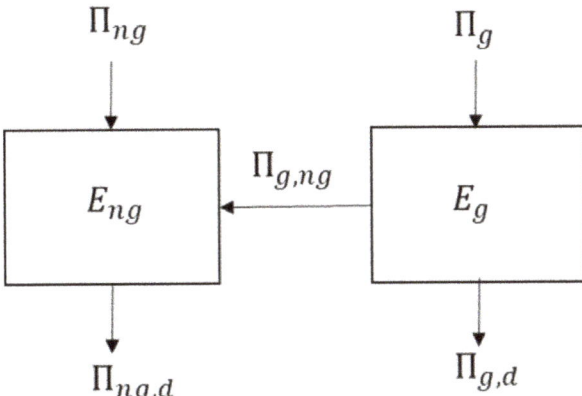

Figure 1. The SEA model.

As the distance r is included in the model, the room acoustic parameters as a function of distance can be calculated. However, in the calculations performed, a representative value of r is used. See Equation (41).

The theoretical background for the model is presented in the next chapter and verifying measurements in Section 5. The new model will hereinafter be referred to as "non-diffuse" and the classical diffuse field model (Sabine) as "diffuse".

3. Theory
3.1. The SEA Model

The sound field in a room with absorbent ceiling treatment is modelled as an SEA system consisting of two subsystems. One subsystem comprises non-grazing waves and the other comprises grazing waves. The term grazing refers to the angle of incidence towards the ceiling absorber. Thus, grazing comprises waves travelling almost parallel to the absorbent ceiling. The coupling loss factor between the two subsystems is related to the energy transfer from the grazing subsystem to the non-grazing subsystem. This energy transfer is most often due to the interior fittings in the room such as furniture, but could also be due to a tilting wall, for example. The back-transfer from the non-grazing to the grazing subsystem is neglected. The SEA model is illustrated in Figure 1.

The power flow into the grazing (g) and non-grazing (ng) subsystem (Π_{ng}, Π_g), as well as the dissipated power $\left(\Pi_{ng,d}, \Pi_{g,d}\right)$, are shown in Figure 1. The total energy in the subsystems are denoted as E_{ng} and E_g, respectively. The power lost by the grazing subsystem to the non-grazing is represented by $\Pi_{g,ng}$. Generally, a weak coupling is assumed, i.e., that the losses related to the coupling between the two system is less than the internal losses in the grazing and non-grazing subsystems [54].

In a room with non-uniform distribution of absorption, such as the rectangular room with a highly absorbent ceiling and the other surfaces almost reflecting, the energy decay is estimated by

$$E(t) = E_{ng}(0)e^{-\omega\eta_{ng}t} + E_g(0)e^{-\omega\eta_g t} \qquad (3)$$

$E_{ng}(0)$ and $E_g(0)$ are the initial energies for the non-grazing and grazing subsystems, respectively. The loss factor in the non-grazing and the grazing subsystems are denoted as η_{ng} and η_g, respectively.

Using $\Pi = \omega\eta E$ and, assuming that the coupling loss factor is negligibly small compared to the internal losses in the two subsystems, the energy ratio is given by

$$E(t) = E_{ng}(0)\left(e^{-\omega\eta_{ng}t} + \frac{E_g(0)}{E_{ng}(0)}e^{-\omega\eta_g t}\right) = E_{ng}(0)\left(e^{-\omega\eta_{ng}t} + \frac{\eta_{ng}\Pi_g}{\eta_g\Pi_{ng}}e^{-\omega\eta_g t}\right) \qquad (4)$$

The condition in Equation (4) above is valid for a rectangular room with absorbent ceiling, but without furniture. Including furniture will lead to the introduction of a coupling loss factor related to the energy transfer from the grazing to the non-grazing sound field, see Figure 1. Replacing η_g in Equation (4) by $\eta_g + \eta_{g,ng}$ where $\eta_{g,ng}$ is the coupling loss factor, the absorbing and scattering effect of furniture can be accounted for. The coupling loss factor is further discussed in Section 3.2.5.

As shown in [55], the ratio Π_g/Π_{ng} is approximately given by N_g/N_{ng}, where N_g and N_{ng} are the number of modes in the grazing and the non-grazing subsystems, respectively.

In geometrical acoustics, sound waves are often represented as rays with a certain sound intensity. Further, in room acoustical calculations, the reverberation time is a well-established parameter and normally the frequency depending on reverberation times are studied in frequency bands, usually octave bands.

By converting Equation (4) into sound intensity, assuming octave band values and using the relation $\Delta\Pi_g/\Delta\Pi_{ng} \approx \Delta N_g/\Delta N_{ng}$, and further introducing the reverberation time T using the relation $\omega\eta = 6\ln(10)/T$, we get

$$I(t) = I_{ng}(0)\left(e^{-13.8t/T_{ng}} + \frac{T_g\Delta N_g}{T_{ng}\Delta N_{ng}}e^{-13.8t/T_g}\right) \qquad (5)$$

The procedure presented for a linear decay by Barron and Lee [27] is applied for the double sloped decay, as given by Equation (5).

The steady-state condition at $t = 0$ gives the power balance

$$W = \omega\eta\frac{I}{c}V \qquad (6)$$

where W is the input power and V is the room volume.

Assuming a point source and a distance r_0 between the source and receiver and further, that the sound field at steady-state is diffuse with a reverberation time T_{ng}, the intensity at steady-state is given by [56]

$$I(0) = I_0 r_0^2 \frac{T_{ng}}{V}\frac{4\pi c}{6\ln(10)} = 312 I_0 r_0^2 \frac{T_{ng}}{V} \qquad (7)$$

where I_0 is the intensity of the direct sound at the distance r_0 from the sound source.

The total (energy) decay, as given by Equation (5), adjusted towards the steady-state intensity in Equation (7) will be given by

$$I(t) = 312 I_0 r_0^2 \frac{T_{ng}}{V(1+C)}\left(e^{-13.8t/T_{ng}} + Ce^{-13.8t/T_g}\right) \qquad (8)$$

where

$$C = \frac{T_g\Delta N_g}{T_{ng}\Delta N_{ng}} \qquad (9)$$

Including the direct sound gives

$$I(t) = I_d + I_{rev} \tag{10}$$

I_{rev} is given by Equation (8) and I_d is the direct sound at distance r given by

$$I_d = \frac{W}{4\pi r^2} \tag{11}$$

where W is the input power.

Following Barron et al. [27], the sound strength G is calculated. The sound strength G is defined as

$$G = L_p - L_{p,10} \tag{12}$$

where L_p is the sound pressure level at the measurement point and $L_{p,10}$ is the sound pressure level at a distance of 10 m in a free field given by

$$L_{p,10} = 10 \log \left(\frac{\rho c}{p_{ref}^2} \frac{W}{4\pi 10^2} \right) \tag{13}$$

where p_{ref} is 2×10^{-5} Pa.

Combining Equations (8), (10) and (12) gives

$$G = 10 \log \left(\frac{100}{r^2} + 31{,}200 \frac{T_{ng}}{V(1+C)} \left(e^{-\frac{13.8t}{T_{ng}}} + Ce^{-\frac{13.8t}{T_g}} \right) \right) \tag{14}$$

Setting $t = r/c$ [27] i.e., the time for the sound wave to propagate r metres, gives the final expression. This implies that the decay starts after the direct sound arrived at the receiver position.

$$G = 10 \log \left(\frac{100}{r^2} + 31{,}200 \frac{T_{ng}}{V(1+C)} \left(e^{-\frac{0.04r}{T_{ng}}} + Ce^{-\frac{0.04r}{T_g}} \right) \right) \tag{15}$$

The received sound energy is divided into three components, the direct sound (d), the early reflected sound i.e., a delay <50 ms (e_{50}), and the late reflected sound i.e., a delay >50 ms (l_{50}). Using Equation (8) normalized to $I_0 = W/(4\pi 10^2)$ gives

$$d = 100/r^2 \tag{16}$$

$$e_{50} = I_n(t) - I_n(t+50) = 31{,}200 \frac{T_{ng}}{V(1+C)} \left[e^{-\frac{0.04r}{T_{ng}}} \left(1 - e^{-\frac{0.691}{T_{ng}}}\right) + Ce^{-\frac{0.04r}{T_g}} \left(1 - e^{-\frac{0.691}{T_g}}\right) \right] \tag{17}$$

$$l_{50} = I_n(t+0.05) = 31{,}200 \frac{T_{ng}}{V(1+C)} \left(e^{-\frac{0.04r+0.691}{T_{ng}}} + Ce^{-\left(\frac{0.04r+0.691}{T_g}\right)} \right) \tag{18}$$

The sound strength G is given by

$$G = 10 \log(d + e_{50} + l_{50}) \tag{19}$$

The speech clarity C_{50} is given by

$$C_{50} = 10 \log \left(\frac{d + e_{50}}{l_{50}} \right) \tag{20}$$

T_{20} is calculated using the logarithmic version of Equations (8) and the -5 to -25 dB dynamical range according to ISO 3382-2.

To calculate T_{20}, C_{50}, and G, the inherent parameters T_{ng}, T_g, and C in Equations (8), (17), and (18) have to be estimated. This is described in the next paragraph.

3.2. Estimation of the Inherent Parameters T_{ng}, T_g and C

This chapter concerns the approach of estimating the inherent parameters in Equations (8), (17), and (18). Estimation of these parameters is of central importance in the model and some detailed explanations are presented in this paragraph. These estimations involve considerations regarding how to define absorption and the number of modes for the grazing and non-grazing sound fields and how to take into account the effect of sound-scattering objects in the room. The method involves defining a grazing and non-grazing region, according to Figure 2. The grazing sector is defined by the grazing angles θ_g. For the non-grazing sector, two approaches were used: a theoretical one and an empirical one. Before we go into the derivation of θ_g and the limits for the non-grazing sector, the calculation of the angle-dependent absorption coefficient will be discussed.

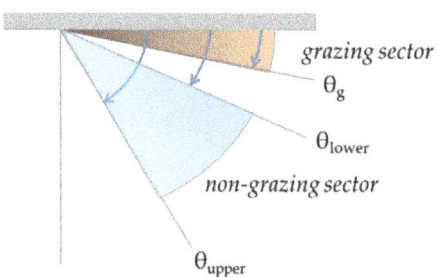

Figure 2. Illustration of the grazing and non-grazing sectors.

3.2.1. The Angle-Dependent Absorption Coefficient

For each sector, representative absorption coefficients (α_{ng}, α_g) and a representative number of modes (ΔN_{ng}, ΔN_g) have to be determined. It is assumed that the surface impedance of the ceiling absorber is known or can be estimated. Several types of commercial software' are available today for calculating the angle-dependent surface impedances [57,58] for different types of absorbers. In this study, only suspended ceilings of porous material were investigated. For porous absorbers, the surface impedance $Z(f,\theta)$ can be calculated by applying empirical models if the air flow resistivity is known. In this case Miki's model was used [59]. An extended reaction is assumed when calculating α_{ng} and α_g. The angle-dependent absorption coefficient for a plane sound wave impinging on a plane infinite surface is given by

$$\alpha(f,\theta) = 1 - \left| \frac{Z(f,\theta)\cos(\theta) - \rho_0 c_0}{Z(f,\theta)\cos(\theta) + \rho_0 c_0} \right|^2 \qquad (21)$$

where $Z(f,\theta)$ is the surface impedance at incidence angle θ, ρ_0 is the density of air, and c_0 is the speed of sound. The surface impedance for an extended reaction is calculated as [60]

$$Z(f,\theta) = \frac{Z_c k}{k_x} \left[\frac{-jZ_0 \cot(k_x d) + Z_c \frac{k}{k_x}}{Z_0 - jZ_c \frac{k}{k_x}\cot(k_x d)} \right] \qquad (22)$$

where k is the wave number in the absorber, $k_x = \sqrt{k^2 - k_0^2 \sin^2(\theta)}$ is the normal component of k, k_0 is the wave number in air, d is the thickness of the absorber, and Z_c is the characteristic impedance of the absorber. The backing impedance Z_0 is given by

$$Z_0(f,\theta) = -j\left(\rho_0 c_0 \frac{k_0}{k_x}\right) \cot(k_0 d_0 \cos\theta) \qquad (23)$$

where d_0 is the depth of the air cavity behind the absorber.

The characteristic impedance for the absorber Z_c is calculated by Miki's model according to

$$Z_c = \rho_0 c_0 \left[1 + 0.070 \left(\frac{f}{\sigma}\right)^{-0.632} - j0.107 \left(\frac{f}{\sigma}\right)^{-0.632}\right] \quad (24)$$

and wave number

$$k = \frac{\omega}{c}\left[1 + 0.109 \left(\frac{f}{\sigma}\right)^{-0.618} - j0.160 \left(\frac{f}{\sigma}\right)^{-0.618}\right] \quad (25)$$

The only material parameter needed for Miki's formula is the air flow resistivity σ of the porous material. Miki's formula is an improvement of the Delany and Bazley model [61]. Another modification of the Delany and Bazley model has been developed by Komatsu [62].

The extended reaction (the angle-dependent impedances) is of particular importance for accurate estimation at low frequencies. This is illustrated in Figure 3 where local and extended reactions are compared for the reverberation time T_{20} measured in a sparsely furnished room with dimensions 7.56 m × 7.30 m × 3.50 m and with a 15 mm thick absorbent ceiling at a mounting height of 200 mm (case 4 in Section 4). The figure shows the results using extended vs. local reaction in the model. Considerable deviation at low frequencies (125 Hz and 250 Hz) appears.

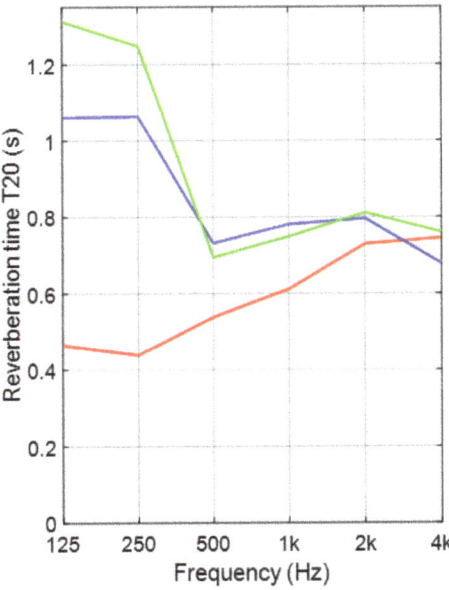

Figure 3. Local vs. extended reaction. Calculations according to the SEA model in a classroom with a 15 mm thick porous ceiling absorber with a mounting height of 200 mm (case 4 in the Section 5). Calculated local reaction (red), measured (blue), calculated and extended reaction (green).

3.2.2. Estimation of α_g and ΔN_g

To calculate the total energy decay, Equation (8) in Section 3.1, the number of modes in each sector and the corresponding reverberation times must be known. In this paragraph and the following paragraph, we will firstly estimate the representative absorption coefficients α_g and α_{ng} for the grazing and non-grazing sectors in Figure 2, as well as the number of modes ΔN_g and ΔN_{ng} in each sector.

The grazing sector is defined by an angle θ_g given by

$$\theta_g = \arccos\left(\frac{c}{4fL_x}\right) \quad (26)$$

The derivation of θ_g is given in Appendix A. The grazing sector in the wavenumber space is illustrated in Figure 4.

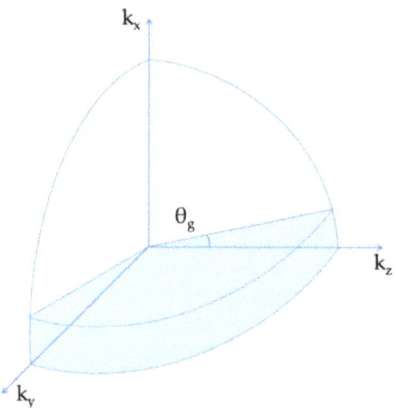

Figure 4. Grazing sector in the wavenumber space.

By knowing the surface impedance of the ceiling, the angle-dependent absorption coefficient can be calculated. The grazing absorption coefficient α_g is then calculated as the average absorption coefficient in the grazing region, i.e., between $\pi/2 - \theta_g$ and $\pi/2$. This absorption is often quite small but not negligible when compared to the total absorption for the grazing field. Equation (26) is a high-frequency estimation. At low frequencies, i.e., at 125 Hz and 250 Hz, the grazing absorption is estimated by

$$\alpha_{g,ceiling} = \pi \rho c A'_{xl} \quad (27)$$

where A'_{xl} is the real part of the admittance for the ceiling absorber. A'_{xl} is given by the real part of $1/Z$, where Z is given by Equation (22), assuming an extended reaction. The derivation of Equation (27) is given in Appendix B.

The number of grazing modes in the frequency band Δf is given by [55]

$$\Delta N_g(\theta_g) = \left[\left(\frac{4\pi f^2 V}{c^3}\right)\cos\left(\frac{\pi}{2} - \theta_g\right) + \left(\frac{2f}{c^2}\right)(\pi L_y L_z + \theta_g(L_x L_z + L_x L_y)) + \left(\frac{1}{c}\right)(L_y + L_z)\right]\Delta f \quad (28)$$

where V is the volume and L_x, L_y, and L_z are height, length, and width of the room, respectively. As $\theta_g \to 0$, the number of grazing modes corresponds to the tangential and axial modes in the yz plan.

3.2.3. Estimation of α_{ng} and ΔN_{ng}

To estimate α_{ng}, an intermediate step was used. This step includes the introduction of a weighted normalised absorption coefficient given by

$$\alpha_n(f,\theta) = \frac{\alpha(f,\theta)\Delta N(f,\theta)}{\max(\alpha(f,\theta)\Delta N(f,\theta))} \quad (29)$$

In this expression, $\Delta N(f, \theta)$ is the number of modes as a function of frequency and angle, as given by Equation (28) replacing θ_g with θ. The absorption coefficient $\alpha(f, \theta)$ is the angle-dependent absorption coefficient given by Equation (21), assuming an extended reaction. The non-grazing absorption coefficient α_{ng} is given by Equation (21) for an angle (θ_{ng}) corresponding to the maximum value in the distribution given by Equation (29).

Examples of this distribution are given for case 1 in Section 4 and for the frequencies 250 Hz and 4000 Hz, see Figure 5. In the classical diffuse field assumption, the angle-dependent absorption coefficient is weighted by the factor $\sin(2\theta)$, according to the Paris formula [63]. For comparison, the diffuse field weighting $\sin(2\theta)$ is also shown. As can be seen in the figure, there is a bias between the classical approach and the distribution, according to Equation (29). The representative angle for the non-grazing absorption coefficient is somewhat higher compared to the $\sin(2\theta)$. For the higher frequency, we see that the classical weighting corresponds to almost 45 degrees, as expected. The irregular shape at 250 Hz is due to the assumption of an extended reaction.

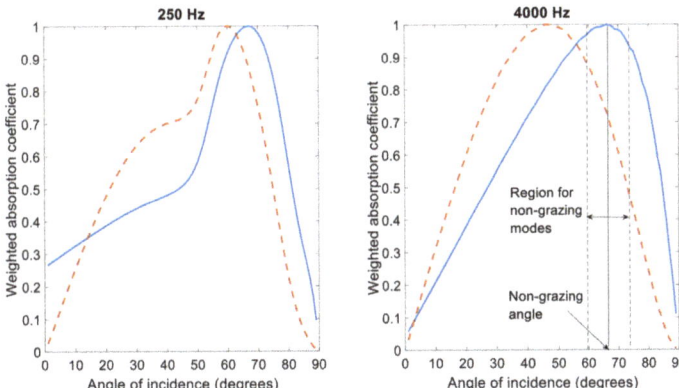

Figure 5. Distribution curves for the normalised weighted absorption coefficient according to Equation (29) for case 1 in Section 4.1. (Red) diffuse model, (blue) non-diffuse model.

Two approaches for determination of ΔN_{ng} were used: a theoretical one and an empirical one. For the empirical approach, the number of non-grazing modes was determined by adjustment towards experimental results for several configurations where room dimensions and acoustical treatment and furnishing were varied. An approach using minimization of a cost function to perform a curve fitting is presented in [64]. In the empirical method, the upper and lower angles defining the non-grazing sector, see Figure 2, are given by $\theta_{ng,\,lower} = \theta_{ng}(1 - \Delta\theta)$ and $\theta_{ng,\,higher} = \theta_{ng}(1 + \Delta\theta)$, where $\Delta\theta$ was estimated by comparison with measurements. Note that the angle of incidence $= \pi/2 - \theta$. Further, θ_{higher} is restricted to be less than $\pi/2$. The values for $\Delta\theta$ is given in Table 1 for the octave bands 125 Hz to 4000 Hz.

Table 1. Empirical determined limit parameter $\Delta\theta$ for defining the non-grazing region.

Frequency Hz	125	250	500	1000	2000	4000
$\Delta\theta$	0.63	0.31	0.14	0.17	0.07	0.08

The number of modes in the non-grazing sector ΔN_{ng} in Figure 2 is given by the repeated use of Equation (28) and is given by

$$\Delta N_{ng} = \Delta N_g \left(\theta_{ng,upper}\right) - \Delta N_g \left(\theta_{ng,lower}\right) \tag{30}$$

The theoretical approach involves calculating the number of modes for the non-grazing sector ΔN_{ng} as

$$\Delta N_{ng} = \frac{1}{\alpha_{ng}} \int_0^{\pi/2} \alpha(f,\theta) N(f,\theta) d\theta \tag{31}$$

where α_{ng} is the absorption coefficient corresponding to the angle defined by the maximum in the weighted normalized absorption coefficient given by Equation (29). By knowing this non-grazing angle, see Figure 5 right, the non-grazing absorption coefficient α_{ng} can be given by Equation (21).

For a room with dimensions 7.56 m × 7.30 m × 3.50 m and with an absorbent ceiling corresponding to case 1 in Table 2, the number of included modes in the non-grazing group given by the empirical and the theoretical approaches are compared in Figure 6. At lower frequencies (125 Hz and 250 Hz), the correspondence is good. At higher frequencies, the theoretical estimation gives significantly higher values compared to the empirical one. The consequence of this discrepancy will be further discussed in Section 5.

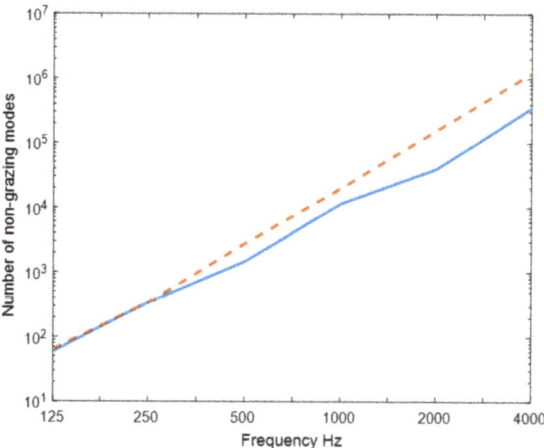

Figure 6. The number of modes in the non-grazing sector estimated by the empirical (solid) and theoretical (dashed) approaches.

The empirical and theoretical approaches described above are used for frequencies of 500 Hz and above. At 125 Hz and 250 Hz, the non-grazing absorption coefficient is estimated in the same way as the grazing one at low frequencies, i.e.,

$$\alpha_{ng,ceiling} = \pi \rho c A'_{xl} \tag{32}$$

where A'_{xl} is the real part of the admittance for the ceiling absorber, see Equation (27).

By knowing the number of modes in each sector, i.e., ΔN_g and ΔN_{ng}, and the representative absorption coefficients α_g and α_{ng}, we can go on and estimate the corresponding reverberation times T_g and T_{ng}.

3.2.4. Estimation of T_g and T_{ng}

The non-grazing reverberation time T_{ng} is given by

$$T_{ng} = \frac{0.161 V}{A_{ng,ceiling} + A_{furniture} + A_{surface} + 4mV} \tag{33}$$

where $A_{ng,ceiling} = \alpha_{ng} S_{ceiling}$ and α_{ng} is the absorption coefficient corresponding to the angle given by the maximum in the weighted normalized absorption coefficient, as described in Section 3.2.3. $A_{furniture}$ is the Sabine equivalent absorption area for the furniture. An estimation of $A_{furniture}$ is given in [49] as $A_{furniture} = V_{furniture}^{2/3}$. $A_{surface}$ is the equivalent absorption area for the walls and floor. Normally, the absorption coefficients for those surfaces are rather small and can be found in tables, e.g., in [49]. The air absorption is taken into account by the term $4mV$ where m is the energy attenuation constant in air and V is the room volume.

Equation (33) is similar to the Sabine formula, but the skewness in the energy distribution is taken into account by using α_{ng}, as described in the paragraph above.

The grazing reverberation time T_g is given by a 2-dim version of Sabine formula [55]

$$T_g = \frac{0.127V}{A_{g,ceiling} + A_{sc} + A_{surface} + \pi mV} \quad (34)$$

where $A_{g,ceiling} = \alpha_g S_{ceiling}$ and α_g represents the absorption coefficient for the grazing sector, as derived in Section 3.2.2. In this formula, we also introduce the parameter equivalent scattering absorption area A_{sc}. This parameter quantifies the absorption and scattering effects of furniture and other objects in rooms with absorbent ceiling treatment. Thus, it also accounts for the directional scattering effects that can appear in these types of rooms, depending on the objects' orientation relative to the absorbent ceiling. The estimation of A_{sc} will be further discussed in the next paragraph. $A_{surface}$ is similar, as in Equation (33). It could be stated that a 2-dimensional statistical absorption coefficient should be used instead of a 3-dimensional one, but as the difference is small [55], $A_{surface}$ is calculated in the same way, as in Equation (33). It is assumed that the contribution of the floor is small and that it can be represented by the statistical absorption coefficient. However, for the air absorption, the distinction between the 2- and 3-dimensional sound fields is accounted for by using πmV instead of $4mV$.

3.2.5. Estimation of A_{sc}

The sound-scattering effects of furniture and other objects in rooms will greatly influence the room acoustic parameters in rooms where the absorbent material is concentrated to the ceiling. Reverberation time T_{20} and speech clarity C_{50} will be particularly affected. Sound strength G will normally be less affected as it is related to the steady-state conditions and thus will not be sensitive to the distribution of the absorbent material. To quantify the scattering effect, the following procedure was used.

In the terminology of SEA, the transfer of energy from the grazing to the non-grazing sound field is expressed in a coupling loss factor $\eta_{g,ng}$. The power flow $\Pi_{g,ng}$ from the grazing to the non-grazing subsystem is given by

$$\Pi_{g,ng} = \omega \eta_{g,ng} E_g \quad (35)$$

where $\eta_{g,ng}$ is the coupling loss factor from the grazing to the non-grazing subsystem and E_g is the energy in the grazing subsystem.

The coupling loss factor $\eta_{g,ng}$ can be estimated in a rectangular room with a highly absorptive ceiling. It is assumed that the two-system SEA model is valid for the sound field in the room, both with and without scattering objects (furniture) present. This is very often the case in rooms with absorbent ceiling treatment, as it is really difficult to create isotropic conditions in these types of rooms.

The coupling loss factor is then given by

$$\eta_{g,ng} = \eta_{g,with\ obj} - \eta_{g,without\ obj} \quad (36)$$

where $\eta_{g,with\ obj}$ is the grazing loss factor with objects in the room and $\eta_{g,\ without\ obj}$ is the grazing loss factor without objects in the room. These loss factors are determined from the reverberation time T_{20}, i.e., the late part of the decay curves in the room with and without objects. The relation between the reverberation time T and the loss factor is given by

$$\eta = \frac{6ln10}{\omega T} \quad (37)$$

In a two-dimensional sound field, an equivalent scattering absorption area can be defined as [55]

$$A_{sc} = \frac{\pi \omega V}{c} \eta_{g,ng} \quad (38)$$

where c is the speed of sound and V is the room volume.

Combining Equations (36)–(38) gives the equivalent scattering absorption area for objects as

$$A_{sc} = 0.127V \left(\frac{1}{T_{20,with}} - \frac{1}{T_{20,without}} \right) \quad (39)$$

where $T_{20, with}$ and $T_{20,without}$ are the reverberation times in the room with ceiling absorber, with and without objects, respectively. Equation (39) assumes that the late reverberation time T_{20} in a room with a highly absorptive ceiling is determined by a two-dimensional sound field. The measure A_{sc} is affected by the sound scattered into the ceiling and by the absorption of the objects. This measure is similar to the equivalent absorption area used in Sabine formula. It is used in the same way in Equation (34).

Of course, the A_{sc} will depend on the ceiling absorption properties. However, if the mean absorption coefficient of the ceiling absorber, for the mid and high frequencies, is larger than about 0.7, we will obtain a reasonable estimation of A_{sc} that can be used in most common situations of rooms with absorbent ceilings [65].

The A_{sc} for the investigated furniture configurations were measured according Equation (39) and are further discussed in the Section 5.

3.3. Summary

By knowing T_g, T_{ng}, ΔN_g, and ΔN_{ng}, the coefficient C in the basic formulas in Section 3.1 can be calculated. It is given by

$$C = \frac{T_g \Delta N_g}{T_{ng} \Delta N_{ng}} \quad (40)$$

It is possible to calculate the distance r between the sound source and the receiver for the actual positions, but in our calculations a representative distance was used given by

$$r = \frac{1}{2}\sqrt{L_y^2 + L_z^2} \quad (41)$$

where L_y and L_z are the width and length of the rectangular room, respectively.

Thus, all parameters are given and can be inserted into Equations (16)–(18) for further calculation of C_{50}, and G. T_{20} is calculated using the logarithmic version of Equation (8).

It should also be mentioned that, as the number of grazing and non-grazing modes are mainly related to the floor area and the volume of the room, it is of interest to investigate the model's applicability for other room shapes than rectangular, as long as the ceiling absorber is parallel to the floor.

4. Measurements and Methods

4.1. Measurement Configurations

The measurements were performed in a mock-up of a classroom with dimension length × width × height = 7.56 m × 7.30 m × 3.50 m, where 3.50 m refers to the height to the soffit. The classroom was sparsely furnished with 10 tables, 19 chairs, and 3 shelves, see Figure 7.

Figure 7. Sparsely furnished classroom mock-up. To the right with wall panels on two adjacent walls.

Two types of suspended ceilings were tested at two mounting heights. One of the suspended ceilings was tested in combination with wall panels on two adjacent walls, see Figure 7 right. The different configurations and specification of the material used are presented in Table 2.

Table 2. Measurement configurations.

Case	Ceiling and Wall Panels	Mounting Height (Depth of Air Cavity) of Suspended Ceiling (mm)	Air Flow Resistivity of Ceiling Absorber (kPas/m^2)
1	50 mm glasswool ceiling absorber *, no wall panels	750	11.8
2	15 mm glasswool ceiling absorber **, no wall panels	785	77.8
3	15 mm glasswool ceiling absorber **, 6.48 m^2 40 mm glasswool wall absorber *** distributed on two adjacent walls and directly mounted on the walls, see Figure 7.	785	NA
4	15 mm glasswool ceiling absorber **	185	77.8
5	50 mm glasswool ceiling absorber *	150	11.8

* Ecophon Industry Modus, ** Ecophon Gedina A, *** Ecophon Wall Panel A. Note: the air flow resistivity is only used as input data for the ceiling absorbers and not for the wall panels. For the wall panels the practical absorption coefficients are used, see Figure 8.

The absorption data for the products used are presented in Figure 8. The absorption coefficients were measured according to ISO 354 [45] and evaluated by ISO 11654 [66]. This presentation of absorption data as a practical absorption coefficient is common practice by manufactures of absorbent ceilings.

4.2. Measurement Method

The room impulse responses were measured using the Dirac system (Dirac type 7841, v.6.0). An exponential sweep signal was fed to an omnidirectional loudspeaker and recorded by an omnidirectional microphone. Two loudspeaker positions at the front of the classroom were used and, for each loudspeaker position, six microphone positions were used throughout the room. No microphone positions were closer than 2 m to the loudspeaker and none were closer than 1 m to any of the room surfaces.

The room acoustic parameters measured were reverberation time T_{20} (s), speech clarity C_{50} (dB), and sound strength G (dB). C_{50} and G are defined in ISO 3382-1. T_{20} was evaluated according to ISO 3382-2 using the interval −5 to −25 dB of the decay curve. The sound strength G was measured using a constant sound power source (Nor278, Norsonic). The sound power source was located in the same positions as the loudspeaker.

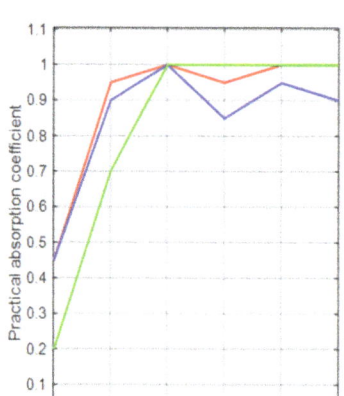

Figure 8. The practical absorption coefficients for the ceiling absorbers and wall panels used in the experiments. (Red) Ecophon Industry Modus 50 mm, (blue) Ecophon Gedina A 15 mm, (green) Wall Panel A 40 mm.

4.3. Repeatability

A repeatability test was performed for the measurement procedure described above. The measurements were repeated five times. Between each measurement, the loudspeaker and the microphone were taken out of the room and reinstalled at different positions. For details see [67].

In Table 3, the uncertainty is given for the measurement procedure.

Table 3. Uncertainty interval related to repeatability, corresponding to a 95% confidence interval, for the measurement procedure used in the experiments.

	G_{avg} (dB)	$C_{50,avg}$ (dB)	$T_{20,avg}$ (s)
125 Hz	±0.61	±0.56	±0.077
250 Hz	±0.30	±0.29	±0.018
500 Hz	±0.40	±0.29	±0.010
1000 Hz	±0.25	±0.27	±0.006
2000 Hz	±0.37	±0.38	±0.010
4000 Hz	±0.36	±0.36	±0.008

The variations in repeated measurements are less noticeable (JND), according to ISO 3382-1 [1]. This supports the discussion of significant differences in the measurements.

4.4. Estimation of the Equivalent Scattering Absorption Area A_{sc}

The A_{sc} for the furniture configurations is estimated by Equation (39). The A_{sc} for the furniture in combination with the two ceiling treatments and for the two mounting heights, given in Table 2, were measured. No wall panels were present during these measurements. The results are presented in Section 5.1. The same number of microphone and loudspeaker positions were used, as for the measurements of the room acoustic parameters.

4.5. Comparison between Measurements and Calculations

The measurements and calculations were compared for the octave band frequencies 125 Hz to 4000 Hz. Calculations of C_{50} and G were performed with the formulas presented in Section 3, according to Equations (19) and (20). T_{20} was calculated by the logarithmic version of Equation (8) using the -5 to -25 dB dynamical range according to ISO 3382-2.

For comparison, calculations according to the Sabine formula were included. The reverberation time T was calculated as

$$T = \frac{0.161V}{A_{ceiling} + A_{furniture} + A_{surface} + 4mV} \tag{42}$$

where $A_{ceiling} = \alpha_p S_{ceiling}$ and α_p is the practical absorption coefficient given in Figure 8. $A_{furniture} = V_{furniture}^{2/3}$, according to the EN 12354-6. For a sparsely furnished room, $V_{furniture}$ is approximately 1–2% of the room volume [29]. $A_{surface}$ and the air absorption were calculated in the same way as in Equation (33).

The absorption coefficients for the floor and walls were estimated from the reverberation time measurements in the empty room, i.e., without an absorbent ceiling. Those values were used both in the diffuse and non-diffuse calculations for calculating $A_{surface}$.

Assuming a linear decay under diffuse field conditions and a reverberation time, given by Equation (42), C_{50} and G are calculated as

$$C_{50} = 10 \log \left(10^{(6/T)0.05} - 1 \right) \tag{43}$$

And

$$G = 10 \log \left(\frac{4}{A} \right) + 31 \tag{44}$$

where $A = 0.16\frac{V}{T}$ is the equivalent absorption area in m^2 sabin and V is the room volume.

5. Results
5.1. Estimation of A_{sc}

The equivalent scattering absorption area A_{sc} for the furniture was measured for configurations 1, 2, 4, and 5 given in Table 2. The A_{sc} is estimated according to Equation (39). In Figure 9, the results are presented together with the averaged values.

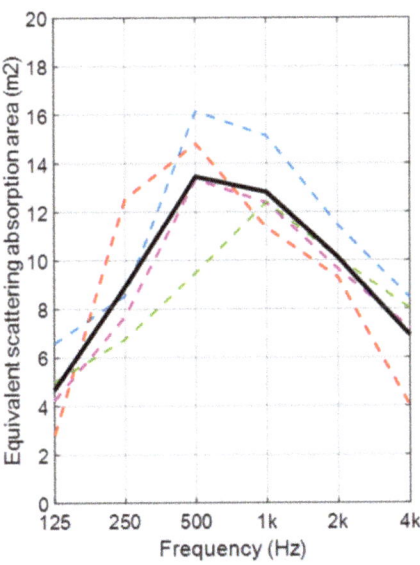

Figure 9. A_{sc} for the furniture estimated from cases 1, 2, 4 and 5 in Table 2, (black) average, (blue) case 4, (red) case 5, (purple) case 1 and (green) case 2.

A_{sc} depends on the absorption of furniture as well as the scattered sound energy transmitted to the non-grazing sound field and mainly absorbed by the ceiling absorber. As can be seen in Figure 9, the frequency behavior is quite similar for the different cases despite the fact that there is a variation of the ceiling absorber concerning airflow resistivity, thickness, and mounting height. This supports the idea that, for ceiling absorbers with a reasonably high absorption, see comment in Section 3.2.5, the correction for furniture absorption and scattering by A_{sc} is justified. For furnishing with tables, chairs, and shelves, the highest values of A_{sc} appears for the mid frequencies, as apparent from Figure 9. In practice, it is also possible to define A_{sc} per m² floor area to obtain a value that can be used for different sizes of rooms. In [29], values of A_{sc} per m2 floor area are suggested for what can be considered as sparse, normal, and dense furnishing. It is noteworthy that, in EN 12354-6, the correction for furniture and other objects in the room is independent of frequency.

5.2. Measurement Results

The measurements results are presented in Figures 10–14, corresponding to the cases 1 to 5 in Table 2. In the figures (a) is the reverberation time T_{20} in seconds, (b) is the speech clarity C_{50} in dB and (c) is the sound strength G in dB. Comparisons are made between measurements, Sabine calculation and the non-diffuse calculation. For the non-diffuse calculation, both the empirical and the theoretical approaches, discussed in Section 3.2.3, are shown.

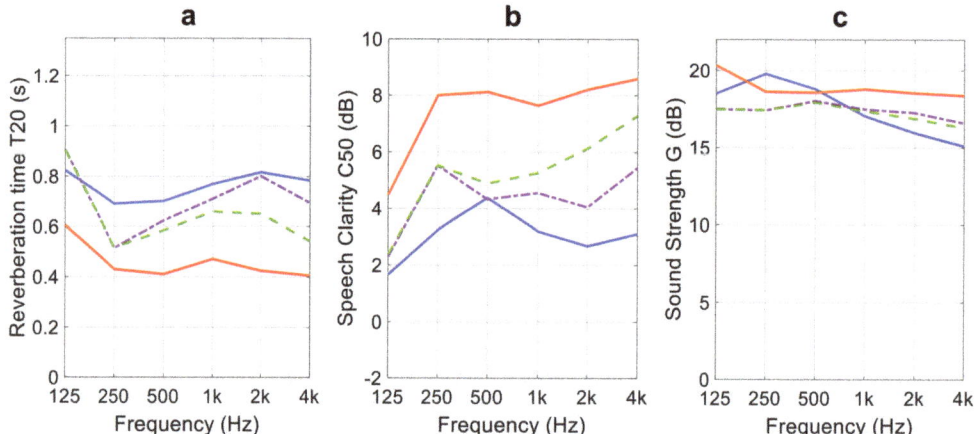

Figure 10. Sparsely furnished room with dimensions 7.35 × 7.50 × 3.50 m. Ceiling treatment: 50-mm glass wool absorber at a mounting height (air cavity behind the absorber) of 750 mm. (**a**) Reverberation time T_{20} in seconds, (**b**) speech clarity C_{50} in dB, (**c**) sound strength G in dB. Curves shown are (red) diffuse calculation (Sabine), (blue) measurement, (dashed) non-diffuse calculation, where the number of non-grazing modes is estimated by Equation (31), and (dash-dotted) non-diffuse calculation, where the number of non-grazing modes is empirically estimated, see Table 1.

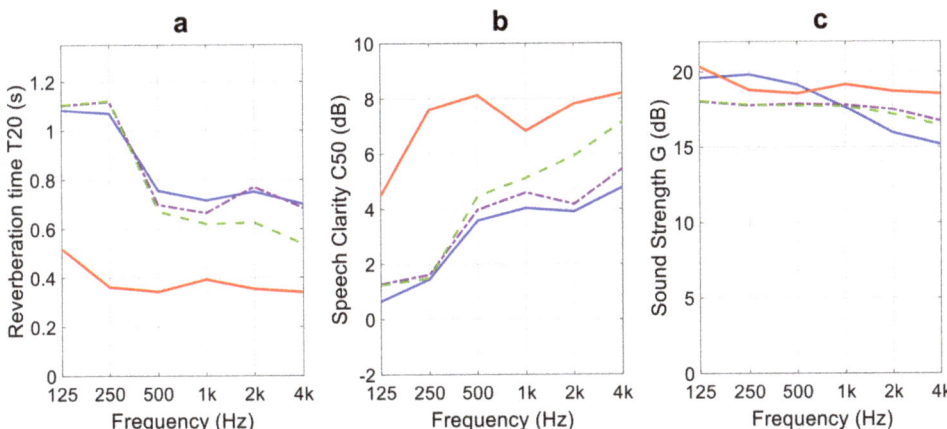

Figure 11. Sparsely furnished room with dimensions 7.35 × 7.50 × 3.50 m. Ceiling treatment: 15-mm glass wool absorber at a mounting height (air cavity behind the absorber) of 785 mm. (**a**) Reverberation time T_{20} in seconds, (**b**) speech clarity C_{50} in dB, (**c**) sound strength G in dB. Curves shown are (red) diffuse calculation (Sabine), (blue) measurement, (dashed) non-diffuse calculation, where the number of non-grazing modes is estimated by Equation (31), and (dash-dotted) non-diffuse calculation, where the number of non-grazing modes is empirically estimated, see Table 1.

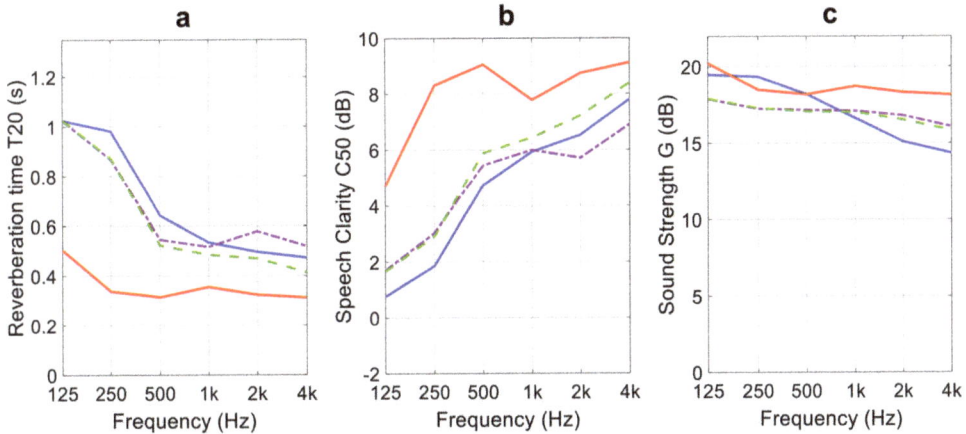

Figure 12. Sparsely furnished room with dimensions 7.35 × 7.50 × 3.50 m. Ceiling treatment: 15 mm glass wool absorber at a mounting height (air cavity behind the absorber) of 785 mm. 6.48 m² 40 mm glass wool wall absorber equally distributed on two adjacent walls and directly mounted on the walls, see Figure 7. (**a**) Reverberation time T_{20} in seconds, (**b**) speech clarity C_{50} in dB, (**c**) sound strength G in dB. Curves shown are (red) diffuse calculation (Sabine), (blue) measurement, (dashed) non-diffuse calculation, where the number of non-grazing modes is estimated by Equation (31), (dash-dotted) non-diffuse calculation, where the number of non-grazing modes is empirically estimated, see Table 1.

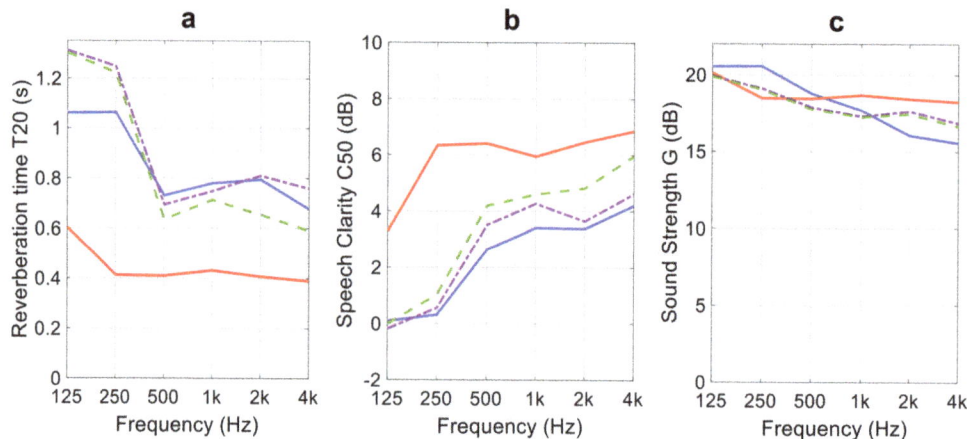

Figure 13. Sparsely furnished room with dimensions 7.35 × 7.50 × 3.50 m. Ceiling treatment: 15 mm glass wool absorber at a mounting height (air cavity behind the absorber) of 185 mm. (**a**) Reverberation time T_{20} in seconds, (**b**) speech clarity C_{50} in dB, (**c**) sound strength G in dB. Curves shown are (red) diffuse calculation (Sabine), (blue) measurement, (dashed) non-diffuse calculation, where the number of non-grazing modes is estimated by Equation (31), (dash-dotted) and non-diffuse calculation, where the number of non-grazing modes is empirically estimated, see Table 1.

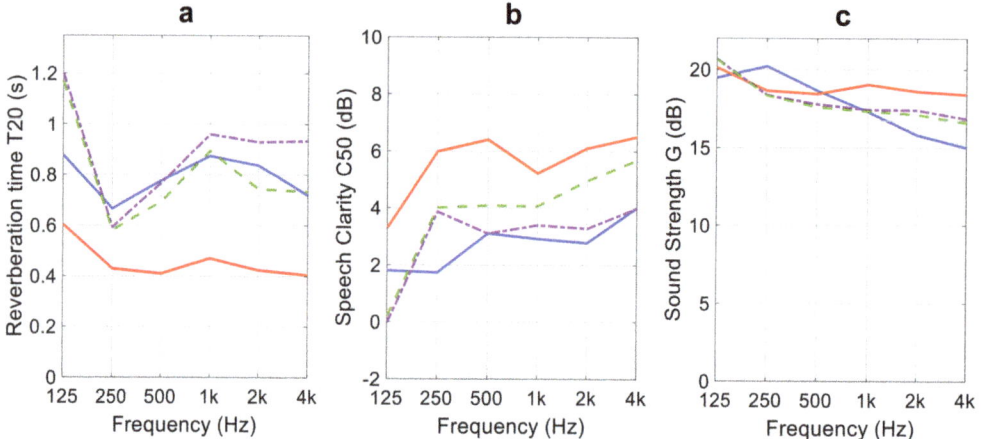

Figure 14. Sparsely furnished room with dimensions 7.35 × 7.50 × 3.50 m. Ceiling treatment: 50 mm glass wool absorber at a mounting height (air cavity behind the absorber) of 150 mm. (**a**) Reverberation time T_{20} in seconds, (**b**) speech clarity C_{50} in dB, (**c**) sound strength G in dB. Curves shown are (red) diffuse calculation (Sabine), (blue) measurement, (dashed) non-diffuse calculation, where the number of non-grazing modes is estimated by Equation (31), and (dash-dotted) non-diffuse calculation, where the number of non-grazing modes is empirically estimated, see Table 1.

Overall, the non-diffuse model fits better with the measurement results than the diffuse model. In particular, the overestimation of the absorption in the diffuse model is reduced in the non-diffuse model. The large differences between the diffuse calculations and the measurement results are typical for sparsely furnished rooms with an absorbent ceiling treatment. The cause of this is the lack of diffusion and the influence of the grazing sound field. Naturally, the empirical estimation given in Table 1 agrees better with measurements than the theoretical approach, according to Equation (31). This is more apparent at the higher frequencies. It is noticeable that the non-diffuse model captures the frequency behavior better than the diffuse one.

An important feature of the non-diffuse model is the reaction to wall panels. The effect of wall panels is the reduction in the energy in the grazing sound field. In sparsely furnished rooms, this largely influences the late reverberation time and the speech clarity. This is clearly shown in Figure 15. The correspondence between the non-diffuse calculation and measurement is good. For the diffuse model, a much smaller effect is noticed. The effect of wall panels on sound strength G is small. As G is a steady-state measurement, it is mainly related to the total absorption in the room, assuming that the sound field is fairly diffuse before the onset of the decay. During the decay, the degeneration of the sound field towards a grazing sound field will affect reverberation times and speech clarity to a great extent, as shown in the experimental results.

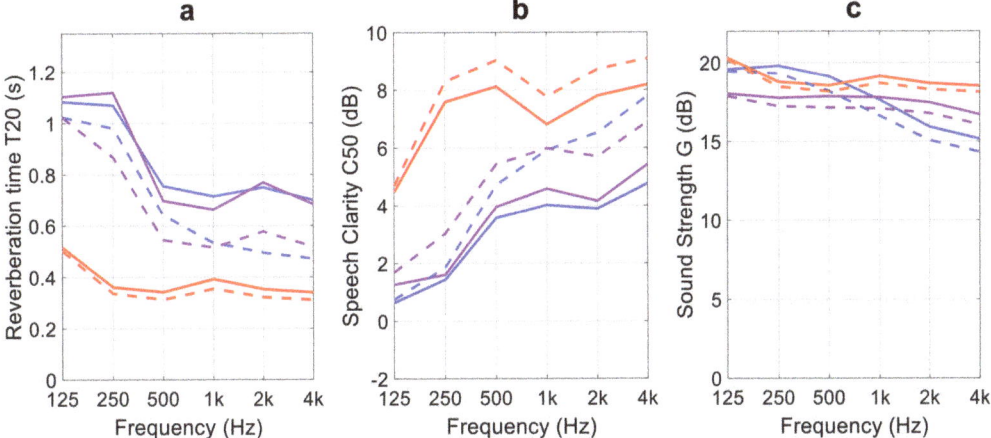

Figure 15. Comparison of case 2 and 3 in Table 2, i.e., the cases with and without wall panels. (**a**) Reverberation time T_{20} in seconds, (**b**) speech clarity C_{50} in dB, (**c**) sound strength G in dB. (Blue solid) measured without wall panels, (blue dashed) measured with wall panels, (purple solid) non-diffuse calculation without wall panels, (purple dashed) non-diffuse calculations with wall panels, (red solid) diffuse calculations without wall panels, and (red dashed) diffuse calculations with wall panels. The empirical approach is used for the non-diffuse calculations, see Section 3.2.3.

Note that the results presented above refer to a sparsely furnished room which is very sensitive to the accuracy of the input data. It is notable that the case with the wall panels decrease the discrepancy between the theoretical and empirical model and also fits better with measurements.

In Figure 16, a comparison of the two ceilings absorbers corresponding to case 1 and 2 in Table 2 is shown. The practical absorption coefficients for these absorbers, as given by the manufactures, is shown in Figure 8. Besides the large difference between the diffuse calculations on the one hand (red curves) and the measurements and non-diffuse calculations on the other (blue and green curves) some other remarks can be made. The practical absorption coefficients, based on ISO 354 measurements, show similar values at 125 Hz for the two absorbers. Accordingly, the diffuse calculations show the same reverberation time at this frequency. However, the non-diffuse calculations give a large difference at 125 Hz which also corresponds to the measurements. It is also noteworthy that, at high frequencies, the non-diffuse calculations and the measurements show contradictory behavior in comparison with the diffuse calculations. The diffuse calculations follow the difference in the practical absorption coefficients which is not the case for the measurements and the non-diffuse calculations. This emphasizes the fact that the absorption coefficients measured under reverberant conditions, as in ISO 354, do not comprise sufficient information for the acoustic design of rooms with ceiling treatment. Other information is needed and, in the model presented, the surface impedance of the ceiling absorber is necessary input data.

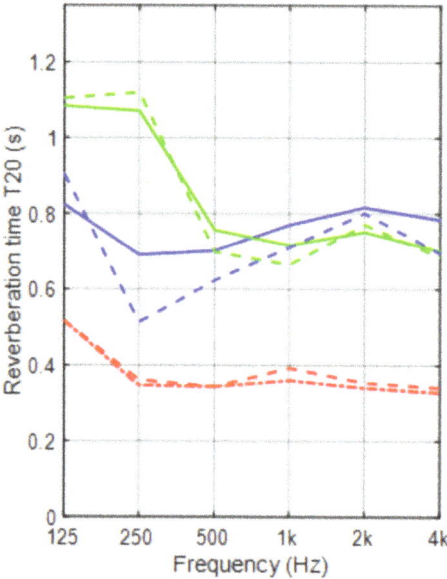

Figure 16. Measured and calculated reverberation times for case 1 and 2 in Table 2. Fifty mm and fifteen mm thick porous ceiling absorbers at a mounting height of 760 mm and 785 mm, respectively, were investigated. (Dashed green) 15 mm absorber, non-diffuse calculation; (solid green) 15 mm absorber, measurement; (dashed blue) 50 mm absorber, non-diffuse calculation; (solid blue) 50 mm absorber, measurement; (dashed red) 50 mm absorber, diffuse calculation; and (dash-dot red) 15 mm absorber, diffuse calculation.

6. Discussion

A model is presented based on a subdivision of the sound field into a grazing and non-grazing subsystem, where grazing refers to sound waves propagating almost parallel to the absorbent ceiling. An advantage of this approach is its interpretation of sound scattering due to interior equipment such as furniture, diffusors, or similar. The scattering effect is quantified in a parameter related to the energy transfer from the grazing to the non-grazing group. The parameter is denoted as the equivalent scattering absorption area A_{sc} and comprises the scattering and absorbent effect of interior objects in a room with a highly absorptive ceiling. Due to the presumption of an absorbent ceiling, the directional scattering effects of objects will appear. It is assumed that the ceiling absorption is much larger than the average absorption for walls and floors. An average absorption coefficient for the ceiling absorber greater than 0.7 for the octave bands ranging from 250 to 4000 Hz seems to be sufficient for most practical situations, but this has to be further investigated [65]. There is an assumption concerning sufficiently great ceiling absorption to ensure that the energy reflected back to the non-grazing field can be neglected. It might also be of future interest to specify the conditions for a laboratory configuration as to how to estimate A_{sc}. The methodology could be used to give input data for typical furnishing scenarios in different segments such as schools, offices, and healthcare premises. The directional effects of diffusors were studied in a classroom configuration by Arvidsson et al. [67].

In the presented model, it is assumed that the surface impedance is known or can be calculated for the suspended ceiling. Other surfaces are dealt with in a normal way using the practical or statistical absorption coefficients. Data for this can be found in handbooks in acoustics or manufactures' websites. The surface impedance is not a parameter normally provided by the manufacturers of absorbent ceilings. However, several examples of commercial software exist today that calculates the surface impedances for different types of absorbers.

The need to include more complex boundary conditions for improved accuracy has also been noted in the development of simulation models [68]. Furthermore, the assumption of local reaction was investigated and the benefits of an extended reaction were shown to improve in accuracy, especially at lower frequencies [69].

In the model, the distance from the sound source to the receiver is a parameter. In this investigation, a representative value, see Equation (41), was used. For open-plan offices, it is of interest to calculate the sound propagation over distances corresponding, e.g., to different workplaces. As the model accounts for the distance and takes into account the angle-dependent absorption of the ceiling absorber, it would also be of interest to investigate this application.

Another application where the non-diffuse sound fields appear are sport halls. A common treatment in such rooms is an absorbent ceiling. The present model clarifies the considerable deviation between diffuse field calculations and measurements that often appear in these rooms. It also shows the importance of a more uniform distribution of absorbent material.

The general assumption of the SEA approach and the method for a subdivision into a grazing and non-grazing sound field can be further improved. It is assumed that the ceiling is the most absorptive area in the type of rooms investigated. However, a more precise description of the non-uniform absorption conditions would be valuable. Comparison with field measurements of different room types would clarify the models applicability and point out opportunities for improvements. Similarly, the limits in the method of estimating the equivalent scattering absorption area must be further investigated. The statistical approach requires a certain minimum room volume for the application of the model. This needs further investigation, but experiences so far indicate a room volume larger than 50 m^3.

In any event, the purpose of the model is to give a direct and reasonably accurate estimation of room acoustic parameters in rooms with absorbent ceiling treatments. The model accounts for the actual mounting height of the ceiling absorber, including both the scattering and absorbing effects of furniture, and reveals the typical characteristic behavior of sound fields in rooms with ceiling treatment, such as the effects of adding wall absorbers.

7. Conclusions

A statistical energy analysis (SEA) model was developed for rooms with absorbent ceiling treatments. The model is based on a subdivision of the sound field into a grazing and non-grazing subsystem where grazing refers to sound waves propagating almost parallel to the absorbent ceiling. The scattering and absorbing effects of furniture and other interior objects is quantified in a measure denoted as the equivalent scattering absorption area A_{sc}. This parameter is related to the energy transfer between the grazing and non-grazing subsystem. The back-transfer from the non-grazing to the grazing subsystem is assumed to be negligible. As a consequence, it is assumed that the ceiling absorption is much greater than the average absorption for walls and floors. An average absorption coefficient for the ceiling absorber greater than 0.7 for the octave bands ranging from 250 to 4000 Hz seems to be sufficient for most practical situations, but this has to be further investigated. In the model, it is assumed that the surface impedance for the suspended ceiling is known or can be calculated. Other surfaces are dealt with in the usual way, using the practical or statistical absorption coefficients. Based on the airflow resistance of the ceiling absorbers investigated, the surface impedances are estimated by the Miki's model, assuming an extended reaction. Thus, the actual mounting height of the ceiling absorber can be accounted for.

The new model was compared with the classical diffuse field model. Experiments were carried out in a classroom mock-up. Two different ceiling absorbers for two different mounting heights were each investigated. One of these cases was also tested in combination with wall panels on two adjacent walls. For all the experiments carried out, the new model shows better agreement with measurements than the classical diffuse field model. The new model reproduces the frequency behaviour of the room acoustic parameters as well as accounting for wall panels in closer agreement with measurements than the diffuse field model.

Further comparison with well-documented field measurements is necessary for the fine-tuning of the model, as well as investigation of the methodology used for estimating the equivalent scattering absorption area.

Author Contributions: Writing—original draft preparation, E.N.; writing—review and editing, E.N.; Experiments and investigations related to Sections 3.2.5 and 5.1, E.N. and E.A. All authors have read and agreed to the published version of the manuscript.

Funding: This research received no external funding.

Institutional Review Board Statement: Not applicable.

Informed Consent Statement: Not applicable.

Data Availability Statement: The data presented in this study are available on request from the corresponding author.

Acknowledgments: The authors want to thank Ecophon R&D and Innovation for providing material, laboratory facilities and equipment. Thanks!

Conflicts of Interest: The authors declare no conflict of interest.

Appendix A

Derivation of the grazing angle at high frequencies.

A more profound argumentation for the theory outlined in this appendix is given in [55,70].

We consider the rectangular room in Figure A1.

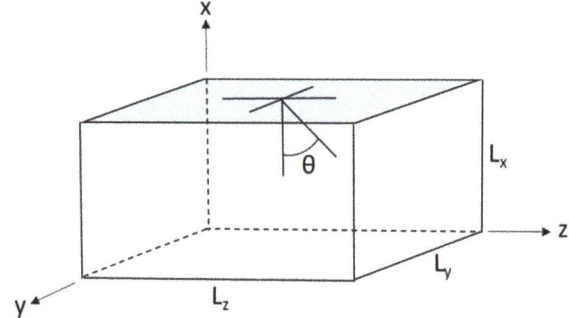

Figure A1. Room with absorbent ceiling.

From the wave equation, the complex wave number is given by

$$\underline{k}^2 = \underline{k}_x^2 + \underline{k}_y^2 + \underline{k}_z^2 \tag{A1}$$

and

$$\cos(\theta) = \frac{k_x}{k} \tag{A2}$$

where k_x and k is the real part of $\underline{k}_x$ and $\underline{k}$, respectively.

Near the absorbing ceiling we expect a phase step of nearly π for grazing incidence. This means that the real component of $\underline{k}_x$ is approximately [55]

$$k_x = \frac{\pi}{L_x}\left(n_x + \frac{1}{2}\right) \tag{A3}$$

where n_x is an integer 0, 1, 2 . . .

Equations (A2) and (A3) gives

$$\cos(\theta) = \frac{c\pi}{\omega L_x}\left(n_x + \frac{1}{2}\right) \tag{A4}$$

θ_g is defined by Equation (A4) for $n_x = 0$. Thus we get

$$\theta_g = \arccos\left(\frac{c}{4fL_x}\right) \tag{A5}$$

By knowing the impedance Z of the ceiling absorber, the grazing absorption coefficient α_g is then calculated as the average absorption coefficient in the grazing region defined by θ_g.

Appendix B

Grazing absorption at low frequencies.

If we consider sound propagation mainly in the yz-plan in Figure A1, the surfaces at $x = 0$ and $x = L_x$ are exposed for the grazing sound field. An expression for a grazing decay constant at low frequencies was derived by Morse and Bolt [71]. An expression of the decay constant at low frequencies is given by

$$\delta = \rho c^2 \left(\frac{A'_{x0} + A'_{xl}}{2l_x} + \frac{A'_{y0} + A'_{yl}}{l_y} + \frac{A'_{z0} + A'_{zl}}{l_z}\right) \tag{A6}$$

where A' is the real part of the admittance. Assuming all the walls and floor in Figure A1 rigid, except for the ceiling, we get

$$\delta = \rho c^2 \frac{A'_{xl}}{2l_x} \tag{A7}$$

For the almost two-dimensional grazing sound field, the contribution from the ceiling to the grazing absorption is given by [55].

$$\eta_{g,ceiling} = \frac{c}{\pi V \omega} S_{ceiling} \alpha_{g,\,ceiling} \tag{A8}$$

The relation between the loss factor η and the decay constant δ is

$$\eta = \frac{2\delta}{\omega} \tag{A9}$$

The grazing ceiling absorption is given by combining Equations (A7)–(A9). We get

$$\alpha_{g,ceiling} = \pi \rho c A'_{xl} \tag{A10}$$

where A'_{xl} is the real part of the admittance for the ceiling absorber. A'_{xl} is given by real part of $1/Z$ where Z is given by Equation (22), assuming an extended reaction.

Equation (A10) is used as an approximation for $\alpha_{g,ceiling}$ for the frequencies 125 Hz and 250 Hz.

References

1. ISO. *ISO 3382-1: 2009, Acoustics—Measurement of Room Acoustic Parameters—Part 1: Performance Spaces*; ISO: Geneva, Switzerland, 2009.
2. ISO. *ISO 3382-2: 2008, Acoustics—Measurement of Room Acoustic Parameters—Part 2: Reverberation Time in Ordinary Rooms*; ISO: Geneva, Switzerland, 2008.
3. Astolfi, A.; Pellerey, F. Subjective and objective assessment of acoustical and overall environmental quality in secondary school clas srooms. *J. Acoust. Soc. Am.* **2008**, *123*, 163–173. [CrossRef] [PubMed]
4. Shield, B.; Conetta, R.; Dockrell, J.; Connolly, D.; Cox, T.; Mydlarz, C. A survey of acoustic conditions and noise levels in secondary school classrooms in England. *J. Acoust. Soc. Am.* **2015**, *137*, 177–188. [CrossRef]

5. Shield, B.M.; Dockrell, J. The effects of environmental and classroom noise on the academic attainments of primary school children. *J. Acoust. Soc. Am.* **2008**, *123*, 133–144. [CrossRef] [PubMed]
6. Shield, B.M.; Dockrell, J. The effects of noise on children at school: A review. *J. Build. Acoust.* **2003**, *10*, 97–116. [CrossRef]
7. Ljung, R.; Israelsson, K.; Hygge, S. Speech Intelligibility and Recall of Spoken Material Heard at Different Sig-nal-to-noise Ratios and the Role Played by Working Memory Capacity. *Appl. Cogn. Psychol.* **2013**, *27*, 198–203. [CrossRef]
8. Yang, W.; Bradley, J.S. Effects of room acoustics on the intelligibility of speech in classrooms for young children. *J. Acoust. Soc. Am.* **2009**, *125*, 922–933. [CrossRef] [PubMed]
9. Bradley, J.S.; Sato, H. The intelligibility of speech in elementary school classrooms. *J. Acoust. Soc. Am.* **2008**, *123*, 2078–2086. [CrossRef]
10. Oberdörster, M.; Tiesler, G. *Akustiche Ergonomie der Schule, Schriftenreihe der Bundesanstalt fur Arbeitsschutz und Arbeitsmedizin, Forshung Fb1071*; Universität Bremen: Bremen, Germany, 2006.
11. Sato, H.; Bradley, J.S. Evaluation of acoustical conditions for speech communication in working elementary school classrooms. *J. Acoust. Soc. Am.* **2008**, *123*, 2064–2077. [CrossRef]
12. Lyberg-Åhlander, V.; Rydell, R.; Löfqvist, A. Speaker's comfort in teaching environments: Voice problems in Swedish teaching staff. *J. Voice* **2011**, *25*, 430–440. [CrossRef]
13. Pelegrin-Garcia, D.; Brunskog, J.; Lyberg-Åhlander, V.; Löfqvist, A. Measurement and prediction of voice support and room gain in school classrooms. *J. Acoust. Soc. Am.* **2012**, *131*, 194–204. [CrossRef] [PubMed]
14. Pelegrin-Garcia, D. Speakers' comfort and voice level variation in classrooms: Laboratory research. *J. Acoust. Soc. Am.* **2012**, *132*, 249–260. [CrossRef]
15. Bistrup, M. *Health Effects of Noise on Children and the Perception of the Risk of Noise*; National Institute of Public Health: Cpenhagen, Denmark, 2001.
16. Bistrup, M. *Children and Noise-Prevention of Adverse Effects*; National Institute of Public Health: Cpenhagen, Denmark, 2002.
17. Waye, K.P.; Fredriksson, S.; Hussain-Alkhateeb, L.; Gustafsson, J.; Van Kamp, I. Preschool teachers' perspective on how high noise levels at preschool affect children's behavior. *PLoS ONE* **2019**, *14*, e0214464. [CrossRef]
18. Ryherd, E.E.; Persson Waye, K.; Ljungkvist, L. Characterizing noise and perceived work environment in a neurological intensive care unit. *J. Acoust. Soc. Am.* **2008**, *123*, 747–756. [CrossRef]
19. Haapakangas, A.; Hongisto, V.; Eerola, M.; Kuusisto, T. Distraction distance and disturbance by noise—An analysis of 21 open-plan offices. *J. Acoust. Soc. Am.* **2017**, *141*, 127–136. [CrossRef]
20. ISO. *ISO 3382-3:2012 Acoustics—Measurement of Room Acoustic Parameters—Part 3: Open Plan offices*; ISO: Geneva, Switzerland, 2012.
21. ISO. *ISO 22955:2021 Acoustics—Acoustic Quality of Open Office Spaces*; ISO: Geneva, Switzerland, 2021.
22. Bradley, J. Review of objective room acoustics measures and future needs. *Appl. Acoust.* **2011**, *72*, 713–720. [CrossRef]
23. Sato, H.; Morimoto, M.; Sato, H.; Wada, M. Relationship between listening difficulty and acoustical objective measures in reverberant fields. *J. Acoust. Soc. Am.* **2008**, *123*, 2087–2093. [CrossRef]
24. Harvie-Clark, J.; Dobinson, N.; Hinton, R. Acoustic Response in Non-Diffuse Rooms. In Proceedings of the Euronoise, Prague, Czech Republic, 10–13 June 2012.
25. Rindel, J.H. Acoustical capacity as a means of noise control in eating establishment. In Proceedings of the Joint Baltic Nordic Acoustics Meeting, Odense, Denmark, 18–20 June 2012.
26. Nijs, L.; Rychtáriková, M. Calculating the optimum reverberation time and absorption coefficient for good speech intelligibility in classroom design using U50. *Acta Acust. United Acust.* **2011**, *97*, 93–102. [CrossRef]
27. Barron, M.; Lee, L.J. Energy relations in concert auditoriums. I. *J. Acoust. Soc. Am.* **1988**, *84*, 618–628. [CrossRef]
28. Barron, M. Theory and measurement of early, late and total sound levels in rooms. *J. Acoust. Soc. Am.* **2015**, *137*, 3087–3098. [CrossRef] [PubMed]
29. Nilsson, E. Input data for acoustical design calculations for ordinary public rooms. In Proceedings of the 24th International Congress on Sound and Vibration, London, UK, 23–27 July 2017.
30. Sabine, W.C. *Collected Papers on Acoustics*; Dover Publications: New York, NY, USA, 1964.
31. Rasmussen, B.; Brunskog, J.; Hoffmeyer, D. Reverberation time in class rooms—Comparison of regulations and classi-fication criteria in the Nordic countries. In Proceedings of the Joint Baltic—Nordic Acoustics Meeting, Odense, Denmark, 18–20 June 2012.
32. IEC. *IEC 60268-16 Sound System Equipment—Part 16: Objective Rating of Speech Intelligibility by Speech Transmission Index*; IEC: Geneva, Switzerland, 2011.
33. Astolfi, A.; Parati, L.; D'Orazio, D.; Garai, M. The new Italien standard UNI 11532 on acoustics for schools. In Proceedings of the 23rd International Congress on Acoustics, Aachen, Germany, 9–13 September 2019.
34. Cremer, L.; Muller, H.A.; Schultz, T.J. *Principles and Applications of Room Acoustics Vol 1*; Applied Science Publishers: London, UK, 1982.
35. Kuttruff, H. *Room Acoustics*, 3rd ed.; Elsevier Science Publishers Ltd.: London, UK, 1991.
36. Joyce, W.B. Sabine's reverberation time and ergodic auditoriums. *J. Acoust. Soc. Am.* **1975**, *58*, 643–655. [CrossRef]
37. Joyce, W.B. Exact effect of surface roughness on the reverberation time of a uniformly absorbing spherical enclosure. *J. Acoust. Soc. Am.* **1978**, *64*, 1429–1436. [CrossRef]
38. Fitzroy, D. Reverberation formulae which seem to be more accurate with non-uniform distribution of absorption. *J. Acoust. Soc. Am.* **1959**, *31*, 893–897. [CrossRef]

39. Neubauer, N.O. Estimation of reverberation time in rectangular rooms with non-uniformly distributed absorption using a modified Fitzroy equation. *Build. Acoust.* **2001**, *8*, 115–137. [CrossRef]
40. Arau-Puchades, H. An improved reverberation formula. *Acustica* **1988**, *65*, 163–180.
41. Cucharero, J.; Hänninen, T.; Lokki, T. Influence of sound-absorbing material placement on room acoustical parame-ters. *Acoustics* **2019**, *1*, 644–660.
42. Sakuma, T. Approximate theory of reverberation in rectangular rooms with specular and diffuse reflections. *J. Acoust. Soc. Am.* **2012**, *132*, 2325–2336. [CrossRef] [PubMed]
43. Bistafa, S.R.; Bradley, J.S. Predicting reverberation times in a simulated classroom. *J. Acoust. Soc. Am.* **2000**, *108*, 1721–1731. [CrossRef]
44. Prodi, N.; Visentin, C. An experimental evaluation of the impact of scattering on sound field diffusivity. *J. Acoust. Soc. Am.* **2013**, *133*, 810–820. [CrossRef] [PubMed]
45. ISO. *ISO 354:2003 Acoustics—Measurement of Sound Absorption in a Reverberation Room*; ISO: Geneva, Switzerland, 2003.
46. Nolan, M.; Berzborn, M.; Fernandez-Grande, E. Isotropy in decaying reverberant sound fields. *J. Acoust. Soc. Am.* **2020**, *148*, 1077–1088. [CrossRef]
47. Nilsson, E. Decay Processes in Rooms with Non-Diffuse Sound Fields Part I: Ceiling Treatment with Absorbing Material. *Build. Acoust.* **2004**, *11*, 39–60. [CrossRef]
48. Nilsson, E. Decay processes in rooms with non-diffuse sound fields—Part II: Effect of irregularities. *Build. Acoust.* **2004**, *11*, 133–143. [CrossRef]
49. EN 12354-6:2004, *Building Acoustics—Estimation of Acoustic Performance of Buildings from the Performance of Elements—Part 6: Sound Absorption in Enclosed Spaces*; CEN: Brussels, Belgium, 2004.
50. Lochner, J.; Burger, J. The influence of reflections on auditorium acoustics. *J. Sound Vib.* **1964**, *1*, 426–454. [CrossRef]
51. Visentin, C.; Pellegatti, M.; Prodi, N. Effect of a single lateral diffuse reflection on spatial percepts and speech intelligi-bility. *J. Acoust. Soc. Am.* **2020**, *148*, 122–140. [CrossRef]
52. Bradley, J.S.; Sato, H.; Picard, M. On the importance of early reflections for speech in rooms. *J. Acoust. Soc. Am.* **2003**, *113*, 3233–3244. [CrossRef]
53. Lyon, R.H.; De Jong, R.G. *Theory and Application of Statistical Energy Analysis*, 2nd ed.; RH Lyon Corp: Cambridge, MA, USA, 1998; p. 02138.
54. Crighton, D.G.; Dowling, A.P.; Ffowcs Williams, J.E.; Heckl, M.; Leppington, F.G. *Modern Methods in Analytical Acoustics*; Springer London: London, UK, 1992; Chapter 8.
55. Nilsson, E. *Decay Processes in Rooms with Non-Diffuse Sound Fields*; Report TVBA-1004; Engineering Acoustics, LTH, Lund University: Lund, Sverige, 1992.
56. Barron, M.; Barron, M. Growth and decay of sound intensity in rooms according to some formulae of geometric acoustics theory. *J. Sound Vib.* **1973**, *27*, 183–196. [CrossRef]
57. Available online: https://alphacell.matelys.com/ (accessed on 16 July 2021).
58. Available online: https://norsonic.se/product_single/norflag/ (accessed on 16 July 2021).
59. Miki, Y. Acoustical properties of porous materials. Modifications of Delany-Bazley models. *J. Acoust. Soc. Jpn. (E)* **1990**, *11*, 19–24. [CrossRef]
60. Jeong, C.-H. Guidline for adopting the local reaction assumption for porous absorbers in terms of random incidence ab-sorption coefficients. *Acta Acoust. United Acust.* **2011**, *97*, 779–790. [CrossRef]
61. Delany, M.; Bazley, E. Acoustical properties of fibrous absorbent materials. *Appl. Acoust.* **1970**, *3*, 105–116. [CrossRef]
62. Komatsu, T. Improvement of the Delaney-Bazley and Miki models for fibrous sound-absorbing materials. *Acoust. Sci. Tech.* **2008**, *29*, 121–129. [CrossRef]
63. Paris, E.T. On the coefficient of sound-absorption measured by the reverberation method. *Philos. Mag.* **1928**, *5*, 489. [CrossRef]
64. Bakoulas, K. Optimization of an Energy-Based Room Acoustics Model that Considers Scattering and Non-Uniform Absorption. Master's Thesis, Department of Electrical Engineering, Technical University of Denmark, Lyngby, Denmark, 2017.
65. Nilsson, E. Sound scattering in rooms with ceiling treatment. In *Nordtest Technical Report TR 606*; Nordic Innovation Centre: Oslo, Norway, 2007.
66. ISO. *ISO 11654:1997 Acoustics—Sound Absorbers for Use in Buildings—Rating of Sound Absorption*; ISO: Geneva, Switzerland, 1997.
67. Arvidsson, E.; Nilsson, E.; Hagberg, D.B.; Karlsson, O.J.I. The effect on room acoustical parameters using a combina-tions of absorbers and diffusors—An experimental study in a classroom. *Acoustics* **2020**, *2*, 505–523. [CrossRef]
68. Marbjerg, G.; Brunskog, J.; Jeong, C.-H.; Nilsson, E. Development and validation of a combined phased acoustical radi-osity and image source model for predicting sound fields in rooms. *J. Acoust. Soc. Am.* **2015**, *138*, 1457–1468. [CrossRef] [PubMed]
69. Pind, F.; Jeong, C.-H.; Engsig-Karup, A.P.; Hesthaven, J.S.; Stromann-Andersen, J. Time-domain room acoustic simula-tions with extended-reacting porous absorbers using the discontinuous Galerkin method. *J. Acoust. Soc. Am.* **2020**, *148*, 2851–2863. [CrossRef] [PubMed]
70. Cremer, L.; Muller, H.A.; Shultz, T.J. *Principles and Applications of Room Acoustics Vol 2*; Applied Science Publishers: London, UK, 1982.
71. Morse, P.M.; Bolt, R.H. Sound waves in rooms. *Rev. Mod. Phys.* **1944**, *16*, 69–150. [CrossRef]

Article

Coherent Image Source Modeling of Sound Fields in Long Spaces with a Sound-Absorbing Ceiling [†]

Hequn Min * and Ke Xu

Key Laboratory of Urban and Architectural Heritage Conservation, Ministry of Education, School of Architecture, Southeast University, 2 Sipailou, Nanjing 210096, China; xuke@seu.edu.cn
* Correspondence: hqmin@seu.edu.cn
† This paper is an extension of a conference paper presented in INTER-NOISE 2014.

Abstract: Sound-absorbing boundaries can attenuate noise propagation in practical long spaces, but fast and accurate sound field modeling in this situation is still difficult. This paper presents a coherent image source model for simple yet accurate prediction of the sound field in long enclosures with a sound absorbing ceiling. In the proposed model, the reflections on the absorbent boundary are separated from those on reflective ones during evaluating reflection coefficients. The model is compared with the classic wave theory, an existing coherent image source model and a scale-model experiment. The results show that the proposed model provides remarkable accuracy advantage over the existing models yet is fast for sound prediction in long spaces.

Keywords: long space; coherent image source method; sound-absorbing boundary; sound field modeling; scale-model experiment

1. Introduction

Sound prediction is very important for design of practical long spaces such as traffic tunnels and subway stations to evaluate acoustical qualities such as speech intelligibility of public address systems [1]. For noise control in such long spaces, it is often the case to apply acoustical liners to the space ceiling for larger noise attenuation. Regarding sound prediction in such spaces, classic room acoustics formulas are unsatisfactory [2,3] because the sound field is not diffused due to the extreme dimensions. The commonly used incoherent geometrical acoustics models [4–8] cannot account for the interference between multiple sound reflections on impedance boundaries, which were experimentally observed to be distinct and can notably affect the sound prediction accuracy in this situation [3,9,10], especially at lower frequencies and in early parts of the impulse response [10].

For the coherent geometrical acoustics models, Li et al. developed a numerical model for coherent sound prediction inside long spaces [3,11] and afterwards applied this prediction model into full-scale tunnels [9] and a long space with impedance discontinuities [12]. It was shown that their coherent prediction model provides much better prediction accuracy than the usual incoherent ones for long spaces with reflective boundaries. However, for applications with sound-absorbing boundaries, the applicability of their model may be limited. The numerical model of Li et al. [3] originated from a coherent image source method by Lemire and Nicolas [13], in which it is implicitly assumed that the wave front shapes remain spherical during each successive reflection of the initial spherical wave radiation [13]. This assumption may hardly hold for reflections on sound-absorbing boundaries.

Recently, Min et al. [14] proposed a coherent image source method for fast yet accurate sound prediction in flat spaces with absorbent boundaries. They proposed different refection coefficients to evaluate the reflections on the absorbent and reflective boundaries, which avoids the prediction difficulties with absorbent boundaries in the method of Lemire

and Nicolas. Unfortunately, their model is currently limited to spaces with two parallel infinite boundaries in theory.

Upon reviewing the studies above, there is still the problem of sound prediction in long spaces with sound-absorbing boundaries for practical noise control. In this paper, a coherent image source model is extended and examined theoretically and experimentally for fast yet accurate sound prediction in long spaces with sound-absorbing boundaries.

2. Theoretical Method

Figure 1 shows the cross-sectional geometry of a long rectangular space with a height of H and width W. For simplicity, four boundaries in this space, the ceiling, ground, and right and left walls, are assumed to be locally reactive with a uniform normalized specific admittance of β_c, β_g, β_r, and β_l, respectively. The ceiling is defined to be sound absorptive with a relatively high sound absorption coefficient, while other boundaries are sound reflective with a relatively low absorption coefficient. The space extends infinitely along the y-direction as a typical case and a point source is located at $(x_S, 0, z_S)$ and a receiver is located at (x_r, y_r, z_r) inside.

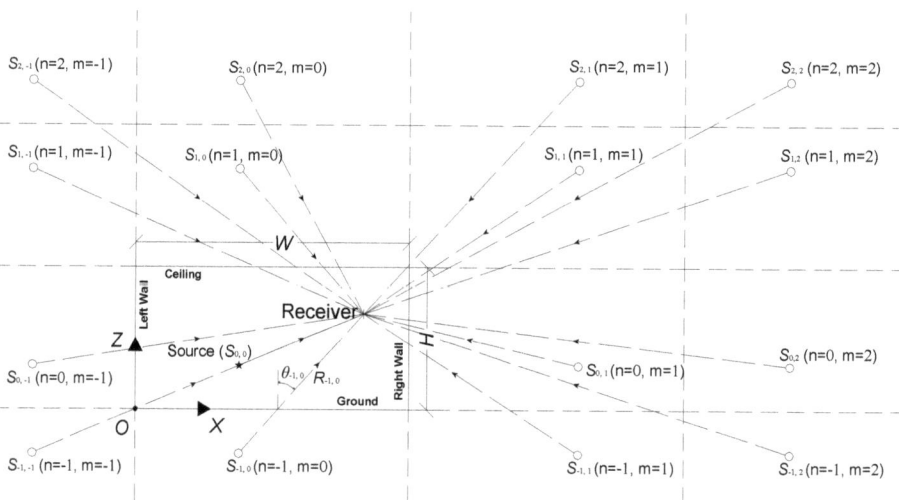

Figure 1. Cross-sectional geometry of a long rectangular space with height H and width W, and the image sources formed by multiple reflections on its four boundaries.

To model the sound field, we first assume that $kW \gg 1$ and $kH \gg 1$ (with k for the wavenumber) so that the boundaries may be considered infinity for each sound reflection on them [13]. The total sound pressure field at receiver can be approximated as a summation of successive sound reflections on four boundaries:

$$P_{tot} \approx \sum_{n=-\infty}^{+\infty} \sum_{m=-\infty}^{+\infty} P_{n,m} \qquad (1)$$

where $n, m = 0, \pm 1, \pm 2, \ldots$, and $P_{n,m}$ represents the sound field contribution from the (n, m)-th order image source, in which a positive n (or m) is for an image source located above the ceiling (or rightwards from the right wall) while a negative n (or m) is for that located below the floor (or leftwards from the left wall), as shown in Figure 1. Particularly, $P_{0,0}$ denotes the direct sound from the real source $S_{0,0}$. Based on the assumption of $kW \gg 1$

and $kH \gg 1$, $P_{n,m}$ can be approximated from the plane wave expansion of a spherical wave as follows [14,15]:

$$P_{n,m} \approx \frac{jk}{8\pi^2} \int_0^{2\pi} \int_0^{\frac{\pi}{2}-j\infty} e^{j\mathbf{k}\cdot\mathbf{R}_{n,m}} [V_g(\theta)]^{n_g} [V_c(\theta)]^{n_c} [V_l(\alpha)]^{m_l} [V_r(\alpha)]^{m_r} \sin\theta \, d\theta \, d\varphi \quad (2)$$

where n_g, n_c, m_l, and m_r are used to count reflection times on the ground, ceiling, and left and right walls in the path from $S_{n,m}$ to the receiver, respectively. They can be determined from the order (n, m) by

$$\begin{aligned} n_{g,c} &= \tfrac{|n|}{2} \mp \tfrac{1}{2}\text{sign}(n)\text{rem}(|n|,2) \\ m_{l,r} &= \tfrac{|m|}{2} \mp \tfrac{1}{2}\text{sign}(m)\text{rem}(|m|,2) \end{aligned} \quad (3)$$

In Equation (2), $\mathbf{R}_{n,m} = (R_{n,m}\sin\theta_{n,m}\cos j_{n,m}, R_{n,m}\sin\theta_{n,m}\sin j_{n,m}, R_{n,m}\cos\theta_{n,m})$ represents the distance vector from $S_{n,m}$ to the receiver, with explicit azimuth angles $\theta_{n,m}$ and $j_{n,m}$. $V_g(\theta)$ and $V_c(j)$ are the plane wave reflection coefficients on the "infinite" ground and ceiling with the incidence angle θ, respectively, while $V_l(\alpha)$ and $V_r(\alpha)$ are those on the left and right walls with the incidence angle $\alpha = \pi/2 - \theta$, respectively. These plane wave reflection coefficients can be correspondingly evaluated by [16]

$$V_{g,c,r,l}(\zeta) = \frac{\cos\zeta - \beta_{g,c,r,l}}{\cos\zeta + \beta_{g,c,r,l}} \quad (4)$$

Through an identical mathematical transformation similar to that from Equations (8)–(12) in Ref. [14], the evaluation of $P_{n,m}$ in Equation (2) can be simplified as

$$P_{n,m} \approx \frac{jk}{8\pi} \int_{-\frac{\pi}{2}+j\infty}^{\frac{\pi}{2}-j\infty} V(\theta)\sin\theta \cdot H_0^1(kr\sin\theta) \, e^{jkR_{n,m}\cos\theta\cos\theta_{n,m}} d\theta \quad (5)$$

where $V(\theta)$ represents the term $[V_g(\theta)]^{n_g}[V_c(\theta)]^{n_c}[V_l(\alpha)]^{m_l}[V_r(\alpha)]^{m_r}$, $r = R_{n,m}\sin\theta_{n,m}$, and $H_0^1(.)$ is the first Hankel function with zero-th order. In Equation (5), the ray field from single reflection on the reflective boundaries, $P_{-1,0}$ for example, can be further evaluated as [13,17]

$$P_{-1,0} = Q_{ref}(S_{-1,0}, R|GB) \cdot \frac{e^{jkR_{-1,0}}}{4\pi R_{-1,0}} \quad (6)$$

where $Q_{ref}(S_{-1,0}, R|GB)$ represents the single reflection coefficient on the reflective ground boundary (GB) and is evaluated as [13,17]

$$Q_{ref}(S_{-1,0}, R|GB) = V_g(\theta_{-1,0}) + [1 - V_g(\theta_{-1,0})]F(w_n), \quad (7)$$

in which

$$F(w_n) = 1 + j\sqrt{\pi} \cdot w_n \cdot g(w_n), \quad (8)$$

$$w_n = \sqrt{kR_{-1,0}} \cdot \frac{1+j}{2} \cdot (\cos\theta_{-1,0} + \beta_g), \text{ and} \quad (9)$$

$$g(w_n) = e^{-w_n^2}\text{erfc}(jw_n). \quad (10)$$

Further analytical approximation of $g(w_n)$ is available in Ref. [18].

It was shown that the wave front shape before and after each reflection on the reflective boundaries can almost remain the same [14,19]. This suggests that single reflection coefficients Q_{ref} shall be weakly dependent on θ and be almost uniform for different spatial parts of incident wave fronts of any shapes [14]. Accordingly, during the ray propagation from $S_{n,m}$ to receiver in Figure 1, the evaluation of each reflection upon one reflective boundary (or each "transmission" through it or its images) can be approximated by once-weighting the ray field with the corresponding single reflection coefficient Q_{ref} on this boundary [14].

Thus, after the ray field being weighted for n_g, m_l, and m_r times due to "transmission" through the reflective ground, left wall (LW) and right wall (RW), and their images, the evaluation of $P_{n,m}$ can be simplified by

$$P_{n,m} \approx [Q_{ref}(S_{n,m}, R|GB)]^{n_g} \cdot [Q_{ref}(S_{n,m}, R|LW)]^{m_l} \cdot [Q_{ref}(S_{n,m}, R|RW)]^{m_r}$$
$$\cdot \frac{jk}{8\pi} \int_{-\frac{\pi}{2}+j\infty}^{\frac{\pi}{2}-j\infty} [V_c(\theta)]^{n_c} \sin\theta \cdot H_0^1(kr\sin\theta) \, e^{jkR_{n,m}\cos\theta\cos\theta_{n,m}} d\theta, \quad (11)$$

where the integral involves only the reflection coefficient on the absorptive ceiling boundary (CB) and can be further evaluated through the second order approximation provided by Brekhovskikh [15] to yield

$$P_{n,m} \approx [Q_{ref}(S_{n,m}, R|GB)]^{n_g} \cdot [Q_{ref}(S_{n,m}, R|LW)]^{m_l} \cdot [Q_{ref}(S_{n,m}, R|RW)]^{m_r}$$
$$\cdot \left\{ V_t(\theta_{n,m}|CB, n_c) - \frac{j[V'_t(\theta_{n,m}|CB, n_c)\cot\theta_{n,m} + V''_t(\theta_{n,m}|CB, n_c)]}{2kR_{n,m}} \right\} \cdot \frac{e^{jkR_{n,m}}}{4\pi R_{n,m}}, \quad (12)$$

where $V_t(\theta_{n,m}|CB, n_c) = [V_c(\theta_{n,m})]^{n_c}$, and $V'_t(\theta_{n,m}|CB, n_c)$ and $V''_t(\theta_{n,m}|CB, n_c)$ are the first and second derivatives of $V_t(\theta_{n,m}|CB, n_c)$ at $\theta_{n,m}$, respectively. This equation may be rewritten as an image source model form as

$$P_{n,m} = Q_{n,m} \frac{e^{jkR_{n,m}}}{4\pi R_{n,m}}, \quad (13)$$

where $Q_{n,m}$ represents a combined reflection coefficient corresponding to the ray with reflection order (n,m) as

$$Q_{n,m} = [Q_{ref}(S_{n,m}, R|GB)]^{n_g} \cdot [Q_{ref}(S_{n,m}, R|LW)]^{m_l} \cdot [Q_{ref}(S_{n,m}, R|RW)]^{m_r} \cdot Q_{abs}(S_{n,m}, R|CB, n_c) \quad (14)$$

in which $Q_{abs}(S_{n,m}, R|CB, n_c)$ represents one reflection coefficient accounting for overall effect from successive reflections on the absorptive ceiling boundary as

$$Q_{abs}(S_{n,m}, R|CB, n_c) \approx V_t(\theta_{n,m}|CB, n_c) - \frac{j[V'_t(\theta_{n,m}|CB, n_c)\cot\theta_{n,m} + V''_t(\theta_{n,m}|CB, n_c)]}{2kR_{n,m}} \quad (15)$$

One can easily expand $Q_{abs}(S_{n,m}, R|CB, n_c)$ for analytical evaluation, and this is not presented here for succinctness. Equations (1), (13), and (14) provide a coherent image source model for long rectangular spaces with a sound-absorbent ceiling.

3. Results and Discussion

Numerical simulations are firstly carried out to validate the proposed coherent image source model. As the classic wave theory is analytically exact in the spaces studied in this paper [16], it is used as a reference method to provide benchmark results in validations. The coherent image source method by Lemire and Nicolas [13] that was widely used in previous studies [3,9,12] is also investigated for comparisons. Numerical implementation of the methods above stays similar to that in Refs. [13,14], except the geometry of four boundaries in Figure 1.

In simulations, a long rectangular space with $W \times H$ = 20 m × 5 m is considered to simulate one city road tunnel with four lanes. For simplicity, four tunnel boundaries are all assumed to be rigidly backed layers of homogeneous porous material. Attenborough's "three-parameter" approximation [14,20] is applied to evaluate surface admittances for these boundaries, in which the boundary media parameters of flow resistivity (σ), porosity (Ω), tortuosity (T), pore shape factor (S_p), and thickness (d) are used for evaluation. The tunnel ceiling is defined as highly sound absorptive, with σ = 10 cgs (where 1 cgs = 1 kPa s m^{-2}), Ω = 1, T = 1, S_p = 0.25, and d = 0.1 m, such as a wool layer. The ground has σ = 10 k cgs, Ω = 0.2, T = 1.4, S_p = 0.5, and d = 0.05 m to represent a compact asphalt pavement layer. The right and left walls have σ = 0.5 k cgs, Ω = 0.1, T = 1, S_p = 0.3, and d = 0.01 m to represent cement plaster over concrete walls. Figure 2 shows the corresponding normal incident absorption coefficients of these four boundaries in simulations.

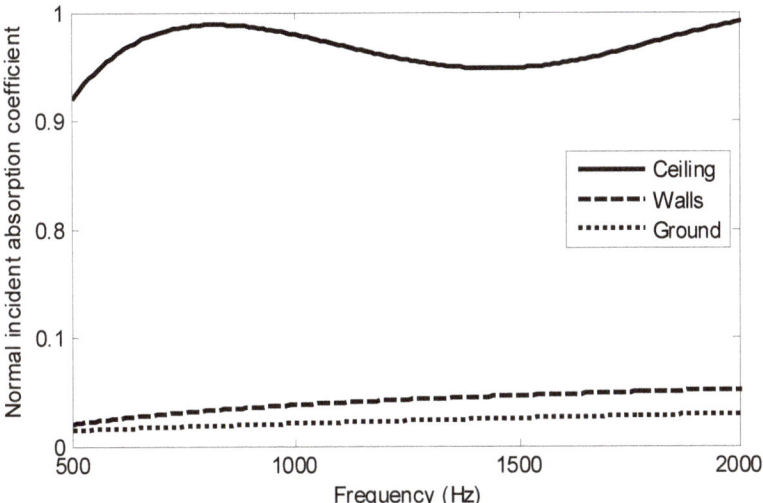

Figure 2. Spectra of the normal incident absorption coefficient on four boundaries of the rectangular long space in numerical simulations.

Two sets of numerical simulations are conducted. In the first set, predictions of sound pressure level (SPL) spectrum at the receiver (6 m, 50 m, 1 m) from a point source at (6 m, 0 m, 1 m) are investigated. Predictions from the proposed method, the wave theory, and the method of Lemire and Nicolas are compared in Figure 3. It is shown that the results from the proposed method agree excellently with those of the wave theory over frequencies from 500 to 2000 Hz, with only small deviations (<1 dB) at few lower frequencies. This suggests the successful extension of the coherent image source method by Min et al. [14] for spaces enclosed by four perpendicular finite boundaries in this paper. It can also be observed from Figure 3 that the predictions with the method of Lemire and Nicolas differ significantly from the benchmark results over frequencies in this situation. This indicates that the existing coherent models [3,9,12] based on the method of Lemire and Nicolas can hardly be accurate in long spaces with absorbent boundaries because the assumption of spherical wave front shapes for each successive reflection is unsatisfied in this situation. All simulations are executed in Matlab 2010b on the same personal computer with a 2.4 GHz Intel Core i5-560M processor and 8 Gbytes of random access memory. Computational time records show that, for results at all the 31 frequencies in Figure 3, evaluation of the proposed model and the method of Lemire and Nicolas takes 50.3 s and 49.7 s, respectively, while the corresponding execution time with the wave theory takes over 2 h. This indicates the remarkable advantage of the proposed model both at accuracy and efficiency for sound predictions in long spaces with absorbent boundaries.

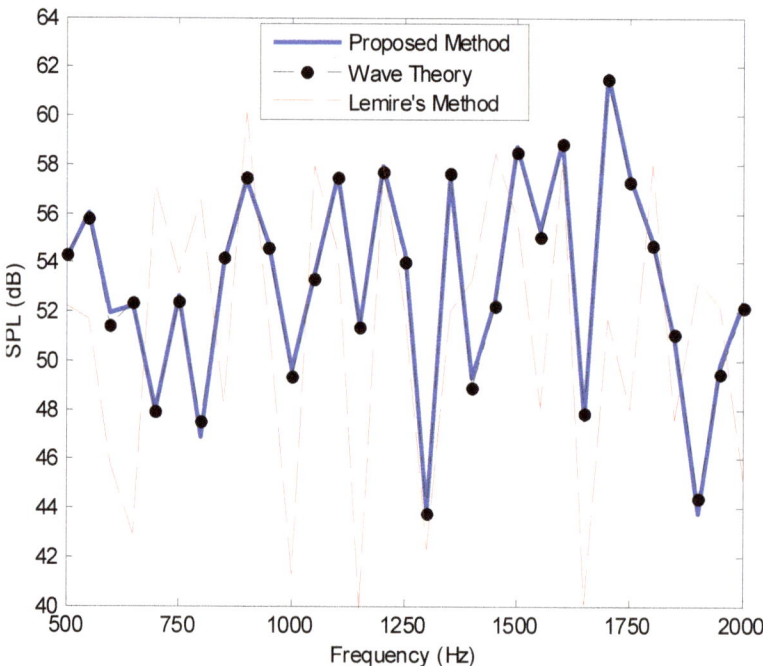

Figure 3. Comparison of predictions on sound pressure level (SPL) spectrum in a long space with 20 m wide and 5 m high in numerical simulations. The source is located at (6 m, 0 m, 1 m), and the receiver is located at (6 m, 50 m, 1 m). The solid line represents the results with the proposed method, the solid circles represent the results with a dot-dash line are those with the wave theory that is considered a benchmark, and the dash line is the result with the method of Lemire and Nicolas (Lemire's method).

In the second simulation set, predictions of the SPL distribution inside the long rectangular space are investigated. Figure 4 presents SPL predictions at frequency of 1000 Hz versus the receiver location along the tunnel extension direction. The source is located at (6 m, 0 m, 1 m) and the receiver is located at (6 m, y_r, 1 m) with y_r moving from 5 m to 200 m along the space length-extending direction. From Figure 2, the absorption coefficients of the ceiling, ground, and walls at a frequency of 1000 Hz are 0.98, 0.02, and 0.04, respectively. Figure 4 shows remarkably good agreement between the proposed model and the wave theory, even at receiver locations far away from the source compared to the space height and width. However, large prediction differences can be found between the method of Lemire and Nicolas and the reference method in this situation. In Figure 4, prediction error from the proposed method increases at a longer source/receiver distance. The reason may be that, when the receiver moves farther away from the source compared to the space cross section dimensions, high-order reflection rays provide relatively higher contributions in the receiver total sound field. In the proposed model, reflection ray field is evaluated through Equations (13) and (14) by approximating each reflection at reflective boundaries as one single reflection. This may accumulate larger errors for higher-order reflection rays.

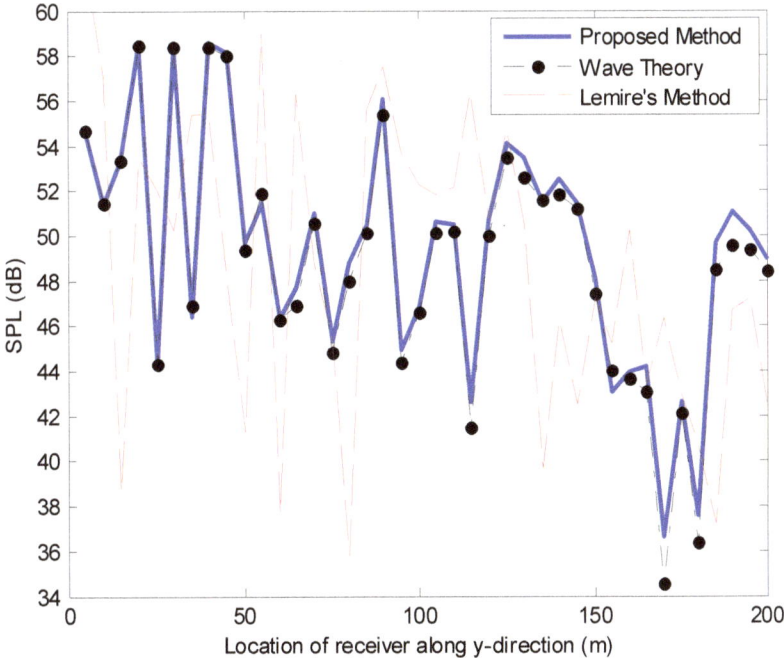

Figure 4. Comparison of predictions on sound pressure level (SPL) at frequency of 1000 Hz vs. the receiver location along the y-direction. The source is located at (6 m, 0 m, 1 m) and the receiver is located at (6 m, yr, 1 m) with yr moving from 5 m to 200 m. The solid line represents the results with the proposed method, the solid circles with a dot-dash line are those with the wave theory that is considered the benchmark, and the dash line is the result with the method of Lemire and Nicolas (Lemire's method).

Figure 5 presents the SPL predictions at frequency of 1000 Hz versus the receiver location along the tunnel width direction. The source is located at (6 m, 0 m, 1 m) and the receiver is located at (x_r, 50 m, 1 m) with x_r moving from 1 m to 19 m. The results show excellent prediction agreement between the proposed model and the wave theory, even at receiver locations close to boundary interaction corners compared to the wavelength. In Figure 5, large discrepancies remain between the method of Lemire and Nicolas, and the reference method. Predictions are also compared on the SPL at 1000 Hz versus the receiver location along the tunnel height direction, with the source at (6 m, 0 m, 1 m) and the receiver at (6 m, 50 m, z_r) with z_r moving from 0.125 m to 4.875 m. The corresponding results are presented in Figure 6. It is shown that results from the proposed model almost overlap those from the wave theory, not only at receiver locations close to the reflective ground but also at those in the vicinity of the absorbent ceiling compared to the wavelength. Predictions with the method of Lemire and Nicolas still bias much from the benchmark results in this situation. The computational time in this simulation set is also recorded for comparison. In Matlab 2010b on the same computer as mentioned above, an evaluation of the proposed model and the method of Lemire and Nicolas takes about 1.8 s and 1.7 s for results at each receiver location in Figure 4 to Figure 6, respectively, while the corresponding calculation with the wave theory takes over 80 s. These show that the proposed coherent image source model can accurately predict the sound fields in long rectangular spaces with an absorbent ceiling, while its computational load stays at a same level with the existing models [3,9,12,13].

A scale-model experiment was carried out to further verify the predictions. One model long rectangular space was built with inner dimensions of 0.7 m width, 0.45 m height, and 10 m length for the measurements, which was scaled with 1:10 to represent a tunnel with $W \times H = 7 \text{ m} \times 4.5 \text{ m}$ in full scale (all of the following dimensions referred to are scaled ones unless otherwise stated). Panels of 20 mm thick high-density fiberboard were used to build the model long space, and the model's inner surfaces were well finished to represent sound reflective ground and wall boundaries. A layer of 50 mm thick fiberglass was used as a liner on the top panel to represent an absorbent ceiling. To minimize the sound reflection on the two ends of the model's long space for infinite extension, liners of 200 mm thick fiberglass were applied onto those two end panels. The specific normalized admittances of the model boundaries were preliminarily measured through an impedance tube kit typed B&K 4206.

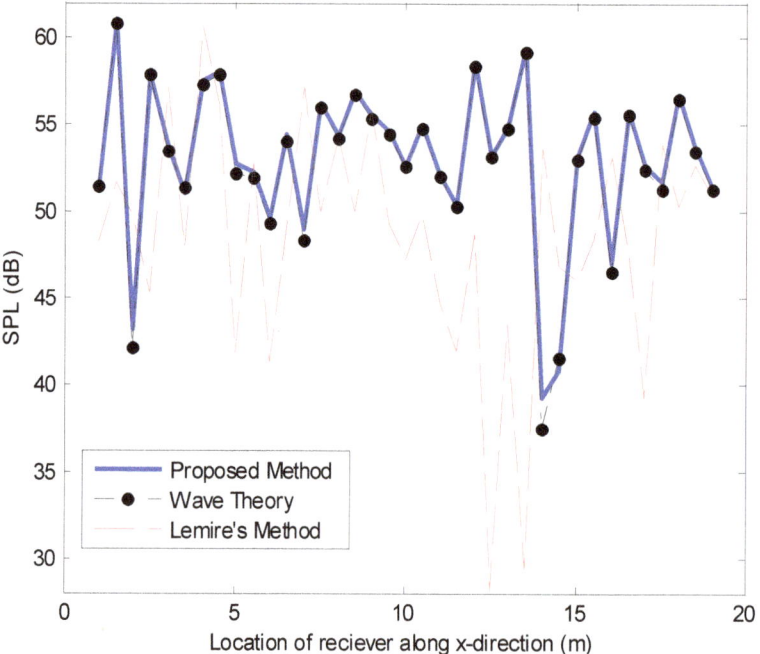

Figure 5. Comparison of predictions on sound pressure level (SPL) at a frequency of 1000 Hz vs. the receiver location along the x-direction. The source is located at (6 m, 0 m, 1 m) and the receiver is located at (x_r, 50 m, 1 m) with x_r moving from 1 m to 19 m. The solid line represents the results with the proposed method, the solid circles are those with the wave theory that is considered as a benchmark, and the dash line is the result with the method of Lemire and Nicolas (Lemire's method).

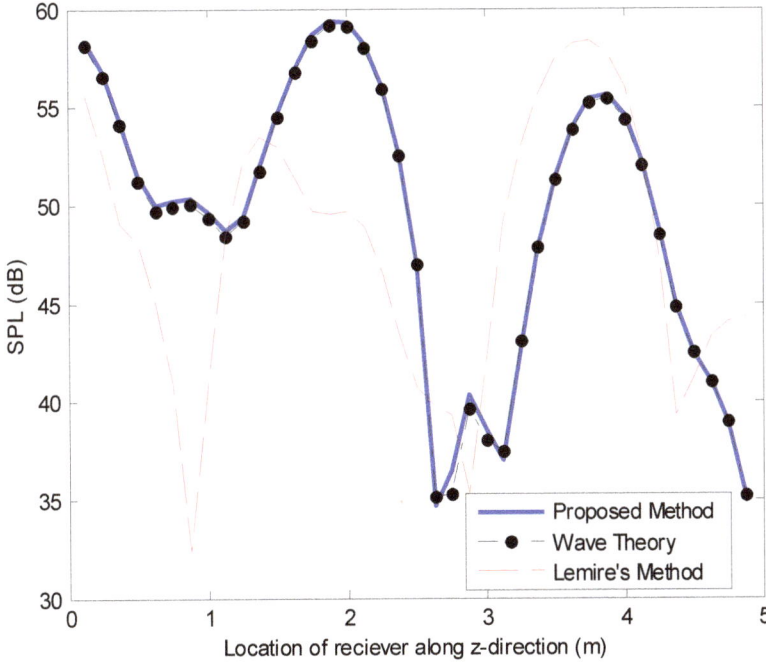

Figure 6. Comparison of predictions on sound pressure level (SPL) at frequency of 1000 Hz vs. the receiver location along the z-direction. The source is located at (6 m, 0 m, 1 m) and the receiver is located at (6 m, 50 m, z_r) with z_r moving from 0.125 m to 4.875 m. The solid line represents the results with the proposed method, the solid circles are those with the wave theory that is considered as a benchmark, and the dash line is the result with the method of Lemire and Nicolas (Lemire's method).

A speaker driver with a tube of internal diameter of 2 cm and length of 1 m was applied to represent a point source [3]. One microphone typed B&K 4190 was used as the receiver. Sound signals were generated and collected through one B&K Pulse system 3560D. In measurements, high enough levels of white noise were generated into the model long space to ensure the steady SPL at most locations inside remained at least 15 dB higher than the background noise. In accordance with coordinates defined in Figure 1, in the experiment, the point source (the speaker tube mouth) was located at (0.35 m, 0 m, 0.2 m) and the receiver was located at (0.35 m, y_r, 0.1 m) with y_r moving from 0.1 m to 8 m to investigate the SPL distribution along the space extension direction. Relative attenuation (RA) was used to present the measured and predicted results in the experiment, which is defined as subtracting the SPL at (0.35 m, 0.1 m, 0.2 m) from that at receiver. The predictions from the wave theory that are used as benchmarks in simulations were firstly compared with the experimental data. Figure 7 presents the comparison results on the RA distribution along the y-direction. The frequency of 1000 Hz was chosen without loss of generality, at which β_c was (1.2068–1.4338i), corresponding to a normal incident absorption coefficient of 0.7, while β_g, β_r, and β_l were (0.0272 + 0.1041i), corresponding to a normal incident absorption coefficient of 0.1. In Figure 7, reasonable agreement is shown between the wave theory predictions and the experimental data. By considering experimental uncertainty and errors such as those from the receiver locations and model tunnel dimensions in measurements, this agreement supports the reliability of the benchmark results used in numerical validations above.

Predictions from the proposed method and the method of Lemire and Nicolas were compared with the experimental data, as presented in Figure 7 as well. It is shown that, although the assumption of $kW \gg 1$ and $kH \gg 1$ can be hardly satisfied in this case with a wavelength of 0.344 m, the predictions from the proposed method can still have reasonable agreement with the benchmark results, which indicates that such a requirement may be relaxed in applying the proposed method. In this case, the method of Lemire and Nicolas can predict reasonably well at the receiver in the vicinity of the source, however deviating far from the benchmarks when source/receiver distance being large compared to the wavelength. These comparison results in the experimental case provide further validations on the proposed method.

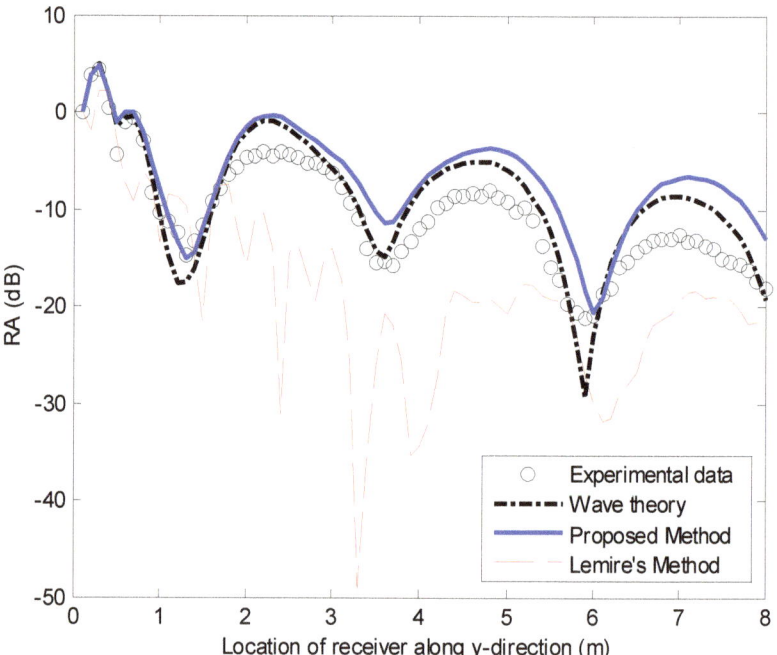

Figure 7. Comparison of measurements and predictions on relative attenuation (RA) at a frequency of 1000 Hz vs. the receiver location along the *y*-direction inside a scale-model long rectangular space. The circles represent the experimental results averaged from several measurements, the dash-dotted line represents the prediction results with the wave theory, the solid line denotes those with the proposed method, and the dash line is the result with the method of Lemire and Nicolas (Lemire's method).

4. Conclusions

In this paper, a coherent image source model is proposed for simple yet accurate sound prediction in long rectangular spaces with a sound absorbing ceiling. Predictions from the proposed model were compared with those from the wave theory, those from the existing coherent image source models, and the measurements in a scaled-model experiment. The results show that the proposed method can predict the sound field in long rectangular spaces with remarkable accuracy advantages over the existing coherent image source models but has computational load at the same level as the latter. The work in this study takes an important step in extending the coherent image source method proposed in Ref. [14] for versatile predictions in enclosed spaces.

Author Contributions: Conceptualization, H.M.; methodology, H.M.; software, H.M.; validation, H.M.; formal analysis, H.M.; investigation, H.M. and K.X.; resources, H.M.; data curation, H.M.; writing—original draft preparation, H.M.; writing—review and editing, H.M. and K.X.; visualization,

H.M.; supervision, H.M.; project administration, H.M.; funding acquisition, H.M. All authors have read and agreed to the published version of the manuscript.

Funding: This research was funded by the Natural Science Foundation of China, grant number 51408113, and the Natural Science Foundation of Jiangsu Province, China, grant number BK20140632.

Conflicts of Interest: The authors declare no conflict of interest. The funders had no role in the design of the study; in the collection, analyses, or interpretation of data; in the writing of the manuscript; or in the decision to publish the results.

References

1. Kang, J. *Acoustics of Long Spaces: Theory and Design Guidance*; Thomas Telford Limited: London, UK, 2002.
2. Kang, J. The unsuitability of the classic acoustical theory in long enclosures. *Architect. Sci. Rev.* **1996**, *39*, 89–94. [CrossRef]
3. Li, K.M.; Iu, K.K. Propagation of sound in long enclosures. *J. Acoust. Soc. Am.* **2004**, *116*, 2759–2770. [CrossRef]
4. Kang, J. A method for predicting acoustic indices in long enclosures. *Appl. Acoust.* **1997**, *51*, 169–180. [CrossRef]
5. Yang, L.; Shield, B.M. The prediction of speech intelligibility in underground stations of rectangular cross section. *J. Acoust. Soc. Am.* **2001**, *109*, 266–273. [CrossRef] [PubMed]
6. Kang, J. Numerical modelling of the speech intelligibility in dining spaces. *Appl. Acoust.* **2002**, *63*, 1315–1333. [CrossRef]
7. Nosal, E.-M.; Hodgson, M.; Ashdown, I. Improved algorithms and methods for room sound-field prediction by acoustical radiosity in arbitrary polyhedral rooms. *J. Acoust. Soc. Am.* **2004**, *116*, 970–980. [CrossRef] [PubMed]
8. Jan, H.; Hopkins, C. Prediction of sound transmission in long spaces using ray tracing and experimental Statistical Energy Analysis. *Appl. Acoust.* **2018**, *130*, 15–33.
9. Li, K.M.; Iu, K.K. Full-scale measurements for noise transmission in tunnels. *J. Acoust. Soc. Am.* **2005**, *117*, 1138–1145. [CrossRef]
10. Yousefzadeh, B.; Hodgson, M. Energy- and wave-based beam-tracing prediction of room-acoustical parameters using different boundary conditions. *J. Acoust. Soc. Am.* **2012**, *132*, 1450–1461. [CrossRef]
11. Li, K.M.; Lam, P.M. Prediction of reverberation time and speech transmission index in long enclosures. *J. Acoust. Soc. Am.* **2005**, *117*, 3716–3726. [CrossRef] [PubMed]
12. Lam, P.M.; Li, K.M. A coherent model for predicting noise reduction in long enclosures with impedance discontinuities. *J. Sound Vib.* **2007**, *299*, 559–574. [CrossRef]
13. Lemire, G.; Nicolas, J. Aerial propagation of spherical sound waves in bounded spaces. *J. Acoust. Soc. Am.* **1989**, *85*, 1845–1853. [CrossRef]
14. Min, H.; Chen, W.; Qiu, X. Single frequency sound propagation prediction in flat waveguides with locally reactive impedance boundaries. *J. Acoust. Soc. Am.* **2011**, *130*, 772–782. [CrossRef]
15. Brekhovskikh, I. *Waves in Layered Media*, 2nd ed.; Academic: New York, NY, USA, 1980; pp. 225–320.
16. Morse, P.M.; Ingard, K.U. *Theoretical Acoustics*; McGraw-Hill: New York, NY, USA, 1968; pp. 492–509.
17. Attenborough, K.; Hayek, S.I.; Lawther, J.M. Propagation of sound above a porous half space. *J. Acoust. Soc. Am.* **1980**, *68*, 1493–1501. [CrossRef]
18. Abramowitz, M.; Stegun, I.A. *Handbook of Mathematical Functions with Formulas, Graphs, and Mathematical Tables*; Dover: New York, NY, USA, 1965; p. 328.
19. Ingard, U. On the reflection of a spherical sound wave from an infinite plane. *J. Acoust. Soc. Am.* **1951**, *23*, 329–335. [CrossRef]
20. Attenborough, K. Ground parameter information for propagation modeling. *J. Acoust. Soc. Am.* **1992**, *92*, 418–427. [CrossRef]

Power Response and Modal Decay Estimation of Room Reflections from Spherical Microphone Array Measurements Using Eigenbeam Spatial Correlation Model

Amy Bastine *,†, Thushara D. Abhayapala † and Jihui (Aimee) Zhang †

Audio & Acoustic Signal Processing Group, The Australian National University, Canberra 2601, Australia; thushara.abhayapala@anu.edu.au (T.D.A.); jihui.zhang@anu.edu.au (J.Z.)
* Correspondence: amy.bastine@anu.edu.au
† These authors contributed equally to this work.

Featured Application: Room Mode Analysis.

Abstract: Modal decays and modal power distribution in acoustic environments are key factors in deciding the perceptual quality and performance accuracy of audio applications. This paper presents the application of the eigenbeam spatial correlation method in estimating the time-frequency-dependent directional reflection powers and modal decay times. The experimental results evaluate the application of the proposed technique for two rooms with distinct environments using their room impulse response (RIR) measurements recorded by a spherical microphone array. The paper discusses the classical concepts behind room mode distribution and the reasons behind their complex behavior in real environments. The time-frequency spectrum of room reflections, the dominant reflection locations, and the directional decay rates emulate a realistic response with respect to the theoretical expectations. The experimental observations prove that our model is a promising tool in characterizing early and late reflections, which will be beneficial in controlling the perceptual factors of room acoustics.

Keywords: reflection power; room response; directional decay rates; room modes; eigenbeam processing; spatial correlation

1. Introduction

In any enclosed acoustic space, the sound received by a listener is the superposition of the direct sound from the source and the reflected sounds from the surrounding surfaces. The numerous reflections termed reverberation cause persistence of sound even after the source ceases, until these reflected waves decay due to absorption by the surrounding surfaces. The intricate sound field generated by these reflected waves provides the sense of acoustic space to the perceived sound. However, severe reverberation can cause spectral distortions and reduce speech intelligibility. The study of reverberation is complicated since it is a product of many factors like sound frequency, room shape, room size, room geometry, source and receiver locations, source and receiver directivity, etc. A comprehensive understanding of the reflection sound field distribution, resonant frequencies, and modal decay rates is necessary to control audible artifacts and achieve desired sound perception quality in room acoustic applications.

Initially, the objective parameters like reverberation time, percentage articulation (PA) [1], decay rates [2], and statistical measures of room impulse responses (RIR) [3] were the only measures of reverberation. However, later studies [4,5] found that these measures vary with the sound frequency and wall surface properties. This necessitated the frequency-dependent spatio-temporal analysis of sound fields for accurate characterization of room acoustics. The existing 3D room acoustic parameter estimation methods either depend on predictions based on computational acoustics or derive the parameters directly from real sound field measurements. The room acoustic analysis using prominent computational models like ray/geometrical [6,7], wave/element [8], statistical energy [9], or synthetic RIR [10,11] methods are computationally complex and applicable to limited frequency ranges. The lack of proper consideration of the source and environment factors, frequency-dependent wave behavior, and precise reflection methods reduce the estimation accuracy of these computational approaches, especially in highly reverberant environments [12]. Furthermore, the analysis of intermediate frequencies using these computational models is complicated because of the dominant diffraction effects and the influence of both wave and ray acoustic behaviors.

The characterization of real acoustic environments requires 3D acoustic scene analysis using spatial sound field measurements. This led to the development of several microphone arrays designs [13–15] and processing methods like sound intensity mapping [16], plane-wave decomposition (PWD) and steered beamforming [17–19], sound intensity vector analysis [20], and multi-channel correlation model [21]. Gover et al. used PWD beamforming in [18] to estimate the angular distribution and anisotropy index of the spatial sound field from the RIRs recorded by a spherical microphone array. The recent works in [22–24] allow similar analysis in terms of isotropy measures and directional energy decays using Schroeder integration [25] and PWD of directional RIRs. However, these methods require a large number of RIR measurements for an accurate analysis of the room acoustic field. This problem was overcome with the introduction of higher-order spherical harmonic (eigenbeam)-based processing of spherical microphone array measurements [12,26–28], which provided higher spatial resolution for analysis compared to the previous methods. Subsequently, more robust techniques [29–31] were developed to achieve efficient parameterization of the spatial sound field using modal decomposition. In [32], the eigenbeam rotational invariance technique (EB-SPRIT) was used to identify room modes and damping parameters from RIRs. In [33,34], Samarasinghe et al. used the spatial correlation of higher-order eigenbeams to estimate the directional characteristics of the reverberant field, and this approach was able to achieve an accurate estimation of direct-to-reverberant energy ratio and dominant reflection directions.

The majority of the existing methods of directional characterization of room reflections derive the parameters from the aggregate sound field formed by the direct and reflected waves. Even though the direct path can be removed from the RIRs, the spatial resolution for directional analysis will be limited by the number of microphones. Moreover, a fine-scale separation of the spatial components of the direct path and reflected path is difficult without the knowledge of the source directivity. Additionally, the lack of incorporation of frequency-dependent surface reflectivities with distinct decay times can cause severe errors in the reflected sound field power distribution estimated by the existing methods [18,24,32]. Hence, a competent room characterization tool should integrate the frequency, time, and spatial dependencies in the formulation of the reflected sound field.

In this paper, we utilize the spatial correlation of higher-order eigenbeams to estimate the directional power response of room reflections by processing the RIR measurements. The proposed technique further facilitates room mode analysis and directional decay rate estimation. In comparison to the previous version of this method in [33,34], we model the reflection power as a function of time, frequency, and direction for comprehending the influence of frequency-dependent wall absorption properties of the room surfaces. This method allows the estimation of the directional features of reflections with higher spatial resolution independent of the direct sound component. The room mode features, directional decay rates and dominant reflection locations generated from the proposed tool can serve many applications like room response equalization, acoustic treatment design, architectural design simulations, room geometry inference, auralization of historic buildings, archaeoacoustics, and other machine hearing technologies.

The remainder of this paper is organized as follows: Section 2 discusses the formulation and implementation procedure of the eigenbeam spatial correlation model for estimating the reflection power response. Section 3 presents the experimental results including the time-frequency spectrum of reflection power, directional decay rates, and dominant reflection directions. Section 4 concludes the paper with a summary of the key findings and mentions the future research plans.

2. Reflection Power Estimation Using Eigenbeam Spatial Correlation Model

In this section, we present the formulation and synthesis of reflection power as a function of time, frequency, and space in the spherical harmonics domain.

2.1. Problem Formulation

Consider a convex room with a single sound source and a spherical microphone array of radius R with Q omnidirectional microphones centered at a location O, as shown in Figure 1. Let the spherical coordinate $\boldsymbol{y}_o = (r_o, \theta_o, \phi_o)$ denote the sound source location with respect to O. Similarly, the qth microphone element is located at $\boldsymbol{x}_q = (R, \theta_q, \phi_q)$ for $q \in \{1, 2, \cdots, Q\}$. In this paper, all the elevation angles are $\in [0, \pi]$ downwards from the Z-axis and the azimuth angles are $\in [0, 2\pi)$ counterclockwise from the X-axis.

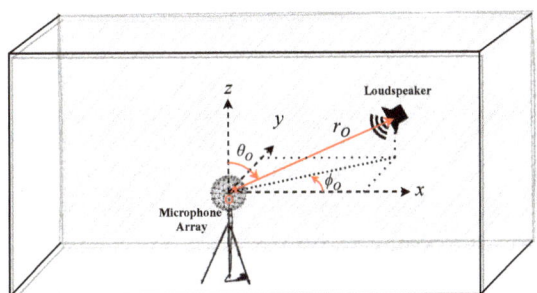

Figure 1. Geometric illustration of the spherical microphone array centered at the coordinate origin and the single sound source located at $\boldsymbol{y}_o = (r_o, \theta_o, \phi_o)$.

We treat the room as a linear time-invariant (LTI) acoustic transmission system whose dynamic behavior is represented by the RIRs derived from the spherical microphone array measurements. Let $H(\boldsymbol{x}_q, \boldsymbol{y}_o, t, k)$ be the room transfer function (RTF), between the source at $\boldsymbol{y}_o$ and the microphone element at $\boldsymbol{x}_q$, obtained from the short-time Fourier transform (STFT) of the RIR. Here, t is the STFT temporal frame index and $k = 2\pi f/c$ is the wavenumber with f and c representing the frequency and speed of sound, respectively. Since the incident sound field at the receiver contains the direct sound and the reflections, we can decompose the RTF $H(\boldsymbol{x}_q, \boldsymbol{y}_o, t, k)$ as

$$H(\boldsymbol{x}_q, \boldsymbol{y}_o, t, k) = H_\mathrm{d}(\boldsymbol{x}_q, \boldsymbol{y}_o, t, k) + H_\mathrm{r}(\boldsymbol{x}_q, \boldsymbol{y}_o, t, k) \qquad (1)$$

where $H_\mathrm{d}(\boldsymbol{x}_q, \boldsymbol{y}_o, t, k)$ and $H_\mathrm{r}(\boldsymbol{x}_q, \boldsymbol{y}_o, t, k)$ are the direct path and reflected path components, respectively.

Assuming that the distance between $\boldsymbol{y}_o$ and $\boldsymbol{x}_q$ is significantly larger than the aperture size of the microphone array, we can represent $H_\mathrm{d}(\boldsymbol{x}_q, \boldsymbol{y}_o, t, k)$ and $H_\mathrm{r}(\boldsymbol{x}_q, \boldsymbol{y}_o, t, k)$ as a composition of plane waves in the spatial domain as

$$H_\mathrm{d}(\boldsymbol{x}_q, \boldsymbol{y}_o, t, k) = G_D(t, k|\boldsymbol{y}_o) e^{ik\hat{\boldsymbol{y}}_o \cdot \boldsymbol{x}_q} \qquad (2)$$

$$H_\mathrm{r}(\boldsymbol{x}_q, \boldsymbol{y}_o, t, k) = \int_{\hat{\boldsymbol{y}}} G_R(t, k, \hat{\boldsymbol{y}}|\boldsymbol{y}_o) e^{ik\hat{\boldsymbol{y}} \cdot \boldsymbol{x}_q} d\hat{\boldsymbol{y}} \qquad (3)$$

where $G_D(t, k|\boldsymbol{y}_o)$ is the direct path gain with respect to O, $\hat{\boldsymbol{y}}_o$ is the unit vector along the source direction, $i = \sqrt{-1}$, $G_R(t, k, \hat{\boldsymbol{y}}|\boldsymbol{y}_o)$ is the gain of the reflected plane wave arriving from the direction $\hat{\boldsymbol{y}} = (1, \theta, \phi)$, and $\int_{\hat{\boldsymbol{y}}} d\hat{\boldsymbol{y}} = \int_0^{2\pi} \int_0^\pi \sin\theta d\theta d\phi$. Here, we have modeled the reflection gain G_R as a non-isotropic directional distribution function that varies with frequency and time to comprehend a real room with inhomogeneous surfaces that have frequency-dependent wall impedance and damping coefficients.

By examining $\mathbb{E}\left\{H_d H_d^*\right\}$ based on (2), where $\mathbb{E}\{\cdot\}$ represents the statistical expectation operator, we can express the direct path power as

$$P_D(t,k|\boldsymbol{y}_o) = \mathbb{E}\left\{|G_D(t,k|\boldsymbol{y}_o)|^2\right\} \tag{4}$$

where $|\cdot|$ denotes the absolute value. Similarly, by examining $\mathbb{E}\left\{H_r H_r^*\right\}$ based on (3), we can write the power of the reflected sound field component incoming from the direction $\hat{\boldsymbol{y}}$ as

$$P_R(t,k,\hat{\boldsymbol{y}}|\boldsymbol{y}_o) = \mathbb{E}\left\{|G_R(t,k,\hat{\boldsymbol{y}}|\boldsymbol{y}_o)|^2\right\}. \tag{5}$$

We aim to estimate the reflection power $P_R(t,k,\hat{\boldsymbol{y}}|\boldsymbol{y}_o)$ from the RTFs $H(\boldsymbol{x}_q,\boldsymbol{y}_o,t,k) \ \forall \ q$ obtained using a spherical microphone array. Since $P_R(t,k,\hat{\boldsymbol{y}}|\boldsymbol{y}_o)$ is a spherical function, we can simplify its estimation using the spherical harmonic decomposition [35] given by

$$P_R(t,k,\hat{\boldsymbol{y}}|\boldsymbol{y}_o) = \sum_{v=0}^{\infty}\sum_{u=-v}^{v} \gamma_{vu}(t,k|\boldsymbol{y}_o) Y_{vu}(\hat{\boldsymbol{y}}) \tag{6}$$

where $\gamma_{vu}(t,k|\boldsymbol{y}_o)$ are the reflection power coefficients and $Y_{vu}(\cdot)$ is the spherical harmonic function of vth order and uth mode. Thus, we can calculate the reflection power for any incoming direction and time-frequency bin once we estimate $\gamma_{vu}(t,k|\boldsymbol{y}_o)$ coefficients.

2.2. Methodology

For determining the $\gamma_{vu}(t,k|\boldsymbol{y}_o)$ coefficients, we utilize the spatial correlation of higher-order spherical harmonic (eigenbeam) coefficients of the incident sound field. The estimation of the reflection power response involves two main steps:

Step 1: Estimating spherical harmonic coefficients of the incident sound field

In this work, since we are interested in characterizing the room response independent of the source power spectrum, we assume a sound source emitting an impulse signal and treat $H(\boldsymbol{x}_q,\boldsymbol{y}_o,t,k)$ as the incident sound field on the spherical microphone array. For deducing the higher-order spherical harmonic coefficients of the incident sound field, we represent $H(\boldsymbol{x}_q,\boldsymbol{y}_o,t,k)$ as the spherical harmonic decomposition of Helmholtz wave equation solution to the interior sound field problem [12] as

$$H(\boldsymbol{x}_q,\boldsymbol{y}_o,t,k) = \sum_{n=0}^{\infty}\sum_{m=-n}^{n} \alpha_{nm}(t,k|\boldsymbol{y}_o) b_n(kR) Y_{nm}(\hat{\boldsymbol{x}}_q) \tag{7}$$

where $\alpha_{nm}(t,k|\boldsymbol{y}_o)$ are the modal coefficients of the spatial sound field, $\hat{\boldsymbol{x}}_q$ is the unit vector in the direction of the qth microphone, and

$$b_n(kR) = \begin{cases} j_n(kR) & \text{for an open array} \\ j_n(kR) - \dfrac{j_n'(kR)}{h_n'(kR)} h_n(kR) & \text{for a rigid array} \end{cases} \tag{8}$$

with $j_n(\cdot)$ and $h_n(\cdot)$ denoting the spherical Bessel and Hankel functions of order n, respectively, and $(\cdot)'$ represents the first derivative operation. From (7), we can estimate $\alpha_{nm}(t,k|\boldsymbol{y}_o)$ coefficients using the orthogonal property of spherical harmonics [36] as

$$\alpha_{nm}(t,k|\boldsymbol{y}_o) = \frac{\sum_{q=1}^{Q} H(\boldsymbol{x}_q,\boldsymbol{y}_o,t,k) Y_{nm}^*(\hat{\boldsymbol{x}}_q)}{b_n(kR)} \tag{9}$$

where $(\cdot)^*$ denotes the complex conjugation operation. Practically, we truncate $\alpha_{nm}(t,k|\mathbf{y}_o)$ to an order N, such that $N = \lceil kR \rceil$ and $Q \geq (N+1)^2$, where $\lceil \cdot \rceil$ denotes the ceiling operation, to avoid errors due to spatial aliasing and high-pass nature of higher-order Bessel functions [36].

Step 2: Estimating reflection gains using the spatial correlation model

We can now estimate $\gamma_{vu}(t,k|\mathbf{y}_o)$ from the $\alpha_{nm}(t,k|\mathbf{y}_o)$ coefficients using the spatial correlation matrix expression [33] given by

$$\underbrace{\begin{bmatrix} \Lambda_{0000} \\ \Lambda_{001-1} \\ \vdots \\ \Lambda_{00NN} \\ \Lambda_{1-100} \\ \vdots \\ \Lambda_{NNNN} \end{bmatrix}}_{\Lambda(t,k|\mathbf{y}_o)} = \underbrace{\begin{bmatrix} \delta_{0000} & d_{000000} & \cdots & d_{0000VV} \\ \delta_{001-1} & d_{001-100} & \cdots & d_{001-1VV} \\ \vdots & \vdots & \vdots & \vdots \\ \delta_{00NN} & d_{00NN00} & \cdots & d_{00NNVV} \\ \delta_{1-100} & d_{1-10000} & \cdots & d_{1-100VV} \\ \vdots & \vdots & \vdots & \vdots \\ \delta_{NNNN} & d_{NNNN00} & \cdots & d_{NNNNVV} \end{bmatrix}}_{B(k,\mathbf{y}_o)} \times \underbrace{\begin{bmatrix} P_D \\ \gamma_{00} \\ \gamma_{1-1} \\ \vdots \\ \gamma_{V-V} \\ \vdots \\ \gamma_{VV} \end{bmatrix}}_{\Omega(t,k|\mathbf{y}_o)} \quad (10)$$

where

$$\Lambda_{nmn'm'} = \mathbb{E}\left\{\alpha_{nm}(t,k|\mathbf{y}_o)\alpha^*_{n'm'}(t,k|\mathbf{y}_o)\right\} \quad (11)$$

$$\delta_{nmn'm'} = 16\pi^2 i^{(n-n')} Y^*_{nm}(\hat{\mathbf{y}}_o) Y_{n'm'}(\hat{\mathbf{y}}_o) \quad (12)$$

$$d_{nmn'm'vu} = 16\pi^2 i^{(n-n')}(-1)^m \sqrt{\frac{(2v+1)(2n+1)(2n'+1)}{4\pi}} W_1 W_2 \quad (13)$$

with $W_1 = \begin{pmatrix} v & n & n' \\ 0 & 0 & 0 \end{pmatrix}$ and $W_2 = \begin{pmatrix} v & n & n' \\ u & -m & m' \end{pmatrix}$ representing the Wigner 3j symbols [37].

The elements in $\Lambda(t,k|\mathbf{y}_o)$ and $B(k,\mathbf{y}_o)$ can be generated from the $\alpha_{nm}(t,k|\mathbf{y}_o)$ coefficients and source direction information, respectively. Now, we can solve (10) to estimate $\Omega(t,k|\mathbf{y}_o)$ by

$$\hat{\Omega}(t,k|\mathbf{y}_o) = B^\dagger(k,\mathbf{y}_o)\Lambda(t,k|\mathbf{y}_o) \quad (14)$$

where $[\hat{\cdot}]$ and $[\cdot]^\dagger$ indicate estimated values and pseudo-inversion operator, respectively. While solving (14), the order of $\gamma_{vu}(t,k|\mathbf{y}_o)$ in $\hat{\Omega}(t,k|\mathbf{y}_o)$ is truncated to $V \leq \lfloor \sqrt{(N+1)^4 - 1} \rfloor$, where $\lfloor \cdot \rfloor$ indicate flooring operation, to avoid an underdetermined system [34]. Once the $\gamma_{vu}(t,k|\mathbf{y}_o)$ coefficients are extracted from $\hat{\Omega}(t,k|\mathbf{y}_o)$, we can generate the reflection power using Equation (6) for different incoming directions $\hat{\mathbf{y}}$ and time-frequency bins. From $P_R(t,k,\hat{\mathbf{y}}|\mathbf{y}_o)$, we can estimate the total reflected power in any time-frequency bin as

$$P_T(t,k|\mathbf{y}_o) = \int_{\hat{\mathbf{y}}} P_R(t,k,\hat{\mathbf{y}}|\mathbf{y}_o) d\hat{\mathbf{y}}. \quad (15)$$

Substituting (6) in (15) and using the symmetrical property of spherical harmonics [35] $P_T(t,k|\mathbf{y}_o) = \gamma_{00}(t,k|\mathbf{y}_o)$. We can now use $P_R(t,k,\hat{\mathbf{y}}|\mathbf{y}_o)$ and $P_T(t,k|\mathbf{y}_o)$ to analyze the reflection power variations with time, frequency, and direction.

3. Experimental Analysis

In this section, we present the analysis of the reflection power response of two rooms from their RIR datasets recorded using an em32 Eigenmike [38], which is a $Q = 32$ element rigid spherical microphone array of radius $R = 0.042$ m. Both the RIR datasets were measured using a source signal generated from a directional loudspeaker. The first RIR dataset available from the work in [39] is for a small audio laboratory room of size $3.54 \times 4.06 \times 2.70$ m, hereafter referred to as Room-1. The second RIR dataset from [40] pertains to a larger classroom of size 6.5 × 8.3 × 2.9 m, hereafter referred to as Room-2. According to these datasets, the reverberation time (T_{60}) of Room-1 and Room-2 are 0.329 s and 1.12 s, respectively. From the datasets, we have selected the RIRs for different source positions in the XY plane, i.e., $\theta_o = 90°$ at different ϕ_o angles, and at 1 m distance from the microphone array center. The direct path component from the source arrives at the receiver around 0.0026 s and 0.0028 s for Room-1 and Room-2, respectively.

From the selected 32-channel RIRs, we obtain $H(x_q, y_o, t, k)$ using the STFT operation with a 1024-sample Hanning window with 50% overlap, 2048-point fast Fourier transform (FFT), and 48 KHz sampling frequency. We then follow the process described in Section 2.2 to generate $P_R(t, k, \hat{y}|y_o)$ for 500 uniformly distributed $\hat{y}$ directions derived from spiral-based sampling [41] $\forall\, t, k$ bins in the frequency band of 20 to 1500 Hz. These 500 spiral sampled directions provide sufficient spatial resolution to assimilate the sound reflectivity variations across the room surfaces at a reasonable computation cost. Finally, we estimate $P_T(t, k|y_o)$ for analyzing the time-frequency spectrum of the reflection power of the two rooms. While dealing with the temporal response in the following sections, the 0 s in the time-index indicates the moment of sound event occurrence. However, the reflection power response is calculated only from 0.01 s which is the center of the first STFT frame. This frame size was selected after considering a reasonable time-frequency resolution for proper spectral and temporal analysis of reflections in both rooms.

3.1. Theoretical Background

Here we discuss important theoretical concepts of room acoustics and room response characteristics according to prevalent literature [5,42–44] to validate the experimental analysis.

3.1.1. Modal Decay

The reverberation field inside a room leads to the persistence of sound even after the source ceases. The duration of this sound persistence, called the reverberation time R_T [5], is the most commonly used measure of room acoustic quality. In practical applications, acousticians calculate R_T as the 60 dB decay time since source cessation and is referred to as T_{60} [43]. Typically, such estimations assume diffuse sound field conditions and average wall absorption and calculate R_T as a single value to characterize the room acoustics. However, in reality, the wall absorption factors change with frequency [5,44], and hence accurate R_T estimates should be frequency-dependent. Furthermore, the room architecture, variations in surface materials, and source-receiver properties affect the reflection path length [44] and magnitude, which, in turn, influence the decay of different frequency components. Therefore, decay times should be a function of frequency and direction. Since an analytical solution to decay rate estimation is complex, we can derive them numerically through reflection sound field analysis.

3.1.2. Room Modes

The sound propagation in any acoustic enclosure follows different wave characteristic phenomena like reflection, scattering, diffraction, and interference. Such a complex interaction of innumerous waves is characterized through the acoustical wave equation [5]. The frequencies corresponding to the eigenvalues of the acoustic wave equation can form standing waves inside the room to create a resonant behavior leading to non-uniform distribution of reflection power and extended reverberation [5,43,44]. These frequencies are often referred to as room modes or eigenfrequencies.

According to [5,43], at low frequency ranges, the number of resonant frequencies will be small, and they can be excited individually. Hence, the room response will be quite irregular and anisotropic for these frequencies. When we move towards the higher frequencies, the eigenvalues are densely spaced, so they cannot be independently excited. Even though the higher frequencies contribute to the reflected sound pressure, the lack of independent resonance combined with increased scattering makes them relatively uniform and less prominent compared to the lower frequencies. Hence, in a typical room response, we expect high reflection powers with some resonant peaks for low (<300 Hz) to mid (300 to 600 Hz) audible frequencies and decaying magnitude towards the high (>600 Hz) frequencies. The cross-over frequency [5,43] that separates the resonant low-frequency response and the high-frequency diffused reflections is termed as Schroeder frequency (ν_S). It can be calculated using the empirical formula

$$\nu_S \approx 2000\sqrt{\frac{T_{60}}{\Delta}} \tag{16}$$

where Δ is the room volume. From the dimensions and T_{60} of the test rooms, (16) gives $\nu_S \approx 184$ Hz and ≈ 169 Hz for Room-1 and Room-2, respectively.

For a rectangular enclosure, we can calculate the eigenvalues of the wave equation [5,42–44] as

$$\nu_{n_x n_y n_z} = \frac{c}{2}\sqrt{\left(\frac{n_x}{l_x}\right)^2 + \left(\frac{n_y}{l_y}\right)^2 + \left(\frac{n_z}{l_z}\right)^2} \tag{17}$$

where $\{n_x, n_y, n_z\}$ are non-negative integers and $l_x \times l_y \times l_z$ are the room dimensions. When two of $\{n_x, n_y, n_z\}$ equals zero, the solution of (17) gives the axial modes which are considered to be stronger with low decay rates compared to other modes [42]. We can calculate the tangential modes with two non-zero integers in $\{n_x, n_y, n_z\}$ and oblique modes by substituting all non-zero integers in $\{n_x, n_y, n_z\}$.

Figure 2 shows the room mode distribution in Room-1 and Room-2. The axial and tangential modes are calculated from (17), and the line heights in Figure 2 represent the number of resonances occurring at a frequency since different $\{n_x, n_y, n_z\}$ combinations can result in the same $\nu_{n_x n_y n_z}$ frequency. The axial modes were given a higher nominal weight [44] while calculating this distribution due to their inherent prominence. Theoretically, an empty rectangular room of the same dimensions should replicate this trend in their frequency response. However, in a real room environment, the interference of normal modes of different decay rates [44] and the influence of inhomogeneous surfaces and source directivity alter the assumptions behind (17). Therefore, the real room response may vary from the predicted distribution.

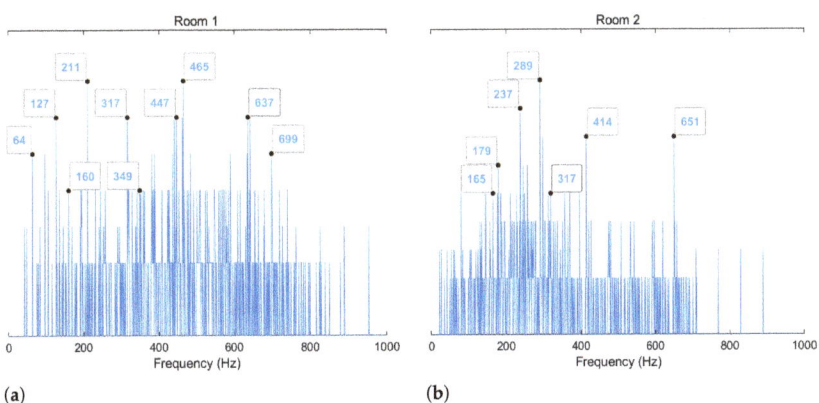

Figure 2. Room mode distribution in (**a**) Room-1 (**b**) Room-2.

For practical validation of the real acoustic phenomenon, we will use the power response generated using the proposed technique to identify the variations in the room mode distribution and modal decays compared to the above theoretical expectations.

3.2. Reflection Power Spectrum

Figures 3 and 4 show the spectrogram of $P_T(t, k|\mathbf{y}_o)$ for different source positions in Room-1 and Room-2, respectively. For both rooms, the lower frequencies show some irregular peaks, and the reflection power of late reverberation clearly decays towards the higher frequencies as we predicted in Section 3.1.2. Additionally, the reflection power is maximum in the initial time instants, and then the power decays with time for all frequencies due to surface absorption. It should be noted that the power decay trend is varying with the frequencies due to the frequency-dependent wall impedance property [5]. Apart from some magnitude variations, the time-frequency spectrum trend is maintained for all source positions in both rooms. In the following sections, we will analyze the reflections power variations with frequency and time in more detail.

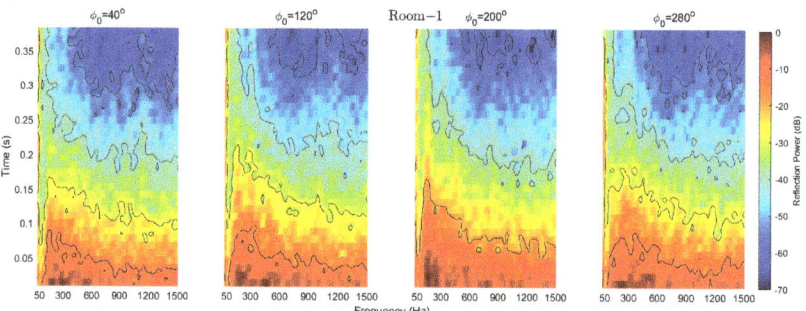

Figure 3. Reflection power response of Room-1 for different source positions.

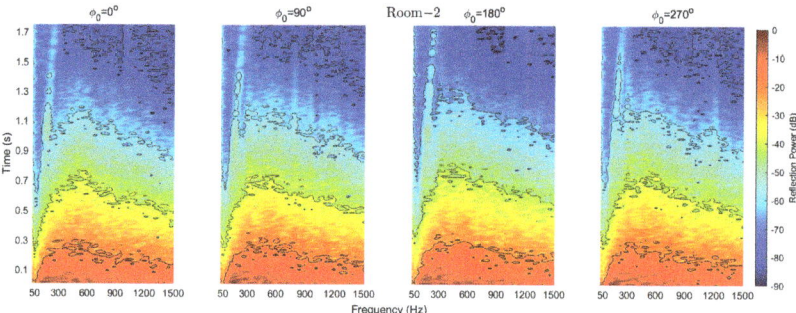

Figure 4. Reflection power response of Room-2 for different source positions.

3.2.1. Frequency Response of Reflection Power

Figures 5 and 6 show the frequency response of time-averaged $P_T(t, k|\mathbf{y}_o)$ for different source positions in Room-1 and Room-2, respectively. These figures provide a clear view of the low-frequency peaks and the decay of power towards the higher frequencies. In Room-1, we can observe high powers around 164 Hz, 211 Hz, and 281 Hz before the onset of the power decay. Compared to Figure 2a, 164 Hz and 211 Hz are closer to the theoretical room modes, whereas many other predicted modes do not appear in the observed response in Figure 5. Similarly, some of the observed peaks in Room-2 around 164 Hz, 304 Hz, 328 Hz, and 492 Hz vary from the theoretical room mode estimates shown in Figure 2b. Additionally, the identification of ν_S is difficult from these responses, but is clearly greater than the predicted ν_S values mentioned in Section 3.1.2. This error is caused by the approximation in (16) by use of frequency-averaged T_{60} and from the influence of source directivity.

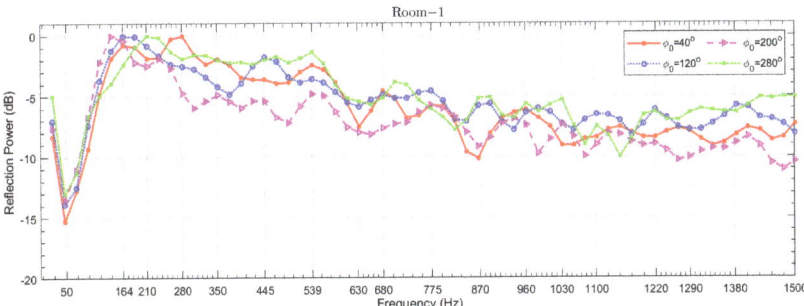

Figure 5. Reflection power with frequency for different source positions in Room-1.

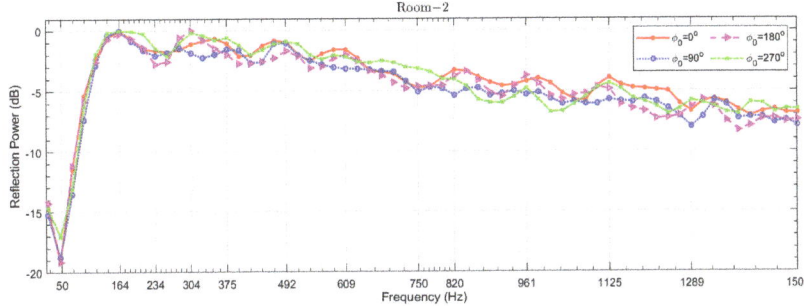

Figure 6. Reflection power with frequency for different source positions in Room-2.

It should also be noted that there are no substantial variations in the frequency response of Room-2 for different source positions. Additionally, in Room-1, the differences are not drastic as should be expected in a smaller room with significant reverberation. This is the result of the formulation of reflection gains with respect to a common listening position (O) and the separation of the direct path component from the reflections. A direct analysis of the frequency response of RIR will show significant differences with the change in source positions. Therefore, the proposed technique can be used to predict the room response behavior independent of the source positions.

3.2.2. Temporal Response of Reflection Power

Figures 7 and 8 show the temporal response of $P_T(t, k|y_o)$ at different frequencies for different source positions in Room-1 and Room-2, respectively. As evident from these figures, the reflection power decays due to surface absorption, and the decay trend is similar for all source positions. Since the damping constants of room surfaces are frequency-dependent, each frequency in Figures 7 and 8 decays at different rates. The lower frequencies like 70 Hz, 141 Hz, and 211 Hz have slower decay rates compared to the other frequencies. As we move from 281 Hz to 633 Hz in Figures 7 and 8, the decay rate stabilizes towards the higher frequencies. Furthermore, the decay of higher frequencies is nearly linear, whereas the lower frequencies (70 Hz to 211 Hz) exhibit a non-linear decay, especially in Room-2. This can be attributed to the highly non-uniform power distribution of the lower frequency resonant modes, which leads to the concentration of sound absorption to certain surfaces [42,43]. In comparison, the high frequencies have more diffused distribution of reflection power, and hence the decay behavior is averaged over broader surface areas.

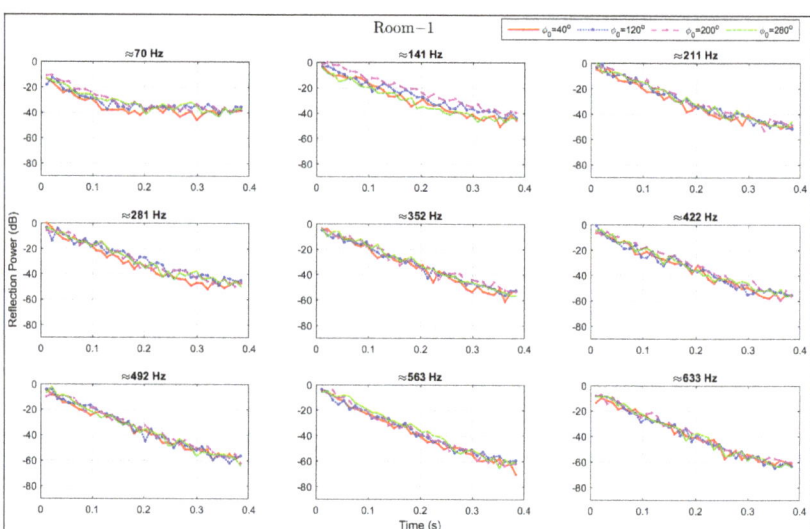

Figure 7. Reflection power with time for different frequencies and source positions in Room-1.

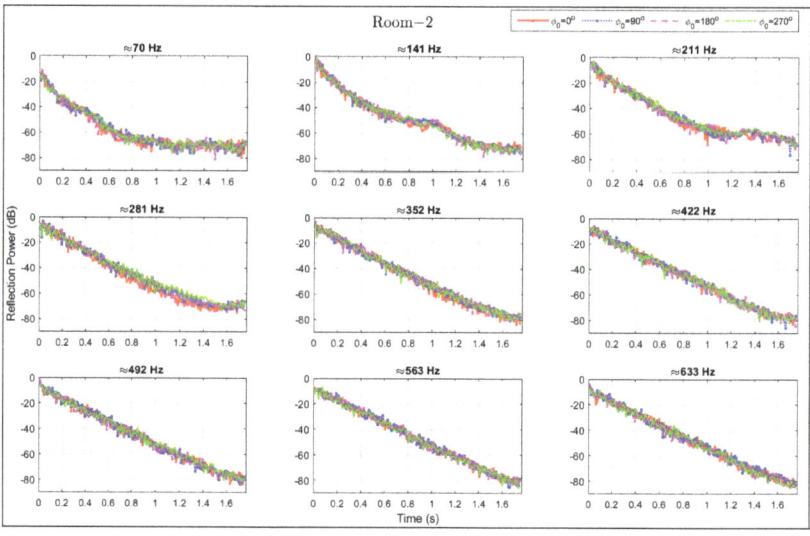

Figure 8. Reflection power with time for different frequencies and source positions in Room-2.

3.3. Decay Time

From the time-frequency spectrum of reflection power, we can estimate the decay time of each frequency to predict the strong room modes in a real room environment. Figures 9 and 10 show the 60 dB decay time of each frequency estimated from the $P_T(t, k|y_o)$ values for different source positions in Room-1 and Room-2, respectively. Even though the temporal response at each frequency in Figures 7 and 8 seems relatively independent of the source positions, the decay times of the frequencies is slightly different for each source position according to Figures 9 and 10. The average decay time, maximum decay time, and the corresponding frequency for each source position in both rooms are summarized under Table 1. We can say that the strongest modes in Room-1 are ≈140 Hz, ≈164 Hz, and ≈258 Hz, which are closer to the peak power frequencies observed in Figure 5. However, in Room-2, the frequencies with maximum decay time are different from the frequencies with maximum reflection power. Hence, we need a deeper insight into the directional variations of power and decay time which we will analyze in the next section.

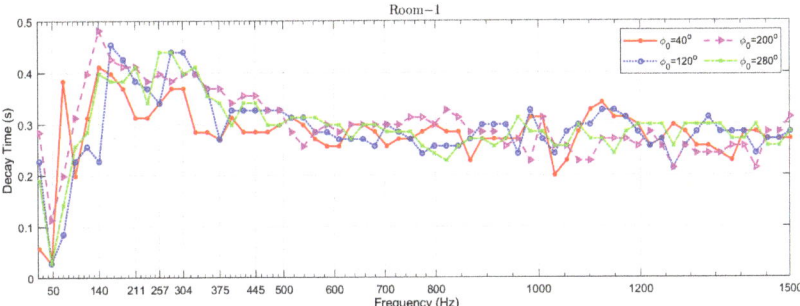

Figure 9. Decay time with frequency for different source positions in Room-1.

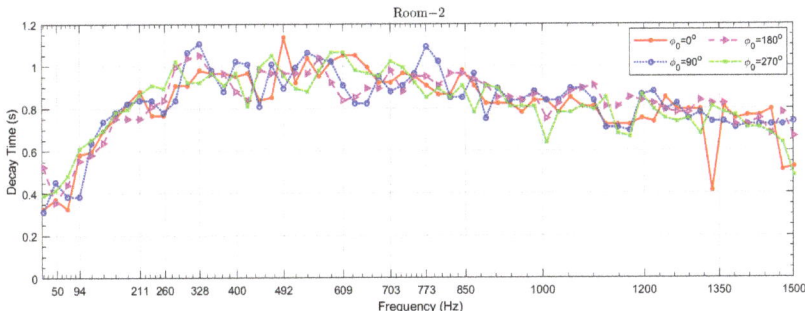

Figure 10. Decay time with frequency for different source positions in Room-2.

Table 1. Maximum and average decay times in Room-1 and Room-2.

In Room	Source Position	Maximum Decay Time (s)	Frequency (Hz) with Maximum Decay Time	Average Decay Time (s)
Room-1	$y_o = (1, 90°, 40°)$	0.4114	140	0.2822
	$y_o = (1, 90°, 120°)$	0.4540	164	0.2899
	$y_o = (1, 90°, 200°)$	0.4823	140	0.2995
	$y_o = (1, 90°, 280°)$	0.4398	258	0.2936
Room-2	$y_o = (1, 90°, 0°)$	1.1349	492	0.8133
	$y_o = (1, 90°, 90°)$	1.1066	328	0.8288
	$y_o = (1, 90°, 180°)$	1.0498	328	0.8341
	$y_o = (1, 90°, 270°)$	1.0640	586	0.8182

3.4. Directional Decays and Dominant Reflection Directions

As we discussed in Section 3.1.1, decay times are a function of frequency and direction. Additionally, from Section 3.3, we found that the modes with higher decay times can be different from the modes with high reflection powers. Therefore, a more comprehensive analysis of the spatial spectrum of these reflections is necessary to identify room surfaces causing the observed behaviors for the frequencies of interest. Figure 11a,b shows the directional decay times of Room-1 for $y_o = (1, 90°, 40°)$ and $y_o = (1, 90°, 120°)$, respectively, obtained from the 60 dB decay time of $P_R(t, k, \hat{y}|y_o)$ in each $\hat{y}$ direction. Figure 12a,b shows the directions with high reflection powers in Room-1 for $y_o = (1, 90°, 40°)$ and $y_o = (1, 90°, 120°)$, respectively. The letters indicated near the locations of highest reflection powers in Figure 12 are coarsely mapped onto the real Room-1 environment in Figure 13. As evident from this figure, the locations around 'A', 'C', 'D', and 'E' have glass surfaces with high reflectivity, and hence the observed dominant power directions are valid. Furthermore, there is no evident pattern between the distributions in Figures 11 and 12 for the given modal frequencies, and hence the feature predictions based on computational room acoustic models can be imprecise. In such cases, we can employ the proposed technique to reproduce authentic spatio-temporal room responses.

According to Figures 11 and 12, the directions of high decay times and dominant reflections are different from each other for every frequency and source position. Even though the dominant reflection locations and directional decay distribution have many common factors of influence, the reflection power in a direction strongly depends on the source directivity and source-to-wall distance, whereas the directional decay is mainly a function of the wall impedance coefficients and reflection paths. Hence, as seen in Figure 11, the directional decay will be different between the frequencies due to wall impedance variations, as well as for different source positions due to change in reflection path. In contrast, if we observe Figures 12 and 13, the dominant reflection locations 'A', 'B', and 'C' have similar azimuth values and source-to-wall distance when the source is at $y_o = (1, 90°, 40°)$. Likewise, the elevation values of the dominant reflection locations 'D' and 'E' in Figure 12b are nearly the same when the source is at $y_o = (1.90°, 120°)$. Additionally, the location of 'F' is in the close vicinity of the source position. Thus, the dominant reflection locations are principally determined by the source position and source directivity. For locations with same source-to-wall distance, the dominant reflections will depend on the reflectivity of the surface materials.

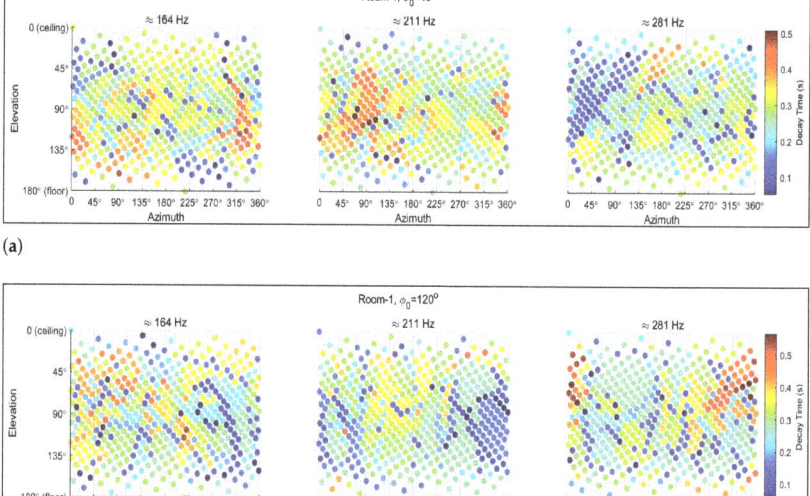

Figure 11. Directional decay times inside Room-1 for the peak frequencies when source is located at (**a**) $y_o = (1, 90°, 40°)$ (**b**) $y_o = (1, 90°, 120°)$.

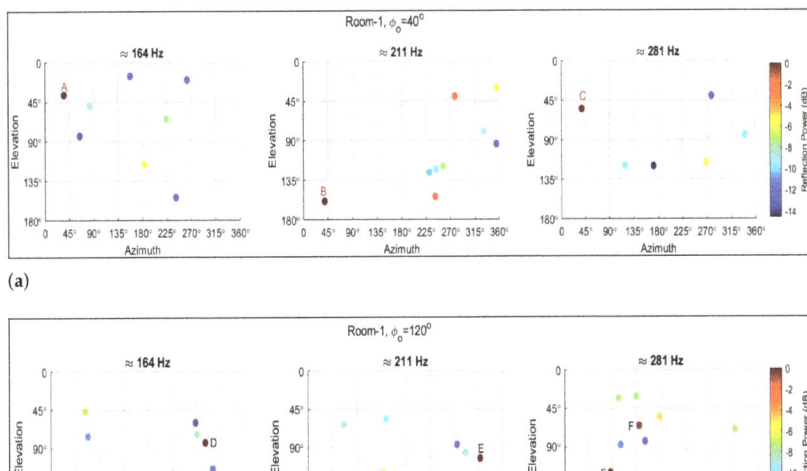

(a)

(b)

Figure 12. Dominant reflection directions inside Room-1 for the peak frequencies when source is located at (**a**) $y_o = (1, 90°, 40°)$ (**b**) $y_o = (1, 90°, 120°)$.

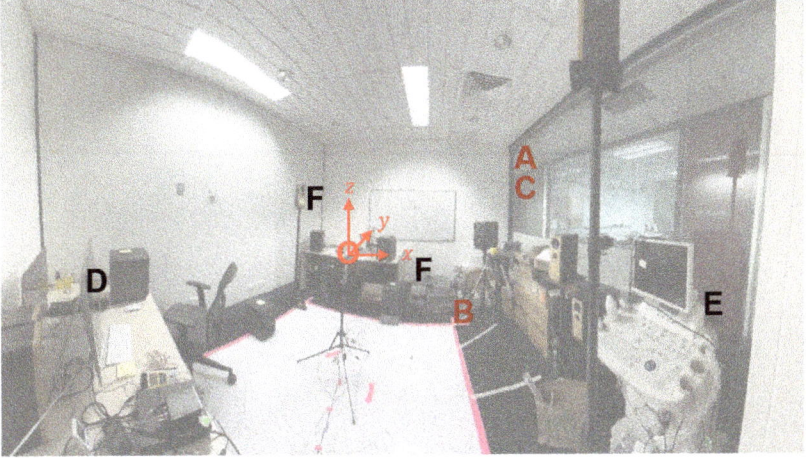

Figure 13. Mapping of dominant reflection directions in Room-1. The letters A to C and D to F represent the directions of highest reflection powers with respect to Figure 12a,b, respectively.

Based on the above observations, the analysis of both directional decay and directional power is essential in characterizing the room reflections. This is particularly important while managing the features of early reflections and late reverberations to achieve desired perception quality. Since the early reflections undergo very few boundary reflections [45], they are mainly defined by the source directivity and source-to-wall distance. Hence, we can use the dominant reflection directions to characterize the behavior of early reflections. The late reverberation undergoes multiple boundary reflections, and they are integrated both spatially and temporally before reaching the receiver [45]. Since the late reverberation characteristics are primarily characterized by the surface absorption and room shape [45,46], we can analyze the directional decay rates to study their behavior. We can further visualize the power spectrum of $P_R(t, k, \hat{\boldsymbol{y}}|\boldsymbol{y}_o)$ across time for an extensive analysis of the variations in the anisotropic spatial properties between the early reflections and late reverberations.

The precise knowledge of frequencies and surfaces contributing to the salient features of these reflections will be useful for defining the perceptual targets for modal control methods [47], optimizing room mode redistribution to improve acoustic quality [48], and devising active [49] and passive [50,51] room acoustic treatment methods.

4. Conclusions

In this paper, we presented a reflection power response estimation technique utilizing the spatial correlation of higher-order eigenbeams derived from spherical microphone array measurements. The formulation of the reflection gain as a function of time, frequency, and direction helps in comprehending a faithful room response for a realistic non-diffuse sound field. The experimental results validate the frequency response and temporal response of the reflection power against the theoretical expectations.

The proposed technique can estimate the resonant frequencies and modal decays caused by directional speakers and complex room environments. Furthermore, the directional decay times and dominant reflection directions facilitate the distinction of early and late reflection features. The insights from this room acoustic evaluation technique will be beneficial in controlling the acoustic quality while designing performance spaces. Particularly, the findings from this method will be more reliable than computational room models while deciding acoustic treatment schemes compatible with the source directivity. Additionally, the room mode features identified from this method can be incorporated in spectral equalization algorithms to improve speech intelligibility and remove audible artifacts. The dominant reflection locations and directional decay spectrum can aid in the inference of room geometry and calibration of the room acoustics in virtual reality-based rendering of heritage sites.

The method can also be adapted for blind estimation of the discussed characteristics from the direct processing of microphone recordings for any arbitrary source signal, since we can separate the reflected power from the direct path power. Moreover, apart from spherical microphone arrays, any arbitrary array designs that can generate accurate spatial sound field coefficients can be integrated with the proposed algorithm. The future work shall expand the method to include multiple sources in noisy environments to conceive more real-world applications.

Author Contributions: Conceptualization, A.B., T.D.A. and J.Z.; Methodology, A.B., T.D.A. and J.Z.; Software, A.B.; Formal analysis, A.B.; Investigation, A.B.; Validation, A.B., T.D.A. and J.Z.; Resources, A.B., T.D.A. and J.Z.; Writing—original draft preparation, A.B.; Writing—review and editing, T.D.A. and J.Z.; Visualization, A.B.; Supervision, T.D.A. and J.Z.; Project administration, T.D.A.; Funding acquisition, T.D.A.; All authors have read and agreed to the published version of the manuscript.

Funding: This research was funded by Australian Research Council (ARC) Discovery Project Grant No. DP180102375.

Institutional Review Board Statement: Not applicable.

Informed Consent Statement: Not applicable.

Data Availability Statement: Not applicable.

Conflicts of Interest: The authors declare no conflicts of interest.

Abbreviations

The following abbreviations are used in this manuscript:

RIR	Room impulse response
PWD	Plane-wave decomposition
EB-SPRIT	Eigenbeam rotational invariance technique
LTI	Linear time invariant

RTF	Room transfer function	
STFT	Short time Fourier transform	
FFT	Fast Fourier transform	

References

1. Morse, P.M.; Bolt, R.H. Sound waves in rooms. *Rev. Mod. Phys.* **1944**, *16*, 69. [CrossRef]
2. Karjalainen, M.; Antsalo, P.; Makivirta, A.; Valimaki, V.; Peltonen, T. Estimation of Modal Decay Parameters from Noisy Response Measurements. *J. Audio Eng. Soc.* **2002**, *50*, 5290.
3. Stewart, R.; Sandler, M. Statistical measures of early reflections of room impulse responses. In Proceedings of the Conference Digital Audio Effects (DAFx-07), Bordeaux, France, 10–15 September 2007; pp. 59–62
4. Long, M. *Architectural Acoustics*; Elsevier: Amsterdam, The Netherlands, 2005.
5. Kuttruff, H. *Room Acoustics*, 6th ed.; CRC Press: Boca Raton, FL, USA, 2017.
6. Allen, J.B.; Berkley, D.A. Image method for efficiently simulating small-room acoustics. *J. Acoust. Soc. Am.* **1979**, *65*, 943–950. [CrossRef]
7. Lehmann, E.A.; Johansson, A.M. Prediction of energy decay in room impulse responses simulated with an image-source model. *J. Acoust. Soc. Am.* **2008**, *124*, 269–277. [CrossRef]
8. Hamilton, B. Finite Difference and Finite Volume Methods for Wave-Based Modelling of Room Acoustics. 2016. Available online: https://www.researchgate.net/profile/Brian-Hamilton-5/publication/310902744_Finite_Difference_and_Finite_Volume_Methods_for_Wave-based_Modelling_of_Room_Acoustics/links/583acf2a08ae3a74b4a01683/Finite-Difference-and-Finite-Volume-Methods-for-Wave-based-Modelling-of-Room-Acoustics.pdf (accessed on 27 May 2021)
9. Lyon, R.H. *Theory and Application of Statistical Energy Analysis*; Elsevier: Amsterdam, The Netherlands, 2014.
10. Kim, H.; Hernaggi, L.; Jackson, P.J.; Hilton, A. Immersive Spatial Audio Reproduction for VR/AR Using Room Acoustic Modelling from 360° Images. In Proceedings of the Virtual Reality 3D User Interfaces, Osaka, Japan, 23–27 March 2019; pp. 120–126
11. Remaggi, L.; Neidhardt, A.; Hilton, A.; Philip, J.B.J. Perceived quality and spatial impression of room reverberation in VR reproduction from measured images and acoustics. In Proceedings of the 23rd International Congress Acoustics, Aachen, Germany, 9–13 September 2019
12. Samarasinghe, P. Modal based Solutions for the Acquisition and Rendering of Large Spatial Soundfields. Ph.D. Thesis, College of Engineering and Computer Science, Australian National University, Canberra, Australia, 2014
13. Schroeder, M.R. Measurement of sound diffusion in reverberation chambers. *J. Acoust. Soc. Am.* **1959**, *31*, 1407–1414. [CrossRef]
14. Broadhurst, A. An acoustic telescope for architectural acoustic measurements. *Acta Acust. United Acust.* **1980**, *46*, 299–310.
15. Yamasaki, Y.; Itow, T. Measurement of spatial information in sound fields by closely located four point microphone method. *J. Acoust. Soc. Jpn. (E)* **1989**, *10*, 101–110. [CrossRef]
16. Merimaa, J.; Lokki, T.; Peltonen, T.; Karjalainen, M. Measurement, Analysis, and Visualization of Directional Room Responses. In Proceedings of the Audio Engineering Society Convention, New York, NY, USA, 21–24 September 2001.
17. Ward, D.B.; Abhayapala, T.D. Reproduction of a plane-wave sound field using an array of loudspeakers. *IEEE Trans. Speech Audio Process.* **2001**, *9*, 697–707. [CrossRef]
18. Gover, B.N.; Ryan, J.G.; Stinson, M.R. Measurements of directional properties of reverberant sound fields in rooms using a spherical microphone array. *J. Acoust. Soc. Am.* **2004**, *116*, 2138–2148. [CrossRef]
19. Park, M.; Rafaely, B. Sound-field analysis by plane-wave decomposition using spherical microphone array. *J. Acoust. Soc. Am.* **2005**, *118*, 3094–3103. [CrossRef]
20. Tervo, S.; Korhonen, T.; Lokki, T. Estimation of reflections from impulse responses. *Build. Acoust.* **2011**, *18*, 159–173. [CrossRef]
21. Hioka, Y.; Niwa, K.; Sakauchi, S.; Furuya, K.; Haneda, Y. Estimating Direct-to-Reverberant Energy Ratio Using D/R Spatial Correlation Matrix Model. *IEEE Trans. Audio Speech Lang. Process.* **2011**, *19*, 2374–2384. [CrossRef]
22. Alary, B.; Massé, P.; Välimäki, V.; Noisternig, M. Assessing the anisotropic features of spatial impulse responses. In Proceedings of the EAA Spatial Audio Signal Processing Symposium, Paris, France, 6–7 September 2019; pp. 43–48
23. Nolan, M.; Berzborn, M.; Fernandez-Grande, E. Isotropy in decaying reverberant sound fields. *J. Acoust. Soc. Am.* **2020**, *148*, 1077–1088. [CrossRef]
24. Berzborn, M.; Nolan, M.; Fernandez-Grande, E.; Vorländer, M. On the directional properties of energy decay curves. In Proceedings of the 23rd International Congress Acoustics, Aachen, Germany, 9–13 September 2019.
25. Schroeder, M.R. New method of measuring reverberation time. *J. Acoust. Soc. Am.* **1965**, *37*, 1187–1188. [CrossRef]
26. Abhayapala, T.D.; Ward, D.B. Theory and design of high order sound field microphones using spherical microphone array. In Proceedings of the 2002 IEEE International Conference on Acoustics, Speech, and Signal Processing, Orlando, FL, USA, 13–17 May 2002; Volume 2, pp. 1949–1952.
27. Poletti, M.A. Three-dimensional surround sound systems based on spherical harmonics. *J. Audio Eng. Soc.* **2005**, *53*, 1004–1025.
28. Lovedee-Turner, M.; Murphy, D. Three-dimensional reflector localisation and room geometry estimation using a spherical microphone array. *J. Acoust. Soc. Am.* **2019**, *146*, 3339–3352. [CrossRef]
29. Rafaely, B.; Balmages, I.; Eger, L. High-resolution plane-wave decomposition in an auditorium using a dual-radius scanning spherical microphone array. *J. Acoust. Soc. Am.* **2007**, *122*, 2661–2668. [CrossRef]

30. Rafaely, B.; Peled, Y.; Agmon, M.; Khaykin, D.; Fisher, E. Spherical Microphone Array Beamforming. In *Speech Processing in Modern Communication: Challenges and Perspectives*; Cohen, I., Benesty, J., Gannot, S., Eds.; Springer: Berlin/Heidelberg, Germany, 2010; pp. 281–305. [CrossRef]
31. Sun, H.; Mabande, E.; Kowalczyk, K.; Kellermann, W. Joint DOA and TDOA estimation for 3D localization of reflective surfaces using eigenbeam MVDR and spherical microphone arrays. In Proceedings of the 2011 IEEE International Conference on Acoustics, Speech and Signal Processing (ICASSP), Prague, Czech Republic, 22–27 May 2011; pp. 113–116. [CrossRef]
32. Kereliuk, C.; Herman, W.; Wedelich, R.; Gillespie, D.J. Modal analysis of room impulse responses using subband ESPRIT. In Proceedings of the International Conference Digital Audio Effects (DAFx-18), Aveiro, Portugal, 4–8 September 2018
33. Samarasinghe, P.N.; Abhayapala, T.D.; Chen, H. Estimating the Direct-to-Reverberant Energy Ratio Using a Spherical Harmonics-Based Spatial Correlation Model. *IEEE/ACM Trans. Audio Speech Lang. Process.* **2016**, *25*, 310–319. [CrossRef]
34. Samarasinghe, P.N.; Abhayapala, T.D. Blind estimation of directional properties of room reverberation using a spherical microphone array. In Proceedings of the 2017 IEEE International Conference on Acoustics, Speech and Signal Processing (ICASSP), New Orleans, LA, USA, 5–9 March 2017; pp. 351–355.
35. Williams, E.G. *Fourier Acoustics: Sound Radiation and Nearfield Acoustical Holography*; Academic Press: Cambridge, MA, USA, 1999.
36. Rafaely, B. *Fundamentals of Spherical Array Processing*; Springer: Berlin/Heidelberg, Germany, 2015; Volume 8.
37. Olver, F.W.; Lozier, D.W.; Boisvert, R.F.; Clark, C.W. *NIST Handbook of Mathematical Functions*; Cambridge University Press: New York, NY, USA, 2010.
38. em32 Eigenmike® Microphone Array Release Notes (v17.0). Available online: https://mhacoustics.com/sites/default/files/ReleaseNotes.pdf (accessed on 6 April 2021).
39. Birnie, L.I.; Abhayapala, T.D.; Samarasinghe, P.N. Reflection Assisted Sound Source Localization through a Harmonic Domain MUSIC Framework. *IEEE/ACM Trans. Audio Speech Lang. Process.* **2019**, *28*, 279–293. [CrossRef]
40. Olgun, O.; Hacihabiboglu, H. METU SPARG Eigenmike em32 Acoustic Impulse Response Dataset v0.1.0. Available online: http://doi.org/10.5281/zenodo.2635758 (accessed on 28 September 2020).
41. Semechko, A. Suite of Functions to Perform Uniform Sampling of a Sphere. 2020. Available online: https://www.mathworks.com/matlabcentral/fileexchange/37004-suite-of-functions-to-perform-uniform-sampling-of-a-sphere (accessed on 20 July 2020)
42. Cox, T.J.; D'Antonio, P.; Avis, M.R. Room sizing and optimization at low frequencies. *J. Audio Eng. Soc.* **2004**, *52*, 640–651.
43. Crocker, M.J. *Handbook of Noise and Vibration Control*; John Wiley & Sons: Hoboken, NJ, USA, 2007.
44. Everest, F.A. Master Handbook of Acoustics. *J. Acoust. Soc. Am.* **2001**, *110*, 1714–1715. [CrossRef]
45. Schimmel, S.M.; Muller, M.F.; Dillier, N. A fast and accurate "shoebox" room acoustics simulator. In Proceedings of the 2009 IEEE International Conference on Acoustics, Speech and Signal Processing, Taipei, Taiwan, 19–24 April 2009; pp. 241–244.
46. Izumi, Y.; Otani, M. Relation between Direction-of-Arrival distribution of reflected sounds in late reverberation and room characteristics: Geometrical acoustics investigation. *Appl. Acoust.* **2021**, *176*, 107805. [CrossRef]
47. Fazenda, B.M.; Stephenson, M.; Goldberg, A. Perceptual thresholds for the effects of room modes as a function of modal decay. *J. Acoust. Soc. Am.* **2015**, *137*, 1088–1098. [CrossRef] [PubMed]
48. Papadopoulos, C.I. Redistribution of the low frequency acoustic modes of a room: A finite element-based optimisation method. *Appl. Acoust.* **2001**, *62*, 1267–1285. [CrossRef]
49. Fazenda, B.; Wankling, M.; Hargreaves, J.; Elmer, L.; Hirst, J. Subjective preference of modal control methods in listening rooms. *J. Audio Eng. Soc.* **2012**, *60*, 338–349.
50. Fuchs, H.; Lamprecht, J. Covered broadband absorbers improving functional acoustics in communication rooms. *Appl. Acoust.* **2013**, *74*, 18–27. [CrossRef]
51. Cox, T.; d'Antonio, P. *Acoustic Absorbers and Diffusers: Theory, Design and Application*; CRC Press: Boca Raton, FL, USA, 2016.

Article

On the Sequence of Unmasked Reflections in Shoebox Concert Halls

Juan Óscar García Gómez [1,*], Oliver Wright [2], Elisabeth van den Braak [3], Javier Sanz [2], Liam Kemp [1] and Thomas Hulland [3]

1. Marshall Day Acoustics, Collingwood 3066, Australia; lkemp@marshallday.com
2. Marshall Day Acoustics, Auckland 1010, New Zealand; oliver.wright@marshallday.co.nz (O.W.); javier.sanz@marshallday.co.nz (J.S.)
3. Marshall Day Acoustics, Wellington 6011, New Zealand; bertie.vandenbraak@marshallday.co.nz (E.v.d.B.); tom.hulland@marshallday.co.nz (T.H.)
* Correspondence: jogarciagomez@marshallday.com

Abstract: Highly appreciated concert halls have their own acoustic signature. These signatures may not often be consciously appraised by general audiences, but they have a significant impact on the appreciation of the hall. Previous research indicates that two of the most important defining elements of a hall's acoustic signature are (i) the reflection sequence and relative reflection levels at the listener position and (ii) the perceptibility of the reflections based on perception thresholds. Early research from Sir Harold Marshall identified the importance of unmasked early reflections to enhance a concert hall's acoustic signature. The authors see an opportunity to extend the existing research by further examining the sequence of unmasked reflections. By analysing the cross-sections of three concert halls, this manuscript quantifies potential links between a hall's architectural form, the resultant skeletal reflections, and the properties of its acoustic signature. While doing so, the manuscript identifies potential masking reflections through visual and analytical assessment of a hall's skeletal reflections. It is hypothesized that the "rhythm" of the reflection sequence could hold key insights into the hall's "personality" and acoustic signature. If so, this could present new design tools and considerations for new concert halls and the diagnosis of underperformance in existing halls.

Keywords: concert hall acoustics; lateral reflections; shoebox typology; spatial impression; perception thresholds; skeletal reflections; reflection sequence; perception thresholds

1. Introduction

Sir Harold Marshall defined Presence as *"a dimension in which cultural phenomena and cultural events become tangible and have an impact on our senses and our bodies"* [1]. How to achieve Presence in a space is a cross-disciplinary mission that involves the input of experts in several fields, acoustics among them. The concept of Presence is similarly described by Salter and Blesser, with the definition of Aural Spaces as places that have an impact on the emotions or the behaviours of the listeners [2]. Concert halls are spaces where thousands of people gather to share one-off experiences each evening and where the concepts of Presence and Aural Spaces become particularly relevant.

Every space has, at least, one purpose which should be engineered by the acoustic architect through a detailed understanding of the nuts and bolts of acoustics [2]. Spaces such as control rooms in recording studios are meant to be acoustically neutral, providing an exceptionally linear acoustic response that avoids any tone colouration between the monitors and the sound engineer. Concert halls, on the other hand, are meant to enhance the sound from the orchestra—nobody would enjoy listening to an orchestra in an anechoic chamber. Concert halls are an extension of the orchestra, with their own character, and like any other musical instrument, should be designed to be rich and unique. High-end violins, for example, are not appraised by their neutral and linear sound, but by their unique sound signature. Indeed, highly appreciated halls such as the Große Musikvereinssaal have a strong and unique sound, also described as a character or acoustic signature that has a significant impact on the appreciation of the hall by an audience.

Concert halls are collective instruments played by the conductor, by the orchestra as an ensemble, and by each musician independently. The best orchestras grasp the hall's character and integrate it as part of the musical expression. Needless to say, the acoustics of a concert hall, along with adding character to the sound, should provide favourable acoustic conditions for the orchestra and the learning curve should be gentle so that visiting orchestras can adapt to its acoustics quickly within a short seating call or warm-up rehearsal.

2. Research Purpose

The authors of this manuscript believe that the foundational concept of reflection sequence and its potential to enhance the acoustic signature of a concert hall has not been explored in enough detail as of yet. With this work, the authors aim to provide further insight into the potential relationship between the reflection sequence and the acoustic signature of a concert hall. The manuscript is structured as follows:

- Literature review;
  - Acoustic signature;
  - Reflection sequence;
  - Perception thresholds;
- Skeletal Analysis of three shoebox concert halls;
- Potential masking diagram;
- Design features to potentially enhance the acoustic signature of a concert hall;
- Conclusions.

3. Literature Review

Four concepts are consistently used in this manuscript: acoustic signature, skeletal reflection, reflection sequence, and perception thresholds. Given that the three concepts are closely related, it is worth defining them first.

- Acoustic signature: the distinctive acoustic properties of an aural object;
- Skeletal reflections: discrete early reflection provided by major architectural elements;
- Reflection sequence: the temporal order and relative level of a set of discrete sound reflections;
- Perception threshold: the weakest sound stimulus that a human can sense.

The following paragraphs provide a brief historical reference and explain their relevance to the design of concert halls

3.1. Acoustic Signature

Humans possess an inherent spatial ability to understand their surroundings by how passive objects modify the acoustics of a space. The frequency content and level of background noises as well as the delays of sounds bouncing off passive objects creates an aural image to a listener. Therefore, to some extent, we can "see" with our ears thanks to our ability to decode spatial attributes using acoustic cues [2].

Shaping these passive objects—that is, the walls, ceilings, and architectural elements of concert halls—is a relatively young science. One of the major advances on the field was the finding of the effect of lateral reflections in 1952, when Meyer identified the apparent extension of a sound source with the presence of a sound reflection [3,4].

In the late 1960s, Marshall and Barron advocated that perceptible early lateral reflections were significant in improving Spatial Impression (SI) [5,6]. Marshall concluded that a sequence of skeletal reflections that are perceptible and unmasked could enhance the room response, providing the room with a *"premium quality of acoustical experience"* [7].

This statement is essential to understand the purpose of this work. While most of the standard acoustic parameters are based on energy integrals, the concept of unmasked reflection sequences provides an innovative approach to the design of concert halls to be used alongside the rest of the acoustic parameters.

In the 1970s, Wettschurek's research on the unmasking of reflections related to the loudness levels signalled a connection between SI and Loudness [8]. This research was followed up by Kahle et al. in 2017, linking the musical dynamic of a hall to the unique reflection sequence at a particular listening position as a reason why concert halls "wake up" differently [9]. A concert hall that "wakes up" as the dynamic increases could be interpreted as a concert hall that reveals its acoustic signature as its sequence of early reflections becomes unmasked as the overall reflection level increases.

Through a set of listening tests in 2011, the Aalto University team found that listeners could recognize a concert hall through its early sound, which is primarily composed of a sequence of early reflections created by the room's shape [10]. This reveals the strong connection between the sequence of early reflections and the recognizability of a concert hall.

Therefore, designing concert halls with a strong acoustic signature could be a strategy to create spaces that left an aural memory on the audience. Indeed, the mere-repeated-exposure effect describes that an individual exposed to familiar and new stimulus objects shows a preference for the familiar object. This could be one of the reasons why the Große Musikvereinssaal has become such an archetypical hall or why Funkhaus Studio 1, with its strong character, has become iconic for classical recordings.

From the literature, two elements with relevant influence on the signature of a concert hall are:

(i) The reflection sequence and relative reflection levels at the listener position.
(ii) The perceptibility of the reflections relative to the perception thresholds.

3.2. Reflection Sequence

For a particular listening position, a series of discrete early reflections are provided by the hall's architectural shape. These principal discrete reflections were identified as *skeletal reflections* (SR) by Marshall [5]. Based on a complex relationship between the arrival times, directions, and levels, certain reflections fall below or above the perception thresholds and are considered as *masked* or *unmasked*, respectively.

In 1981, Barron and Marshall attested that sound reflection levels add incoherently for SI—in short, the temporal order of reflections from different directions was deemed of little importance for SI [3,11]. As an objective criterion to measure SI, Barron developed the Lateral Energy Fraction (LF), which has been proven to correlate highly with subjective listener preferences. Probably due to the success of the LF criteria and other acoustic parameters, further design techniques related to the sequence of unmasked reflections have not been explored yet even though the applications of both concepts are complementary.

The acoustic parameters included in the ISO 3382-1:2009 "Acoustics Measurement of room acoustic parameters—Part 1: Performance spaces" provide a rich framework to interpret the acoustics of a concert hall. However, as suggested by Bradley, further research has to be done to fully understand missing elements such as the preferred design criteria or the most relevant range of frequencies for each parameter [12]. From this set of parameters, it is not clear which would allow distinguishing between "good" and "excellent" acoustics [10] or what makes a hall recognizable from other halls [13].

Furthermore, despite the Reverberation Time (RT), which is largely constant through a room, many other parameters are location-dependent, making it difficult to describe a room by a single number. For example, Clarity values have been proven to vary beyond the standard uncertainty even between two adjacent seats [14].

The early reflections are a sonic representation of the hall's architectural shape. Halls with a simple shape such as the Große Musikvereinssaal create a "simple" reflection sequence, while halls with a complex architecture generally provide a greater quantity of early reflections, creating a more complex acoustic signature. Halls including large amounts of acoustic diffusion or *visual noise* will gravitate towards a blurred or less obvious acoustic signature given that the early reflections are highly weakened, likely below the perception threshold.

Halls should be shaped to provide the best acoustic conditions for their intended use. Certain concert halls would benefit from a stronger acoustic signature to enhance the aural experiences, whereas other halls such as black box-type halls intended for amplified sound will benefit from more neutral acoustics.

3.3. Perception Thresholds

The perception of sound reflections in a concert hall does not only correspond to a binary audible/inaudible criterion but to a gradient of perceptibility with three main thresholds [15]:

(i) Detection threshold below which sound reflections are inaudible. In this work, the perception threshold term is used to reference this threshold.
(ii) Image-shift threshold above which reflections are audible but contribute to some spatial effect such as image-shift or localization blur.
(iii) Echo-threshold above which two separate auditory events are audible.

Acoustic perception thresholds were first investigated by Seraphim using speech signals for several source and reflection directions [16]. These are the perception thresholds that were used in the well-known 1967 note to the editors by Marshall, "A note on the importance of room cross-section in concert halls" [5].

A year later, Marshall published a second letter to the editors, "Levels of Reflection Masking in Concert Halls" after Schubert's research on the audibility of musical reflections in a concert hall [17,18]. This included Schubert's thresholds for a reflection at $\alpha = 30°$ using a choral motif, not a significant change compared to his previous paper but just a more appropriate use of detection thresholds. Schubert's data indicated that Seraphim's thresholds were unlikely to be applicable to the case of a concert hall, since the musical masking is, in any case, sustained much more than the masking of speech. This is because the listener perception is generally less sensitive to reflections if the sound signal is music instead of speech. Moreover, the reverberation in a concert hall renders the detection of a sound reflection even more difficult. Schubert also showed that the masking threshold level is approximately 10 dB lower for a lateral reflection normal to the direct sound than it is for a reflection having approximately the same direction as the direct sound [17,18].

In 1970, Barron published "The Subjective Effects of First Reflections in Concert Halls—The Need for Lateral Reflections" [6]. This paper tested the threshold for a single side reflection at $\alpha = 40°$ with a Mozart motif at a mean level of 81 dB. The results of this test are well known to acousticians and are often referenced in the literature.

Twenty-two years later, Olive and Toole continued the research on thresholds for audibility and image shift. Their research showed that the more reverberant the environment is, the higher the audibility threshold becomes [15,19]. From an acoustic point of view, the consequence of this finding is that if lateral reflections are to be perceived, the reflections should be strong and well above the threshold of audibility. Even surfaces with low levels of scattering or residual acoustic absorption could drag these reflections below the audibility threshold in a reverberant field, making them imperceptible.

As part of their work, Olive and Toole prepared a complete comparison of audibility thresholds, which can be found in [19]. For reference, Schubert's "Handel Concerto Grosso" in that comparison was the threshold used in Marshall's second note and in this manuscript. Schubert's work on perception thresholds is even cited in relevant books such as Kuttruff's Room Acoustics [20].

Strong lateral reflections have been shown to enhance musical dynamics and the dynamic responsiveness in concert halls [21,22]. The literature also shows that in most cases, image-shift thresholds are significantly higher than detection thresholds. This means that reflections must be substantially louder than just detectable to mislead localization [15].

Image shift produced by early reflections is likely due to a combination of a strong reflection arriving after a weakened direct sound and/or low energetic early reflections. In the authors' previous experience, cases of image shift are often found on the voice signal of opera singers, likely accentuated by the lack of stage enclosure to fill with the gap between the direct sound and the first hall reflection.

While scattered reflections below the audibility threshold have the potential to increase acoustic parameters values such as LF or Clarity, Lokki et al. indicated that sound reflections from scattering surfaces will result in Temporal Envelope Distorting (TED) reflections, making it difficult for these reflections to be fused with the direct sound [10].

3.3.1. Overall Listening Levels and Angle of Arrival

The perception thresholds also depend on the overall listening levels. This was first investigated by Wettschurek in 1976 [8]. Forty years after the original publication, Green et al. revived the topic and demonstrated how reflections from different directions and arrival times are masked or unmasked at different overall listening levels [22].

From Wettschurek's work, the main conclusion of interest for this manuscript is that the relative perception threshold for lateral and frontal reflections is dependent on the overall listening levels. The perception threshold for a lateral reflection at *pianissisimo* is up to 8 dB higher than for a *forte*. Reflections that are felt under the perception threshold at lower listening levels would be perceived during *crescendos*. Green et al. refer to this effect as dynamic responsiveness or how a concert hall "wakes up" [22].

3.3.2. Note on Perception Thresholds

As shown above, several authors have shed light on perception thresholds, greatly improving the general knowledge under different listening conditions and sources. However, the reader should note the these are highly case-dependent, and this effect is magnified when considering the circumstances and listening experience of each member of an audience.

Similar to the precautions provided by Barron in 1971 [6], the perception thresholds should be interpreted as a baseline for the design of concert halls. Despite their relative "*objectiveness*" provided in the literature and this manuscript, the reader should be aware that the thresholds are highly case-dependent and should be understood as such.

4. Skeletal Reflections Analysis

4.1. Introduction

The Skeletal Reflections Analysis is a visually descriptive medium to evaluate and compare halls. It also enables the identification of potential relationships between a reflection sequence and the acoustic signature of a concert hall.

This analysis is generally applied to the cross-section of shoebox concert halls. For further typologies, a 3D analysis would be better suited.

This method was first introduced by Marshall [5]. The analysis was applied to the wall boundaries of two theoretical hall shapes to shed light on the importance for lateral reflections to arrive earlier than ceiling reflections in order to avoid masking.

While the original procedure did not include sound reflectors or galleries, this work includes the architectural elements, which can provide strong Temporal Envelope Preserving (TEP) reflections [10]. These elements are mostly flat or slightly curved and large enough to reflect the 125 Hz frequency band, which roughly corresponds to surfaces larger than 2.7 m. For corner reflections, the total surface considers the vertical and horizontal components.

The power of this analysis lies in its simplicity, which can be executed using the back-of-an-envelope approach and still provide great insights, which can be used for decision making at any stage of a design.

For the set of concert halls included in this work, the SRs are manually found in the cross-section. The automatization of the procedure by the use of the Image-Source method is an option; however, a critical selection of reflections should be made. The inclusion of non-relevant reflections would complexify the analysis.

4.2. Methodology

The Skeletal Reflections Analysis in this manuscript is consistent with the procedure followed by Marshall. The reader should refer to the original paper for an in-detail description of the procedure [5]. In short, the procedure as applied in this work is the following:

- A simplified cross-section of a shoebox concert hall is used;
- A source position S on stage is assumed. The receiver R is considered at 18.3 m (60 ft) from the stage located 1 m off-axis from the centre line of the hall;
- The cross-section represents the concert hall in the middle point between the stage and the source position;
- The cross-section is then placed on the same plane as the source, as shown in Figure 1. This assumes that all reflecting surfaces in the plane with the source are angled such that they provide a reflection to the receiver. The relocation of the cross-section may be counterintuitive, but it is consistent with the original procedure. It is the subject of discussion and may be modified in future analysis;
- The skeletal reflections points are identified and represented as numbered black dots. The image sources are calculated and identified as grey numbered dots;
- The levels of the reflections are based on the path length and the effect of grazing incidence and scattering surfaces. The effect of grazing incidence to a seat in the centre of the audience is accounted for as
    - -15 dB for a flat floor configuration [23],
    - -10 dB for a gentle audience rake (5–6°) (*), and
    - -5 dB for a steeper audience rake (10–11°) (*);
    - The effect of scattering surfaces is accounted as -5 dB per bounce [24];

(*) author's assumptions.

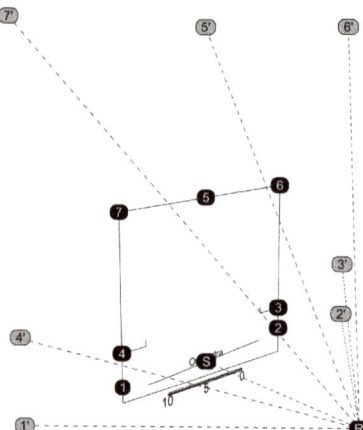

Figure 1. Skeletal reflections procedure, graphical explanation. The source position is named S, and the receiver is R. The numbers in black dots indicate the sound reflection position, grey dots indicate image sources positions. The procedure and graphs were produced using a custom script in Grasshopper for Rhino3d.

- Data from Schubert are used to draw the reflections' absolute threshold and probable threshold for useful contribution [18];
- To facilitate the reading of the echograms, the reflections have been numbered from the bottom up. Zero (0) corresponds to the direct sound. One (1) and two (2) represent the two grazing lateral reflections. The rest of the reflections are case-dependent.

In reality, concert halls have numerous reflections, and an orchestra with multiple directivity patterns can hardly be assumed to be represented by a single point source as in the Skeletal Analysis in Marshall's procedure. However, the simplified echogram based on SRs presented by Marshall and explored in this manuscript serves as a preliminary tool at the early phases of the design. During the design phase, multiple sources and seat locations with varying overall levels should be compared.

4.3. Case Studies

A set of concert halls with common characteristics has been selected for comparing SRs. The selection criteria are:

- Shoebox typology;
- Similar capacity;
- The halls should have proven successful acoustics.

The list of concert halls along with an overview of their capacity, dimension, and pictures are shown in Table 1 and Figure 2.

Table 1. Data for selected concert halls.

Concert Hall	Code	Capacity (N)	Volume (V) [m³]	V/N [m³/seat]	$RT_{500-1000,occ}$ [s]	Floor Rake at Stalls
Große Musikvereinssaal	GMV	1680	15,000	8.9	2.0	0°
Perth Concert Hall	PCH	1729	15,650	9.0	2.1	10–11°
Stavanger Konserthuset [25]	SK	1500	22,168	14.8	2.2	5–6° [26]

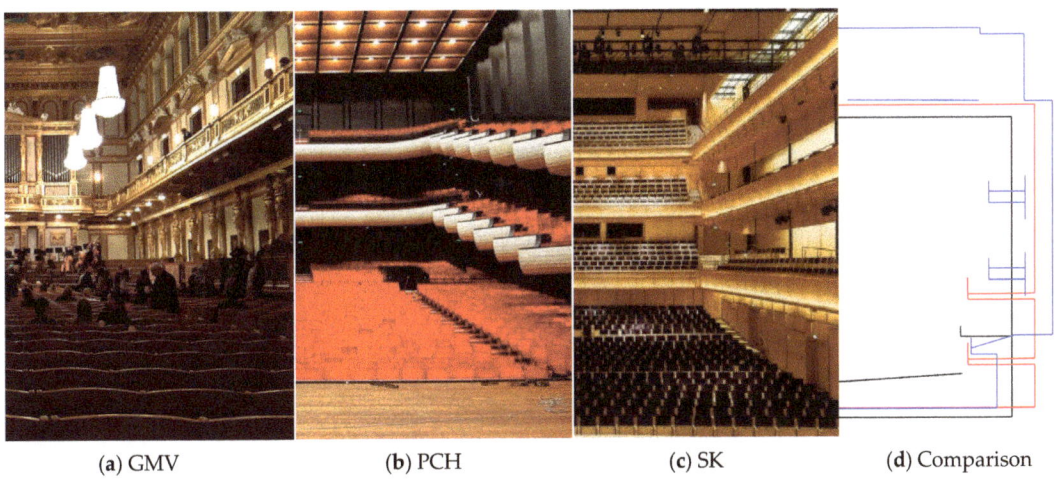

(a) GMV   (b) PCH   (c) SK   (d) Comparison

Figure 2. (**a**) Große Musikvereinssaal (photo: J.O. García); (**b**) Perth Concert Hall (photo: Marshall Day Acoustics); (**c**) Stavanger Konserthuset (photo: https://sso.no/en/seating-plan-fartein-valen/ accessed on 24 August 2021); (**d**) Comparison of concert hall's cross-section: (**Black**) Große Musikvereinssaal, (**Red**) Perth Concert Hall, (**Blue**) Stavanger Konserthuset [25].

4.4. Analysis

The Skeletal Analysis is presented in the form of an echogram. The vertical lines indicate the arrival of an SR, which is the height of the line related to the relative level to the theoretical unobstructed direct sound. The numbers above each vertical line represent the location where the reflection happens, indicated in a short section adjacent to the echogram. Two perception thresholds are shown in the SR echogram in Figure 3 (dashed line), absolute threshold of perceptibility after Schubert (solid line), probable threshold for useful contribution. The absolute arrival time and the Initial Time Delay Gap (ITDG) are shown in the horizontal axis of the echogram.

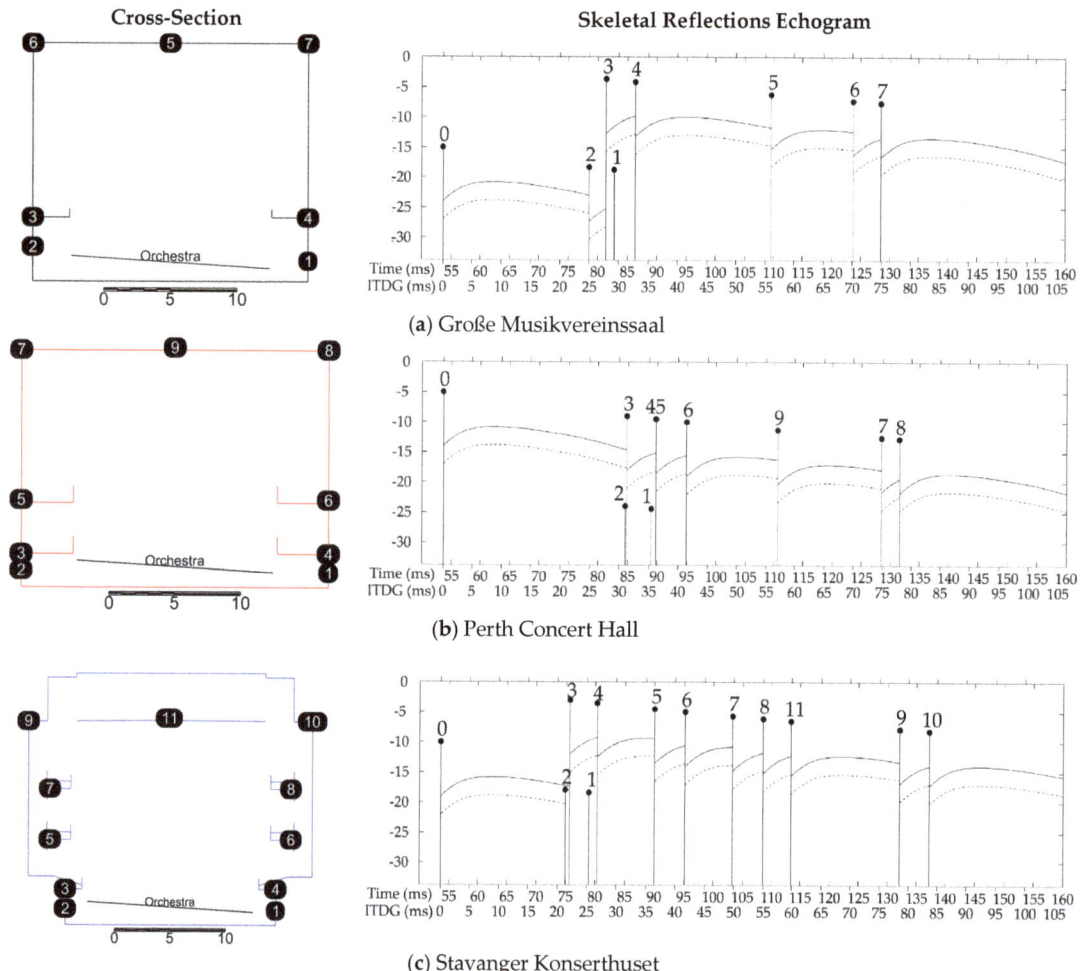

Figure 3. Skeletal Reflections Analysis for three concert halls: (**a**) Große Musikvereinssaal, (**b**) Perth Concert Hall, (**c**) Stavanger Konserthuset. Time indicates the absolute time of arrival, and Initial Time Delay Gap (ITDG) is the arrival time after the direct sound. The dashed line indicates the absolute threshold of perceptibility after Schubert. The solid line is the probable threshold for useful contribution.

The reader should note that reflections from the stage enclosure fill the time gap between the direct sound and the first lateral but are not shown in this analysis. The stage enclosure reflections, due to their direction and the arrival time, are generally masked by the direct sound but contribute to filling in the gap between the direct sound and the first halls reflection.

When analysing the SRs in Figure 3, the reader should note that as discussed above, the perception thresholds for lateral reflections are lower than from any other direction. Therefore, in reality, if two reflections arrive simultaneously from two different directions, it is likely that the more lateral (perpendicular to the ear) reflection will predominate and mask the other. Relative reflection levels in the analysis below do not compensate for direction nor overall listening levels.

Regarding the attenuation of the direct sound, halls with a flat floor will be the most affected by audience attenuation. In the original analysis, Marshall used a 15 dB attenuation in excess of the inverse square law based on data from Schulz and Watters [23]. The steeper audience rakes are the less affected ones. The following convention has been followed: flat floor −15 dB (GMV); gentle audience rake −10 dB (SK); pronounced audience rake −5 dB (PCH).

4.5. Results

The analysis of the cross-section is divided into three main groups: Lower room, sidewalls, and ceiling. The rhythm and density of SRs are not related to the *quality* of the acoustic signature. The presence of a recognizable acoustic signature is important for the subjective quality of a concert hall. However, the properties of this signature can and should be different, depending on the hall's musical and architectural purpose.

4.5.1. Lower Room

Reading the echograms from left to right will tell us which hall provides the quickest early reflection of the three halls, which is related to ITDG and Acoustic Intimacy.

In halls with flat floor or gentle audience rake, very early overhead reflections are critical to provide reflections that enhance the direct sound which is very much obstructed by the audience. In this case, SK due to its narrow parterre and the first balcony soffit (reflections 3–4) provides the earliest lateral overhead reflections.

In halls with full-width stalls of 23–25 m, the first lateral reflection will only arrive at 30–40 ms after the direct sound, as is the case of PCH. Even though the energy from a first early reflection arriving at that time contributes to enhancing Clarity and Early Energy, other important criteria such as Acoustic Intimacy would benefit from the shortest ITDG possible.

In the set of studied concert halls, PCH, which has a wider parterre also has a steeper audience rake. This allows for an unobstructed direct sound that compensates for the "late" first early reflection. In addition to ITDG, Lokki et al. suggest the spectrum of the sound reflections is important to enhance intimacy; low and high-frequencies enhance it more than mid-frequencies [27]. Therefore, the use of large or fine-scale scattering applied to the sidewalls could reduce the subjective impression of intimacy. In the three halls in this manuscript, the use of fine-scale scattering is residual. Even in the GMV, which is believed to be highly diffused, Marshall pointed out that *"inspection shows that there is far more plane surface than one would think"* [28].

Generally, finding strategies to physically narrow the room will improve acoustic and visual intimacy by creating earlier first lateral reflections, resulting in the audience feeling closer to the stage. This design approach was also employed for the Philarmonie de Paris. In this case, the balconies and clouds were narrowed to create an acoustic and visual intimate room within a larger reverberant volume [29–32].

4.5.2. Sidewalls

There are two clear design approaches in this set of concert halls. First, GMV and PCH were designed to provide quick second-order overhead lateral reflections from relatively low balcony undersides, which arrive virtually simultaneously with the sidewall reflections. This results in enhancing the direct sound, increasing Apparent Source Width, and creating envelopment. Above the side balconies, the relatively flat walls go all the way up to the ceiling. These large upper wall surfaces are providing a *breath,* or reflection-free time window between the first reflections and the ceiling reflection. The Skeletal Analysis of GMV and PCH shows a clear separation between the direct sound, first lateral reflections, ceiling reflection, and late-ish early reflections. This distinction between the sound fields could align with Lokki's impressions on GMV, indicating that this distinction would preserve articulation and reduce masking the onset of notes [27].

PCH includes two overhead reflections from the undersides of the second balcony (5–6). These reflections arrive barely 5–10 ms after the first balcony reflections (3–4), increasing the risk of temporal masking.

In contrast to the low balconies and flat empty upper walls of GMV, SK exhibits a rational balcony height distribution and a denser short section. This feature is meant to deliver a great amount of early energy to the audience to enhance the early lateral sound, providing high clarity in a larger volume—SK is 7000 m^3 larger than GMV and PCH.

The rhythm of reflections shown in the echograms is immediately obvious. Halls like SK show a denser echogram between 20 ms and 60 ms after the direct sound. This is informed by the number of early reflections provided by the balconies soffits compared to GMV. Based on the Skeletal Analysis, the presence of multiple reflections from side balconies would increase the chances of masking the ceiling reflection, which is harder to perceive due to a higher perception threshold for zenith reflections as shown by Wettschurek [8].

4.5.3. Ceiling

A ceiling reflection has a great potential to "open" the room, i.e., to enhance the room perception and the feeling of being in a large volume. However, if this reflection arrives too late, or if due to limited or no lateral reflections, the preceding reflection-free period would be too long, it could make the ceiling reflection be perceived as an echo. In the case of an excessively late ceiling reflection for major reasons, providing a strong reflection sequence beforehand could help to reduce the risk of echo by renovating the precedence effect.

The risk of echo perception motivates the use of highly diffusive ceilings that do avoid echoes but also misses all the above-mentioned aural benefits. Instead of designing to avoid risks, the ceiling reflections can be fine-tuned by:

- Limiting the ceiling height to 18 m, a rule-of-thumb limit for large shoeboxes;
- Creating enough lateral reflections;
- Accounting for the higher perception threshold for zenith reflections.

The three halls analysed show a similar approach to ceiling reflections. PCH and SK include some sort of diffusiveness to the ceiling when compared to GMV—PCH a coffered ceiling and SK convex reflector panels—but all the halls allow for relatively strong ceiling reflections, which might be important for their success. GMV and PCH provide a considerable reflection-free time before the ceiling reflection, GMV almost 20 ms, PCH 15 ms. The reflection-free time provided by SK is variable between 5–15 ms, depending on the movable ceiling position. This reflection-free period was first mentioned by Marshall, and given the quality of these concert halls, it is an acoustic feature that should be at least considered for the development of a new hall to avoid potential masking.

5. Potential Masking Diagram

When analysing the cross-section of a concert hall, one of the main goals is to identify which reflections may be masked. Once identified, it is up to the acoustician to modify the shape of the room to avoid potential masking. Of interest for this work is the creation of aural cues that could provide the room with an acoustic signature.

Complementary to the Skeletal Analysis, the Potential Masking diagram is proposed by the authors as a technique to visualize which architectural elements provide simultaneous sound reflections.

Figure 4 shows the Potential Masking diagram applied to the cross-section of the three concert halls. Each cross-section incorporates a series of circles centred at the source position. Each circle corresponds to the arrival of the first reflection after the direct sound from a source on the stage edge to a receiver at 18.3 m—the same as used for the Skeletal Masking Analysis. The analysis is straightforward: architectural elements provide reflections to the audience, which share a curve that could lead to masking sound reflections.

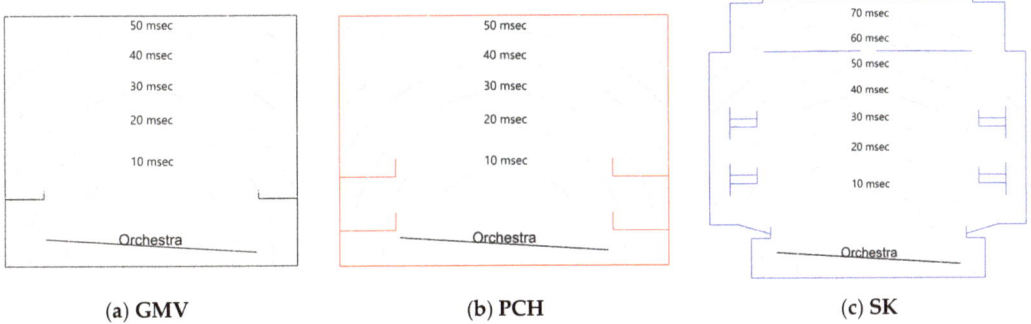

Figure 4. Potential Masking diagram for three concert halls: (**Black**) Große Musikvereinssaal, (**Red**) Perth Concert Hall, (**Blue**) Stavanger Konserthuset.

From Figure 4, the influence of the parterre width can be seen —the narrower parterre (SK) provides the first lateral reflections around 20–25 ms—whereas in the wider ones, the first lateral reflections only reach the receiver around 30–35 ms.

It is also common for various halls to have a balcony providing strong second-order reflections that is on the same circle as the ceiling. This could lead to masking the ceiling reflection (given that the threshold for a lateral reflection is lower than for the ceiling), which could lead to a room that "opens" less easily. The use of curved surfaces, as is the case of the vertical soffits in SK, mitigates the strength of the lateral reflections that could mask the ceiling reflection. Note these are potential effects that could be important cues for the acoustician when designing or fine-tuning a concert hall.

As a rule of thumb, avoiding architectural elements that create SRs being located on the same arc may increase the chances for the reflections to be unmasked and therefore meaningful for the acoustic signature of the room.

6. Conclusions

6.1. Potential Room Signature Enhancers

From the Skeletal Reflections Analysis of three successful venues, a series of design techniques have emerged as potential enhancers of the concert hall acoustic signature based on the cross-section. These are summarised as follows:

- Avoiding *bunched* SRs;
- Providing a reflection-free period in the order of 10–15 ms before key SRs;
- Fine-tuning the energy of the SRs—the perception thresholds can be used as context to evaluate the sound energy;
- Creating a sequence of "strong" SR to reduce the risk of image shift and echoes.

The reader should note these features are hypotheses, and further research, including listening tests, is required.

6.2. General

Early research from Marshall identified the importance of unmasked early reflections. The authors have sought to further this research and address a perceived gap in the existing literature.

The authors consider the sequence of reflections a key feature of the hall geometry that has significant influence on the hall's acoustic signature. Through analysing the cross-section of three concert halls, this manuscript quantifies potential links between a hall's architectural form, the resultant SRs, and the properties of its acoustic signature.

While doing so, the manuscript identifies potential masking reflections through visual and analytical assessment of a hall's SRs.

The "rhythm" (bunching and spaces) of the discrete reflection sequence could hold key insights into the hall's acoustic signature. If so, this could hold ramifications on the design of new concert halls and the diagnosis of underperformance in existing concert halls.

This work so far has focussed on theoretical investigations, and further research is required. The authors hope to continue this research and test the resultant hypotheses through a series of listening tests using 3D impulse responses.

Author Contributions: Conceptualization, J.Ó.G.G., O.W., E.v.d.B., J.S. and T.H.; methodology, J.Ó.G.G.; software, J.Ó.G.G.; validation, J.Ó.G.G., O.W., E.v.d.B., J.S. and T.H.; formal analysis, J.Ó.G.G.; investigation, J.Ó.G.G.; resources, J.Ó.G.G.; data curation, J.Ó.G.G.; writing—original draft preparation, J.Ó.G.G.; writing—review and editing, O.W., E.v.d.B., J.S., T.H. and L.K.; visualization, J.Ó.G.G.; supervision, J.Ó.G.G., O.W., E.v.d.B., J.S. and T.H.; project administration, J.Ó.G.G.; funding acquisition, J.Ó.G.G.; All authors have read and agreed to the published version of the manuscript.

Funding: This research was funded by Marshall Day Acoustics Ltd.

Acknowledgments: This research would not have been possible without the innovative research by Harold Marshall. His enthusiasm and scientific contribution to the development of concert halls is a source of inspiration. His guidance and exchanges during the development of this research are greatly valued.

Conflicts of Interest: The authors declare no conflict of interest.

References

1. Marshall, A.H. Acoustical dimensions for production of presence in music spaces. In Proceedings of the IOA Auditorium Acoustics, Hamburg, Germany, 4–6 October 2018.
2. Salter, L.R.; Blesser, B. *Spaces Speak, Are You Listening? Experiencing Aural Architecture*; MIT Press: Cambridge, MA, USA, 2009.
3. Marshall, A.H.; Barron, M. Spatial responsiveness in concert halls and the origins of spatial impression. *Appl. Acoust.* **2001**, *62*, 91–108. [CrossRef]
4. Richardson, E.G.; Meyer, E. Technical aspects of sound. *Phys. Today* **1963**, *16*, 82–84. [CrossRef]
5. Marshall, A.H. A note on the importance of room cross-section in concert halls. *J. Sound Vib.* **1967**, *5*, 100–112. [CrossRef]
6. Barron, M. The subjective effects of first reflections in concert halls—The need for lateral reflections. *J. Sound Vib.* **1971**, *15*, 475–494. [CrossRef]
7. Marshall, A.H. Acoustical determinants for the architectural design of concert halls. *Arch. Sci. Rev.* **1968**, *11*, 81–87. [CrossRef]
8. Wettschurek, R. Über die Abhängigkeit Raumakustischer Wahrnehmung von der Lautstärke. Ph.D. Thesis, Technical University of Berlin, Berlin, Germany, 1976.
9. Kahle, E.; Green, E.; Knauber, F.; Wulfrank, T.; Jurkiewicz, Y. How (and why) does every Concert Hall "wake up" differently? *J. Acoust. Soc. Am.* **2017**, *141*, 3599. [CrossRef]
10. Lokki, T.; Pätynen, J.; Tervo, S.; Siltanen, S.; Savioja, L. Engaging concert hall acoustics is made up of temporal envelope preserving reflections. *J. Acoust. Soc. Am.* **2011**, *129*, EL223–EL228. [CrossRef] [PubMed]
11. Barron, M. Basic design techniques to achieve lateral reflctions in concert halls. In Proceedings of the International Symposium on Room Acoustics, Amsterdam, The Netherlands, 15–17 September 2019.
12. Bradley, J.S. Review of objective room acoustics measures and future needs. *Appl. Acoust.* **2011**, *72*, 713–720. [CrossRef]
13. Haapaniemi, A.; Lokki, T. Identifying concert halls from source presence vs. room presence. *J. Acoust. Soc. Am.* **2014**, *135*, EL311–EL317. [CrossRef] [PubMed]
14. Witew, I.; Dietrich, P.; de Vries, D.; Vorländer, M. Uncertainty of room acoustic measurements—How many measurement positions are necessary to describe the conditions in auditoria? In Proceedings of the International Symposium on Room Acoustics, Melbourne, Australia, 29–31 August 2010.
15. Sheaffer, J. From Source to Brain: Modelling Sound Propagation and Localisation in Rooms. Ph.D. Thesis, University of Salford, Salford, UK, 2013.
16. Seraphim, H.P. Über die wahrnehmbarkeit mehrerer Ruckwurfe von Sprachschall. *Acta Acust. United Acust.* **1961**, *11*, 80–91.
17. Marshall, A.H. Levels of reflection masking in concert halls. *J. Sound Vib.* **1968**, *7*, 116–118. [CrossRef]
18. Schubert, P. Untersuchungen über die Wahrnehmbarkeit von Einzelrückwürfen bei Musik. *Technische Mitteilung RFZ* **1966**, *3*, 124–127.
19. Olive, S.E.; Toole, F.E. The detection of reflections in typical rooms. *J. Audio Eng. Soc.* **1989**, *539*, 539–553.
20. Kuttruff, H. *Room Acoustics*; Applied Science Publishers: London, UK, 1979.
21. Pätynen, J.; Tervo, S.; Robinson, P.W.; Lokki, T. Concert halls with strong lateral reflections enhance musical dynamics. *Proc. Natl. Acad. Sci. USA* **2014**, *111*, 4409–4414. [CrossRef] [PubMed]
22. Green, E.; Kahle, E. Dynamic spatial responsiveness in concert halls. *Acoustics* **2019**, *1*, 549–560. [CrossRef]
23. Schulz, T.; Watters, B.C. Propagation of sound across audience seating. *J. Acoust. Soc. Am.* **1964**, *36*, 885–896. [CrossRef]

24. Ryu, J.K.; Jeon, J.Y. Subjective and objective evaluations of a scattered sound field in a scale model opera house. *J. Acoust. Soc. Am.* **2008**, *124*, 1538–1549. [CrossRef]
25. Kahle, E.; Möller, H.; Ognedal, T. The New Concert Hall in Stavanger. In Proceedings of the Institute of Acoustics, Reading, UK, 10–11 April 2008.
26. Kahle, E.; Jurkiewicz, Y.; Katz, B.F. Stavanger concert hall, acoustic design and measurement results. In Proceedings of the Institute of Acoustics, Paris, France, 29–31 October 2015.
27. Lokki, T.; Pätynen, J.; Tervo, S.; Kuusinen, A.; Tahvanainen, H.; Haapaniemi, A. The secret of the Musikverein and other shoebox concert halls. In Proceedings of the Institute of Acoustics, Paris, France, 29–31 October 2015.
28. Marshall, A.H. On the architectural implications of "diffusing surfaces". In Proceedings of the IOA Auditorium Acoustics, Hamburg, Germany, 4–6 October 2018.
29. Marshall, A.H.; Day, C.W. The conceptual acoustic design for La Philharmonie de Paris, Grande Salle. In Proceedings of the Institute of Acoustics, Paris, France, 29–31 October 2015.
30. Barron, M.; Marshall, A.H. Spatial impression due to early lateral reflections in concert halls: The derivation of physical measure. *J. Sound Vib.* **1981**, *77*, 211–232. [CrossRef]
31. Kahle, E. Halls without qualities—Or the effect of acoustic diffusion. In Proceedings of the Institute of Acoustics: Auditorium Acoustics, Hamburg, Germany, 4–6 October 2018.
32. Kirkegaard, L.; Gulsrud, T. In search of a new paradigm: How do our parameters and measurement techniques constrain approaches to concert hall design? *Acoust. Today* **2011**, *7*, 7–14. [CrossRef]

Article

Comfort Distance—A Single-Number Quantity Describing Spatial Attenuation in Open-Plan Offices

Valtteri Hongisto * and Jukka Keränen

Acoustics Laboratory, Turku University of Applied Sciences, Joukahaisenkatu 3, FI-20520 Turku, Finland; jukka.keranen@turkuamk.fi
* Correspondence: valtteri.hongisto@turkuamk.fi

Abstract: ISO 3382-3 is globally used to determine the room acoustic conditions of open-plan offices using in situ measurements. The key outcomes of the standard are three single-number quantities: distraction distance, r_D, A-weighted sound pressure level of speech, $L_{p,A,S,4m}$, and spatial decay rate of speech, $D_{2,S}$. Quantities $L_{p,A,S,4m}$ and $D_{2,S}$ describe the attenuation properties of the office due to room and furniture absorption and geometry. Our purpose is to introduce a new single-number quantity, comfort distance r_C, which integrates the quantities $L_{p,A,S,4m}$ and $D_{2,S}$. It describes the distance from an omnidirectional loudspeaker where the A-weighted sound pressure level of normal speech falls below 45 dB. The study explains why the comfort criterion level is set to 45 dB, explores the comfort distances in 185 offices reported in previous studies. Based on published data, the r_C values lie typically within 3 m (strong attenuation) and 30 m (weak attenuation). Based on this data, a classification scheme was proposed. The new quantity could benefit the revised version of ISO 3382-3.

Keywords: open-plan offices; spatial decay; ISO 3382-3; room absorption; office noise; speech

1. Introduction

Office noise and lack of speech privacy are among the environmental factors causing the largest dissatisfaction in open-plan offices [1]. One of the main reasons for this might be that work performance in concentration-demanding tasks has been found to improve with reducing intelligibility of irrelevant speech [2]. This is supported by the finding that disturbance due to noise was lower in offices having lower speech intelligibility [3]. Behavioral means can significantly affect the amount of irrelevant speech in offices, such as reducing speech effort, using high-quality headsets during phone meetings, or preferably to move to another room during such calls. Likewise, one can try to avoid the adverse effects of noise by moving to a silent environment during concentration-demanding work tasks. Room acoustic treatment can also reduce office noise. The disturbance caused by remote speech can be reduced by simultaneous application of sound absorbers (e.g., ceiling, walls, screens, and furniture), blocking of sound propagation (e.g., screens, and furniture), and electroacoustic sound masking [4]. Virjonen et al. [5] have shown that open-plan offices can significantly differ from each other with respect to acoustic quality. Therefore, the potential of solving noise problems in offices with room acoustic means is large.

ISO 3382-3 standard [6] was published in 2012 to promote the room acoustic design of offices. It describes a method for determining the room acoustic properties of open-plan offices using acoustic measurements. The measurement reports five single-number quantities (SNQs) that together fully describe the room acoustic performance of an open-plan office:

- the spatial decay rate of A-weighted sound pressure level (SPL) of speech, $D_{2,S}$ [dB], i.e., the reduction of A-weighted SPL of speech when the distance to the speaker is doubled (Figure 1),

- the A-weighted SPL of speech at 4 m distance from the speaker, $L_{p,A,S,4m}$ [dB] (Figure 1),
- distraction distance, r_D [m], i.e., the distance from the speaker where Speech Transmission Index, STI, falls below 0.50,
- privacy distance, r_P [m], i.e., the distance where Speech Transmission Index, STI, falls below 0.20, and
- A-weighted SPL of the background noise of an unoccupied office, $L_{p,A,B}$.

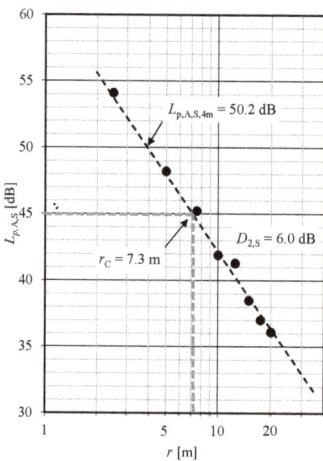

Figure 1. A-weighted SPL, $L_{p,A}$, as a function of distance, r, to the speaker (black circles) and a linear fit over the data (dashed line). Definitions for $D_{2,S}$, $L_{pA,S,4m}$, and r_C are given in Section 1. The data is not related to this study.

ISO 3382-3 [6] defines the SPL of normal effort speech to be used in the determination of the abovementioned SNQs. This guarantees that different operators obtain similar measurement results from the same office as shown by the Round Robin test of Hongisto et al. [7]. D'Orazio et al. [8] reported measurement results from the office where the background noise level, $L_{p,A,B}$, varied even 13 dB within a single measurement path. ISO 3382-3 [6] states that the mean of $L_{pA,B}$ values along the measurement positions shall be used in the position-dependent STI determinations. In such special cases, the uncertainty of r_D and r_P may be higher than in the Round Robin test of Ref. [7] where the spatial distribution of background noise was smooth.

ISO 3382-3 [6] was largely based on the method described by Hongisto et al. [9], who studied 15 different open-plan offices. An extended version involving 16 offices was published later by Virjonen et al. [5]. They suggested the abovementioned SNQs that deal with spatial decay instead of temporal decay of sound since reverberation time was not associated with spatial decay rate in a non-diffuse sound field. Therefore, reverberation time did not belong to the reported SNQs of ISO 3382-3. Furthermore, they showed that the A-weighted SPL of speech was usually linearly associated with logarithmic distance. Because speech is the main noise source in offices, it was justified to focus on the spatial decay of A-weighted SPL of speech.

ISO 14257 [10] was an important role model in the development of ISO 3382-3 because the new quantities were revolutionary at that time when most room acousticians were used to measuring reverberation time and background noise levels in the first place. ISO 14257 [10] was among the first acoustic standards that focused on a non-diffuse sound field. It involved two SNQ's that were considered during the standardization of ISO 3382-3: rate of spatial decay of SPL per distance doubling, DL_2 [dB], and excess of SPL, DL_f [dB]. The latter describes how much the spatial decay rate deviates from the free field. However, these quantities were determined in octave bands. Such a large amount of reported outcomes did not serve the purpose of ISO 3382-3 of providing simple and scarce SNQs as the main outcomes. Because it was evident from office surveys that speech is the main noise source, and that speech has a standardized spectrum shape and overall level, the approach of using A-weighted SPL of speech was justified. This led to the definition of $D_{2,S}$ as a primary quantity describing the spatial decay rate. However, it was not alone sufficient to describe the spatial decay since the sound attenuation in the nearfield varies a lot between offices due to different room height, screen height, and room absorption. Therefore, $L_{p,A,S,4m}$ was chosen to be used as an anchor point for $D_{2,S}$ slope instead of DL_f, since the former was easier to understand and determine. It should be noted that the SNQs of ISO 3382-3 had to be understandable also among non-acousticians involved with office design, such as building owners, workplace designers, material and furniture providers, authorities, facility managers, occupational physicists, ergonomists, HR people, managers, and office users.

During the standardization process, which lasted from 2009 to 2012, Nilsson and Hellström [11] proposed an alternative option to $L_{p,A,S,4m}$ and DL_f: the distance of comfort, d_C [m]. It was the distance, where an acceptable A-weighted SPL of speech was achieved. It should be noted that d_C is not an alternative quantity of distraction distance r_D since d_C is purely based on spatial attenuation of speech and it ignores the background noise level of the room, unlike r_D. However, their approach did not gain support at that time since there was too little published evidence about the suitable d_C values, and it was also based on DL_f which was already discarded in ISO 3382-3. Furthermore, there was already some uncertainty about the acceptance of r_D among acousticians and non-acousticians. It was found safer to limit the distance-related SNQs to r_D and r_P, which were derived from the spatial decay of STI.

Authors' interactions with non-acousticians have learned that privacy-related SNQs, i.e., r_D and r_P, have been well understood. An important reason for this was a study, which showed that cognitive performance deteriorates with increasing STI, i.e., with reducing speech privacy [12]. A later important reason was a cross-sectional study showing that shorter r_D was associated with a lower probability of being highly disturbed by office noise [3]. Against expectations, $D_{2,S}$ did not show any association with that probability. The most probable reason is that the latter ignores the effect of background noise (masking).

Authors' experience has been that the attenuation-related SNQs, i.e., $D_{2,S}$ and $L_{p,A,S,4m}$, have been more difficult to understand by non-acousticians. The reason for this is that both quantities have the same unit but different definitions. It would be useful to have a simpler attenuation-related SNQ to facilitate communication with non-acousticians.

Seddigh et al. [13] described the room acoustic properties of their open-plan offices by comfort distance as introduced by Nilsson and Hellström [11]. They defined the comfort distance as the distance where the A-weighted SPL of speech falls below $L_{p,A,C}$ = 48 dB. However, the comfort criterion level, $L_{p,A,C}$, was not based on a thorough analysis of the existing measurement data.

The A-weighted SPL of speech, $L_{p,A,S}$, depends linearly on logarithmic distance, r, from the speaker. Therefore, $L_{p,A,S}$ can be determined from the linearly fitted SNQ values of ISO 3382-3 by

$$L_{p,A,S} = L_{p,A,S,4m} + 2D_{2,S} - \frac{D_{2,S}}{\log_{10}(2)} \cdot \log_{10}(r) \quad (1)$$

If $L_{p,A,S}$ equals the comfort criterion level, $L_{p,A,C}$, the distance r_C, where this is achieved, i.e., comfort distance, gets a general form:

$$r_C = 2^{(L_{p,A,S,4m} - L_{p,A,C} + 2 \cdot D_{2,S})/D_{2,S}} \quad (2)$$

This form was recently used by Hongisto et al. [7]. They set the comfort criterion level to $L_{p,A,C}$ = 45 dB. However, they did not describe the origin of that choice. Most importantly, the comfort distance can be calculated by the SNQs which are already determined in ISO 3382-3. Some countries already have mandatory target values or voluntary classification systems for the room acoustic quality of open-plan offices using the SNQs of ISO 3382-3 [14,15]. It would be useful to find suitable limiting values for the acoustic classes A–D also for comfortable distance. Because r_C belongs among the key SNQs in the draft international standard ISO DIS 3382-3 [16], the elaboration of the scientific basis of r_C is justified.

The purpose of our study is to present the scientific basis of comfort distance to better introduce it as a new SNQ in the revised version of ISO 3382-3 [6,16]. The second purpose was to compare the ISO 3382-3 [6] data reported in previous studies to calculate the range of typical comfort distance values using Equation (2). The third purpose was to propose limit values for the classification of comfort distance based on all available data.

2. Materials and Methods

We utilized the measurement data of $D_{2,S}$ and $L_{p,A,S,4m}$ of Keränen and Hongisto [17], which represents well the range of values where the SNQs of ISO 3382-3 could usually lie. They reported altogether 26 measurements in acoustically different open-plan offices (Table 1, Figure 2). Each measurement corresponds to a single path in one direction.

Table 1. The data of the 26 offices of Ref. [17] used in our study. L is the length of the office in the direction of the measurement path. The other quantities were defined in Section 1. The notation (both numbers and letters) is adopted from Ref. [17].

Office ID	L [m]	$L_{p,A,B}$ [dB]	$L_{p,A,S,4m}$ [dB]	$D_{2,S}$ [dB]	r_D [m]
1	16	39	53.8	4.0	14.2
2	27	45	57.2	4.2	18.5
3	16	42	52.5	4.6	9.5
4	60	41	49.4	5.7	5.6
5	18	35	50.9	6.0	15.4
6	36	44	52.6	6.2	5.4
7	19	31	47.5	6.3	13.8
8	19	39	52.4	6.4	10.3
9	42	40	54.4	6.7	15.3
10	23	39	43.4	9.0	5.5
11	34	35	48.3	9.2	9.9
12	32	37	49.4	9.4	9.3
13	36	31	46.5	11.4	9.5
14	35	31	47.1	11.5	6.2
15	70	31	49.0	11.7	8.1
16	27	33	49.9	12.4	10.0
A	18	34	47.4	4.9	16.2
B	33	32	49.1	6.0	15.3
C	69	29	44.0	6.4	11.4
D	17	38	50.4	6.4	11.9
E	23	34	47.9	7.8	8.8
F	16	35	51.5	8.2	11.1
G	36	32	50.3	9.3	14.0
H	28	38	50.3	9.4	6.0
I	30	38	53.9	9.0	9.7
J	33	39	49.3	11.6	9.3

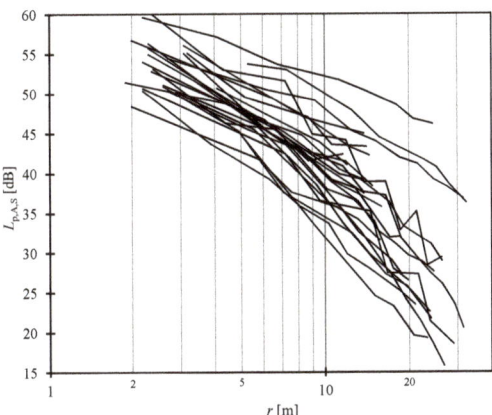

Figure 2. Measured A-weighted SPL of speech, $L_{p,A,S}$, as a function of distance, r, from the omnidirectional loudspeaker, for the 26 open-plan offices of Table 1. The number of measurement positions per office ranged from 4 to 13. The required minimum number of positions is four.

We determined the comfort distance for the 26 offices of Table 1 using 21 different values for the comfort criterion level, $L_{p,A,C}$. The values ranged from 30 to 50 dB in 1-dB steps. The calculation was made using Equation (2). The method is depicted in Figure 3 for office ID 1. This way, each office was assigned by 21 different comfort distances. Simple statistics (mean, minimum, maximum 68% confidence intervals) were determined at every comfort criterion level for the distribution of comfort distances over the 26 offices.

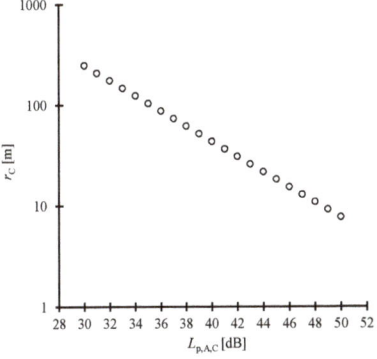

Figure 3. Example of the comfort distances, r_C, obtained for office ID 1 for different comfort criterion levels, $L_{p,A,C}$, from 30 to 50 dB.

It was justified to presume in general that comfort distance should not be larger than the length of the office. Therefore, we calculated for every $L_{p,A,c}$ value the probability P that the comfort distance r_C was larger than the room length within the sample, by

$$P = \frac{N_0}{N} \qquad (3)$$

where N_0 is the number of offices (out of 26 offices in question) fulfilling the adverse criterion $r_C > L$ and N is the total number of offices (26). The room length L of each office is given in Table 1. The desirable situation is $P = 0$. It indicates a high probability that the comfort distance is shorter than a room in most offices beyond the sample of Table 1 since the sample of Table 1 represents a broad range of acoustically different offices.

Our second purpose involved a comparison between previous studies. Some important previous studies are described in Table 2.

Table 2. Eight studies I–VIII reporting measurement data according to ISO 3382-3. N is the number of reported paths.

Study	ID	N	Country	Comment
Keränen and Hongisto (2013) [17]	I	26	Finland	a
Haapakangas et al. (2017) [3]	II	21	Finland	b
Selzer and Schelle (2018) [14]	III	34	Germany	c
Wenmaekers and van Hout (2019) [18]	IV	4	Laboratory	d
Cabrera et al. (2018) [19]	V	20	Australia	e
Yadav et al. (2019) [20]	VI	36	Australia	f
Lüthi and Desarnaulds (2020) [21]	VII	22	Switzerland	g
Keränen et al. (2020) [4]	VIII	22	Laboratory	h

a. 26 separate offices, one path per office; b. 21 separate offices, one path per office; c. 13 offices with 2 to 4 paths; d. Conditions were built by researchers in a real office, why it is called as a laboratory setup; e. 20 separate offices, one path per office; f. 27 offices with one path, 5 offices with two paths, 2 offices with three paths; g. 22 separate offices; h. $L_{p,A,B}$ and r_D were disregarded since background noise was adjustable. Mean of two paths. Six conditions with r_C > 45 m were ignored.

3. Results

The comfort distances of the 26 offices of Ref. [17] for comfort criterion levels ranging from 30 to 50 dB are shown in Figure 4. The corresponding probabilities that the comfort distances exceeded the length of the office are shown in Figure 5.

Figure 5 clearly shows that the probability P reaches zero when $L_{p,A,C}$ > 45 dB. Therefore, this value was chosen as the comfort distance criterion. Further justification for this choice is given in Section 4.

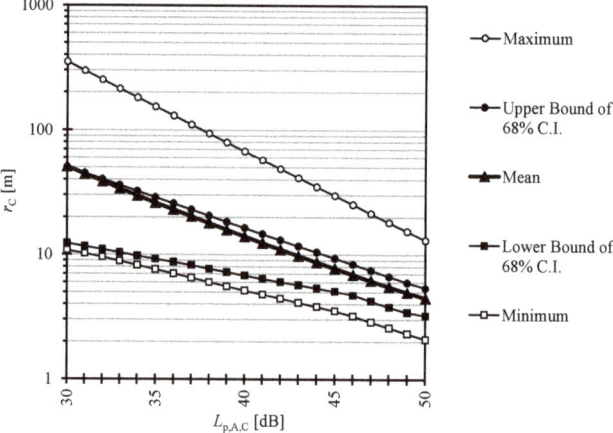

Figure 4. The range of comfort distance values, r_C, as a function of the comfort criterion level, $L_{p,A,C}$, for the 26 offices of Ref. [17] calculated by Equation (2). Mean, maximum, minimum, and 68% confidence interval (C.I.) within the sample of 26 offices are shown.

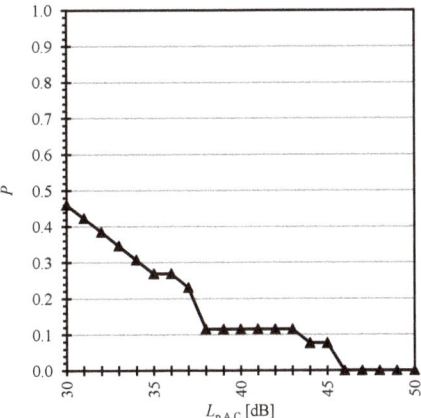

Figure 5. The probability P among the 26 offices of Ref. [17] that the comfort distance, defined by comfort criterion level $L_{p,A,C}$, was larger than the room length.

Figure 6 presents a statistical overview of the single-number values of ISO 3382-3 for the eight studies of Table 2 and the comfort distance calculated by Equation (2). The average of all 179 comfort distances was 9.3 m. The lower and upper bounds of the 68% and 95% confidence intervals were 4.7, 3.5, 13.8, and 25.3 m, respectively.

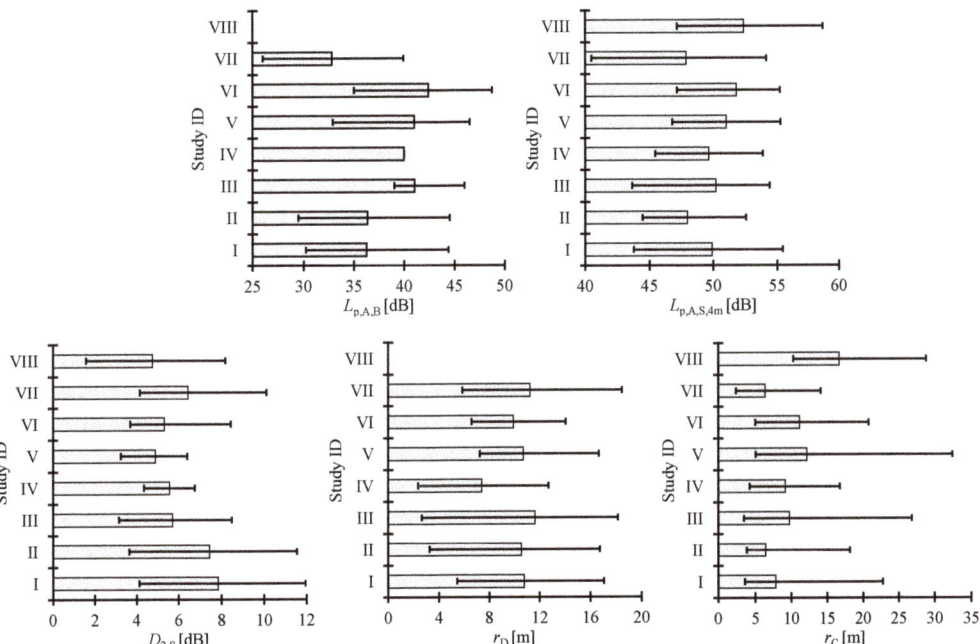

Figure 6. Distribution of measurement results according to ISO 3382-3 standard in the eight studies of Table 2. Bars are the means and whiskers are the 95% confidence intervals.

Figure 7 presents an analysis of how the acoustic classes A–D of comfort distance could be set in a balanced way for the 26 offices of Table 1. We paid attention to three criteria: each class involves at least two offices, the classes are equally spaced, and some offices (two worst ones) can remain unclassified. The limit values for classes A to D became 5, 7, 9, and 11 m for r_C. The ranges for classes A–D are [0–5) m, [5–7) m, [7–9) m, and [9–11) m, respectively. Values of 11 m and higher are unclassified.

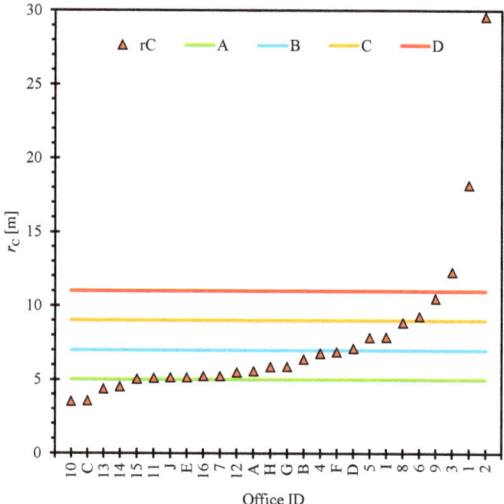

Figure 7. Rank ordered comfort distance, r_C, as a function of office ID (red triangles) for the 26 offices of Table 1. The proposed limits for classes A–D are indicated by horizontal lines. The figure expects that $L_{p,A,C}$ = 45 dB.

Using the same criteria, the proposed limit values for classes A to D became 11, 9, 7, and 5 dB for $D_{2,S}$, 47, 49, 51, and 53 dB for $L_{p,A,S,4m}$, and 6, 8, 10, and 12 m for r_D.

4. Discussion

As expected, comfort distance increased strongly when the comfort criterion level $L_{p,A,C}$ was reduced. Noise annoyance of broadband steady-state noise is usually low when the level is below 35 dB L_{Aeq} [22]. Using this level as a comfort criterion level is not justified since the comfort distance would exceed 50 m in most offices of our sample. On the other hand, setting the comfort criterion level to 48 dB, as Seddigh et al. [13] did, is not justified since such a high level is probably no longer perceived as comfortable. For example, Veitch et al. [23] and Hongisto et al. [24,25] suggested that the level of sound masking should not exceed 45 dB L_{Aeq} to avoid the triggering of noise annoyance due to masking sound itself. It is also notable that Bottalico et al. [26] showed that people start to raise voice effort due to the Lombard effect when the background noise level exceeds 43.3 dB L_{Aeq}. This supports the use of a comfort criterion level lower than 48 dB. The mean levels during the workday are usually 48–59 dB $L_{Aeq,8h}$ according to a major survey of offices [27]. This supports also that 48 dB might not be comfortable since it exceeds the average activity noise level. The probability of comfort distance being larger than room length reached zero when the comfort criterion level was 46 dB or larger. Thus, setting the comfort criterion level higher than 45 dB is not supported from this practical viewpoint. In conclusion, it is feasible to set the comfort criterion level at most to $L_{p,A,C}$ = 45 dB. Among the 26 open-plan offices of Ref. [17], the mean value of r_C was 7.8 m and the values ranged from 3.5 to 30.0 m, when $L_{p,A,c}$ = 45 dB.

Figure 6 involves a broad perspective over the eight studies of Table 2. If laboratory study VIII is ignored due to the small room size, the rest of the studies indicate a somewhat similar distribution of comfort distances as Study I [17], where the classification scheme was based upon. All eight studies suffer from selection bias: offices have not been randomly selected from the building population. Because of that, none of the studies declare that their data represents the general distribution of acoustic quality in their country. In this light, Figure 6 also suggests that the distribution of room acoustic properties of open-plan offices do not drastically differ from each other in different countries. Figure 6 also represents the best available knowledge on the room acoustics of open-plan offices at the global level. It would be important to systematically analyze the target values and measurement results also from other countries to see the progress in room acoustic design at a global level. This would help in the development of research, business, design guidelines, and target values in the future.

The classification scheme was created using three criteria explained in Section 3. The scheme may look demanding with respect to the distribution shown in Figure 6 since only a minority of offices can reach class A. For example, an office representing the mean of the 26 offices of Ref. [17], i.e., r_C = 7.8 m, reaches only class C. Informative (non-mandatory) annex of international standard draft ISO DIS 3382-3 [16] describes that "Typical values of r_C with poor and good room acoustic conditions are $r_C > 11$ m and $r_C < 5$ m, respectively". This description is supported by our proposal.

5. Conclusions

The scientific basis of comfort distance was introduced. Comfort distance was calculated using the single-number values of $D_{2,S}$ and $L_{p,A,S,4m}$ determined according to in situ measurements by ISO 3382-3 [6]. Comfort distance describes the distance where A-weighted SPL of normal effort speech falls below 45 dB. The mean value of comfort distance was 7.8 m in our database containing 26 offices. The values ranged from 3 to 30 m.

A classification scheme was presented according to which the best class (A) is reached when comfort distance is shorter than 5 m. The worst class (D) is reached when the comfort distance is between 9–11 m. Values above 11 m are unclassified.

Comfort distance could be used as an option in the revised ISO 3382-3 standard to facilitate the comparison of open-plan offices with respect to speech attenuation performance and to facilitate the communication of measurement results with non-acousticians. Furthermore, comfort distance enables the classification of speech attenuation performance using a single quantity instead of two quantities.

Author Contributions: Conceptualization, V.H.; methodology, V.H. and J.K.; validation, J.K.; formal analysis, V.H.; investigation, V.H. & J.K.; resources, V.H.; data curation, V.H. and J.K.; writing—original draft preparation, V.H. and J.K.; writing—review and editing, V.H. and J.K.; visualization, V.H.; supervision, V.H.; project administration, V.H.; funding acquisition, V.H. Both authors have read and agreed to the published version of the manuscript.

Funding: This research was funded by ACADEMY OF FINLAND (grant No. 314788) and Turku University of Applied Sciences.

Institutional Review Board Statement: Not applicable.

Informed Consent Statement: Not applicable.

Conflicts of Interest: The authors declare no conflict of interest. The funders had no role in the design of the study; in the collection, analyses, or interpretation of data; in the writing of the manuscript, or in the decision to publish the results.

References

1. Frontczak, M.; Schiavon, S.; Goins, J.; Arens, E.; Zhang, H.; Wargocki, P. Quantitative relationships between occupant satisfaction aspects of indoor environmental quality and building design. *Indoor Air* **2012**, *22*, 119–131. [CrossRef]
2. Haapakangas, A.; Hongisto, V.; Liebl, A. The relation between the intelligibility of speech and cognitive performance—A revised model based on laboratory studies. *Indoor Air* **2020**, *30*, 1130–1146. [CrossRef] [PubMed]
3. Haapakangas, A.; Hongisto, V.; Eerola, M.; Kuusisto, T. Distraction distance and disturbance by noise—An analysis of 21 open-plan offices. *J. Acoust. Soc. Am.* **2017**, *141*, 127–136. [CrossRef] [PubMed]
4. Keränen, J.; Hongisto, V.; Hakala, J. The effect of sound absorption and screen height on spatial decay of speech in open-plan offices. *Appl. Acoust.* **2020**, *166*, 107340. [CrossRef]

5. Virjonen, P.; Keränen, J.; Hongisto, V. Determination of acoustical conditions in open-plan offices—Proposal for new measurement method and target values. *Acta Acust. Acust.* **2009**, *95*, 279–290. [CrossRef]
6. *International Standard ISO 3382-3:2012 Acoustics—Measurement of Room Acoustic Parameters—Part 3: Open Plan Offices*; International Organization for Standardization: Geneve, Switzerland, 2012.
7. Hongisto, V.; Keränen, J.; Labia, L.; Alakoivu, R. Precision of ISO 3382-2 and ISO 3382-3—A Round-Robin test in an open-plan office. *Appl. Acoust.* **2021**, *175*, 107846. [CrossRef]
8. D'Orazio, D.; Rossi, E.; Garai, M. Comparison of different in situ measurements techniques of intelligibility in an open-plan office. *Build. Acoust.* **2018**, *25*, 111–122. [CrossRef]
9. Hongisto, V.; Virjonen, P.; Keränen, J. Determination of acoustical conditions of open offices—Suggestions for acoustic classification. In Proceedings of the 19th International Congress on Acoustics, Madrid, Spain, 2–7 September 2007.
10. *International Standard ISO 14257:2001. Acoustics—Measurement and Parametric Description of Spatial Sound Distribution Curves in Workrooms for Evaluation of Their Acoustical Performance*; International Organization for Standardization: Geneve, Switzerland, 2001.
11. Nilsson, E.; Hellström, B. Room acoustic design in open-plan offices. In Proceedings of the Euronoise, Edinburgh, UK, 26–28 October 2009.
12. Hongisto, V. A model predicting the effect of speech of varying intelligibility on work performance. *Indoor Air* **2005**, *15*, 458–468. [CrossRef] [PubMed]
13. Seddigh, A.; Berntson, E.; Jönsson, F.; Bodin Danielson, C.; Westerlund, H. Effect of variation in noise absorption in open-plan office: A field study with a cross-over design. *J. Environ. Psychol.* **2015**, *44*, 34–44. [CrossRef]
14. Selzer, J.; Schelle, F. Practical aspects of measuring acoustics in German open plan offices. In Proceedings of the Euronoise, Crete, Greece, 27–31 May 2018; pp. 1919–1924.
15. Hongisto, V.; Keränen, J. Open-plan offices—New Finnish room acoustic regulations. In Proceedings of the Euronoise, Crete, Greece, 27–31 May 2018.
16. *Draft International Standard. ISO DIS 3382-3:2021 Acoustics—Measurement of Room Acoustic Parameters—Part 3: Open Plan Offices*; International Organization for Standardization: Geneve, Switzerland, 2021.
17. Keränen, J.; Hongisto, V. Prediction of the spatial decay of speech in open-plan offices. *Appl. Acoust.* **2013**, *74*, 1315–1325. [CrossRef]
18. Wenmaekers, R.; Van Hout, N. How ISO 3382-3 acoustic parameter values are affected by furniture, barriers and sound absorption in a typical open plan office. In Proceedings of the 23rd International Congress on Acoustics, Aachen, Germany, 9–13 September 2019.
19. Cabrera, D.; Yadav, M.; Protheroe, D. Critical methodological assessment of the distraction distance used for evaluating room acoustic quality of open-plan offices. *Appl. Acoust.* **2018**, *140*, 132–142. [CrossRef]
20. Yadav, M.; Cabrera, D.; Love, J.; Kim, J.; Holmes, J.; Caldwell, H.; de Dear, R. Reliability and repeatability of ISO 3382-3 metrics based on repeated acoustic measurements in open-plan offices. *Appl. Acoust.* **2019**, *150*, 138–146. [CrossRef]
21. Lüthi, G.; Desarnaulds, V. Analysis of open plan acoustic parameters based on Swiss and international databases of in situ measurements. In Proceedings of the ICSV27, Prague, Czech Republik, 12–16 July 2020.
22. Virjonen, P.; Hongisto, V.; Radun, J. Annoyance penalty of periodically amplitude-modulated wide-band sound. *J. Acoust. Soc. Am.* **2019**, *146*, 4159–4170. [CrossRef] [PubMed]
23. Veitch, J.; Bradley, J.; Legault, L.; Norcross, S.; Svec, J. *Masking Speech in Open-Plan Offices with Filtered Pink Noise Noise: Noise Level and Spectral Composition Effects on Acoustic Satisfaction*; Internal Report IRC-846; Institute for Research in Construction: Ottawa, ON, Canada, 1 April 2002.
24. Hongisto, V.; Oliva, D.; Rekola, L. Subjective and Objective Rating of Spectrally Different Pseudorandom Noises—Implications for Speech Masking Design. *J. Acoust. Soc. Am.* **2015**, *137*, 1344–1355. [CrossRef] [PubMed]
25. Hongisto, V.; Varjo, J.; Oliva, D.; Haapakangas, A.; Benway, E. Perception of water-based masking sounds—Long-term experiment in an open-plan office. *Front. Psychol.* **2017**, *8*, 1117. [CrossRef] [PubMed]
26. Bottalico, P.; Passione, I.I.; Graetzer, S.; Hunter, E.J. Evaluation of the Starting Point of the Lombard Effect. *Acta Acust. Acust.* **2017**, *103*, 169–172. [CrossRef] [PubMed]
27. Yadav, M.; Cabrera, D.; Kim, J.; Fels, J.; de Dear, R. Sound in occupied open-plan offices: Objective metrics with a review of historical perspectives. *Appl. Acoust.* **2021**, *177*, 107943. [CrossRef]

Article

A Trial Acoustic Improvement in a Lecture Hall with MPP Sound Absorbers and FDTD Acoustic Simulations

Matteo Cingolani [1], Giulia Fratoni [1], Luca Barbaresi [1], Dario D'Orazio [1], Brian Hamilton [2] and Massimo Garai [1,*]

1. Department of Industrial Engineering, University of Bologna, 40136 Bologna, Italy; matteo.cingolani6@unibo.it (M.C.); giulia.fratoni2@unibo.it (G.F.); luca.barbaresi@unibo.it (L.B.); dario.dorazio@unibo.it (D.D.)
2. Acoustics and Audio Group, University of Edinburgh, Edinburgh EH8 9DF, UK; brian.hamilton@ed.ac.uk
* Correspondence: massimo.garai@unibo.it

Citation: Cingolani, M.; Fratoni, G.; Barbaresi, L.; D'Orazio, D.; Hamilton, B.; Garai, M. A Trial Acoustic Improvement in a Lecture Hall with MPP Sound Absorbers and FDTD Acoustic Simulations. *Appl. Sci.* **2021**, *11*, 2445. https://doi.org/10.3390/app11062445

Academic Editors: Nikolaos M. Papadakis, Massimo Garai and Stavroulakis Georgios

Received: 23 January 2021
Accepted: 3 March 2021
Published: 10 March 2021

Publisher's Note: MDPI stays neutral with regard to jurisdictional claims in published maps and institutional affiliations.

Copyright: © 2021 by the authors. Licensee MDPI, Basel, Switzerland. This article is an open access article distributed under the terms and conditions of the Creative Commons Attribution (CC BY) license (https:// creativecommons.org/licenses/by/ 4.0/).

Abstract: Sound absorbing micro-perforated panels (MPPs) are being increasingly used because of their high quality in terms of hygiene, sustainability and durability. The present work investigates the feasibility and the performance of MPPs when used as an acoustic treatment in lecture rooms. With this purpose, three different micro-perforated steel specimens were first designed following existing predictive models and then physically manufactured through 3D additive metal printing. The specimens' acoustic behavior was analyzed with experimental measurements in single-layer and double-layer configurations. Then, the investigation was focused on the application of double-layer MPPs to the ceiling of an existing university lecture hall to enhance speech intelligibility. Numerical simulations were carried out using a full-spectrum wave-based method: a finite-difference time-domain (FDTD) code was chosen to better handle time-dependent signals as the verbal communication. The present work proposes a workflow to explore the suitability of a specific material to speech requirements. The measured specific impedance complex values allowed to derive the input data referred to MPPs in FDTD simulations. The outcomes of the process show the influence of the acoustic treatment in terms of reverberation time (T_{30}) and sound clarity (C_{50}). A systematic comparison with a standard geometrical acoustic (GA) technique is reported as well.

Keywords: acoustics; micro-perforated panels; FDTD simulation; speech intelligibility

1. Introduction

Nowadays, the interest in sound absorbing materials is growing due to the variety of their possible applications, from room acoustics [1] to environmental noise control [2,3]. Porous and fibrous absorbers [4–6] have until now been the most used materials in noise control application because of their high performance-to-cost ratio in the frequency band of interest. In the last decades, new requirements have become important, such as durability, recyclability, hygienic problems, environmental sustainability and optical transparency, which are no longer suitable for porous and fibrous materials. In order to satisfy these requirements, specific classes of sound absorbing materials have been proposed: among them, micro-perforated panels (MPPs) [7–11]. During the 1970s, the first MPP acoustic model proposed by Maa [7] defined the absorbers as a combination of a thin panel with sub-millimetric holes, an air cavity and a rigid wall. The air cavity is required to perform a Helmholtz-type resonance. Moreover, an equivalent fluid (EF) model was theorized by Atalla and Sgard [12]. In the last decades, the applications, the improvements and the theoretical developments of such materials have been extensively studied and MPP multiple-layers have been introduced to provide wide-band absorption, creating more efficient sound absorbing systems [13–15].

MPPs can be made of various materials, including plywood, glass and sheet metal. Therefore, they are extremely attractive from an ecological point of view, especially for architectural applications [16,17]. Among the potential applications of MPPs, there is their use as an acoustic treatment in existing lecture rooms to enhance the verbal communication conditions [18]. Reducing the reverberant field in a specific frequency range contributes to decreasing the vocal effort of the speaker and the distraction of the students [19–21]. Since in the last years the acoustical comfort of teachers and students is one of the most debated topics [22–24], the possibility to choose a sustainable and high-performance material could meet the need of improving the acoustics in existing lecture halls.

In this work, three steel MPP specimens were designed with specific constitutive geometrical parameters. The sound absorption mathematical equations and the electro-acoustic analogies were used to simplify complex mechanical issues into equivalent electrical circuits. The specimens were constructed using 3D additive printing and the manufacturing issues encountered during the process are described and shown in detail. Experimental measurements made with the transfer-function method [25] in an impedance tube [26] are reported and compared with predictive models. An optimization of MATLAB implementations was carried out taking into account the practical issues encountered during the measurements. After the experimental phase, the performance of MPPs was evaluated focusing on the application to large-scale rooms. In particular, the effects of MPPs as an acoustic treatment in an existing lecture hall previously surveyed [27] were explored by means of full-spectrum finite-difference time-domain (FDTD) simulations [28].

2. Modeling the Sound Absorption of Micro-Perforated Panels (MPP)

A micro-perforated panel consists of a thin panel with a specific perforation ratio made by a distribution of sub-millimeter holes backed by an air cavity and a rigid wall, as shown in Figure 1.

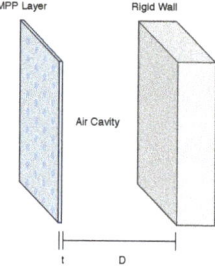

Figure 1. Schematic representation of a single-layer micro-perforated panel (MPP) and its dimensional parameters.

The acoustic complex impedance of an MPP (Z) is the result of different contributions: the real part of the impedance that needs to be matched to the air impedance (Z_0) and the imaginary part of the impedance provided by the air cavity and the perforations. The values of the perforation diameter d, the thickness of the panel t, the porosity ϕ and the air cavity thickness D are the four constitutive parameters that influence the range of frequencies absorbed and the bandwidth as well. Taking into account Maa's definition [8] and according to Cobo's notation [13], it is possible to define the input complex impedance of a single-layer (SL) MPP Z_1 as follows:

$$Z_{1,SL} = Z_{holes} + Z_{edge} + Z_c. \tag{1}$$

The impedance Z_{holes} defines the viscous dissipation within the holes, Z_{edge} the distortion of the flow in the perforation edges and Z_c is the resonance in the air cavity:

$$Z_{holes} = \frac{\Delta p}{u} = i\frac{\omega \rho_0}{\phi}\left[1 - \frac{2}{s\sqrt{-i}}\frac{J_1(s\sqrt{-i})}{J_0(s\sqrt{-1})}\right]^{-1} \tag{2}$$

$$Z_{edge} = R_s + iX_m = 2\sqrt{2\eta\omega\rho_0} + i\frac{\omega\rho_0 0.85d}{F(\epsilon)} \tag{3}$$

where Δp is the pressure difference at both sides of the tubes, u is the particle velocity in the tube, ρ_0 is the air density, η is the air dynamic viscosity, ϕ is the porosity, $s = d\sqrt{\frac{\omega\rho_0}{4\eta}}$ is the perforation constant (d being the diameter of the holes), J_1 and J_0 the Bessel functions of first-class and order 1 and 0, respectively, and $F(\epsilon)$ is the Fok function, a correction factor of the mass reactance [13] and $\epsilon = \sqrt{\phi}$. Considering all the parameters introduced so far, it is possible to study the equivalent electro-acoustic system for the MPP, as shown in Figure 2.

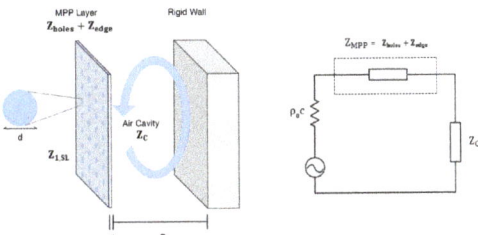

Figure 2. Single-layer (SL)-MPP schematic representation (**left**) and the corresponding equivalent electrical system (**right**).

In practice, multiple-layer MPPs are usually preferred because of their extended absorption band.

A double-layer MPP (DL-MPP) consists of two MPPs with impedances $Z_{MPP,1}$ and $Z_{MPP,2}$ and two air cavities with impedances Z_{c1} and Z_{c2} (see Figure 3). Considering the sound waves passing through the DL-MPP system from left to right at normal incidence, the input impedance to the DL-MPP system is:

$$Z_{1,DL} = Z_{holes,1} + Z_{edge,1} + Z_0 \frac{Z_{2,DL}\cos(kD_2) + iZ_0\sin(kD_1)}{Z_0\cos(kD_1) + iZ_{2,DL}\sin(kD_2)} \tag{4}$$

with

$$Z_{2,DL} = Z_{edge,2} + Z_{holes,2} + Z_{c2}. \tag{5}$$

Therefore, the absorption of a DL-MPP globally depends on eight constitutive and geometrical parameters, four for the first layer and four for the second layer: the diameter of the holes (d_1, d_2), the thickness (t_1, t_2), the distance (D_1, D_2), and the porosity (ϕ_1, ϕ_2). For this reason, it is difficult to predict the acoustic performance of a DL-MPP and to find, a priori, a combination of these parameters providing the maximum absorption in a specific frequency range.

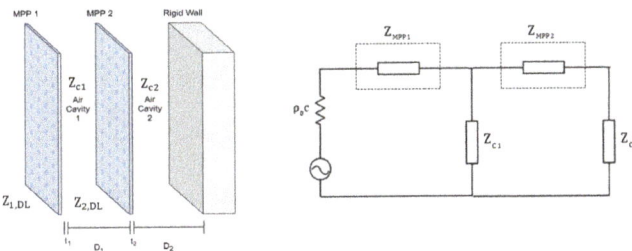

Figure 3. Double-layer (DL)-MPP schematic representation (**left**) and the corresponding equivalent electrical system (**right**).

Atalla and Sgard [12] introduced the so-called equivalent fluid (EF) model following the Johnson–Champoux–Allard approach with an equivalent tortuosity [5]: they assumed an MPP coupled at both sides to a semi-infinite fluid. All of the phenomena involved are recalled in Figure 4. In the EF model, the viscous boundary within the perforations and around the edges is represented by the resistive part of the normal surface impedance, and the movement of the air cylinder—the length of which is greater than the panel thickness—is taken into account in the reactive part of the MPP impedance. In addition to this, a new length correction is introduced in order to consider the increase of air mass vibrating inside the cylinder. Allard demonstrated that the viscous and thermal lengths (Λ and Λ', respectively) can be considered equal to the hydraulic radius of the perforations in case of straight cylindrical pores.

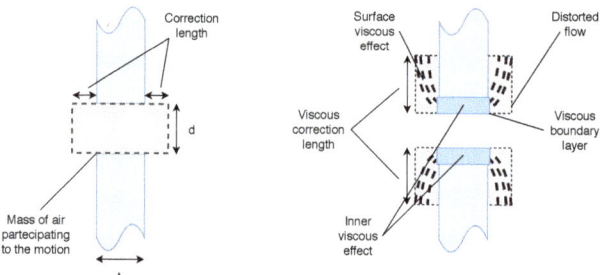

Figure 4. Physical phenomena involved in an MPP: surface viscous effects and inner viscous effects.

The EF model introduces the effective density $\tilde{\rho}_e$, which considers viscous and inertial effects that govern the front face impedance, inside the perforations, defined as:

$$\tilde{\rho}_e = \rho_0 \left[1 + \frac{\sigma\phi\left(1 + \frac{4\omega\rho_0\eta}{\sigma^2\phi^2 r^2}\right)^{1/2}}{i\omega\rho_0} \right] \tag{6}$$

where σ is the flow resistivity, defined as $\sigma = \dfrac{8\eta}{\phi r^2}$. Thus, the impedance of the holes can be rewritten as:

$$Z_{holes} = i\omega\tilde{\rho}_e \frac{t}{\phi}. \tag{7}$$

The edge effects are introduced in the EF model through the geometrical tortuosity α_{inf}. In this work, the definition of α_{inf} provided by Atalla and Sgard [12] is used, meaning that not only the intrinsic properties of the material and its micro-geometry but also the media in contact with the panel will be considered. In case of a panel radiating on both sides:

$$\alpha_{inf} = 1 + \frac{2\epsilon_e}{t} \tag{8}$$

where t is the thickness of the panel, and $\epsilon_e = 0.48\sqrt{\pi r^2}(1 - 1.14\sqrt{\phi})$ (valid when $\phi < 0.4$). In the EF model, the tortuosity replaces the Fok function introduced in Maa's model [9]. Thus, the panel impedance for a SL-MPP using the EF model is:

$$Z_{MPP,EF} = i\frac{\omega\rho_0 \alpha_{inf} t}{\phi}\left[1 + \frac{\sigma\phi}{i\omega\rho_0 \alpha_{inf}}\left(1 + i\frac{4\omega\rho_0 \alpha_{inf}^2 \eta}{\sigma^2\phi^2 r^2}\right)^{1/2}\right] \tag{9}$$

$$Z_{1,EF} = Z_{MPP,EF} + Z_c \tag{10}$$

A double-layer MPP system can be studied with the EF model as well, defining for the second layer of the MPP structure $Z_{MPP2,EF}$ as:

$$Z_{MPP2,EF} = i\frac{\omega\rho_0\alpha_{\inf,2}t_2}{\phi_2}\left[1 + \frac{\sigma_2\phi_2}{i\omega\rho_0\alpha_{\inf,2}}\left(1 + i\frac{4\omega\rho_0\alpha_{\inf,2}^2\eta}{\sigma_2^2\phi_2^2r_2^2}\right)^{1/2}\right] \quad (11)$$

$$Z_{2,EF} = Z_{MPP2,EF} + Z_{c2} \quad (12)$$

Additionally, as done earlier with the Maa model, the entire double-layer surface impedance $Z_{1,EF}$ is obtained as follows:

$$Z_{EF} = Z_{1,EF} + Z_0\frac{Z_{2,EF}\cos(kD_2) + iZ_0\sin(kD_1)}{Z_0\cos(kD_1) + iZ_{2,EF}\sin(kD_2)}. \quad (13)$$

3. MPP Samples

The theoretical models have been used as a reference to design and develop different samples of micro-perforated panels. The purpose was to find the best configurations of the double-layer MPP's parameters in order to predict and simulate the behavior of sound absorbing structures that match the characteristic curve of speech, in view of the application to an university lecture hall.

3.1. Samples Manifacturing

The first step for the realization of an MPP layer was the material choice and, consequently, the standard thickness of the samples. Taking into consideration the sub-millimeter perforation diameter and the high quality standard in terms of durability and endurance, the choice was stainless steel. The aim was to obtain six samples with circular cross-sections and a diameter of ∼40 mm in order to allow the measurements in the impedance tube according to the technical standard ISO 10534-2 [26]. These three types of samples allowed to obtain different combinations between different samples, including double-layer structures with two equal samples. Considering the required perforation size (<1 mm) and geometry (circular cross-section), the stainless steel samples were manufactured using a 3D additive printing technique. A 316L grade stainless steel powder was used, specific for laser powder bed fusion (LPBF). The 3D printing process parameters are reported in Table 1.

Table 1. 3D printing process parameters. LPBF, laser powder bed fusion.

Parameter	Value
Stainless steel powder grade	316L (LPBF)
Powder grains diameter	30–40 μm
Powder layer height	60 μm
Laser rated power	360 W
Laser rated diameter	50 μm (max 80 μm)
Energy density	130 J/m^3
Material specific weight	7.98 g/cm^3

This choice allowed us to print the three different types of samples in a single cylinder of stainless steel with a diameter of ∼40 mm and a height of ∼30 mm as shown in Figure 5, respecting the working restrictions of the printer. Once the stainless steel was printed, the problem was to find a specific cutting technique to make samples of 1 mm thickness. The only possible choice was to use a wire electrical discharge machining (WEDM). The WEDM technique works without increasing the heat during the cutting.

Thanks to this technique, two samples of each type were obtained from the cylinder. The thickness of each sample was respected with a tolerance of ≈ ±0.3–0.6 mm, as shown in Figure 6.

Considering the scale of the perforation size, the real shapes of the perforations were not as expected and the real dimensions were bigger than expected: this was due to the limited accuracy of the processing and manufacturing machinery. Thus, all of the resulting perforations had irregular cross-sections: Ning et al. [29] demonstrated that increasing the specific surface area of a perforation could increase the sound energy dissipation and expand the sound absorption bandwidth, when the inscribed and circumscribed circle are assumed to be unchanged. In fact, keeping constant the outer diameter of the hole, an irregular cross-section increases the length of the hole perimeter and can add sharp edges; in this situation, viscous dissipation increases the sound absorption and the bandwidth.

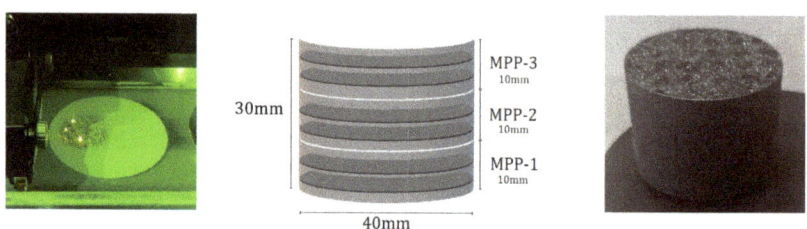

Figure 5. 3D metal printing (**left**); 3D sketch of the cylinder (**center**) and real photograph (**right**) of the cylinder. Two equal samples for three different kinds of specimen were obtained from the cylinder.

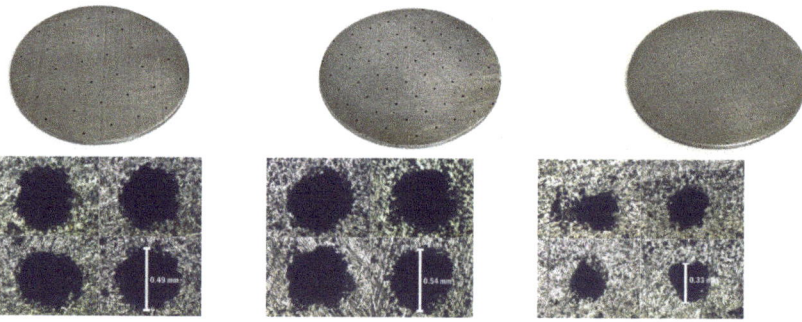

Figure 6. The three MPP samples (**top**) and the effective visualisation of perforations seen at the microscope (**bottom**).

3.2. Measuring Equipment

The use of the impedance tube with two microphone locations and a digital frequency analysis system for the determination of the sound absorption coefficient and acoustical surface impedance for normal sound incidence is shown in the standard ISO 10534-2 [26]. Considering a sample rate of 192 kHz and the speed of sound in air at $T = 18.6$ °C, $c_0 = 343$ m/s, the effective dimensions of the tube were calculated and are reported in Figure 7.

The working frequency range of the impedance tube is 300–4400 Hz: the lower frequency is limited by the accuracy of the signal processing equipment; the upper frequency depends only on the physical dimensions of the tube. The measurements were made with the so-called one microphone method: recording the tube response using one microphone in two different locations. This choice eliminates phase mismatch between microphones. According to ISO 10534-2 [26], the transfer function method was developed in MATLAB [30], using the tube impedance measurements scripts of the ITA-Toolbox [31].

The measurement chain consisted of a loudspeaker for the signal generation, a power supply, an audio device working with the audio stream input output (ASIO) drivers, a microphone and a battery power signal conditioner, as reported in Table 2.

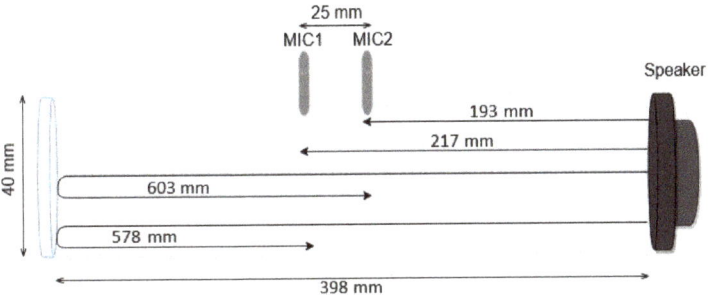

Figure 7. Effective dimensions (**top**) and actual photograph (**bottom**) of the impedance tube used in the ISO 10534-2 measurements.

Table 2. Equipment used for the impedance tube measurements.

Device	Model	Specifics
Microphone	PCB Piezoeletronics 130E20	Free-Field, 20–10 k Hz
Loudspeaker	SICA Z000795	200–10 k Hz
Signal Conditioner	PCB Piezoeletronics 480B21	Output Current 3 mA
Signal Amplifier	Tracopower TXL 100–125	$SNR = 96$ dBS/N
Soundcard	RME Fireface 800	Sample Rate 44,100 Hz
Software	MATLAB R2019a	ITA-Toolbox scripts

The output signal is an exponential sweep [32] converted to an analog signal by the digital-to-analog converter (DAC) of the audio device. The exponential sine sweep used for the experimental measurements was in the range of 250–5280 Hz, according to the working frequency range of the impedance tube (300–4400 Hz). The signal pressure coming from the microphone was converted to a digital audio object through an AD converter. All of the digital signal processing was developed in MATLAB with the help of the ITA-Toolbox impedance tube calculation scripts. The schematic is reported in Figure 8.

The recorded audio file of length 2.97 s was sampled with a sample rate of 44,100 Hz at 24-bit depth, and a time data file of 131,072 samples was obtained.

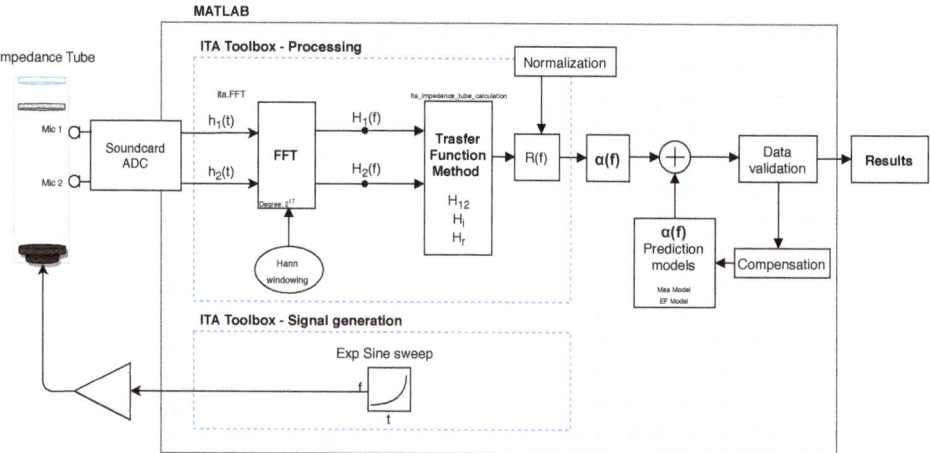

Figure 8. Data flow: excitation signal generation, acquisition of the impulse responses $h_1(t)$ and $h_2(t)$, processing, comparison with models, output of results.

3.3. Models Compensation

The experimental results for the MPP configurations showed some discrepancies between the measured data and both theoretical models decribed in Section 2. There were two important differences in common for each configuration:

- A small frequency shift of the sound absorption peaks;
- Absorption bandwidths larger than expected.

The first issue is connected to the samples' mounting procedure (Figure 9).

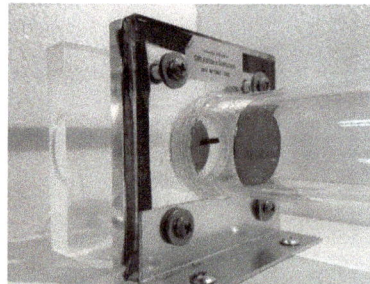

Figure 9. Details of the sample MPP02 mounted inside the impedance tube.

All samples were rounded and had a diameter slightly smaller than the sample holder. This did not guarantee the samples to be firmly mounted inside the tube. In order to avoid the side gaps between the specimen and the tube and to respect the mechanical boundary conditions, small strips of adhesive tape were applied to clamp the edges as much as possible. This caused small vibrations of the samples and a part of the energy was dissipated through the adhesive strips, increasing the absorption bandwidth.

The absorption bandwidth is due to the variability of the geometrical parameters of the samples: mainly, the irregular shapes and sizes of the perforations are the reasons why the measured data bandwidths appear to be larger than the predicted ones [33,34]. In particular, the smaller the perforations were, the bigger were the discrepancies. The perforations' perimeter was larger than expected and, consequently, a bigger absorption bandwidth occured. Instead of acting on the sample mounting procedure, a new compensating impedance has been added to the models, taking into consideration the discrepancies.

Considering the electro-acoustic analogy of the single-layer MPPs, the following changes were applied:

- R_{comp}, a resistance in series with the MPP impedance, associated to the dissipative losses due to the irregular perforations;
- L_{comp}, a new inductance in parallel with the MPP surface impedance, associated with the displacement of air along the boundary of the sample and the sample vibration [35].

This is still valid for the double-layer configurations: furthermore, the mounting inaccuracy is emphasized by the doubling of the MPP layer. In a DL configuration, two different resistances in series with the MPP impedance and two inductances in parallel with the MPP impedances have been considered, one for every MPP layer. Thus, in order to consider the geometrical dimension issues of the steel samples and the main problem connected to the sample mounting procedure, some changes to the equivalent circuits were made, as shown in Figure 10.

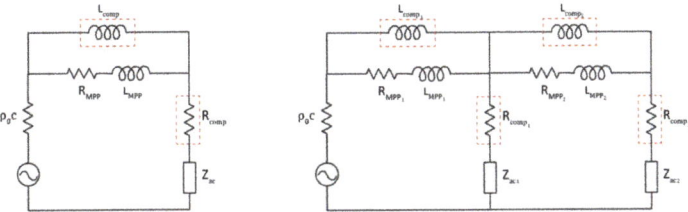

Figure 10. Electro-acoustical equivalent circuit for SL-MPP and DL-MPP configurations: the inductance L_{comp} is in parallel with the MPP surface impedance Z_{MPP}, and the resistance R_{comp} is in series with Z_{MPP}.

The explicit equation of the surface impedance in the case of a DL-MPP configuration after compensation is the following:

$$Z_{1,EF,comp} = (Z_{MPP1,EF} L_{comp,1}) / (Z_{MPP1,EF} + L_{comp,1}) + R_{comp,1} + Z_c \qquad (14)$$

$$Z_{2,EF,comp} = (Z_{MPP2,EF} L_{comp,2}) / (Z_{MPP2,EF} + L_{comp,2}) + R_{comp,2} + Z_c \qquad (15)$$

$$Z_{EF,comp} = Z_{1,EF,comp} + Z_0 \frac{Z_{2,EF,comp} \cos(kD_2) + iZ_0 \sin(kD_1)}{Z_0 \cos(kD_1) + iZ_{2,EF,comp} \sin(kD_2)} \qquad (16)$$

In the present work, the values of R_{comp} and L_{comp} for the DL-MPP combinations were estimated by trial and error, trying to match the experimental data.

3.4. Specimen Properties

Three different SL configurations were measured, one for every type of sample (MPP01, MPP02, MPP03). A different value of the air cavity thickness was chosen for every MPP sample, in order to obtain a good sound absorption peak around 1000 Hz. The same was done for four different DL configurations, composed of the type of samples mentioned above. Two layers of MPP provide two different resonance peaks: the choice was to try to find DL configurations with an absorption peak at 1000 Hz and a second peak in a lower frequency range. The three samples had the effective geometrical parameters reported in Table 3.

The comparison between the equivalent fluid model and the experimental measurements is reported in terms of sound absorption coefficient and normalized surface impedance in Figure 11 for three SL configurations.

The three samples were used in four different combinations of DL configurations with air cavities reported in Table 4, and a comparison was made between the EF model and the experimental measurements in terms of sound absorption coefficient and normalized surface impedance in Figure 12. In contrast to the SL comparisons, the DL experimental measurements showed bigger discrepancies from the model: this is due to the mounting procedure inaccuracies and the irregular perforations of two different samples, both doubled. The model compensation is not enough to completely compensate for the non-idealities, mainly in terms of normalized surface impedance.

Table 3. Expected (exp.) and effective (eff.) geometrical parameters of the three MPP specimens.

Specimen	MPP01		MPP02		MPP03	
	Exp.	Eff.	Exp.	Eff.	Exp.	Eff.
Thickness (mm)	1	0.94 ± 0.03	1	0.95 ± 0.01	1	0.92 ± 0.02
Porosity (%)	0.5	0.5	1	1	0.5	0.5
Perforation diameter (mm)	0.5	0.483 ± 0.04	0.5	0.482 ± 0.03	0.3	0.33 ± 0.05

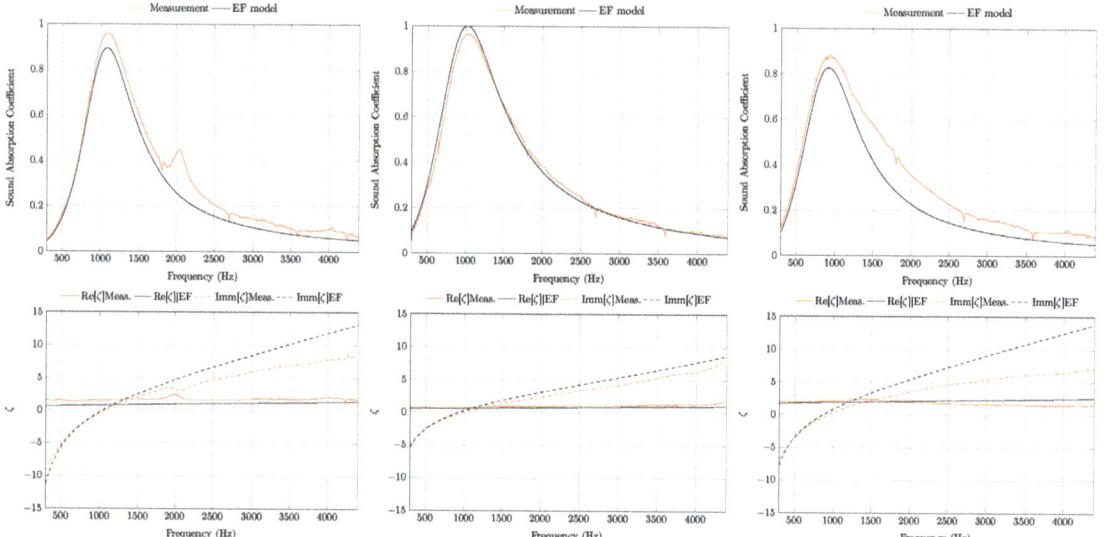

Figure 11. Sound absorption coefficient and normalized surface impedance of single-layer MPPs: MPP01 with an air cavity of $D = 14.8$ mm, MPP02 with an air cavity of $D = 30$ mm and MPP03 with an air cavity of $D = 20$ mm.

Table 4. Specific normalised impedances corresponding to the four double-layer configurations. Octave band data were fitted based on the outcomes of the compensation model described in the text.

	1st Layer	1st Cav. mm	2nd Layer	2nd Cav. mm	ζ					
					125 Hz	250 Hz	500 Hz	1 kHz	2 kHz	4 kHz
A	MPP01	40	MPP01	30	$1.3 - 16.4i$	$1.3 - 6.8i$	$1.8 - 0.9i$	$2.2 + 2.0i$	$1.6 + 8.4i$	$4.4 + ai$
B	MPP01	50	MPP02	50	$1.4 - 12.3i$	$1.5 - 4.9i$	$1.9 - 0.1i$	$2.5 + 2.7i$	$1.8 + 9.0i$	$3.1 - ai$
C	MPP02	45	MPP03	45	$0.4 - 14.2i$	$0.4 - 6.2i$	$0.7 - 1.4i$	$1.7 - 0.4i$	$0.5 + 3.5i$	$1.8 + 8.5i$
D	MPP01	35	MPP03	35	$1.4 - 18.1i$	$1.4 - 7.7i$	$1.5 - 1.6i$	$3.1 + 2.3i$	$1.8 + 7.5i$	$2.1 + 15.7i$

$a \to \infty$: the function diverges to infinity or minus infinity.

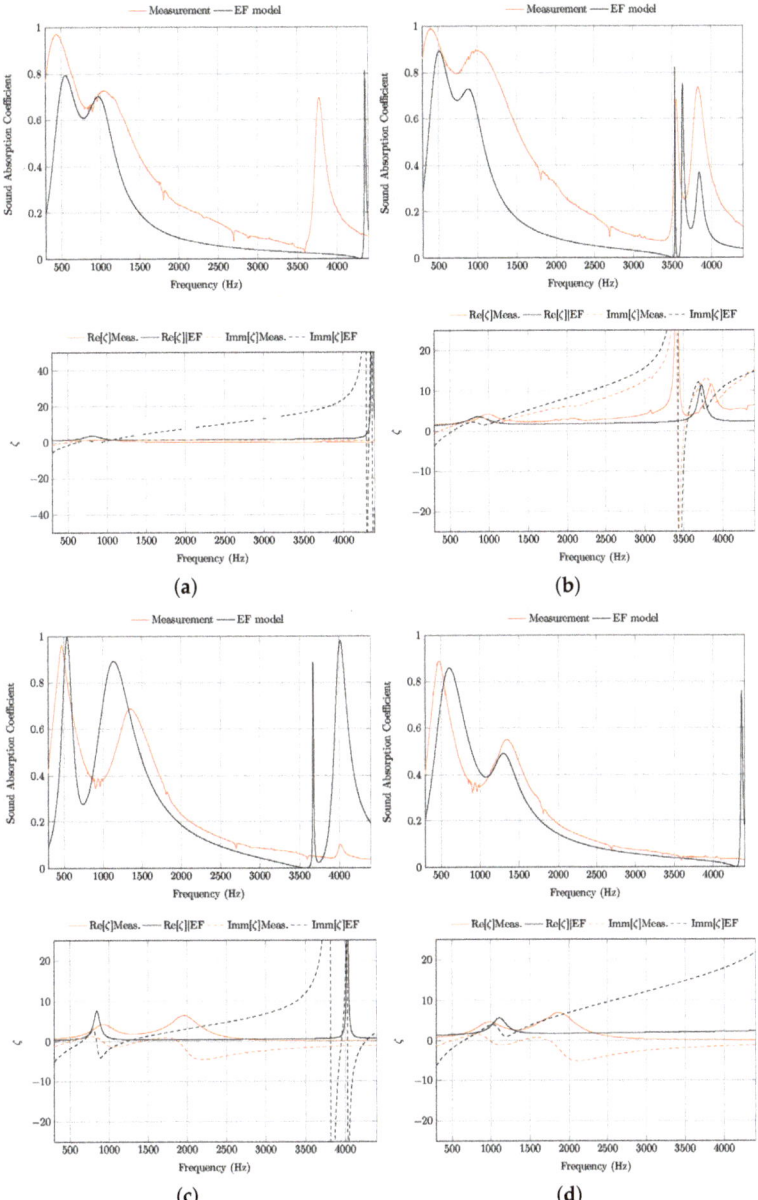

Figure 12. Sound absorption coefficient and normalized surface impedance for four DL configurations. (**a**) Configuration A. (**b**) Configuration B. (**c**) Configuration C. (**d**) Configuration D.

4. Finite-Difference Time-Domain Simulations

In room acoustics, multiple simulation methods aim to directly solve the wave equation in homogeneous form:

$$\left(\nabla^2 - \frac{1}{c_0^2}\frac{\partial^2}{\partial t^2}\right)p = 0 \qquad (17)$$

where ∇^2 is the 3D Laplacian operator, c_0 is the speed of sound at $T = 20\,°C$ and RH= 50%, and $p = p(\mathbf{x},t)$ is the pressure (deviation from ambient). The finite-difference time-domain (FDTD) method is among the oldest numerical methods to solve time-dependent partial differential equations (PDEs), like the wave equation [36]. In FDTD methods, space and time continuous domains are typically discretized with regular Cartesian grids and updating recursions occur in the nodal points to calculate the acoustic properties such as the pressure $p(\mathbf{x},t)$ or the particle velocity potential $u(\mathbf{x},t)$ [37]. The solution $p(\mathbf{x},t)$ of the wave equation, with $\mathbf{x} \in \mathbb{R}^3$, is approximated by a grid function $p_{l,m,p}^n$ at spatiotemporal points $x = lh, y = mh, z = ph$ and $t = nk$, with l, m, n and p representing integer numbers, h the grid spacing, and k the time step [38,39]. A large variety of explicit FDTD methods follows the same general scheme:

$$\delta_t^2 p_{l,m,p}^n = \lambda^2[(\delta_x^2 + \delta_y^2 + \delta_z^2) + a(\delta_x^2\delta_y^2 + \delta_x^2\delta_z^2 + \delta_y^2\delta_z^2) + b(\delta_x^2\delta_y^2\delta_z^2)]p_{l,m,p}^n \quad (18)$$

where λ is the dimensionless quantity defined as the Courant number $\lambda = ck/h$ and a and b are the specific coefficients of a general family of compact explicit schemes [40]. The operators δ_t^2 and δ_x^2 act on $p_{l,m,p}^n$ as follows:

$$\delta_t^2 p_{l,m,p}^n = p_{l,m,p}^{n+1} - 2p_{l,m,p}^n + p_{l,m,p}^{n-1}, \quad \delta_x^2 p_{l,m,p}^n = p_{l+1,m,p}^n - 2p_{l,m,p}^n + p_{l-1,m,p}^n \quad (19)$$

and similarly for δ_y^2 and δ_z^2. The value of the Courant number is closely correlated to the stability condition of the system, which is guaranteed avoiding the exponential growth of the numerical solution (see, e.g., [41]). In the simple case of the seven-point scheme (a = b = 0), the stability condition is expressed as:

$$\lambda \leq \frac{1}{\sqrt{3}} \quad (20)$$

which is the so-called Courant–Friedrichs–Lewy (CLF) condition [36].

However, in this paper the cubic close-packed (CCP) scheme (see Figure 13), with $a = 0.25$, $b = 0$ and $\lambda = 1$, is employed for its favorable numerical dispersion properties [41]. The consequent maximal time step, for a certain grid spacing, is k < h/c. As is the case for any 3D FDTD scheme, the computational cost is proportional to h^{-3}, which can quickly lead to large simulation grids. Then, such grid-based simulations are natural candidates for parallel acceleration on graphics processing units (GPUs) [42].

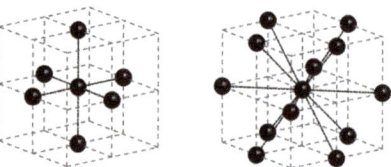

Figure 13. Two possible stencils: standard 7-point (**left**) and cubic close-packed (CCP) 13-point (**right**). Image credit: see reference [28].

5. Application in a Case Study: Use of MPPs as a Ceiling Acoustic Treatment

A wide employment of sustainable materials such the multi-layer MPPs analyzed in this study is expected to occur in several typologies of enclosed environments [35,43], replacing or integrating the common sound absorbing treatments. MPP absorbers are theoretically expected to return the same acoustic behavior regardless of their constitutive material. Therefore, concerning sustainability aspects, they can be made of any green material, reducing the environmental footprint of the whole process. The acoustic simulation of a wide application in a 3D virtual enclosure is a useful tool in a preliminary step in the assessment of their performance [44–46].

With this purpose, the acoustic condition of a large university lecture hall has been simulated with a ceiling-mounted system of double-layer MPPs. Such a kind of large rooms shows the highest values of reverberation time at 500 and 1000 Hz (see Figure 4 in [27]), the same frequency range occupied by the human voice [47]. As too high values of reverberation time deteriorate speech intelligibility, a material whose sound absorbing properties are mostly centered in such a frequency range may return useful outcomes in acoustic treatments of existing university halls. Otherwise, an acoustic treatment with porous or fiber materials would provide the most significant absorbing contribution at higher frequencies [48], sometimes entailing too dry conditions at 2000 Hz and 4000 Hz.

The room chosen as a case study is a historical lecture hall in the University of Bologna (see Figure 14). A previous campaign of acoustic measurements [27,49] in the hall allowed to collect the main room criteria [50] and intelligibility indexes [47]. The setting of sound source and receiver locations chosen for the measurements campaign is also provided in Figure 14.

The 3D virtual model of the hall-about 900 m^3 was built with Sketchup [51] according to consolidated approximation guidelines (see Figure 15), i.e., with a certain degree of geometrical approximation and with a proper division in macro-layers depending on the materials [52,53]. The 3D model was modeled using a small group of different materials (see Table 5) to reduce the uncertainties underlying the assignment of material properties to each surface. With regard to the absorption area distribution, it should be noted that, generally, the seats are the most sound absorbing objects in a lecture hall in unoccupied state [23,44–46]. The remaining layers of the model (walls, floor, ceiling) are made up of rather hard and reflective surfaces, and thus they show low values in the whole frequency range. It should also be highlighted that the random incidence absorption coefficients at low frequencies of wooden parts are due to the air cavity behind them (see dataset in Table 5).

The state-of-the-art procedure to simulate the sound field of a large lecture hall would imply the use of geometrical acoustics (GA) techniques [46,54]. Nevertheless, MPPs are not suitable to be characterized by the energy-based parameters, i.e., the absorption coefficients, usually involved in GA practice. Therefore, the need arose to simulate the acoustic behavior by using a wave-based algorithm, such as a FDTD model, that, additionally, assures the correct computation of the diffraction effects due to the objects' edges [55].

 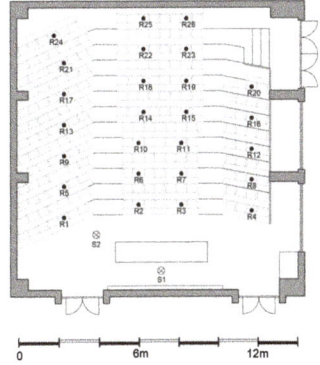

Figure 14. Interior view (**left**) and 2D plan (**right**) of the lecture hall under study. Sound sources (S1, S2) and the spatial grid of microphone receivers (R1, R2, ..., RN) used in the measurements campaign are reported.

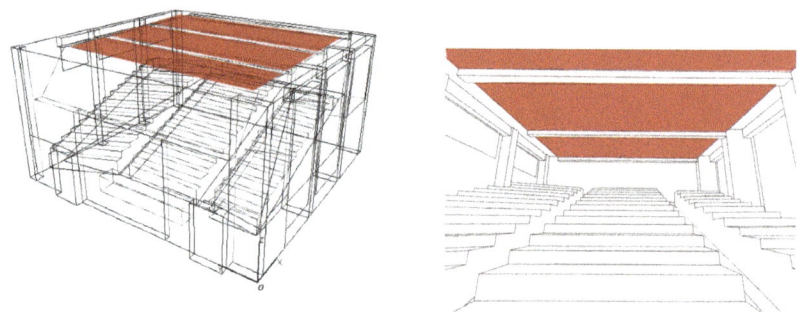

Figure 15. Exterior (**left**) and interior (**right**) view of the acoustic treatment virtually introduced into the CAD model. The ceiling-mounted MPPs are highlighted in red.

Table 5. Calibration setup (without MPP): random incidence absorption coefficients used in geometrical acoustics (GA) simulations and considered in the backward optimization process to obtain the acoustic admittances for the finite-difference time-domain (FDTD) simulation. The macro-layers used in the present work divide the materials into: hard/reflective surfaces (plaster floor), elements absorbing slightly at low frequencies (wood, windows) and the most sound absorbing area (seats).

Materials	125 Hz	250 Hz	500 Hz	1000 Hz	2000 Hz	4000 Hz
Plaster floor	0.01	0.02	0.03	0.03	0.04	0.06
Wood	0.15	0.18	0.04	0.04	0.04	0.04
Windows	0.10	0.10	0.08	0.04	0.04	0.04
Seats	0.40	0.37	0.26	0.19	0.17	0.16

With this purpose, in the 3D virtual model the edges of the blocks containing the seats and the long tables were modeled (see Figure 16b). The contour map shows the edge diffraction due to the seats, which are typically the only irregular element in a lecture hall contributing to the sound field diffusion. Figure 16 also shows a qualitative comparison with the standard GA simulation that involves a simpler modeling of the seats' blocks and a scattering coefficient (see Figure 16a).

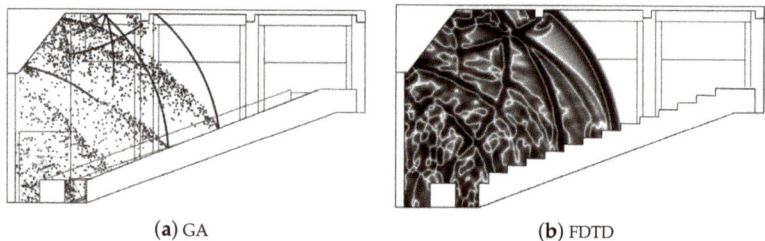

(**a**) GA (**b**) FDTD

Figure 16. Qualitative visualization of sound wave propagation within the lecture hall by means of GA (**a**) and FDTD (**b**) simulations throughout the longitudinal section. The sound source is located at the teacher's position (S1 in Figure 14).

The DL-MPP corresponding to Configuration A in Table 4 was introduced in the model replacing most of the part of the existing false ceiling, as can be seen in Figure 15. Typically, the acoustic impedances needed as boundary conditions for an FDTD numerical simulation derive from the large amount of random incidence absorption coefficient datasets available through optimization processes [56]. Indeed, uncertainties inherent in energy parameters (absorption coefficients)—also due to ISO 354 limits [57]—are propagated in the conversion to non-unique corresponding complex acoustic impedances [58]. In this case, it has been possible to avoid some of those typical uncertainties in the workflow by directly starting with complex acoustic impedances values as input boundary conditions (see Figure 17) [59,60].

Concerning the locally reactive absorption properties of the MPPs, the outcomes of the compensation model described in Section 3.3 were used to fit complex impedances to a boundary impedance model comprising a parallel network of second-order series RLC circuits [61]. The fitting procedure optimized non-negative circuit parameters (resistances, inductances and capacitances) in order to minimize the Euclidean norm of the error between the complex admittance of the circuit network and that of the data. A similar process was carried for known absorption coefficients of the remaining materials, following [62]. The complex admittance data and the model approximation to be used in the FDTD simulation are shown in Figure 17. As can be seen in the figure, the filter model described in [61,62] and used in the FDTD simulation faithfully reproduced the complex admittance of the MPP material.

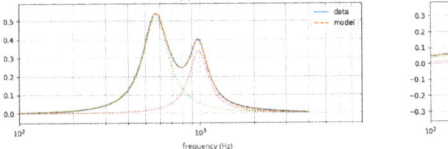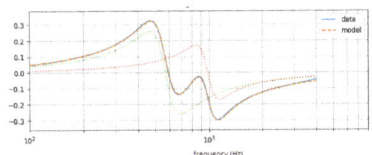

Figure 17. Real (**left**) and imaginary (**right**) parts of complex specific admittances from data for MPP material (solid line) along with fitted filter model (dashed line) used in the FDTD simulation. Dotted lines show underlying resonances in the filter approximation.

The FDTD method chosen simulated up to 8000 Hz with 6.75 points per wavelength (Cartesian grid spacing $h = 6.35$ mm and time step $k = 18.5$ µs), which required approximately 2 h per sound source using four Nvidia Titan X (Maxwell) GPUs. It should be noticed that even though the FDTD model chosen in this work can switch to a ray-tracing algorithm over a certain frequency [63], in this case it was possible to run a full wave-based simulation thanks to the moderate complexity of the geometry and the availability of high computational power.

Before introducing the double-layer MPP into the virtual model of the lecture hall, the calibration process was carried out based on the main room criteria acquired from the acoustic measurements. The 3D model was tuned in terms of reverberation time (T_{30}) considering the sound source in location S1 and averaging the values over all the receiver points shown in Figure 14. Table 6 reports the trend in frequency of the measured values, the equivalent values derived from the calibrated FDTD model ("without MPP") and the variations due to the introduction of MPPs ("with MPP"). During the tuning of the 3D model, the tolerance range for the discrepancies between measured and simulated values was kept equal to twice the common JND [50], considering 10% of the measured values according to recent remarks [64]. Table 6 also provides the T_{30} mean values derived from the corresponding GA simulations to keep the comparison with the standard simulation procedure. Certainly in the GA model all of the temporal delays and acoustic behaviors due to the peculiarity of MPPs are approximated by energy parameters (absorption coefficients). With regard to the results in terms of reverberation time, the main contribution of the acoustic treatment is relevant at 500–1000 Hz, as expected, and still significant at 250 and 2000 Hz, due to the broadband performance obtained with DL configurations. The effects of treating a lecture hall with such a material instead of a common porous material are positive not only because they cover the frequency range occupied by the human voice, but also because they compensate for the typical excessive reverberation that undermines the verbal communication in historical rooms.

Table 6. Trend in frequency of T_{30} mean values corresponding to the results of the measurements ("Meas"), the equivalent values derived from the calibrated FDTD model ("without MPP") and the variations due to the introduction of MPPs ("with MPP"). The T_{30} mean values derived from the corresponding GA simulations are provided as a term of comparison.

T_{30} (s)	125 Hz	250 Hz	500 Hz	1000 Hz	2000 Hz	4000 Hz
Without MPP (Meas.)	1.48	1.36	1.60	1.84	1.83	1.51
Without MPP (FDTD)	1.44	1.25	1.54	1.73	1.70	1.45
Without MPP (GA)	1.43	1.25	1.62	1.79	1.78	1.52
With MPP (FDTD)	1.32	0.92	0.88	1.22	1.52	1.44
With MPP (GA)	1.33	1.06	0.87	1.00	1.34	1.50

Concerning the room criteria related to intelligibility, it is quite intuitive that in an optimal condition for the verbal communication their values should be as uniform as possible throughout the space. Therefore, the spatial behavior of sound clarity (C_{50}) [50] is provided versus the distance between the source and the receiver (see Figure 18). The same five conditions already seen in Table 6 are reported in the graph: the measured values, the calibration outcomes and the effects of the acoustic treatment according to FDTD and GA results. On average, it is possible to quantify an increase of C_{50} values at mid frequencies higher than 4 dB, corresponding to four times the JND of the sound clarity. The acoustic treatment allowed to significantly increase the early-to-late ratio ($C_{50,M} \approx 2$ dB) from a poor acoustic condition ($C_{50,M} \approx -2.5$ dB). Figure 18 also shows a good match among all the slopes of the linear regressions involved. This is probably due to the fact that the calibration was achieved by using an energy parameter, the reverberation time, that is derived from the energy decay curve as the sound clarity.

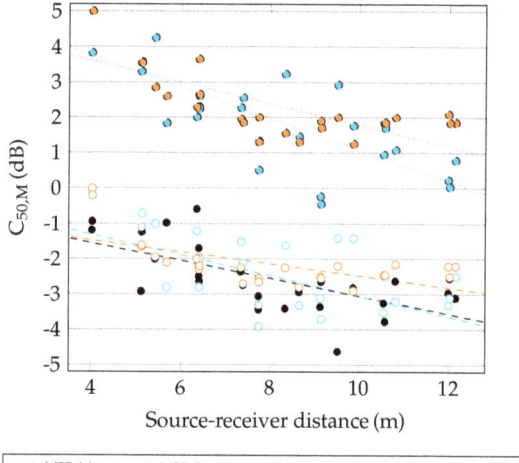

Figure 18. Trend of sound clarity (C_{50}) as a function of the sound source-receiver distance (values and linear regressions). "M" indicates that the values have been averaged over 500 and 1000 Hz. Results of the measurements ("Meas"), the equivalent values derived from the calibrated FDTD model ("w/o MPP") and the variations due to the introduction of MPPs ("w MPP") are provided by using both FDTD and GA techniques.

Finally, an overview is provided about other recent works that handle the microperforated panels with wave-base simulation methods. Table 7 helps to place the present work in a framework of similar research studies to highlight similarities and methodological differences. Concerning the boundary conditions, it should be highlighted that only in reference [65] an extended reaction model for SL-MPP absorbers was used, including the incident angle dependence of surface impedance. In references [66,67] and in the present work the local reaction is assumed. To the best of the authors' knowledge, at the time of writing there are several issues in handling extended reaction in FDTD and concurrently keeping the stability of the system. Indeed, in future developments further efforts will be made to improve the locally reactive simplified model used in the present work with the angle-dependent model.

Table 7. Framework of recent works focused on wave-based simulations of micro-perforated panels. The frequency range managed by the authors, the calculation method, the typology of boundary condition employed and the output parameters obtained are reported.

Reference	Range (Hz)	Calculation Method	Boundary Condition	Output
Liu and Herrin, 2010 [68]	100–5000	BEM	Transfer Impedance	SPL, Insertion Loss
Okuzono and Sakagami, 2015 [66]	30–6000	FEM	Surface Impedance	–
Okuzono and Sakagami, 2018 [65]	125–1000	FEM	Surface Impedance	T_{30}, SPL
Naderyan et al., 2019 [69]	710–1400	FEM	Surface Impedance	Power dissipation
Toyoda and Eto, 2019 [67]	31.5–8000	FDTD	Surface Impedance	SPL, Insertion Loss
Mondet et al., 2020 [56]	100–4500	FDTD	Surface Impedance *	Conversion method
Present work	125–4000	FDTD	Surface Impedance	T_{30}, C_{50}

* Values derived from real-valued absorption coefficients.

6. Conclusions and Outlook

Sound absorbers based on MPPs could be an attractive alternative to the conventional porous and fibrous absorbers when following requirements such as durability, recyclability, cleanliness and environmental sustainability. Since MPP absorbers are theoretically expected to return the same acoustic behavior regardless of their constitutive material, in principle they can be made of any green material, provided that it is hard enough to support micro-perforation. In particular, stainless steel guarantees a long service life, being an unalterable material when applied indoors, and a good hygiene, being resistant to mold and fungi. A possible continuation of this research could be a detailed LCA analysis of a selection of materials usable to make MPP. The present work aims to outline a method for the design and the numerical validation of specific sound absorbers where both active and reactive acoustical properties have to be considered. Thanks to a full-spectrum FDTD simulation method the effects of this material on the intelligibility criteria were explored. First, three custom MPP specimens were designed and manufactured using a 3D additive metal printing process. The analytical predictive models of MPP in single- and double-layer configurations were validated through experimental measurements with the impedance tube obtaining the acoustical properties in terms of sound absorption coefficients and normalized surface impedances. Then, wider surfaces of double-layer MPPs were simulated in a calibrated 3D room corresponding to a real lecture hall. Results showed that MPPs mainly operate in the central octave band of 500 and 1000 Hz with a further useful contribution at 250 and 2000 Hz. Since this frequency range is the one most affecting speech intelligibility and at the same time the most undermined in historical lecture halls, MPPs seem to be a high-performance solution in cases similar to the present one. The university lecture hall taken as a case study is intended to show the positive effect of MPPs as acoustic treatment to the enhancement of speech intelligibility. In terms of time-dependence, the finite-difference time-domain model allowed us to better analyze both the reactive effects of the treatment and the scattering effects of the lecture hall's geometry. Finally, for all these reasons the results of the present work should be indicate a robust method to experimentally measure and test the performance of specific materials, especially considering that the demand for high-performance and sustainable materials, such as the micro-perforated panels, is expected to increasingly grow in the sector of sound absorbing treatments.

Author Contributions: Conceptualization: M.G., D.D.; data curation: M.C.; formal analysis: D.D., M.G.; funding acquisition, M.G.; investigation, M.C. and L.B.; methodology, M.G., D.D.; project administration, M.G.; resources, L.B. and M.G.; MATLAB software: M.C. and M.G.; FDTD simulation, B.H. and G.F.; supervision,M.G, and D.D.; validation, M.C., D.D.; visualization, M.C. and G.F.; writing—original draft preparation, M.C. and D.D.; writing—review and editing, D.D., G.F., B.H. and M.G. All authors have read and agreed to the published version of the manuscript.

Funding: This research was funded by the Ministero dell'Istruzione dell'Università della Ricerca (Italy), in the framework of the project PRIN 2017, grant number 2017T8SBH9: "Theoretical modeling and experimental characterization of sustainable porous materials and acoustic metamaterials for noise control".

Acknowledgments: The authors would like to thank Costantino Marmo, Donatella Alvisi and Nadia Perri, who kindly allowed us to carry out the acoustic measurements in the lecture hall taken as a case study.

Conflicts of Interest: The authors declare no conflict of interest.

Sample Availability: Samples of 3D MPPs are available from the authors.

References

1. Pan, L.; Martellotta F. A Parametric Study of the Acoustic Performance of Resonant Absorbers Made of Micro-perforated Membranes and Perforated Panels. *Appl. Sci.* **2020**, *10*, 1581. [CrossRef]
2. Asdrubali, F.; Pispola G. Properties of transparent sound-absorbing panels for use in noise barriers. *J. Acoust. Soc. Am.* **2007**, *121*, 214–221. [CrossRef]
3. Sakagami, K.; Okuzono, T. Space sound absorbers with next-generation materials: Additional sound absorption for post-pandemic challenges in indoor acoustic environments. *UCL Open Environ. Prepr.* **2020**. [CrossRef]
4. Johnson, D. L.; Koplik J.; Dashen, R. Theory of dynamic permeability and tortuosity in fluid-saturated porous media. *J. Fluid. Mech.* **1987**, *176*, 379–402. [CrossRef]
5. Allard, J.F.; Champoux, Y. New empirical equations for sound propagation in rigid frame fibrous materials. *J. Acoust. Soc. Am.* **1992**, *91*, 3346–3353. [CrossRef]
6. Cucharero, J.; Hänninen, T.; Lokki, T. Angle-Dependent Absorption of Sound on Porous Materials. *Acoustics* **2020**, *2*, 753–765. [CrossRef]
7. Maa, D. Theory and design of microperforated panel sound-absorbing constructions. *Sci. Sin.* **1975**, *18*, 55–71.
8. Maa, D. Microperforated-panel wideband absorbers. *Noise Control Eng. J.* **1987** *29*, 77–84. [CrossRef]
9. Maa, D. Potential of microperforated panel absorber. *J. Acoust. Soc. Am.* **1998** *104*, 2861–2866. [CrossRef]
10. Tayong, R.; Leclaire, P. Holes Interaction Effects under high and medium Sound Intensities for Micro-perforated panels design. In Proceedings of the 10th Congress Francais d'Acoustique, Lyon, France, 12–16 April 2010.
11. Melling, T.H. The acoustic impendance of perforates at medium and high sound pressure levels. *J. Sound Vib.* **1973**, *29*, 1–65. [CrossRef]
12. Atalla, N.; Sgard, F. Modeling of perforated plates and screens using rigid frame porous models. *J. Sound Vib.* **2007**, *303*, 195–208. [CrossRef]
13. Cobo, P.; Simon, F. Multiple-Layer Microperforated Panels as Sound Absorbers in Buildings: A Review. *Buildings* **2019**, *9*, 53. [CrossRef]
14. Lee, D.H.; Kwon, Y.P. Estimation of the absorption performance of multiple-layer perforated panel systems by transfer matrix method. *J. Sound Vib.* **2004**, *278*, 847–860. [CrossRef]
15. Sakagami, K.; Morimoto, M.; Koike, W. A numerical study of double-leaf microperforated panel absorbers. *Appl. Acoust.* **2019**, *2006*, 67–609. [CrossRef]
16. Hoshi, K.; Hanyu, T.; Okuzono, T.; Sakagami, K.; Yairi, M.; Harada, S.; Takahashi, S.; Ueda, Y. Implementation experiment of a honeycomb-backed MPP sound absorber in a meeting room. *Appl. Acoust.* **2020**, *157*, 107000. [CrossRef]
17. Tsay, Y.S.; Yeh, C.Y. A Machine Learning Based Prediction Model for the Sound Absorption Coefficient of Micro-Expanded Metal Mesh (MEMM). *Appl. Sci.* **2020**, *10*, 7612. [CrossRef]

18. Arvidsson, E.; Nilsson, E.; Hagberg, D.B.; Karlsson, O.J. The Effect on Room Acoustical Parameters Using a Combination of Absorbers and Diffusers—An Experimental Study in a Classroom. *Acoustics* **2020**, *2*, 505–523. [CrossRef]
19. Choi, Y. Evaluation of acoustical conditions for speech communication in active university classrooms. *Appl. Acoust.* **2020**, *159*, 107089. [CrossRef]
20. D'Orazio, D.; De Salvio, D.; Anderlucci, L.; Garai, M. Measuring the speech level and the student activity in lecture halls: Visual-vs blind-segmentation methods. *Appl. Acoust.* **2020**, *169*, 107–448. [CrossRef]
21. Brill, L.C.; Smith, K.; Wang, L.M. Building a sound future for students: Considering the acoustics of occupied active classrooms.*Acoust. Today* **2018**, *14*, 14–22.
22. Hodgson, M.; Rempel, R.; Kennedy, S. Measurement and prediction of typical speech and background-noise levels in university classrooms during lectures. *J. Acoust. Soc. Am.* **1999**, *105*, 226–233. [CrossRef]
23. Puglisi, G.E.; Bolognesi, F.; Shtrepi, L.; Warzybok, A.; Kollmeier, B.; Astolfi, A. Optimal classroom acoustic design with sound absorption and diffusion for the enhancement of speech intelligibility. *J. Acoust. Soc. Am.* **2017**, *141*, 3456–3457. [CrossRef]
24. Visentin, C.; Prodi, N.; Cappelletti, F.; Torresin, S.; Gasparella, A. Using listening effort assessment in the acoustical design of rooms for speech. *Build Environ.* **2018**, *136*, 38–53. [CrossRef]
25. Chung, J.Y.; Blaser, D.A. Transfer function method of measuring induct acoustic properties. I. Theory. *J. Acoust. Soc. Am.* **1980**, *68*, 907–913. [CrossRef]
26. EN ISO 10534–2:2001. *Acoustics-Determination of Sound Absorption Coefficient and Impedance in Impedance Tubes-Part.2: Transfer-Function Method*; International Organization for Standardization: Geneva, Switzerland, 2001.
27. Fratoni, G.; D'Orazio, D.; De Salvio, D.; Garai, M. Predicting speech intelligibility in university classrooms using geometrical acoustic simulations. In Proceedings of the 16th IBPSA, Rome, Italy, 2–4 September 2019.
28. Hamilton, B.; Webb, C.J. Room acoustics modelling using GPU-accelerated finite difference and finite volume methods on a face-centered cubic grid. In Proceedings of the Digital Audio Effects (DAFx), Maynooth, Ireland, 2–6 September 2013; 336–343.
29. Ning, J.F.; Ren, S.W.; Zhao, G.P. Acoustic properties of micro-perforated panel absorber having arbitrary cross-sectional perforations. *Appl. Acoust.* **2016**, *111*, 135–142. [CrossRef]
30. *MATLAB, Version 9.6.0.1072779 (R2019a)*; The MathWorks Inc.: Natick, MA, USA, 2019.
31. ITA-Toolbox, Open source MATLAB toolbox for acoustics developed by the Institute of Technical Acoustics of the RWTH Aachen University, Neustrasse 50, 52056, Aachen, Germany. In Proceedings of the DAGA 2017, Kiel, Germany, 6–9 March 2017.
32. Corredor-Bedoya, A.C.; Acuña, B.; Serpa, A.L.; Masiero, B. Effect of the excitation signal type on the absorption coefficient measurement using the impedance tube. *Appl. Acoust.* **2020**, *171*, 107659. [CrossRef]
33. Sakagami, K.; Kusaka, M.; Okuzono, T.; Nakanishi, S. The Effect of deviation due to the manufacturing accuracy in the parameters of an MPP on its acoustic properties: Trial production of MPPs of different hole shapes using 3D printing. *Acoustics* **2020**, *2*, 605–606. [CrossRef]
34. Randeberg, R.T. *Perforated Panel Absorbers with Viscous Energy Dissipation Enhanced by Orifice Design*; Ph.D. Dissertation, Department of Telecommunication, Norwegian University of Science and Technology, Trondheim, Norway, June 2000.
35. Sakagami, K.; Morimoto, M.; Yairi, M. A note on the effect of vibration of a microperforated panel on its sound absorption characteristics. *Acoust. Sci. Technol.* **2005**, *26*, 204–207. [CrossRef]
36. Courant, R.; Friedrichs, K.; Lewy, H. Über die partiellen Differenzengleichungen der mathematischen Physik. *Math. Annal.* **1928**, *100*, 32–74. [CrossRef]
37. Botteldooren, D. Finite-difference time-domain simulation of low-frequency room acoustic problems. *J. Acoust. Soc. Am.* **1995**, *98*, 3302–3308. [CrossRef]
38. Bilbao, S.D. *Numerical Sound Synthesis*; John Wiley & Sons: Hoboken, NJ, USA, 2009.
39. Savioja, L. Real-time 3D finite-difference time-domain simulation of low-and mid-frequency room acoustics. In Proceedings of the 13th International Conference on Digital Audio Effects, Maynooth, Ireland, 6–10 September 2010.
40. Kowalczyk, K.; Van Walstijn, M. Room acoustics simulation using 3-D compact explicit FDTD schemes. *IEEE Trans. Audio Speech Lang. Process.* **2011**, *19*, 34–46. [CrossRef]
41. Hamilton, B. Finite Difference and Finite Volume Methods for Wave-Based Modelling of Room Acoustics. Ph.D. Dissertation, University of Edinburgh, Edinburgh, UK, 2016.
42. Micikevicius, P. 3D finite difference computation on GPUs using CUDA. In Proceedings of the 2nd Workshop on General Purpose Processing on Graphics Processing Units, Washington, DC USA, 8 March 2009; pp. 79–84.
43. Arenas, J.P.; Sakagami, K. *Sustainable Acoustic Materials*; MDPI: Basel, Switzerland, 2020; p. 6540.
44. Bistafa, S.R.; Bradley, J.S. Predicting speech metrics in a simulated classroom with varied sound absorption. *J. Acoust. Soc. Am.* **2001**, *109*, 1474–1482. [CrossRef] [PubMed]
45. Reich, R.; Bradley, J. Optimizing classroom acoustics using computer model studies. *Can. Acoust.* **1998**, *26*, 15–21.
46. Bistafa, S.R.; Bradley, J.S. Predicting reverberation times in a simulated classroom. *J. Acoust. Soc. Am.* **2000**, *108*, 1721–1731. [CrossRef]
47. IEC 60268-16. *Sound System Equipment–Part 16: Objective Rating of Speech Intelligibility by Speech Transmission Index*; IEC: Geneva, Switzerland, 2020.
48. Garai, M.; Pompoli, F. A simple empirical model of polyester fibre materials for acoustical applications. *Appl. Acoust.* **2005**, *66*, 1383–1398. [CrossRef]

49. Martellotta, F. Optimizing stepwise rotation of dodecahedron sound source to improve the accuracy of room acoustic measures. *J. Acoust. Soc. Am.* **2013**, *134*, 2037–2048. [CrossRef]
50. ISO 3382-2. *Acoustics-Measurement of Room Acoustic Parameters-Part 2: Reverberation Time in Ordinary Rooms*; International Organization for Standardization: Geneva, Switzerland, 2008.
51. *SketchUp, Trimble Inc., Version Make*; 935 Stewart Drive, Sunnyvale CA (USA) 94085, 2017.
52. Vorländer, M. *Auralization*; Springer: Aachen, Germany, 2008.
53. Postma, B.N.; Katz, B.F. Perceptive and objective evaluation of calibrated room acoustic simulation auralizations. *J. Acoust. Soc. Am.* **2016**, *140*, 4326–4337. [CrossRef]
54. Choi, Y. J. Effects of periodic type diffusers on classroom acoustics. *Appl. Acoust.* **2013**, *74*, 694–707. [CrossRef]
55. Okuzono, T.; Shadi, M.; Sakagami, K. Potential of Room Acoustic Solver with Plane-Wave Enriched Finite Element Method. *Appl. Sci.* **2020**, *10*, 1969. [CrossRef]
56. Mondet, B.; Brunskog, J.; Jeong, C.H.; Rindel, J.H. From absorption to impedance: Enhancing boundary conditions in room acoustic simulations. *Appl. Acoust.* **2020**, *157*, 106884. [CrossRef]
57. Vercammen, M. *On the Revision of ISO 354, Measurement of the Sound Absorption in the Reverberation Room*; ISO/WD: Geneva, Switzerland, 2018.
58. Jeong, C.H. Converting Sabine absorption coefficients to random incidence absorption coefficients. *J. Acoust. Soc. Am.* **2013**, *133*, 3951–3962. [CrossRef]
59. Takumi, Y.; Okuzono, T.; Sakagami, K. Locally implicit time-domain finite element method for sound field analysis including permeable membrane sound absorbers. *Acoust. Sci. Technol.* **2020**, *41*, 689–692.
60. Cheng, Y.; Cheng, L.; Pan, J. Absorption of oblique incidence sound by a finite micro-perforated panel absorber. *J. Acoust. Soc. Am.* **2013**, *133*, 201–209.
61. Bilbao, S.; Hamilton, B.; Botts, J.; Savioja, L. Finite volume time domain room acoustics simulation under general impedance boundary conditions. In *IEEE-ACM Transactions on Audio, Speech, and Language Processing*; IEEE: Piscataway, NJ, USA, **2016**; Volume 24, pp. 161–173.
62. Hamilton, B.; Webb, C.J.; Fletcher, N.D.; Bilbao, S. Finite difference room acoustics simulation with general impedance boundaries and viscothermal losses in air: Parallel implementation on multiple GPUs. In Proceedings of the Int. Symp. Musical Room Acoust. La Plata, Argentina, 11–13 September 2016.
63. D'Orazio, D.; Fratoni, G.; Rovigatti, A.; Hamilton, B. Numerical simulations of Italian opera houses using geometrical and wave-based acoustics methods. In Proceedings of the 23rd International Congress on Acoustics, Aachen, Germany, 9–13 September 2019.
64. Vorländer, M. Computer simulations in room acoustics: Concepts and uncertainties. *J. Acoust. Soc. Am.* **2013**, *133*, 1203–1213. [CrossRef] [PubMed]
65. Okuzono, T.; Sakagami, K. A frequency domain finite element solver for acoustic simulations of 3D rooms with microperforated panel absorbers. *Appl. Acoust.* **2018**, *129*, 1–12. [CrossRef]
66. Okuzono, T.; Sakagami, K.A finite-element formulation for room acoustics simulation with microperforated panel sound absorbing structures: Verification with electro-acoustical equivalent circuit theory and wave theory. *Appl. Acoust.* **2015**, *95*, 20–26. [CrossRef]
67. Toyoda, M.; Eto, D. Prediction of microperforated panel absorbers using the finite-difference time-domain method. *Wave Motion* **2019**, *86*, 110–124. [CrossRef]
68. Liu, J.; Herrin, D.W. Enhancing micro-perforated panel attenuation by partitioning the adjoining cavity. *Appl. Acoust.* **2010**, *71*, 120–127. [CrossRef]
69. Naderyan, V.; Raspet, R.; Hickey, C.J.; Mohammadi, M. Acoustic end corrections for micro-perforated plates. *J. Acoust. Soc. Am.* **2019**, *146*, EL399–EL404. [CrossRef]

Article

Mechanism Analysis of the Influence of Seat Attributes on the Seat Dip Effect in Music Halls

Hequn Min * and Yitian Liao

Key Laboratory of Urban and Architectural Heritage Conservation, Ministry of Education, School of Architecture, Southeast University, 2 Sipailou, Nanjing 210096, China; 220180134@seu.edu.cn
* Correspondence: hqmin@seu.edu.cn

Abstract: The seat dip effect (SDE) is an acoustic phenomenon of low-frequency band attenuation that occurs in the music halls when the sound of the music passes at a near grazing incidence over the seats. In this paper, the numerical simulations on the basis of the finite element method are conducted to study the influence of seat attributes (seat height, seat spacing and seat absorption) on the SDE and the corresponding mechanism. The mapping of sound spatial distribution related to the SDE is employed to observe the behavior of sound between the seats. The results show that the dip frequency of the SDE can be shifted to frequencies lower than theoretical values when the seat height is smaller than the seat spacing. Additionally, the SDE attenuation can be distinctly suppressed in a sequence from the front seats to the rear seats with an absorption improvement to the seat back or cushion, and the seat back absorption is more effective than the cushion absorption. A mechanism analysis reveals that the SDE is highly associated with standing waves inside the seat gaps and with the "diffusion" effect on the grazing incident waves by energy flow vortexes around the top surfaces of the seats.

Keywords: seat dip effect; seat height; seat spacing; sound absorption; mechanism

1. Introduction

In music halls, when sound passes at a near grazing incidence over the seats, there is a phenomenon of excessive attenuation in the low-frequency band. Different from the sound absorption attenuation that usually occurs at high frequencies caused by sound being absorbed by the audience and the surfaces of the seats, this attenuation phenomenon, known as the seat dip effect (SDE), occurs at about 100 Hz and seems typical in music halls despite the difference made to the sound by the audience.

The effect was first discovered by two studies through investigations in concert halls and scale models [1,2], early in 1964. Subsequently, in situ or scale-model measurements [1–7], and theoretical calculations [8] were extensively conducted to analyze the SDE. It was shown that the SDE caused the excessive attenuation of sound at low frequencies within the hall, which significantly affected spatial impression [9], clarity [3], and timbre [4,10] by affecting the early reflection of the sound, and people's perception of bass frequencies was significantly stronger in concert halls with higher SDE attenuation frequencies [11]. Recently, computer simulations were applied to study the SDE with the wave-based geometrical acoustics method, the finite element method (FEM), the boundary element method (BEM) [12] and the finite-difference time-domain method [13]. The parameters with a possible influence on the SDE were investigated, such as the floor inclination [5], sound incident angle [4,14], and ceiling height [4,7]. Based on these investigations, some guidelines were proposed to reduce the SDE, such as: (1) increasing the ceiling reflection or the side reflections [15,16]; (2) installing ground sound absorbers between the seat rows [3,8]; and (3) sloping the floor or the stage [5,14], etc.

However, for the SDE suppression, choosing the appropriate seat attributes of height, spacing and absorption is a more implementable way than raising the floor slope or

adding reflectors, and seat attribute adjustment can be implemented easily in renovations which affect the hall construction. Regarding the influence of seat attributes on the SDE and possible mechanism, early studies [1,2] showed that the frequency of the maximum attenuation mainly depended on the height of the seat. Later, Ishida's measurements [17] investigated the influence on the SDE from the seat underpass and observed that the dip frequency increased for higher source positions when the height of the seat underpass decreased. Measurements and practical calculations suggested that the SDE was affected by the diffraction on the seat top [3] and by the reflected sound on the auditorium seats [14]. Davies and Cox [12] compared the SDE attenuation with different seat absorptions and seat underpass shapes using the BEM simulation and a scale model and reported that changing the seat absorption had an audible influence on reducing the SDE in concert halls. Tahvanainen et al. [18] conducted scale-model experiments to study the effect of the seat underpass and floor inclination on the SDE, and found that the dip frequency depended on the seat back height and on the obstruction degree of the seat underpass. However, little investigation and discussion on the possible mechanism was conducted in [12,18]. Takahashi [7] investigated the SDE phenomena via scale model measurements and derived a simple analytical model for attempting to explain the SDE mechanism. In his study, the SDE attenuations with different seat back absorptions were investigated and it was found that the seat backs affected the SDE attenuation and dip frequency shift. However, it was difficult for Takahashi to further explore the mechanism due to the limited simulation techniques of that time and only assumption was made that the SDE was highly related to the interference between a direct wave and reflected waves from the seats. At present, sound wave behaviors around the seats regarding the mechanisms of influence from the seat attributes of height, spacing, and absorption on the SDE are not yet clear in the literature; however, this influence is important for the reduction in the SDE by choosing appropriate seat attributes.

Given the literature review above, there is a distinct lack of detailed investigation and mechanism discussion focused on the influence of seat attributes on the SDE, especially from the analysis of the corresponding sound wave behavior around the seats. In this paper, the quantitative attenuation and mapping of sound spatial distribution related to the SDE is evaluated in simulations with the FEM to study the influence of seat height, seat spacing and seat absorption on the SDE, as well as the corresponding mechanism.

2. Simulation

From previous studies [7,15], the reflections from walls and ceilings in a room can mask the SDE phenomenon, and the SDE is substantially related to the sound reflections from the seats and the floor. To simplify the problem and highlight the SDE phenomenon, a two-dimensional geometry with simplified acoustic boundary conditions, shown in Figure 1, was established to represent the section plane of classical music halls in the present study. The simulations were carried out with the acoustic FEM in the frequency domain and were implemented with the commercial software, COMSOL Multiphysics® [19].

In the numerical model geometry in Figure 1a, the room is defined as 24.5 m long and 4.5 m high with four boundaries (ground, ceiling, front wall, and back wall), containing 16 rows of simplified seats with the height h and the spacing w (representing the seat pitch in actual halls). A point source S is 4.5 m away from the front wall, at a height of 2.5 m above the ground, and radiates with a strength of 1 W/m². The receiving points R1-R6 are in the middle of the seat gap, at a height of 1.1 m above the ground. To simplify the problem, we assumed the ground to be perfectly reflecting with infinite impedance, and assumed the front wall, back wall and ceiling to be perfectly absorbing without sound reflections from these boundaries (which were implemented as Perfectly Matched Layers (PML) in numerical models) ensuring that the reflections were only from the ground and the seats to emphasize the SDE. The seats were simplified to vertical slats in the model. This configuration was used in previous scale models [2,4] and calculation models [7], while the seat sound absorption could not be considered in the former plywood models and the sound field distribution and acoustic behavior between the seat gaps was not observed in the latter calculation model. In the numerical models in this paper, the seat back or the floor can be defined as rigid or with a different acoustic impedance. For the investigations of the seat height and seat spacing, the seats were set as rigid boundaries to separate the influence from the seat sound absorption. For investigations on the seat-back sound absorption, the seats were set with different acoustic impedances. It should be noted that the influence from the seat cushion sound absorption on the SDE was investigated by applying different impedances on the ground between the seats in the simplified model. In the present study, the surface sound absorption was illustrated by the normal incidence sound absorption coefficient α (seat back absorption by α_b and the seat cushion absorption by α_c).

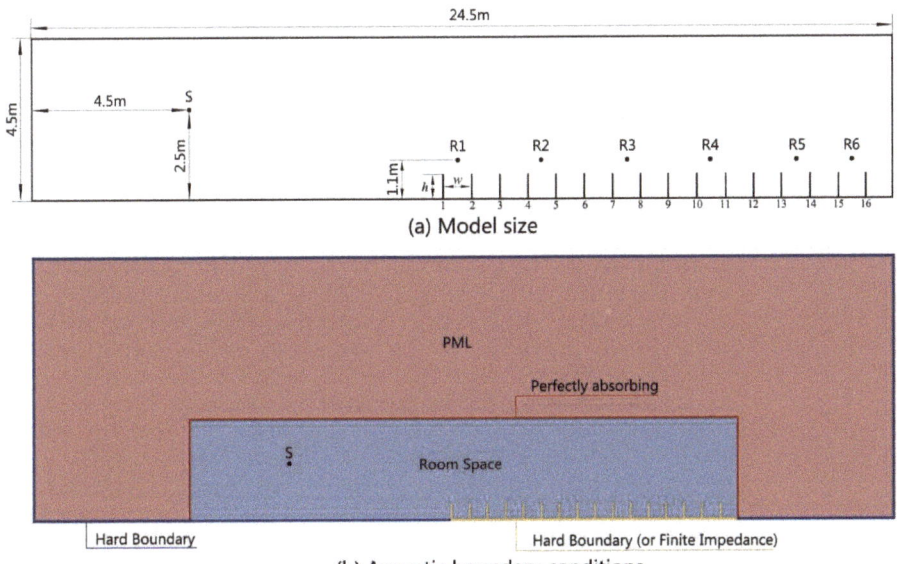

Figure 1. Setup of the simulation model: (**a**) geometry and (**b**) acoustic boundary conditions.

A quantity of Relative Level (*RL*) was applied for the investigations on the characteristics of the SDE, which are defined as:

$$RL = SPL_2 - SPL_1 \qquad (1)$$

where SPL_2 and SPL_1 are the Sound Pressure Levels (*SPLs*) at each receiver when there are seats and no seats, respectively.

At first, the numerical model was compared with the measurements of a 1/10 scale-model from Sessler and West [2] for validation. In the validation case, the geometry dimensions in the simulation were set to equal values of those in the scale-model (as shown in Figure 9 in Ref. [2]) and the sound absorption arrangement of the seats and floor in the simulation was set according to the scale-model conditions [2,8], where the stage and the ground were set as rigid surfaces, and the seats were set as absorptive surfaces with a specific acoustic admittance of 0.01. The *RL* results of the simulation and measurements at a receiver of 1.6 m from the sound source in this validation case were compared in Figure 2. A good agreement can be observed between the simulation and measurements, except that in the measurements the *RL* dip was not identified as clearly as it was in the simulation due to the limited measurement data. This shows the validity of numerical models for the further SDE investigations in this paper.

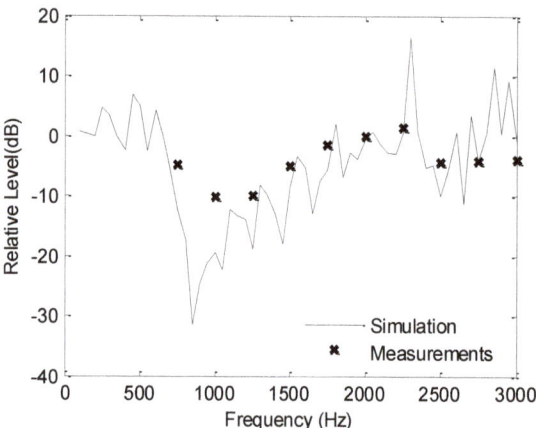

Figure 2. Comparison between the *RL* spectra of simulation and measurements in the validation case. X-marker: measurements in the 1/10 scale model (after Sessler and West [2]). Solid line: simulation with the geometry and boundary conditions in the scale model.

Figure 3a,b show the initial results of the *RL* spectra and SPL spatial distribution associated with the SDE when initially $h = 0.7$ m, $w = 0.8$ m, and the seats and ground are set as rigid boundaries in simulations. In Figure 3a, the dip minimum of the *RL* is close to -30 dB at the dip frequency of 105 Hz, indicating an SDE phenomenon which is the same as the previous observations [1,2]. In Figure 3b, the color map of SPL spatial distribution at the dip frequency of 105 Hz shows that the SDE phenomenon appears around the top opening of seat gaps (emphasized in dotted boxes) with distinctly low SPLs, which should be the focus area on the SDE in acoustic design. In the previous studies mentioned above, the SDE was usually analyzed only through *RL* spectra similar to those shown in Figure 3a. In this study, the color maps similar to Figure 3b at interesting frequencies were applied to observe the two-dimensional SPL attenuation associated with the SDE for further mechanism analysis. In the following, the relationships between three factors (seat height h, seat spacing w, and the seat sound absorption coefficient α) and the SDE are investigated through the dip frequency f_0, the dip minimum RL_0 and the SPL spatial distribution for a further detailed analysis of the SDE mechanism, and for the possible elimination methods for the SDE in the acoustic design stage.

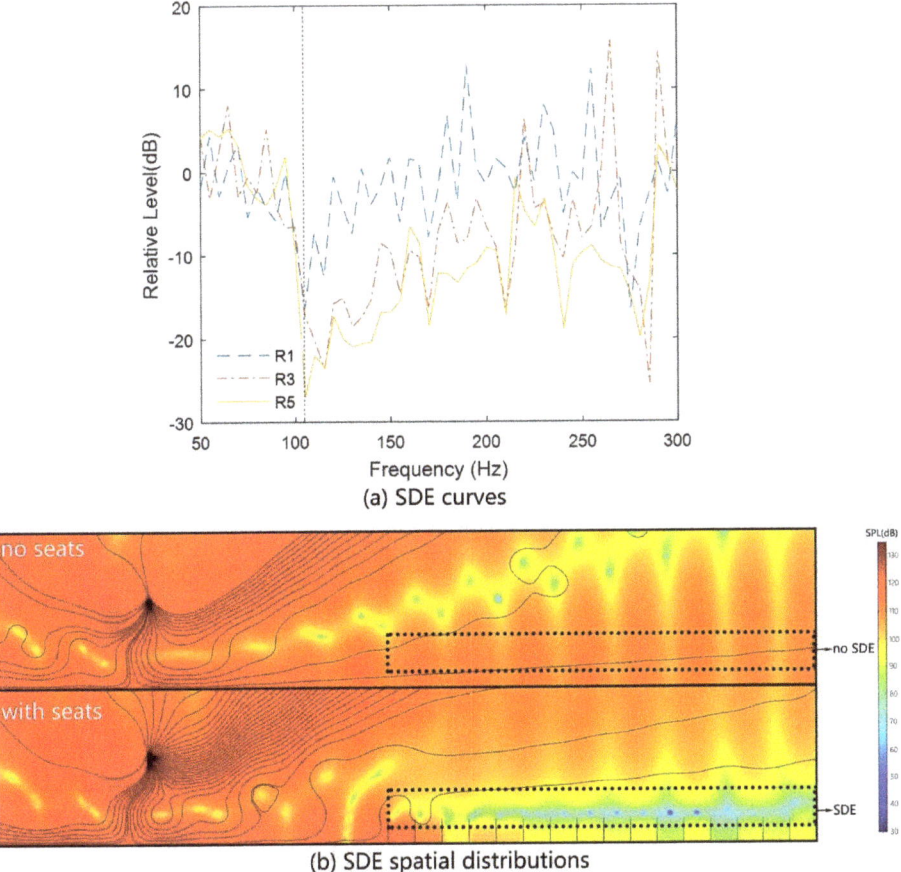

Figure 3. Illustration of the SDE: (**a**) relative level spectra; (**b**) color maps of the SPL spatial distributions at a dip frequency f_0 of 105 Hz. Seats and floor are rigid.

3. Results and Discussion

3.1. Seat Height and Seat Spacing

In previous classical studies [1,2], when $h \approx \lambda/4$, where λ represents the sound wavelength at a specific frequency, the SPLs at the receivers well above the surface of the seat top were measured to be lowest and the corresponding specific frequency was just the dip frequency f_0. This was caused by the classical interference between the grazing direct sound and the vertically reflected sound waves inside the seat gaps near the surface of the seat top. In the first numerical case, investigations were carried out with different seat heights with a fixed seat spacing of 0.8 m to check the SDE phenomenon. Figure 4a presents the corresponding results of the average RL at receivers R4–R6, where the SDE phenomenon was more distinct. When h is 1.1 m, 0.9 m, 0.7 m and 0.5 m, the theoretical f_0 should be 80 Hz, 95 Hz, 120 Hz and 170 Hz, respectively, according to the quarter wavelength theory above. For comparisons, as shown in Figure 4a, the corresponding f_0 in the simulations was about 80 Hz, 90 Hz, 110 Hz and 130 Hz, respectively, and was close to the theoretical values but had a noticeable shift to lower frequencies when the seat height h became smaller than the seat spacing w.

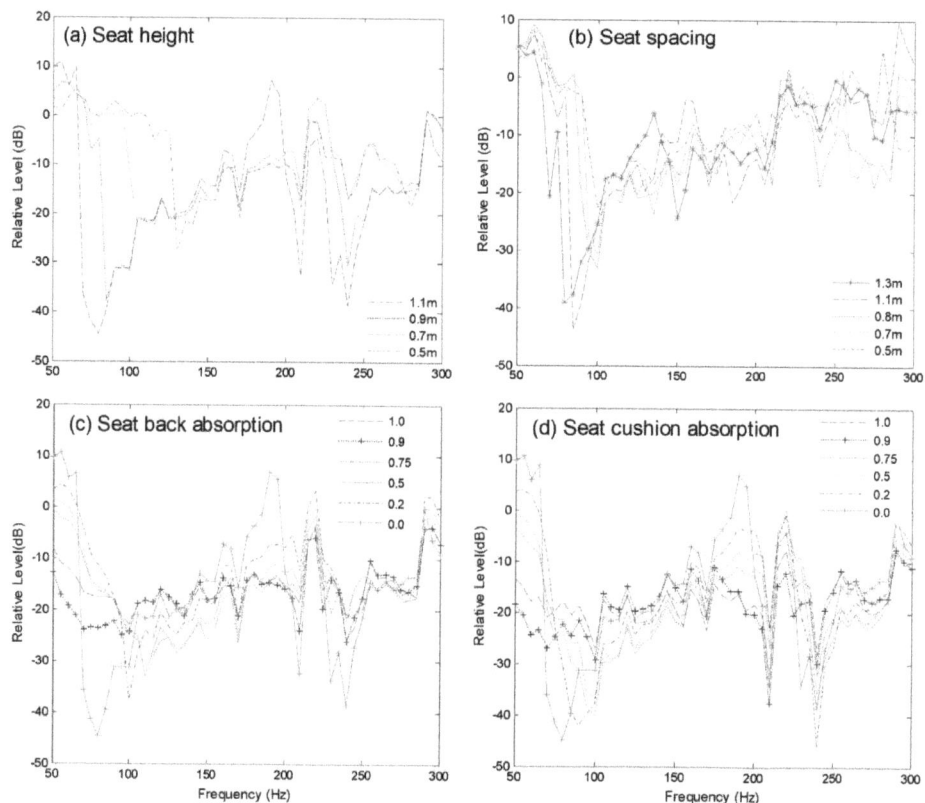

Figure 4. Spectra of average RL at receivers R4-R6 for investigation of influence on the SDE from (**a**) seat heights (w = 0.8 m, seats and floor are rigid); (**b**) seat spacing (h = 0.8 m, seats and floor are rigid); (**c**) seat back absorption (h = 1.1 m and w = 0.8 m) and (**d**) seat cushion absorption (h = 1.1 m and w = 0.8 m).

Another numerical case was conducted on the SDE phenomenon at a different seat spacing w with a fixed seat height of 0.8 m, results of which are presented in Figure 4b. It was shown that, when w was 0.5 m, 0.7 m and 0.8 m, the simulated f_0 was almost the same with a theoretical value of 105 Hz for the seat height h of 0.8 m, but when w was 1.1 m or 1.3 m, and thus larger than the seat height, a noticeable shift to a lower frequency of 85 Hz or 80 Hz was observed, respectively, as well as those in the first numerical case above. This phenomenon indicates that, when the seat height h becomes smaller than the seat spacing w, the dip frequency f_0 may be shifted to lower frequencies compared to values from the quarter wavelength theory [1,2]. This may be caused by the interference between the horizontal reflections of the oblique direct sound on the seat backs. As the seat height becomes smaller or the seat spacing larger, the oblique direct sound from the source that can travel into the seat gaps becomes stronger, and leads to stronger horizontal reflections on the seat backs, and thus a stronger interference between those reflections, whose interference frequency is controlled by the seat spacing w. When h is smaller than w, the interference between the horizontal reflections becomes stronger than the interference between the grazing direct sound and vertical reflections, where the frequency of the former interference is clearly lower than that of the latter interference. This leads to the observed lower frequency shift of dip frequency f_0 in this situation, and can also explain the RL_0 decreasing with w increasing, as shown in Figure 4b.

3.2. Seat Absorption

In the numerical cases above, the SDE phenomenon is investigated with rigid seats and a rigid floor. It is interesting to investigate what occurs when the seats become sound absorbing. Davies and Cox [12] conducted a rough observation on the SDE with several seat absorption coefficients but did not analyze in detail the corresponding strength and characteristic changes of the SDE. In this section, numerical cases are carried out for the SDE with a different seat back absorption coefficient α_b and seat cushion absorption coefficient α_c where $h = 1.1$ m and $w = 0.8$ m. The corresponding RL spectrum results are presented in Figure 4c,d, respectively. It can be observed that, through increasing the seat back or cushion absorption coefficient from 0.0 to 1.0, the dip minimum RL_0 gradually changes from -46 dB to around -23 dB, demonstrating that the SDE attenuation is remarkably weakened by the seat absorption. It is also shown that, the dip frequency f_0 may increase to higher frequencies with a higher seat absorption, especially at the seat back. A similar phenomenon was observed by Takahashi [7], whose complete mechanism is unclear yet and can be related to sound distribution around the vertical slats that are assumed as simplified geometry of seat backs [7].

For a direct visual inspection on the influence of seat absorption, the color maps of the SPL spatial distribution at the corresponding dip frequencies of the SDE with different α_b or α_c, as shown in Figure 4c,d, are presented in Figure 5. Figure 5a,b show the situations of seat back absorption and seat cushion absorption, respectively. In the color maps, the color represents the SPL, and the streamline is derived from the gradient of the calculated sound intensity. As shown in Figure 5, there is no SDE when no seat is in the room. When the rigid seats are set, there is a distinct area of the SDE concentrated at almost the entire surface of the seat tops, with a continuous attenuation of sound energy, and the SDE area becomes larger at seat locations farther away from the source. In Figure 5a, as α_b increases, the concentration strength of the sound energy attenuation is gradually weakened and the area of the sound energy attenuation gradually shifts backwards from the front seats. This shows that the diminishing tendency of the SDE by seat absorption is from the front seats to the rear seats. Therefore, the SDE attenuation investigated at the R4-R6 receiving points (whose positions are marked by black stars in each color map) decreases as α_b increases from 0.0 to 0.9, as shown in Figure 4c. Meanwhile, the streamlines in Figure 5a also gradually extend into the seat gaps, showing that more sound energy flows into the seating area with a higher seat absorption. When $\alpha_b = 1.0$, the SDE attenuation around the surface of the seat tops becomes much weaker in comparison to the situation with no seats.

Figure 5b presents a similar diminishing tendency of the SDE by seat cushion absorption, but the corresponding SDE reduction is observed to not be as distinct as the reduction by the seat back absorption shown in Figure 5a. Through a cross comparison between the results in Figure 5a,b, it is shown that the appropriate sound absorption treatment of the seat back or cushion can play a certain role in suppressing the SDE in music halls, and the seat back absorption is more effective than the seat cushion absorption. From other extensive simulation cases, when h is not larger than w, similar results (not presented here for conciseness) also show a greater beneficial effect on the SDE from the seat back absorption rather than from the seat cushion absorption, and this observation is consistent with discussion presented in Ref. [12]. The above results also suggest that common sound absorption does not have a distinct SDE reduction effect in rear seat areas, but the high sound absorption treatment does. For real applications, it is thought that a high sound absorption treatment at low frequencies on seats could be achieved by filling the appropriate metamaterial absorbers [20,21] into the seat back or cushion. Additionally, in real music halls with audience, it is also suggested that the occupied seats with equivalent absorption should have suppression effects on the SDE. However, for the influence of seat occupancy on the SDE, further study is needed for a confirmation with human models on the seats and will be reported in the future.

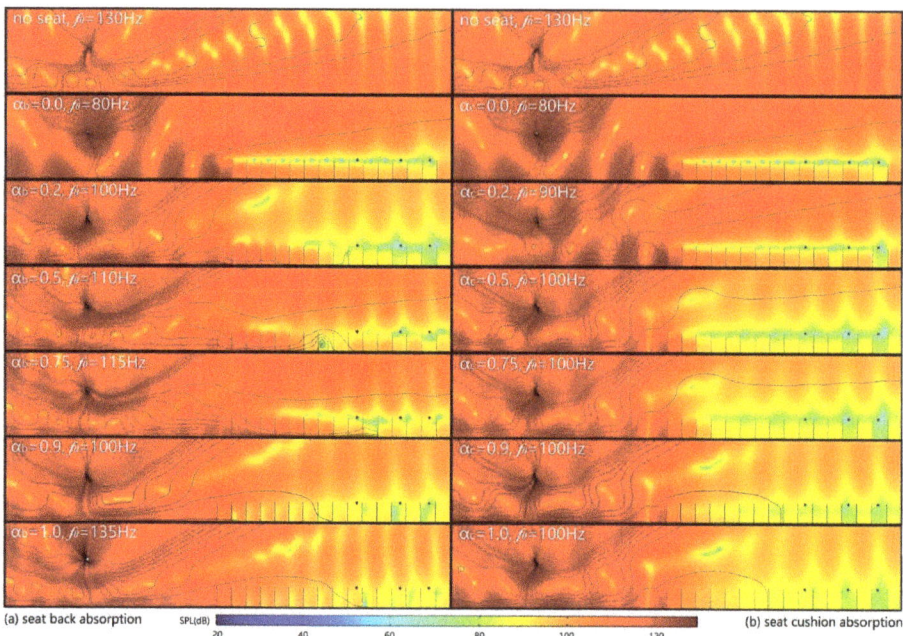

Figure 5. Comparison of SPL distributions at corresponding different dip frequencies with different sound absorption coefficients of seat backs (**a**): no seat; $\alpha_b = 0.0$; $\alpha_b = 0.2$; $\alpha_b = 0.5$; $\alpha_b = 0.75$; $\alpha_b = 0.9$; $\alpha_b = 1.0$; and of seat cushions (**b**): no seat; $\alpha_c = 0.0$; $\alpha_c = 0.2$; $\alpha_c = 0.5$; $\alpha_c = 0.75$; $\alpha_b = 0.9$; $\alpha_c = 1.0$.

3.3. Possible Mechanism

To clarify the influence mechanism of the seat attributes on the SDE mentioned above, two sets of detailed investigations on sound wave behavior between the seats were conducted in simulations on the possible mechanism related to the SDE. For the first investigation set, the color map results of the SPL and sound intensity (SI) distribution at different frequencies inside the rigid seat gaps are presented in Figure 6a–c where $h = 0.8$ m, $w = 1.1$ m and in Figure 6d–f where $h = 1.1$ m, $w = 0.8$ m. The color represents the SPL, and the arrows represent the SI magnitude and direction. In Figure 6a–f, the first phenomenon is that the SI magnitudes gradually decrease from the outside to the inside of the seat gaps, and the SI directions rotate clockwise between the seats and create energy flow vortexes showing the presence of standing waves, as reported by Schultz [1]. The lowest SPL position always appears at the gap center with the height of $\lambda/4$, which is the center point of the energy flow vortex shown in Figure 6a–e with heights of 0.81 m, 0.65 m, and 0.57 m at the frequencies of 105 Hz, 130 Hz, 150 Hz, respectively. It is worth noting that Figure 6f features the two lowest SPL positions at the gap center with heights of 0.35 m and 1.06 m, which are equal to the $\lambda/4$ and $3\lambda/4$ of the investigated frequency, 240 Hz, respectively. This phenomenon of the lowest SPL positions inside the seat gaps located at the gap center, with the height $((2k + 1)\,\lambda)/4$ ($k = 0, 1, 2...$), shows the first influence mechanism on the SDE from seat height and spacing. The second phenomenon is that, when the energy flow vortexes appear near the top surface of the seats, for example the vortexes near the slats' upper ends, as shown in Figure 6a,b,f, the grazing incident waves are observed to be "diffused" by the energy flow vortexes and a continuous attenuation can be created around the top surface of the seats due to this complex interference. This can be another influence mechanism on the SDE from seat height and spacing. These phenomena come not only from the interference between the grazing direct sound and the vertically reflected sound waves inside the seat gaps, but also from those between the oblique sound incidence and horizontal sound reflections inside the seat gaps.

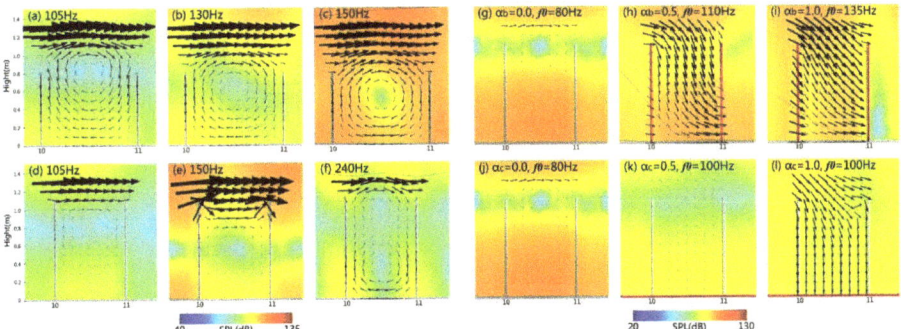

Figure 6. The SPL and SI distribution inside seat gaps: (**a**–**c**) seats are rigid, h = 0.8 m, and w = 1.1 m; (**d**–**f**) seats are rigid, h = 1.1 m, and w = 0.8 m; (**g**–**i**) α_c = 0.0, h = 1.1 m, and w = 0.8 m; (**j**–**l**) α_b = 0.0, h = 1.1 m, and w = 0.8 m.

For the second investigation set, the color map results of the SPL and SI distribution at the corresponding dip frequencies are presented in Figure 6g–l, where α_c = 0.0, h = 1.1 m and w = 0.8 m for the influence of the seat back absorption and Figure 6j–l, where α_b = 0.0, h = 1.1 m, and w = 0.8 m for the influence of the seat cushion absorption. The absorptive seat boundaries (backs or cushions) are marked with thick red lines, while the rigid ones are marked with thick grey lines. It is shown that seat absorption has a clear guiding effect on the SI directions, where sound energy flows are observed to be guided by the absorption to become horizontal into seat backs, in Figure 6g–l, or to become vertical into seat cushions, in Figure 6j–l. Additionally, inside the seat gaps, the horizontal reflections are distinctly reduced by the seat back absorption, as shown in Figure 6h–i, and vertical reflections are distinctly reduced by the seat cushion absorption, as shown in Figure 6k–l. The wave guiding effect and standing wave suppression from the seat absorption remarkably level out the energy distribution and weaken the energy flow vortexes due to the complex interference inside the aforementioned seat gaps, and then reasonably suppress the SDE. This could be the influence mechanism on the SDE from the seat back or cushion absorption. Moreover, since the SDE is governed by the interference among the grazing direct sound and vertical reflection when h is larger than w, as discussed in Section 3.1, from this mechanism it is explicable that the seat back absorption is more effective than the cushion absorption on the SDE reduction in this situation, as shown in Figure 5. The reason for this could be that the seat back absorption guides many vertical energy flows to become horizontal and then effectively weakens the interference governing the SDE in this situation.

4. Conclusions

In this paper, the influence of seat attributes (seat height, spacing, and seat absorption) on the SDE in music halls, as well as its possible mechanism, are studied in FEM simulations with a simplified geometry. The mapping of the SPL and SI distribution is employed to observe the spatial characteristics of the SDE for analysis. The results show that, when the seat height is smaller than the seat spacing, there is a noticeable shift of the dip frequency from the theoretical values to the lower ones. It is also shown that, with the absorption improvement of the seat back or cushion, the SDE attenuation can be suppressed in sequence from the front seats to the rear seats, and the seat back absorption is more effective than the cushion absorption. The mechanism analysis, by observing in detail the wave behavior between the seats, reveals that the standing waves inside the seat gaps act as the main cause of the SDE and the "diffusion" effect on the grazing incident waves by the energy flow vortexes around the top surface of the seats may act as another cause. The high sound absorption treatment of the seats can distinctly weaken the SDE phenomenon for their effective use of reducing the standing waves. In the future work, human models will be added to further study the influence of seat occupancy on the SDE. The work in this paper can provide useful insights and suggestions for the better understanding and suppression of the SDE in actual music halls.

Author Contributions: Conceptualization, H.M.; methodology, H.M. and Y.L.; software, H.M. and Y.L.; validation, H.M. and Y.L.; formal analysis, H.M.; investigation, H.M. and Y.L.; resources, H.M.; data curation, H.M.; writing—original draft preparation, H.M. and Y.L.; writing—review and editing, H.M.; visualization, Y.L.; supervision, H.M.; project administration, H.M.; funding acquisition, H.M. All authors have read and agreed to the published version of the manuscript.

Funding: This research was funded by the Natural Science Foundation of China, grant number 51408113 and the Natural Science Foundation of Jiangsu Province, China, grant number BK20140632.

Institutional Review Board Statement: Not applicable.

Informed Consent Statement: Not applicable.

Data Availability Statement: Not applicable.

Conflicts of Interest: The authors declare no conflict of interest. The funders had no role in the design of the study; in the collection, analyses, or interpretation of data; in the writing of the manuscript, or in the decision to publish the results.

References

1. Schultz, T.J.; Watters, B.G. Propagation of Sound across Audience Seating. *J. Acoust. Soc. Am.* **1964**, *36*, 885–896. [CrossRef]
2. Sessler, G.M.; West, J.E. Sound transmission over theatre seats. *J. Acoust. Soc. Am.* **1964**, *36*, 1725–1732. [CrossRef]
3. Davies, W.; Lam, Y. New attributes of seat dip attenuation. *Appl. Acoust.* **1994**, *41*, 1–23. [CrossRef]
4. Bradley, J.S. Some further investigations of the seat dip effect. *J. Acoust. Soc. Am.* **1991**, *90*, 324. [CrossRef]
5. West, J.E.; Sessler, G.M. Model Study of the Sound Transmission over Raked Theatre Seats. *J. Acoust. Soc. Am.* **1966**, *40*, 1246. [CrossRef]
6. Greenberg, D. Seat-dip phenomenon. *J. Acoust. Soc. Am.* **1994**, *96*, 3267. [CrossRef]
7. Takahashi, D. Seat dip effect: The phenomena and the mechanism. *J. Acoust. Soc. Am.* **1997**, *102*, 1326–1334. [CrossRef]
8. Ando, Y.; Takaishi, M.; Tada, K. Calculation of the sound transmission over theater seats and methods for its improvement in the low-frequency range. *J. Acoust. Soc. Am.* **1982**, *72*, 443–448. [CrossRef]
9. Barron, M.; Marshall, A. Spatial impression due to early lateral reflections in concert halls: The derivation of a physical measure. *J. Sound Vib.* **1981**, *77*, 211–232. [CrossRef]
10. Davies, W.J.; Cox, T.J.; Lam, Y.W. Subjective Perception of Seat-dip Attenuation. *Acta Acust. United Acust.* **1996**, *82*, 784–792.
11. Tahvanainen, H.; Pätynen, J.; Lokki, T. Studies on the perception of bass in four concert halls. *Psychomusicology Music. Mind Brain* **2015**, *25*, 294–305. [CrossRef]
12. Davies, W.J.; Cox, T.J. Response to "Comment on 'Reducing seat dip attenuation". *J. Acoust. Soc. Am.* **2001**, *110*, 1261–1262. [CrossRef]
13. LoVetri, J.; Mardare, D.; Soulodre, G. Modeling of the Seat-dip Effect Using the Finite-difference Time-domain Method. *J. Acoust. Soc. Am.* **1996**, *100*, 2204–2212. [CrossRef]
14. Sakurai, Y.; Morimoto, H.; Ishida, K. The reflection of sound transmission at grazing angles by auditorium seats. *Appl. Acoust.* **1993**, *39*, 209–227. [CrossRef]
15. Pätynen, J.; Tervo, S.; Lokki, T. Analysis of concert hall acoustics via visualisations of time-frequency and spatiotemporal responses. *J. Acoust. Soc. Am.* **2013**, *133*, 842–857. [CrossRef]
16. Tahvanainen, H.; Haapaniemi, A.; Lokki, T. Perceptual significance of seat-dip effect related direct sound coloration in concert halls. *J. Acoust. Soc. Am.* **2017**, *141*, 1560–1570. [CrossRef] [PubMed]
17. Ishida, K. The Measurement and Prediction of Sound Transmission over Auditorium Seats. Ph.D. Thesis, University of Cambridge, Cambridge, UK, 1993.
18. Tahvanainen, H.; Lokki, T.; Jang, H.-S.; Jeon, J.-Y. Investigating the influence of seating area design and enclosure on the seat-dip effect using scale model measurements. *Acta Acust.* **2020**, *4*, 15. [CrossRef]
19. COMSOL. Available online: cn.comsol.com (accessed on 3 May 2021).
20. Min, H.; Guo, W. Sound absorbers with a micro-perforated panel backed by an array of parallel-arranged sub-cavities at different depths. *Appl. Acoust.* **2019**, *149*, 123–128. [CrossRef]
21. Liu, C.R.; Wu, J.H.; Ma, F.; Chen, X.; Yang, Z. A thin multi-order Helmholtz metamaterial with perfect broadband acoustic absorption. *Appl. Phys. Express* **2019**, *12*, 084002. [CrossRef]

Article

A Parametric Study of the Acoustic Performance of Resonant Absorbers Made of Micro-perforated Membranes and Perforated Panels

Lili Pan [1] and Francesco Martellotta [2],*

[1] School of Architecture, South China University of Technology, Guangzhou 510640, China; panlilychinese@outlook.com
[2] Dipartimento di Scienze dell'Ingegneria Civile e dell'Architettura, Politecnico di Bari, Via Orabona 4, 70125 Bari, Italy
* Correspondence: francesco.martellotta@poliba.it; Tel.: +39-080-596-3631

Received: 27 January 2020; Accepted: 24 February 2020; Published: 26 February 2020

Abstract: Sound absorbing surfaces are being increasingly requested for the acoustical treatment of spaces, like offices and restaurants, where high aesthetic standards are requested. In these cases, perforated and micro-perforated panels may represent the ideal solution in terms of low maintenance, durability, and mechanical resistance. In addition, such a solution might be conveniently realized while using optically transparent panels, which might offer extra value, as they could ensure visual contact, while remaining neutral in terms of design. The paper first investigates the reliability of prediction models by comparison with measured data. Subsequently, while taking advantage of a parametric optimization algorithm, it is shown how to design an absorber covering three octave bands, from 500 Hz to 2 kHz, with an average sound absorption coefficient of about 0.8.

Keywords: sound absorption; perforated panels; micro-perforated panels; resonant absorbers

1. Introduction

Investigating noise annoyance problems in places where users are the main sound source and spaces are strictly confined, like open-plan offices [1], restaurants [2–4], and call centers and markets [5], has become a frequent research topic in recent years. In these cases, adding sound absorbing treatments proved to be an easy and efficient method for limiting the problem of high sound pressure level and controlling the speech intelligibility [6]. Among the various sound absorbing materials that are available for the purpose, with different finishing and different acoustic behavior, aesthetic factors combined with ease of maintenance often play a major role in the selection process [7]. Perforated and micro-perforated panels, when compared to more conventional porous absorbers, may offer significant advantages in terms of hygiene, fire-proofing, and durability. Moreover, their acoustic properties could be precisely targeted, achieving a strong absorption at specified frequency band, while being optimized and predicted by computational methods [8–10]. In addition, by using proper materials, such absorbers might easily become eco-friendly, which is an ever increasing requirement among sound absorbers [11,12].

In practical applications, perforated and micro-perforated panels are normally placed in front of a rigid surface with an air cavity between them, forming a series of parallel Helmholtz resonant cells. The sound is absorbed by the viscous loss and specific acoustic impedance of the air in holes and cavity. Accordingly, their sound absorption coefficients are dominated by the diameter and depth of the hole, the perforation rate, and the cavity thickness [10]. After decades of development, the fundamental laws underlying the design of perforated and micro-perforated panel absorbers have been clarified [8–10,13,14] and several proposed prediction models have been validated against measurements [15–19]. For the predictions of homogeneously perforated absorbers, the theoretical models that were proposed by Maa [16], Zwikker, and Kosten [17], are better used to model absorbers with circular perforations, while for other cross-sectional perforations, the models of Stinson [18] or Atalla and Sgard [19] can be used instead. Maa and other researchers have also developed a method to model multi-layer perforated absorbers, while using a transfer matrix solution, taking each layer in turn [8,9,20]. Finite Element Method (FEM) [21,22], Equivalent Circuit Method (ECM) [23], Parallel Transfer Matrix Method (PTMM) [24,25], and Admittance Sum Method (ASM) [26] are other computing methods used to model heterogeneous perforated absorbers. Carbajo et al. [27] explored the adequacy of these four different methodologies in the case of heterogeneous absorbers with isolated or shared cavities, and found that only ECM and FEM yield correct results for both conditions.

Depending on the resonance system, single-leaf perforated panel absorbers always have a limited frequency band, showing only one sharp-narrow absorption peak. Many researchers proposed the use of more complex perforated absorbers, including multi-layer systems [28], three-dimensional (3D) micro-perforated panels [29], combinations of micro-perforated panels with membranes [30,31], arranging the parallel micro-perforated panel absorbers with different cavity depths [32], or combining perforated panels with micro-perforated partitions [33], to achieve a wider absorption band. Sakagami and colleagues studied the theory behind the acoustical property of multi-leaf membranes [34], double-leaf micro-perforated panel [35], their combinations [26,31], and the combination with porous materials [36]. Ayub et al. [37] investigated the sound absorption coefficient of multiple perforated panel systems that were composed of coir fiber and one air gap. The results showed that using such combinations of multi-layer perforated panels and coir fiber could further enhance the sound absorption coefficient in a wider frequency range. Similarly, Shen et al. [38] demonstrated that combining microperforated panels with porous metal could further extend the frequency response of the absorber towards the low frequencies.

However, when investigating the possibility of using those perforated absorbers in modern spaces, like restaurants, open offices, or call centers, the sound absorption material needs to fulfill both acoustic and aesthetic requirements [7]. In this situation, transparent panels may be an optimal choice, because this allows for better visual contact between occupants and is architecturally less intrusive, as demonstrated by the several commercial solutions currently available. Meanwhile, a number of other limitations may be observed: thickness, restricted by space availability, and materials, as to preserve its transparency a resistive layer could not be used in this device. Consequently, a significant contribution may be given by micro-perforation installed behind the perforated panel in order to improve the acoustic performance despite such limitations. In this way, the reduced opening dimension generates more viscous losses inside individual holes. Thus, it makes achieving absorption over a wider frequency band while using a thinner air cavity possible. Therefore, the sound absorbing potential of multiple-leaf perforated panel with micro-perforated membrane was investigated in this paper.

The proposed absorbing elements, in this work, consisted of one or two perforated panels layers, with micro-perforated membranes laying at different distances between perforated panels or between the panels and rigid surface. An analytical model that was based on the theory of Maa [8,9,16,39] and other researchers [13], combined together while using the transfer matrix approach [10,25] was used to investigate the effects of different cavity thickness, hole diameter. and hole spacing, and to find the optimal combination for each type of absorber. The reliability and efficiency of the prediction model was validated by comparison with measured results from single and multiple-leaf and, then, once the method was validated, a triple-leaf panel was developed by means of numerical optimization techniques.

2. Materials

A 5 mm thick transparent Methyl Methacrylate sheet was used to manufacture the perforated panels (PP) in this study. The surface mass of the panel was 6 kg/m^2. This material has high optical transmittance when compared with the conventional glass, and great strength and good toughness.

The panels were laser cut by a computer controlled process. Figure 1 shows the design of the perforated panels used for the preliminary investigation. Four different perforated panels, named A, B, C, and D were made, with their hole diameter and hole spacing, as given in Table 1. Two perforation rates were calculated. The first was the actual rate that was obtained by dividing the hole area by the tube diameter (only considering the 10 cm diameter, corresponding to the "low frequency" tube, see Section 3.1), while the second was calculated assuming the hole pattern to be reproduced on an ideally infinite surface, and it was obviously higher than the first one.

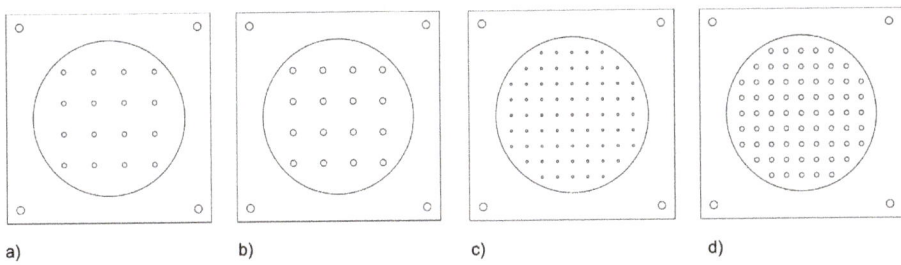

Figure 1. Different layout of the four perforated panels: (**a**) panel A; (**b**) panel B; (**c**) panel C; and, (**d**) panel D.

Table 1. Summary of the characteristics of different panels under investigation.

Panel ID	Hole Diam. (mm)	Spacing (mm)	Hole Number	Perf. Rate (%)	Perf. Rate$_\infty$ (%)
A	3.0	20	16	1.44	1.77
B	4.0	20	16	2.56	3.14
C	1.5	10	69	1.55	1.77
D	4.0	10	69	11.04	12.57

The micro-perforated membrane (MPM) absorbers that were studied in this paper were made while using a polyester transparent film, having a thickness of 0.09 mm and a surface density of 137 g/m^2. The holes in MPM were made by micro-driller that was mounted on a three-axis numerically controlled system. A proper drill bit was chosen, as the application of theoretical models suggested that the best performance could be obtained by using a hole diameter of 0.30 mm, so that the actual resulting diameter was as close as possible to the desired value. However, microscopic analysis (Figure 2) showed unavoidable fluctuations in actual hole diameters that spanned from 0.20 to 0.35 mm. Therefore, taking advantage of a segmentation algorithm that was applied to microscopic images, the actual hole area was calculated, and the "effective" circular hole diameter was found. According to the actual frequency distribution of the effective diameters, the average value of 0.267 mm was considered in the subsequent calculations. Three MPM, named M4, M5, and M6, were prepared with the hole spacing being set accordingly at 4 mm, 5 mm, and 6 mm. Their perforation rates were 0.36%, 0.23%, and 0.16%, respectively. No significant difference appeared between actual and "infinite-extension" perforation rate, given the hole dimension. Sample M6 resulted from the optimization process and was, consequently, only used in the preparation of the triple-layer panel.

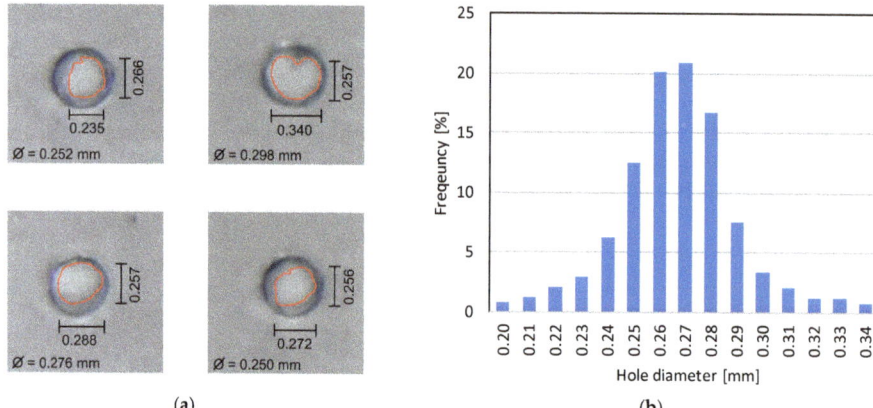

Figure 2. (a) Magnified image of a sample of the holes of the micro-perforated panels with major dimensions and effective diameter (in mm). Red line highlights the hole perimeter obtained by means of the segmentation algorithm; and, (b) Frequency distribution of effective hole diameters.

Laying the selected specimens and spacers in series formed the multi-layer structures that were studied in this paper. The different layers could be tightly combined together through fixing screws at the panel corners. The spacers were made of the same material as the perforated panels mentioned above, which were adapted to form an air cavity of desired thickness. The spacers had different diameters, so that they fitted the tube diameter, in order to ensure that air volume behind each hole corresponded to the expected value. Plasticine and adhesive tape was used to seal every visible gap to avoid that measurements could be affected by sound escaping through air gaps between different parts. It should be noted that the last air cavity, the one right in front of the rigid surface, was formed while using the measuring tube termination, so that its thickness could be conveniently controlled. One sample for each configuration was made, but several measurements were carried out. For the sake of clarity, standard deviations among the measurements were not included in plots if they were negligible.

Several configurations of multi-layer resonant structures were investigated in the study to test the reliability of the prediction methods. Table 2 summarizes the analyzed combinations, starting from the layer nearest to the sound source to the farthest one. The combinations could be divided into four classes (Figure 3): double perforated panels (PP-PP), one perforated panel and a micro-perforated membrane (PP-MPM), double micro-perforated membranes (MPM-MPM), and a triple leaf panel with one PP and two MPM. All of the layers will be described, starting from the outer layer (exposed to sound) to the backing rigid surface, for the sake of clarity and consistence throughout the paper.

Table 2. Configurations of multi-leaf resonant absorbers used in the preliminary test.

Identifier	Panel Arrangement: From the Structure Surface to the Backing Rigid Surface
C/10/B/50	C + 10 mm cavity + B + 50 mm cavity + rigid surface
B/20/M5/40	B + 20 mm cavity + M5 + 40 mm cavity + rigid surface
M4/25M5/25	M4 + 25 mm cavity + M5 + 25 mm cavity + rigid surface
D/10/M4/15/M5/30	D + 10 mm cavity + M4 + 15mm cavity + M5 + 30 mm cavity + rigid surface

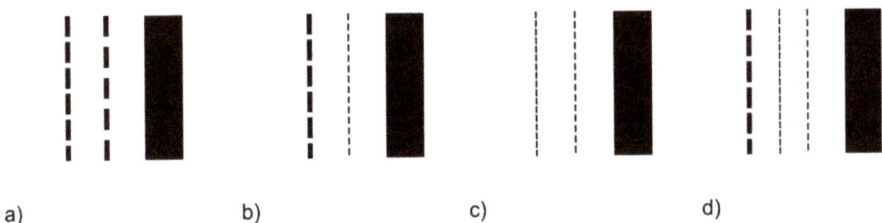

Figure 3. Schematic configuration of the different samples under investigation: (**a**) double leaf perforated panels (PP); (**b**) double leaf one perforated panel and a micro-perforated membrane (PP-MPM); (**c**) double leaf double micro-perforated membranes (MPM-MPM); (**d**) triple leaf optimized triple-layer absorber (PP-MPM-MPM). Black box represents the rigid surface.

3. Methods

3.1. Impedance Tube Method

The two-microphones impedance tube method was applied in this paper to measure the acoustic absorption of resonant absorbers, according to the process that is detailed in ISO 10534-2:1998 standard [40]. In this method, the plane sound wave is generated by a loudspeaker at one end of the tube and is reflected by the specimen at the other end. Two microphones that are set at different positions on the internal surface of the tube are used to record the sound energy decay. The reflection coefficient R of the specimen is calculated from the corrected complex acoustic transfer function. Subsequently, the normal incidence sound absorption coefficient α is calculated as $\alpha = 1 - |R|^2$. The frequency range in which the measurements are reliable depends on the length of the tube for the minimum frequency and on the tube diameter and the microphone spacing for the maximum frequency. Two impedance tubes with different internal diameters (10 cm and 4 cm) are needed in order to cover the whole spectrum range in which the specimens are expected to absorb. These tubes are made of transparent Methyl Methacrylate, 5 mm thick and 85 cm long. At the sending end of the larger tube, a 11 cm loudspeaker is sealed into a wooden case and properly isolated from the tube structure by means of an elastic pad, while a 5 cm loudspeaker is used for the smaller tube (Figure 4). On the other end, there is a shorter tube with a movable rigid termination, which is suitable to form the last air cavity of the test structure. All of the gaps between the interfaces are carefully sealed with plasticine. In the measurements, two microphones (Core Sound) with a flat frequency response from 20 Hz to 20 kHz are used. In the large tube, two different spacings are available, one at 6 cm and one at 20 cm, which correspond to a low frequency limit, respectively, of 283 Hz and 85 Hz, and a high frequency limit of 2 kHz and 770 Hz, respectively. Thus, the large tube covers one-third octave bands from 100 Hz to 1.6 kHz. In the smaller tube a 3 cm spacing between microphones is used, resulting in a frequency range from 566 Hz to 4.9 kHz, allowing for measurements up to the 4 kHz one-third octave band.

Figure 4. The 10 cm standing wave tube used to measure sound absorption coefficients in the frequency range from 100 Hz to 1.6 kHz. One of the triple layer samples is mounted in place.

The whole measurement system is controlled by a MATLAB (2018, Mathworks, Natick, MA, USA) graphic user interface, which generated and played a 5 s linear sweep from 70 Hz to 3 kHz, used in the combination with the larger impedance tube, and from 500 Hz to 5 kHz used in combination with the smaller one. In the measurements, the temperature and the relative humidity were monitored with a precision electronic thermo-hygrometer and were subsequently applied in the calculation process.

3.2. Analytical Prediction Formulation

In order to calculate the overall acoustic impedance, and hence the corresponding sound absorption coefficient, the transfer matrix method was used [10], according to which, if the bottom of layer i has an impedance of z_{si}, and the layer i has a characteristic acoustic impedance z_i, and then the impedance at the bottom of layer $i + 1$ is:

$$z_{si+1} = \frac{-j z_{si} z_i \cot(k_{xi} d_i) + z_i^2}{z_{si} - j z_i \cot(k_{xi} d_i)}, \tag{1}$$

where j is the imaginary unit, k_{xi} is the wavenumber for layer i, and d_i the corresponding thickness. The above formula can be recursively applied to calculate the surface impedance of a multi-layered absorbent, as many researchers have demonstrated [10,25]. In the subsequent sections, its application to the specific case of a series of perforated and micro-perforated panels will be further validated. Obviously, as the final value of the impedance results from the recursive application of Equation (1), it is not possible to obtain a simple and straightforward expression to relate sound absorption and the different input parameters.

The impedance for the different layers can be calculated according to theoretical expressions briefly summarized below. For an air layer (with thickness d), in front of a rigid surface (for which z_{si} is infinite), Equation (1) yields:

$$z_{si+1} = -j \rho c \cot(kd), \tag{2}$$

where ρ is the air density (assumed equal to 1.21 kg/m^3) and c is the speed of sound in air (assumed equal to 340 m/s).

The presence of a perforated panel behaving as a Helmholtz resonator in the front of the air cavity alters its impedance by the addition of mass ($j\omega m$) and resistance (r_m) terms, so that:

$$z_p = r_m + j[\omega m - \rho c \cot(kd)], \tag{3}$$

m being the mass of the vibrating plug of air within the perforations and ω the angular frequency.

The mass of the vibrating plug must take the effect of the radiation impedance and of the mutual interactions between neighboring elements into account, resulting in the addition of an "end correction" to the thickness of the panel. A typical expression for the mass is [10]:

$$m = \rho/\varepsilon \; [t + 2\delta a + (8\nu/\omega \; (1+t/2a))^{0.5}], \tag{4}$$

where ε is the perforation rate, t is the panel thickness, a is the hole radius, ν is the kinematic viscosity of air (assumed equal to $1.5 \cdot 10^{-7}$ m^2/s), and δ is the end correction factor. The latter can be calculated by means of different formulas, among which one that is suitable for more open structures [10] assumes that $\delta = 0.8(1 - 1.47\varepsilon^{1/2} + 0.47\varepsilon^{3/2})$.

The resistance term r_m is responsible for the losses within the device and, consequently, of the absorption as a function of frequency. For normal holes (not sub-millimeter in size), and if no additional porous layers are used, the term can be expressed as:

$$r_m = \rho/\varepsilon \; (8\nu\omega)^{0.5} \; (1 + t/2a), \tag{5}$$

Finally, for microperforated panels the losses occur because of viscous boundary layer effects in the perforations, provided that the diameter is sub-millimeter and comparable to the layer thickness.

Maa mostly developed the theoretical formulation for this device [8,9] and starts by determining the specific acoustic impedance of a cylindrical hole:

$$z_c = j\omega\rho t\left(1 - \frac{2J_1 k\prime\sqrt{-j}}{k\prime\sqrt{-j}J_0(k\sqrt{-j})}\right)^{-1}, \quad (6)$$

Where J_0 and J_1 are the Bessel functions of the first kind, respectively, of zero and first order, t is the tube length (i.e., the layer thickness), while k' depends on the tube diameter a and on the viscosity of the air η (assumed equal to $1.85\cdot10^{-5}$ Pa·s), according to the relation $k' = a\,(\rho\omega/\eta)^{1/2}$.

z_c must be divided by the perforation rate ε and added to the impedance of the cavity in order to get the surface impedance of the Helmholtz resonator (Equation (2)), to the radiation resistance and the end correction:

$$z_{mp} = z_c/\varepsilon - j\rho c\cot(kd) + (2\omega\rho\eta)^{0.5}/(2\varepsilon) + j1.7\omega\rho a/\varepsilon, \quad (7)$$

Once the values of the surface impedances are determined, any combination of different layers can be handled while using the transfer matrix approach. In the subsequent sections, the theoretical model was first validated by comparison with the experimental values and finally used to optimize the design of a broad-band sound absorber.

4. Results and Discussion

4.1. Single-Leaf Resonant Absorber

Only the larger tube was used and, consequently, the frequencies of interest were limited to the one-third octave bands from 100 Hz to 1600 Hz in order to validate the prediction models. Figure 5a shows the predicted and experimental values of sound absorption coefficient for single-leaf perforated panels backed by a 20 mm thick cavity. It was observed that the measured sound absorption coefficients for panel A, B, and C were in a reasonable agreement with their predictions in terms of the maximum absorption coefficient and the resonant frequency. In particular, while using actual perforation rates instead of the infinite-pattern values yielded better results and the agreement was even better for panel C, the one with hole distribution closer to the "ideal" one, showing the smallest differences between the resonant frequencies, as well as the peak sound absorption coefficient value. In fact, the measured peak frequency was at 630 Hz ($\alpha = 0.92$), while the predicted peak was at 620 Hz with $\alpha = 0.93$ when the actual perforation rate was used (while it moved to 660 Hz with $\alpha = 0.89$ when the "infinite" pattern value was used). The deviations of sound absorption coefficient values at the remaining frequencies were almost everywhere below 0.05. For panel A and panel B, the resonant frequency that was measured in laboratory occurred at a frequency a little lower than that predicted, while the peak absorption value was somewhat higher than the corresponding predictions, especially for panel B. The misalignment that was observed between measurements and predictions could be attributed to the influence of further border effects, as the cavity cell behind the holes near the sample edge (close to the boundary) might show a different behavior than the cavity cells of the other holes (close to the center).

Figure 5b presents the validation of single layer micro-perforated membranes, showing the sound absorption coefficient of sample M4 as a function of varying cavity thickness. The measured resonant frequencies of membrane cases were in general agreement with those that were given by the analytical method and calculated while assuming a hole diameter of 0.267 mm. Some measured results had lower absorption than the corresponding predicted values. This deviation could be explained as a consequence of the small variations in hole diameters, as well as of the possible vibration of the membrane, thus modifying the actual behavior of the panel. As expected, when compared with perforated panels, the micro-perforated membranes showed an improvement of sound absorption, with broader bandwidth, higher absorption peak, and extended high frequency response.

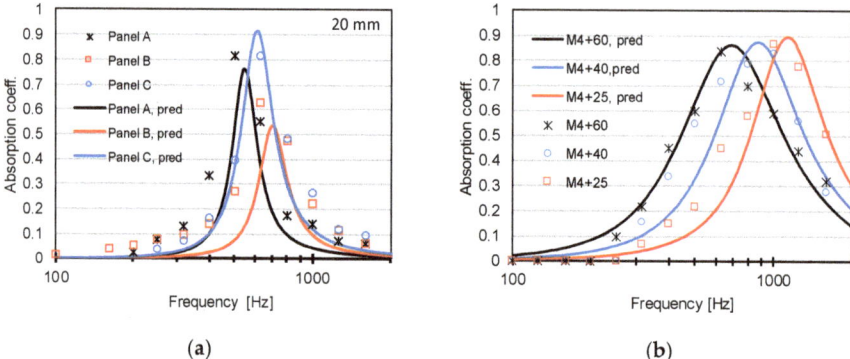

Figure 5. Comparison of the predicted and measured sound absorption coefficients for: (**a**) different perforated panels with a 20 mm cavity; (**b**) micro-perforated membrane M4 with different cavity depth: 25 mm, 40 mm, 60 mm thick.

Overall, for single-leaf resonant absorbers, the results that were measured in experiments showed good agreement with those that were predicted in terms of resonant frequency and peak sound absorption coefficient, thus confirming that the prediction method in this case was reliable and robust.

4.2. Double-Leaf Resonant Absorber

Different combinations of perforated panels and micro-perforated membranes were analyzed, in terms of double-panel, double-membrane, and panel-membrane combinations, in order to achieve better absorbing performance over a wider frequency range and validate the prediction model under more complex conditions (testing the use of the transfer matrix method to calculate the absorption).

First, a double-panel combination that consisted of panel C (backed by 10 mm cavity) and panel B (backed by 50 mm cavity, close to rigid surface) was studied by experiments and analytic method. As a consequence of the addition of a second PP (Figure 6a), two peaks appeared at 300 Hz ($\alpha = 0.82$) and 1320 Hz ($\alpha = 0.98$), with predicted absorption coefficients fitting well with the measured results, with only few deviations occurring at the frequency below 300 Hz with a maximum error being 0.16 (at 200 Hz). The agreement of two results indicates the reliability of the prediction method for double-panel combination.

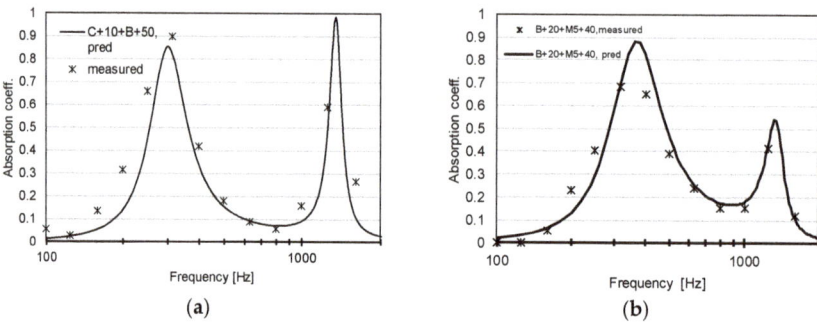

Figure 6. Comparison of the predicted and measured sound absorption coefficients for: (**a**) double-panel structure: perforated panel C followed by a 10 mm cavity combined with panel B followed by a 50 mm cavity, (**b**) perforated panel B followed by a 20 mm cavity combined with M5 followed by a 40 mm cavity.

The absorption characteristic of one combination consisting of perforated panel (B) and micro-perforated membrane (M5) was investigated in order to check the validity of the prediction model for a combination of microperforated membrane and panel (Figure 6b). Even in this case, the second layer caused a new peak to appear. The measured results were in good agreement with the predictions, as the one-third octave band values reasonably matched the predicted curve.

For the case of a double-leaf micro-perforated membrane, a combination of samples M4 (surface layer) and M5 both backed by a 25 mm cavity was taken into account (Figure 7a). The predicted sound absorption coefficient showed a sufficient agreement with the measured values, with the most evident difference appearing on the second peak where the absorption was lower than the predicted value by approximately 0.2. In the comparison with individual micro-perforated membrane (M4 and M5), the combined sample showed a much more efficient absorption capability, with two distinct peaks and a broader bandwidth spanning from 420 Hz to 1940 Hz (the absorption values in this range being always over 0.5). This time the absorption coefficients in the "valley" between the two peaks were higher, unlike previous cases, in which two layers were combined, confirming that micro-perforated panels behave better than normal perforated panels.

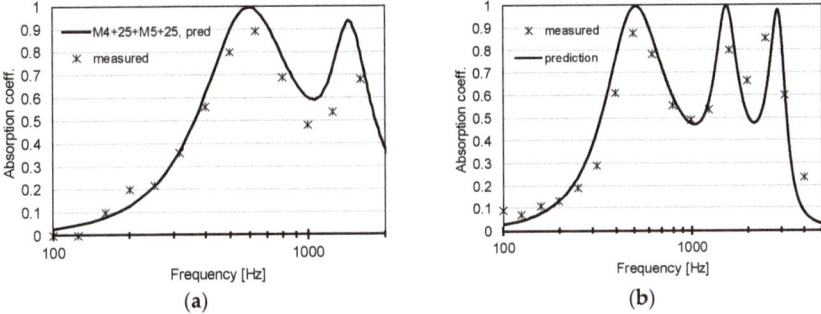

Figure 7. Comparison of the predicted and measured sound absorption coefficients for (**a**) double-membrane structure: micro-perforated membrane M4 followed by a 25 mm cavity combined with M5 followed by a 25 mm cavity. (**b**) triple-membrane structure: panel D, 10 mm cavity, micro-perforated membrane M4 followed by a 15 mm cavity, combined with M5 followed by a 30 mm cavity.

Finally, the configuration with three layers was investigated to validate the prediction model. In this case, measurements were carried out using both tubes given the extended frequency range of the absorber, thus including frequencies up to 4 kHz one-third octave band. For testing purposes, the following assembly was considered to ensure that the peaks were evenly spaced over a wider range and well identifiable: the outer layer was panel D, followed by a 10 mm air gap, and then micro-perforated membrane M4, 15 mm air gap, micro-perforated membrane M5, and a final 30 mm cavity. The results, as shown in Figure 7b, demonstrated that the agreement between measurements and predictions was generally good, with the largest variations appearing around 400 Hz, where the measured absorption was lower, and around 2 kHz, where measurements outperformed predictions.

5. Parametric Study and Panel Optimization

Given the complex relationship between the characteristics of each individual layer, air gaps and the overall absorption, and while considering the good agreement observed with analytical relations, a parametric study was carried out to better understand which combinations of parameters may maximize sound absorption. The parameter chosen as the target value of the optimization was the mean of the sound absorption coefficients that were calculated over the one-third octave bands from 400 Hz to 2.5 kHz (α_{avg}). No conventional single number descriptor was used to underline the complete customization of the process. Among the possible optimized solutions, the one that ensured an average α not differing by more than 2% from the maximum and having the smallest thickness was selected. The overall thickness of the multi-layer combination was arbitrarily limited to 70 mm, because beyond that threshold there are several conventional (and likely cheaper) solutions that could be used. The thickness of the perorated layers was kept constant at 5 mm in order to ensure good mechanical resistance for the external finishing, while the thickness of the microperforated layers was kept at 0.09 mm to avoid changing the available materials.

First, the case of a double leaf perforated panel was considered, obtaining an α_{avg} of 0.45 that resulted from a first panel with 1 mm holes with a 8 mm spacing, over a 15 mm air gap, and a second panel with 1 mm holes with a 4 mm spacing, over a second air gap 25 mm thick. The analysis of the best and worst combinations showed that the essential condition that was needed to obtain high absorption was to keep hole diameter to 1 mm for both layers, with a 4 mm spacing for the outermost layer (resulting in a 4.9% perforation rate). Conversely, the lowest average absorption (below 0.05) was obtained when the spacing between the holes of the outermost layer exceeded 14 mm, while the hole diameter was 1 mm. It can be observed (Figure 8a) that the distance between the two peaks increases by increasing the hole spacing of the innermost layer, mostly because the lowest frequency peak moves towards lower frequencies at a faster pace than the high frequency peak. An increase in the spacing of the outermost layer (Figure 8b) moves the low frequency peak towards even lower frequencies, and at the same time significantly reduces the high frequency peak, thus resulting in the observed poor performance. Similarly, by increasing the thickness of the first air gap (or by decreasing that of the second air gap) the two peaks move away and out of the desired frequency range.

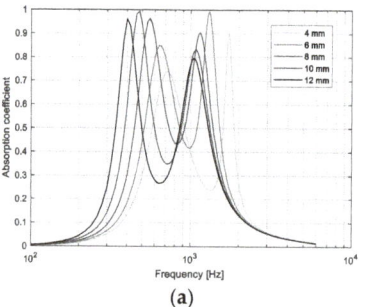
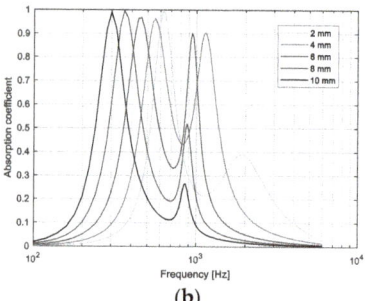

Figure 8. Sound absorption coefficient of a double layer perforated panel for different parameter modifications: (**a**) distance between holes of the innermost layer; and, (**b**) distance between holes of the outermost layer.

A second optimization exercise was carried out with reference to the combination of one perforated and one micro-perforated layer, while assuming the first to be the outermost layer. In this case the best performance yielded an α_{avg} of 0.63 when the first air gap (closest to wall) was 35 mm, the distance between holes in the MPM was 5 mm, the second air gap was 25 mm, and the perforated panel had 1 mm holes with a 2 mm spacing. The analysis of the best performing configurations showed that a spacing of 5 or 6 mm in the MPM was an essential condition, while the hole diameter had to be 1 or 2 mm in the perforated layer, while their spacing had to be 2 or 6 mm, respectively. As in the previous case, the worst performance was attributed to combinations where the perforated layer had a hole diameter of 1 mm, and a spacing exceeding 14 mm. Parametric analysis showed that increasing the distance between holes in the MPM (Figure 9a) caused the low frequency peak to shift towards even lower frequencies, while the high frequency peak remained more or less in the same position but its magnitude significantly decreased. The analysis of the variations as a function of the perforated layer parameters showed (Figure 9b) that using a 1 mm hole determined, despite the slightly higher average value, a quite unbalanced response with increased absorption towards the lower frequencies. Conversely, using the 2 mm hole with 6 mm spacing resulted in a more evenly distributed sound absorption.

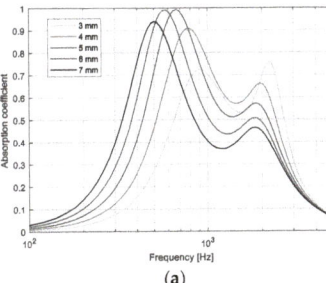
(a)

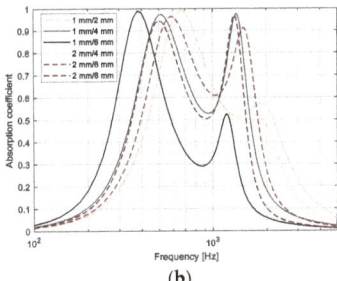
(b)

Figure 9. Sound absorption coefficient of a double layer PP-MPM panel for different parameter modifications: (**a**) distance between holes of the innermost layer (microperforated); and, (**b**) hole diameter and distance between holes of the outermost layer.

The third scenario that was investigated, following the previous discussion, was that both the layers could be micro-perforated. In this case, only air gap thickness and hole spacings were changed given the constraints applied to layer thickness and hole dimensions. The best result, according to previously stated criteria, was given by a first micro-perforated layer with a 5 mm spacing and a 30 mm air gap, followed by the second micro-perforated layer with a 3 mm spacing and a 25 mm air gap. The resulting α_{avg} was about 0.72. The analysis of the best performing combinations showed that all of them shared a 3 mm spacing for the outermost layer, while, for the other one, the spacing varied between 4 and 6 mm. The parameters that minimized average α were, as for the previous cases, a large spacing (exceeding 14 mm) for the holes in the outermost layer. Parametric analysis showed that, by increasing the spacing between holes in the inner layer (Figure 10a), the two resonant peaks in the absorption curve moved towards the lower frequencies, obtaining the maximum values around 5 mm spacing. Similarly, the increase in the hole spacing for the external layer (Figure 10b) caused a more evident shift towards lower frequencies, together with a significant drop in the sound absorption coefficient pertaining to the second peak. It is interesting to point out that, starting from the optimized configuration, a change in the thickness of the air gaps caused a shift of the peaks with no significant reduction of the peaks of absorption (Figure 10 c,d). Thus, the effect on the average absorption was generally negligible, provided that the peaks fell within the desired frequency range.

The final optimization was carried out by adopting a three layer scheme, with two inner micro-perforated layers and the perforated layer being used as the exterior finishing surface, to offer more protection to the MPM. The boundary conditions for the optimization were the same used in the previous cases, with the only difference being the presence of the extra layer. The best possible result that could be achieved in this case was an α_{avg} of about 0.81, but the overall thickness was approximately 10 cm, which was well beyond the 7 cm limit. However, accepting a 2% reduction in the mean α it was nonetheless possible to obtain a suitable panel configuration with an overall thickness of 70 mm and a sound absorption coefficient never falling below 0.7 in the preferred frequency interval. The panel arrangement included a perforated layer with 4 mm holes that were spaced at 10 mm, forming a 20 mm air gap with the subsequent layer being made of a micro-perforated membrane with 4 mm spacing over a 20 mm air gap, and, finally, a second micro-perforated layer with 6 mm spacing with a 25 mm air gap.

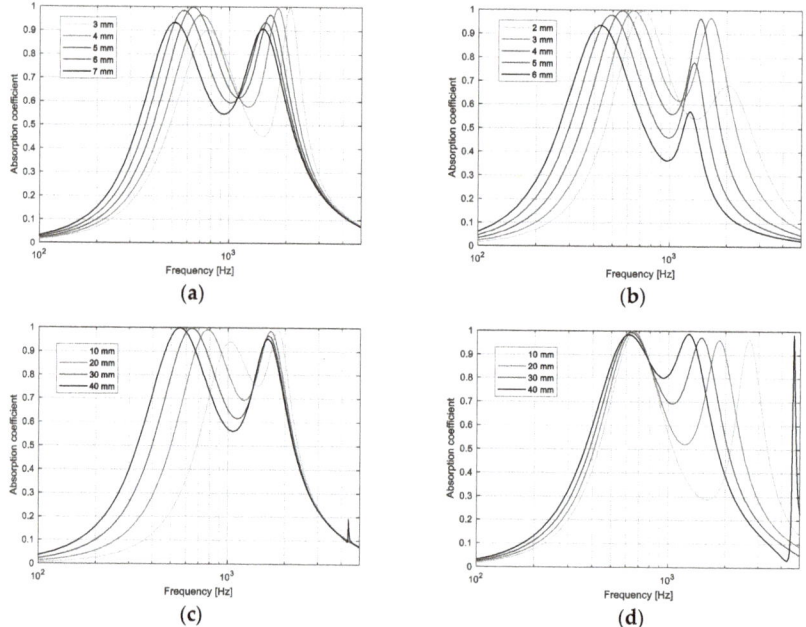

Figure 10. Comparison of the predicted sound absorption coefficients for double-membrane structure. (**a**) distance between holes of the innermost layer; (**b**) distance between holes of the outermost layer; (**c**) thickness of first air gap; and, (**d**) thickness of second air gap.

The analysis of the best performing parameter combinations (Figure 11) showed that, once again, the spacing in the micro-perforated layers played a major role, as in the first one it varied between 5 and 6 mm, while in the second one it was mostly equal to 3 or 4 mm. With reference to the PP, the optimal hole diameter varied between 3 and 6 mm provided that the spacing varied accordingly between 6 and 12 mm (so that the ratio remained optimally around 2.3 in most of the cases and the perforation rate was about 13%). The air gaps had to vary between 20 and 30 mm, with the one closest to wall, which could span over a much larger interval. Conversely, the worst combinations resulted from a very small perforation rate (below 1%) for the outermost layer, independent of the other parameters.

The effect of their variations was investigated starting from the selected best performing combination of parameters. The increase in hole spacing in the first MPM (Figure 12a) yielded a general shifting of the peaks towards lower frequencies, with significant changes in their amplitude. In particular, a higher perforation rate made the high frequency peaks less effective, while decreasing the perforation rate (by increasing the distance) made the low frequency peak less effective. Similarly, increasing hole spacing in the second MPM (Figure 12b) resulted in shifting the peaks towards lower frequency (except the central peak which remained substantially stable around 1.25 kHz), with the interesting effect that, when the distance was larger than 4 mm, the third peak started merging with the second one.

The variations in hole diameter and spacing for the outermost layer were analyzed as a function of hole diameter and, consequently, in terms of the perforation rate (Figure 12c), although viscous losses due to actual hole dimension may also affect the result. A low perforation rate implies a shifting of the peaks towards low frequencies and a significant depression of the second and third peaks, thus dramatically reducing the average sound absorption coefficient, as shown in Figure 3c, in agreement with the results of the optimization. Conversely, an increase in perforation rate beyond the optimal value only affects the high frequency peak, while the absorption due to the MPM remains substantially unchanged.

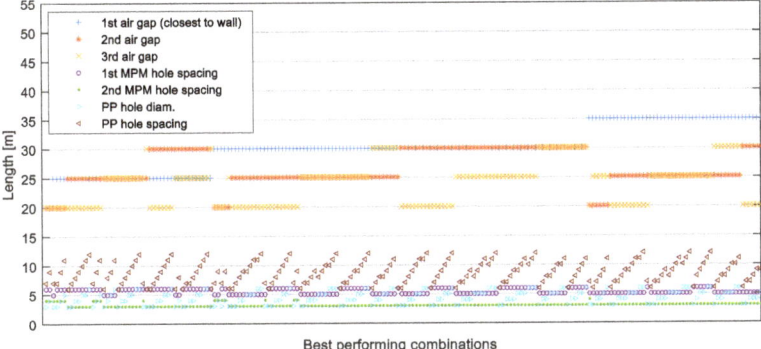

Figure 11. Plot of best 200 combinations of the seven input parameters returned by the optimization and yielding an average absorption coefficient (calculated over the one-third octave bands from 400 Hz to 2.5 kHz) within 2% of the maximum, equal to 0.81.

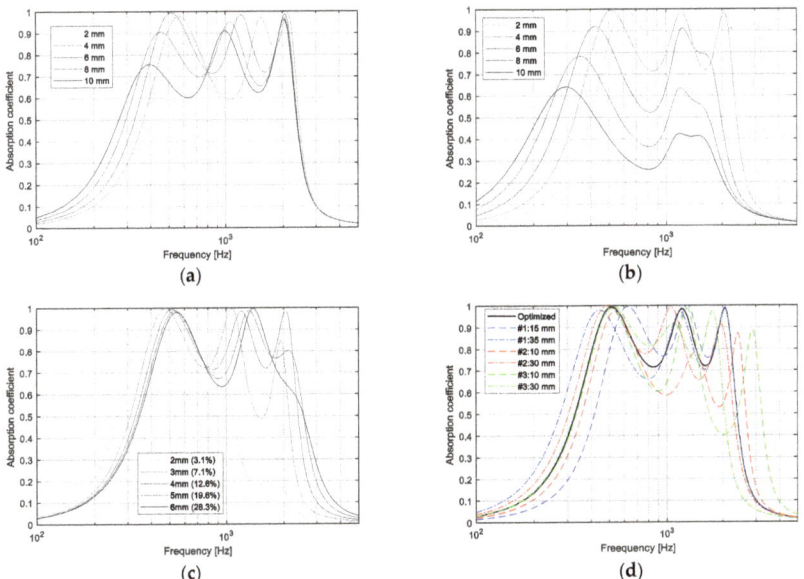

Figure 12. Comparison of the predicted sound absorption coefficients for triple-membrane structure as a function of individual parameter changes: (**a**) spacing of the first MPM; (**b**) spacing of the second MPM; (**c**) perforation rate of PP; and, (**d**) air gaps thickness. Reference parameters were: 25 mm back cavity; 1st MPP with 6 mm spacing and 0.27 mm holes; 20 mm cavity; 2nd MPP with 4 mm spacing and 0.27 holes; 20 mm cavity; PP with 4 mm holes with 10 mm spacing.

Finally, the effect of the air gaps was investigated (Figure 12d). Increasing the thickness of the layer closest to the rigid surface mostly shifted the first peak towards the lowest frequencies, without affecting the maximum absorption, except for a small reduction of the value in the "valley" between the two peaks. It is interesting to notice that the high frequency peak remained unchanged, independent of the air gap value. When the intermediate air gap was increased, it was the second peak (and to a smaller extent the third one) that showed the largest variations, moving towards low frequencies and changing amplitude. A too thin air gap caused the peaks to drop dramatically, thus making the absorber ineffective. However, a reduction in the absorption values also appeared when the air gap was too large. Finally, an increase in thickness of the air gap closer to the exterior layer had no influence at all on the first peak, while it shifted the second and particularly the third one towards lower frequencies. Again, a variation in peaks amplitude appeared, but they remained around and above 0.8, with larger variations appearing in the valleys.

Once the best configuration was identified through the optimization procedure, the resulting information was used to assemble the panel (Figure 13a) and measure its sound absorption coefficient in the impedance tube according to the previously described methods. The results showed (Figure 13b) fairly good agreement between measurements and predictions. The predicted absorption peaks occurred at 500 Hz, 1200 Hz, and 2100 Hz, respectively, and the peak values were all very close to unity, while the lowest absorption values in the valleys were always above 0.7. The measured results showed slightly lower maximum values, particularly for the first peak, compensated by higher values that were observed in the second valley (around 1.25 kHz). This combination shows a clear advantage in terms of absorption range as well as in terms of α_{avg} when compared with previous results corresponding to single and double layers. In addition, under diffuse field conditions, the panel will likely be able to provide an even smoother response over the selected frequency range, as demonstrated by Asdrubali and Pispola [28].

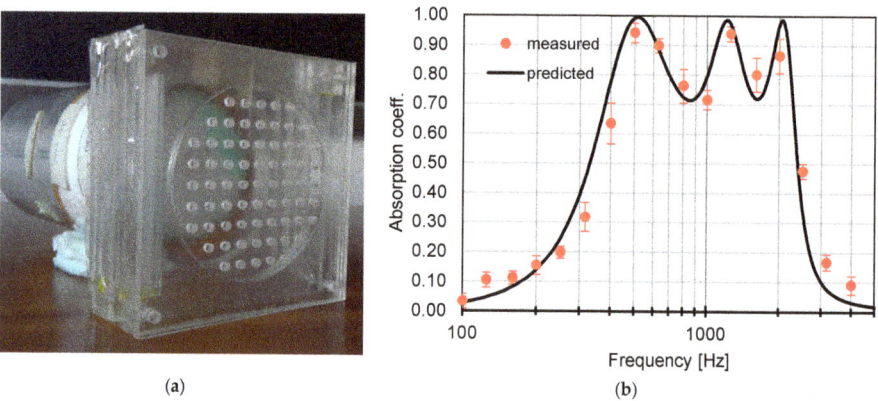

Figure 13. (**a**) The optimized three-layers panel mounted on the sample holder of the standing wave tube. (**b**) Sound absorption coefficient of a triple-leaf resonant absorber, formed by a perforated panel D (with a 20 mm backing cavity, first layer), a micro-perforated membrane with a 4 mm spacing and a 20 mm backing cavity, and a micro-perforated membrane with a 6 mm spacing and a 25 mm backing cavity, close to rigid surface. Error bars represent measured standard deviation.

6. Conclusions

In this paper, the potential of multiple layers of optically transparent sound absorbing materials was investigated. Perforated panels and micro-perforated membranes were first used to validate the predictive models, and the latter were then used to carry out a parametric study of the influence of the different variables and finally optimize an absorber to be effective in the one-third octave bands from 400 Hz up to 2500 Hz while keeping an overall thickness below 7 cm. The optimized triple-layer absorber (PP-MPM-MPM) provided a sound absorption curve that was characterized by three high peaks (close to unity) and two relatively shallow valleys. This type of absorber also had the most uniform sound absorption coefficient over the frequency range of interest, with an average absorption of about 0.80. Accordingly, by arranging perforated panels and micro-perforated membranes in a correct way, it could be possible to obtain the same absorption provided by combinations, including the use of porous materials, properly targeting the desired absorption in the preferred frequency range, while keeping a reduced thickness of the absorber, and, using proper materials, also its transparency. Further studies are under way to extend the number of layers, so as to control sound absorption over an even broader frequency range.

Author Contributions: Conceptualization, F.M.; Methodology, F.M.; Software, F.M.; Formal Analysis, L.P. and F.M.; Investigation, L.P. and F.M.; Data Curation, L.P.; Writing—Original Draft Preparation, L.P.; Writing—Review and Editing, F.M.; Visualization, L.P and F.M. All authors have read and agreed to the published version of the manuscript.

Funding: This research received no external funding.

Conflicts of Interest: The authors declare no conflict of interest.

References

1. Yadav, M.; Kim, J.; Cabrera, D.; de Dear, R. Auditory distraction in open-plan office environments: The effect of multi-talker acoustics. *Appl. Acoust.* **2017**, *126*, 68–80. [CrossRef]
2. Novak, C.C.; La Lopa, J.; Novak, R.E. Effects of Sound Pressure Levels and Sensitivity to Noise on Mood and Behavioral Intent in a Controlled Fine Dining Restaurant Environment. *J. Culinary Sci. Techn.* **2010**, *8*, 191–218. [CrossRef]
3. Hodgson, M.; Steininger, G.; Razavi, Z. Measurement and prediction of speech and noise levels and the Lombard effect in eating establishments. *J. Acoust. Soc. Am.* **2007**, *121*, 2023–2033. [CrossRef]
4. Bottalico, P. Lombard effect, ambient noise, and willingness to spend time and money in a restaurant. *J. Acoust. Soc. Am.* **2018**, *144*, EL209. [CrossRef]
5. Della Crociata, S.; Simone, A.; Martellotta, F. Acoustic comfort evaluation for hypermarket workers. *Build. Environ.* **2013**, *59*, 369–378.
6. Nijs, L.; Saher, K.; den Ouden, D. Effect of room absorption on human vocal output in multitalker situations. *J. Acoust. Soc. Am.* **2008**, *123*, 803. [CrossRef]
7. Adams, T. *Sound Materials*; Frame Pub.: NY, USA, 2017.
8. Maa, D.Y. Theory and Design of Microperforated-Panel Sound-Absorbing Construction. *Sci Sinica* **1975**, *18*, 55–71.
9. Maa, D.Y. Practical single MPP absorber. *Int. J. Acoust. Vib.* **2007**, *12*, 3–6. [CrossRef]
10. Cox, T.J.; d'Antonio, P. *Acoustic Absorbers and Diffusers: Theory, Design, and Application*; Taylor & Francis: New York, NY, USA, 2009.
11. Martellotta, F.; Cannavale, A.; De Matteis, V.; Ayr, U. Sustainable sound absorbers obtained from olive pruning wastes and chitosan binder. *Appl. Acoust.* **2018**, *141*, 71–78. [CrossRef]
12. Rubino, C.; Aracil, M.B.; Gisbert-Payá, J.; Liuzzi, S.; Stefanizzi, P.; Cantó, M.Z.; Martellotta, F. Composite eco-friendly sound absorbing materials made of recycled textile waste and biopolymers. *Materials* **2019**, *12*, 4020. [CrossRef]
13. Ingard, U. On the theory and design of acoustic resonators. *J. Acoust. Soc. Am.* **1953**, *25*, 1037–1061. [CrossRef]
14. Toyoda, M.; Mu, R.-L.; Daiji, T. Relationship between Helmholtz-resonance absorption and panel-type absorption in finite flexible microperforated-panel absorbers. *Appl. Acoust.* **2010**, *71*, 315–320. [CrossRef]
15. Laly, Z.; Atalla, N.; Meslioui, S.-A. Acoustical modeling of micro-perforated panel at high sound pressure levels using equivalent fluid approach. *J. Sound Vib.* **2018**, *427*, 134–158. [CrossRef]
16. Maa, D.Y. Microperforated-panel wideband absorbers. *Noise Control. Eng. J.* **1987**, *29*, 77–84. [CrossRef]
17. Zwikker, C.; Kosten, C.W. *Sound Absorbing Materials*; Elsevier: New York, NY, USA, 1949.
18. Stinson, M.R. The propagation of plane sound waves in narrow and wide circular tubes, and generalization to uniform tubes of arbitrary cross-sectional shape. *J. Acoust. Soc. Am.* **1991**, *89*, 550–558. [CrossRef]
19. Atalla, N.; Sgard, F. Modeling of perforated plates and screens using rigid frame porous models. *J. Sound Vib.* **2007**, *303*, 195–208. [CrossRef]
20. Guignouard, P.; Meisser, M.; Allard, J.F.; Rebillard, P.; Depollier, C. Prediction and measurement of the acoustical impedance and absorption coefficient at oblique incidence of porous layer with perforated facings. *Noise Control Eng. J.* **1991**, *36*, 129–135. [CrossRef]

21. Wang, C.; Huang, L. On the acoustic properties of parallel arrangement of multiple micro-perforated panel absorbers with different cavity depths. *J. Acoust. Soc. Am.* **2011**, *130*, 208–218. [CrossRef]
22. Okuzono, T.; Sakagami, K. A frequency domain finite element solver for acoustic simulations of 3D rooms with microperforated panel absorbers. *Appl. Acoust.* **2018**, *129*, 1–12. [CrossRef]
23. Pieren, R.; Heutschi, K. Modelling parallel assemblies of porous materials using the equivalent circuit method. *J. Acoust. Soc. Am.* **2015**, *137*, 131–136. [CrossRef]
24. Allard, J.F.; Atalla, N. Propagation of sound in porous media. In *Modelling Sound Absorbing Materials*; Wiley Chichester: Chichester, UK, 2009.
25. Verdière, K.; Panneton, R.; Elkoun, S.; Dupont, T.; Leclaire, P. Transfer matrix method applied to parallel assembly of sound absorbing materials. *J. Acoust. Soc. Am.* **2013**, *134*, 4648–4658. [CrossRef]
26. Sakagami, K.; Nagayama, Y.; Morimoto, M.; Yairi, M. Pilot study on wideband sound absorber obtained by combination of two different microperforated panel (MPP) absorbers. *Acoust. Sci. Technol.* **2009**, *30*, 154–156. [CrossRef]
27. Carbajo, J.; Ramis, J.; Godinho, L.; Amado-Mendes, P. Assessment of methods to study the acoustic properties of heterogeneous perforated panel absorbers. *Appl. Acoust.* **2018**, *133*, 1–7. [CrossRef]
28. Asdrubali, F.; Pispola, G. Properties of transparent sound-absorbing panels for use in noise barriers. *J. Acoust. Soc. Am.* **2007**, *121*, 214–221. [CrossRef]
29. Toyoda, M.; Sakagami, K.; Okano, M.; Okuzono, T.; Toyoda, E. Improved sound absorption performance of three-dimensional MPP space sound absorbers by filling with porous materials. *Appl. Acoust.* **2017**, *116*, 311–316. [CrossRef]
30. Gai, X.; Li, X.; Zhang, B.; Xing, T.; Zhao, J.; Ma, Z. Experimental study on sound absorption performance of microperforated panel with membrane cell. *Appl. Acoust.* **2016**, *110*, 241–247. [CrossRef]
31. Sakagami, K.; Fukutani, Y.; Yairi, M.; Morimoto, M. A theoretical study on the effect of a permeable membrane in the air cavity of a double-leaf microperforated panel space sound absorber. *Appl. Acoust.* **2014**, *79*, 104–109. [CrossRef]
32. Mosa, A.I.; Putra, A.; Prasetiyo, I.; Ramlan, R.; Esraa, A.A. Absorption Coefficient of Inhomogeneous MPP Absorbers with Multi-Cavity Depths. *Appl. Acoust.* **2019**, *146*, 409–419. [CrossRef]
33. Carbajo, J.; Ramis, J.; Godinho, L.; Amado-Mendes, P. Perforated panel absorbers with micro-perforated partitions. *Appl. Acoust.* **2019**, *149*, 108–113. [CrossRef]
34. Kiyama, M.; Sakagami, K.; Tanigawa, M.; Morimoto, M. A Basic Study on Acoustic Properties of Double-leaf Membranes. *Appl. Acoust.* **1998**, *54*, 239–254. [CrossRef]
35. Sakagami, K.; Nakamori, T.; Morimoto, M.; Yairi, M. Double-leaf microperforated panel space absorbers: A revised theory and detailed analysis. *Appl. Acoust.* **2009**, *70*, 703–709. [CrossRef]
36. Sakagami, K. Sound absorption systems with the combination of a microperforated panel (mpp), permeable membrane and porous material: Some ideas to improve the acoustic performance of mpp sound absorbers. *Int. J. Recent Res. Appl. Stud.* **2015**, *24*, 59–66.
37. Ayub, M.; Hosseini Fouladi, M.; Ghassem, M.; Mohd Nor, M.J.; Najafabadi, H.S.; Amin, N.; Zulkifli, R. Analysis on Multiple Perforated Plate Sound Absorber Made of Coir Fiber. *Int. J. Acoust. Vib.* **2014**, *19*, 203–211. [CrossRef]
38. Shen, X.; Bai, P.; Yang, X.; Zhang, X.; To, S. Low Frequency Sound Absorption by Optimal Combination Structure of Porous Metal and Microperforated Panel. *Appl. Sci.* **2019**, *9*, 1507. [CrossRef]
39. Maa, D.Y. Potential of microperforated panel absorber. *J. Acoust. Soc. Am.* **1997**, *104*, 2861–2866. [CrossRef]
40. ISO 10534-2:1998 *Acoustics—Determination of Sound Absorption Coefficient and Impedance in Impedance Tubes—Part 2: Transfer-function Method*; International Organization for Standardization: Geneva, Switzerland, 1998.

© 2020 by the authors. Licensee MDPI, Basel, Switzerland. This article is an open access article distributed under the terms and conditions of the Creative Commons Attribution (CC BY) license (http://creativecommons.org/licenses/by/4.0/).

Article

Listeners Sensitivity to Different Locations of Diffusive Surfaces in Performance Spaces: The Case of a Shoebox Concert Hall †

Louena Shtrepi *, Sonja Di Blasio and Arianna Astolfi

Department of Energy, Politecnico di Torino, 10129 Turin, Italy; sonja.diblasio@polito.it (S.D.B.); arianna.astolfi@polito.it (A.A.)
* Correspondence: louena.shtrepi@polito.it; Tel.: +39-011-090-4545
† Part of this work was presented at EURONOISE 2018, Crete, Greece.

Received: 27 May 2020; Accepted: 23 June 2020; Published: 25 June 2020

Abstract: Diffusive surfaces are considered as one of the most challenging aspects to deal with in the acoustic design of concert halls. However, the acoustic effects that these surface locations have on the objective acoustic parameters and on sound perception have not yet been fully understood. Therefore, the effects of these surfaces on the acoustic design parameters have been investigated in a real shoebox concert hall with variable acoustics (Espace de Projection, IRCAM, Paris, France). Acoustic measurements have been carried out in six hall configurations by varying the location of the diffusive surfaces over the front, mid, and rear part of the lateral walls, while the other surfaces have been maintained absorptive or reflective. Moreover, two reference conditions, that is, fully absorptive and reflective boundaries of the hall have been tested. Measurements have been carried out at different positions in the hall, using an artificial head and an array of omnidirectional microphones. Conventional ISO 3382 objective acoustic parameters have been evaluated in all conditions. The results showed that the values of these parameters do not vary significantly with the diffusive surface location. Moreover, a subjective investigation performed by using the ABX method with auralizations at two listening positions revealed that listeners are not sensitive to the diffusive surface location variations even when front-rear asymmetric conditions are compared. However, some of them reported perceived differences relying on reverberance, coloration, and spaciousness.

Keywords: shoebox concert hall; diffusive surfaces; diffusers location; acoustical parameters; variable acoustics; subjective investigation; auralization

---

1. Introduction

The definition of materials for absorptive and diffusive surfaces is the main design issue once the shape and the volume of an auditoria have been determined. These surfaces can be used by acousticians and architects to reach the desired sound field and achieve a trade-off with the aesthetical architectural aspects [1]. In performance spaces, the absorptive surfaces are usually hidden by layers of perforated panels or textiles. Conversely, the diffusive surfaces are commonly visible and become an important part of the design of the interior space. Their effects have been intensively investigated in the last decade and are usually related to corrections of the acoustic glare, echoes, focusing of sound, and enhancement of the uniformity of the sound field [1–3]. Depending on the combination with the absorptive surfaces, they can also generate negative effects, such as the reduction of sound level and reverberation time [4]. Diffusive surfaces are considered one of the most critical aspects in the acoustic design and renovation of concert halls since there is a lack of knowledge on how their effects on the sound field are related to practical design choices, that is, their location and extension. Thus, this experimental study aims to give more insight on the former aspect, by investigating the effects of diffusive surface location on the objective acoustic parameters used in the design process. Moreover, the sensitivity of listeners to variations in the diffusive surfaces location is investigated.

It has been highlighted that the direct relation between the diffusive surfaces and any objective acoustic parameter is not as immediate as the absorptive surfaces related to the reverberation time [5]. Therefore, more adequate diffuser design and evaluation tools for acousticians and architects are needed since the preliminary phases of the design process to promote the use of sound diffusers. In order to better understand the diffusive surfaces effects, several case studies have been used for objective and subjective investigations through measurements in real halls [4,6–8], physical-scale models [4,8–12], and simulations of performance spaces [12–16].

Different investigations have focused on the ISO 3382-1 [17] parameters since these are used as design parameters at a larger scale. Ryu and Jeon [4] found that hemispherical and polygonal diffusers installed on the sidewalls close to the proscenium arch, the sidewalls of stalls, and balcony fronts of a shoebox-horseshoe plan hall decrease sound pressure level (SPL), reverberation time (RT) and early decay time (EDT) at most seats, compared to reflective surfaces. Furthermore, these surfaces affect clarity (C_{80}) and the interaural cross-correlation coefficient (1-$IACC_E$) by increasing and decreasing their values at the front and the rear seats, respectively. Other investigations on the effects of hemispherical diffusers applied to 1:50 scaled rectangular and fan-shaped hall surfaces confirmed the decreasing effects of diffusers on RT and SPL [9]. In this study, the halves of the lateral walls closest to the stage have been judged as the most effective areas for diffuser installation since they reduce the spatial deviation of the acoustic parameters and minimize the decrease of RT and listening level (LL). This was mainly valid for shoebox halls rather than fan-shaped halls. Moreover, large and sparse diffuser profiles resulted as more effective on the acoustic results. Jeon et al. [18] made measurements in real reverse fan-shaped and rectangular halls and found that saw-tooth and cubic shaped diffusers installed on lateral walls do not have any significant effect on the acoustic parameters. However, their presence improves the spatial uniformity of the sound energy. Based on simulations in a fan-shaped hall with two different hall volumes (3600 m^3 and 7300 m^3), Shtrepi et al. [16] showed that the ISO 3382 objective parameters are mostly affected when the diffusive surfaces with a scattering coefficient higher than 0.70 are located on the ceiling, lateral walls and rear wall simultaneously. These effects are more evident in the smaller volume and are reduced when the rear wall only is treated independently of the volume. Jeon et al. [19] have suggested the use of another objective parameter, namely the number of reflection peaks (Np) in an impulse response, which describes the spatial and temporal variation of the sound field. They considered a scaled model of a shoebox hall with polygon- and hemisphere-type diffusive surfaces applied to the lateral walls and ceiling, as well as a real reverse fan-shaped recital hall with diffusive front halves lateral walls closest to the stage. Their measurements showed an increase in the Np at higher frequency bands and no significant differences for the other ISO 3382 parameters. In addition, Jeon et al. [12] showed differences below the just noticeable difference (JND) for the ISO acoustic parameters through simulations in 12 performance halls of various shapes (shoebox, fan-shape, and other complex shapes) and with increasing scattering coefficient of the walls and ceiling. In a second part of the study based on measurements in a scale model of a vineyard-shape hall, they noticed that the periodic diffusers installed over the sidewalls and balcony decrease RT and G (strength), while increase C_{80}. However, this was mainly attributed to the absorption added by the diffusers.

Besides the objective investigations, also the perceptual differences between different surface treatments have been the object of continuous research. Torres et al. [20] showed that changes in diffusion characteristics of the surfaces are audible in a wide frequency region and depend on the input signals, i.e., sustained signals make the perception of the differences easier than impulsive signals. Takahashi and Takahashi [21] and Shtrepi et al. [7] showed that perceptual differences between reflective and diffusive surfaces are related to the listening distance from the surface itself. Moreover, they are related to the difference of scattering coefficient between the compared surfaces [13,15]. Singh et al. [22] found that the perceived diffuseness is related to the interaural cross-correlation coefficient ($IACC$), which is an important parameter in the design process. Furthermore, Jeon et al. [19] showed that the perceived diffuseness could be quantified in terms of the number of reflected peaks (Np), which is correlated to the listener preference. In another study, Ryu and Jeon [4] showed that the preference of the diffusive surface presence highly correlates with the perceived loudness (SPL) and reverberance (EDT). Other studies reported that changes in diffusive surfaces characteristics are mainly perceived in terms of coloration and spaciousness variations [7,20,21,23]. Jeon et al. [12] showed that despite small changes in the objective parameters, the presence of the diffusers made a clear and positive contribution to the overall impression of the listeners, which was mainly related to intimacy and envelopment.

Although these results highlight the importance of the location of the diffusive surfaces and their configuration combined to the size and shape of the hall, there is still need for clear and generalized guidelines useful for acousticians and practitioners alike. Since the scattering properties of these surfaces can be easily assessed by using the ISO 17497-1, -2 [24,25], the application of diffusive surfaces based on scientific investigations, and not only on the architectural and design preferences, should be a common practice for modern concert hall designers. Moreover, the subjective data, i.e., the listeners' sensitivity, would help to determine the measurement accuracy needed for the characterization of these surfaces [26].

However, very little research on this aspect has been carried out in real concert halls due to both technical and economic issues. Therefore, the present study attempts to clarify the influence of diffusive surface location on the objective and subjective aspects by means of both in-situ measurements and perceptual listening tests. Since both technical and economic issues would limit the research, a flexible environment—the hall Espace de Projection at IRCAM (Paris)—has been involved. Six configurations have been created by varying the location of the diffusive surfaces over the front, mid and rear part of the lateral walls, while the other surfaces have been maintained absorptive or reflective. Moreover, two reference conditions, that is, fully absorptive and reflective boundaries of the hall have been tested. The ISO 3382 objective acoustic parameters, such as reverberation time (T_{30}), early decay time (EDT), clarity (C_{80}), definition (D_{50}), center time (T_s), and interaural cross-correlation (IACC) have been estimated from the measured impulse responses. Furthermore, subjective investigations have been performed in order to identify the detectable differences between different locations of the diffusive surfaces.

2. Method

2.1. Objective Measurements

2.1.1. Hall Description

A variable-acoustic environment, the Espace de Projection (ESPRO) at IRCAM in Paris (Figure 1), has been used for in-field measurements in order to investigate how the location of diffusive surfaces can influence the generated sound field. Table 1 provides the architectural and acoustical details of the variability of ESPRO based on Peutz [27,28]. The hall characteristics have been extensively described in Shtrepi et al. [7,13,14] and here only a brief overview is given in order to help the reader understand the context of the experiment.

The ESPRO is a modern facility with variable passive acoustics, which is achieved through the variation of room geometry and surface acoustic properties: the former is reached by moving the ceiling height from 3.5 m up to 10 m, while the latter is controlled by acting on independently pivoting prisms. The prisms are grouped in panels of three and have three faces with different acoustic properties that are reflective, diffusive, and absorptive (Figure 1). The frequency-dependent absorptive and scattering properties of the surfaces have been shown in [7], while diffusion polar distributions have been presented in [13,14]. Based on these references, the data at 500–1000 Hz for the absorptive surfaces present a mean absorption coefficient of a = 0.80, while the diffusive surfaces are characterized by a mean scattering coefficient of s = 0.75 and a diffusion coefficient of $d_{45°}$ = 0.52. The rotation is automated and managed from a control room. Only the eye-level panels, i.e., the first row from the floor level, are controlled manually and could be set in either absorptive or reflective conditions. The floor is a hard-reflective surface.

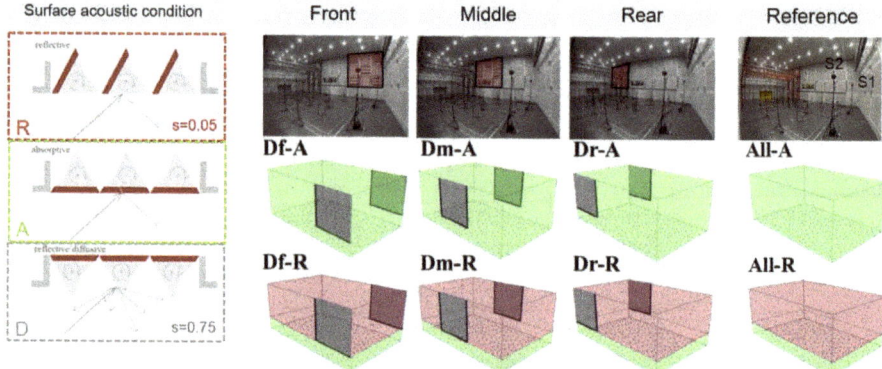

Figure 1. Surface acoustic conditions absorptive, reflective, and diffusive (A, R, and D). Interior view and simplified models of the eight configurations of the hall. Six configurations tested with three different locations of the diffusive surfaces (Df-A, Dm-A, Dr-A, Df-R, Dm-R, Dr-R) and two reference conditions (All-A and All-R).

Table 1. Architectural and acoustical details of the ESPRO based on Peutz [27,28].

Characteristic	Details
Use	Multipurpose
Plan type	Shoebox
Dimensions	$H_{variable}$: H_{min} = 3.5 m; H_{max} = 10 m; W = 15.5 m; L = 24.0 m
Volume	V_{min} = 818.4 m^3; V_{max} = 3720 m^3
Seats	Variable: N_{max} = 350
Ceiling	Variable panels: N_{vp} = 54
Long lateral walls	Variable panels: N_{vp} = 49; Fixed panels: N_{fp} = 12
Short front/rear walls	Variable panels: N_{vp} = 42; Fixed panels: N_{fp} = 12
Reverberation time (500–1000 Hz)	$T_{60, min}$ = 0.4 s; $T_{60, max}$ = 4 s

2.1.2. Hall Acoustic Conditions and Measurements

Six hall configurations have been considered in this study by varying the location of the diffusive surfaces over the lateral walls within two different main acoustic conditions of the overall surfaces of the hall: absorptive (-A) and reflective (-R) (Figure 1). Three conditions of the diffusive surfaces (Figures 1 and 2) have been tested by shifting their location over the front, mid, and rear part of each lateral wall (hereafter labeled Df, Dm, Dr, respectively). Moreover, two reference conditions, that is, all variable surfaces set in the absorptive (All-A) and reflective (All-R) mode have been considered in order to investigate the overall absolute effect of the presence of a diffusive surface.

The absorptive condition was chosen for the eye-level fixed panels in all the measurements in order to avoid the strong reflections from the lower parts of the walls. The ceiling was set at the maximum operative height of 10 m, i.e., leading to a room volume of 3720 m^3.

ISO 3382-1 [17] objective parameters have been measured in the unoccupied room conditions. A detailed description of the measurement set-up is given in [7], while here a brief overview is given in order to help the reader understand the main elements. Measurements have been carried out using the ITA-Toolbox, an open-source toolbox for Matlab [29]. Monaural and binaural measurements have been performed with twenty-four omnidirectional microphones (Sennheiser KE-4) and two artificial heads (ITA Head), respectively (Figure 2). The microphones have been set at a height of 3.7 m in a crossed array that extended to one of the two halves of the audience area (Figures 1 and 2). This height was chosen in order to reach the center of the first level of variable panels. Additionally, the artificial heads (Head 1 and Head 2) have been placed in the middle of the microphone array in order to be representative of the largest number of receiver positions and adjusted at an ear height of 3.7 m from the floor level as the omnidirectional microphones. Head 1 was located close to the central symmetry axis of the room and Head 2 at the midway between the axis and the lateral wall. The impulse responses at these positions have been used for the auralization introduced in the listening test session. Two omnidirectional sound sources have been positioned at the front part of the room. Each source consisted of a three-way system of low, medium, and high-frequency sources, which were positioned at different heights, that is, at 0.40, 3.70, and 3.90 m, respectively [7]. The excitation signal was an exponential sine sweep with a sampling rate of 44.1 kHz, a length of 16.8 s, and a frequency range separated for each speaker of the sources. Two repetitions have been performed for each configuration; however, given the high S/N ratio no averaging was applied [30]. Three Octamic II by RME (Haimhausen, Germany) have been used as microphone preamps and an ADA8000 Ultragain Pro-8 by Behringer (Willich, Germany) served as DA-converter. Loudspeaker, artificial head, and amplifier were custom made devices by the Institute of Technical Acoustics, Aachen, Germany.

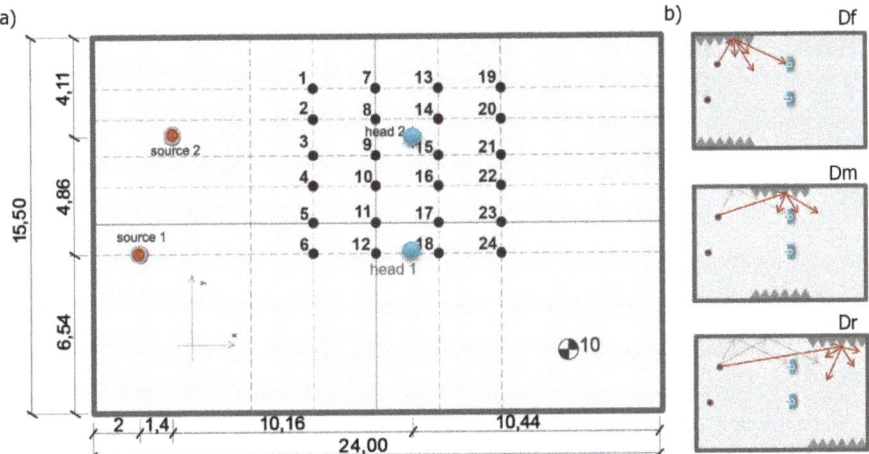

Figure 2. (a) Measurement positions (source, microphones, and artificial head) and main dimensions of the room (metric scale). (b) Schematic plan view of the diffusers location with respect to the source and artificial head positions.

2.1.3. Objective Analyses

The ISO 3382-1 [17] parameters, that is, reverberation time (T_{30}), early decay time (EDT), clarity (C_{80}), definition (D_{50}), center time (T_s), interaural cross-correlation ($IACC$) have been assessed by using the functions of ITA-Toolbox. Specifically, these parameters have been considered as a measure of reverberance and liveness (T_{30} and EDT), clarity and balance between early and late energy, or the balance between clarity and reverberance (C_{80}, D_{50}, and T_s), and perceived spaciousness ($IACC$). This last parameter has been evaluated only for the binaural measurements at the head locations.

Averaged values, as suggested in ISO 3382-1 [17], have been calculated over the 500 Hz and 1000 Hz octave bands, while the $IACC$ values were averaged over 500 Hz, 1000 Hz and 2000 Hz octave band results since these frequencies concern the subjectively most important range. Besides the $IACC$ for the full length of the impulse responses, the early-arriving (0–80 ms) and late-arriving (80 ms-inf) sound have been considered separately in the evaluation of $IACC_E$ and $IACC_L$, respectively. The JND values of each parameter have been used to compare the results for different configurations (Table 2).

Appl. Sci. **2020**, *10*, 4370

Table 2. Objective acoustic parameters and respective JND values.

Parameters	EDT	T_{30}	C_{80}	D_{50}	T_s	$IACC_E$	$IACC_L$	$IACC$
Units	(s)	(s)	(dB)	(%)	(s)	(-)	(-)	(-)
JND	5%	5%	1 dB	5%	0.010	0.075	0.075	0.075

2.2. Subjective Investigation

An auditory experiment has been conducted to investigate the listener's ability to perceive variations of the diffusive surfaces location by using the ABX method [31]. The test also allowed to evaluate the effects of different source and listener positions and type of music/signal passages (Figure 3).

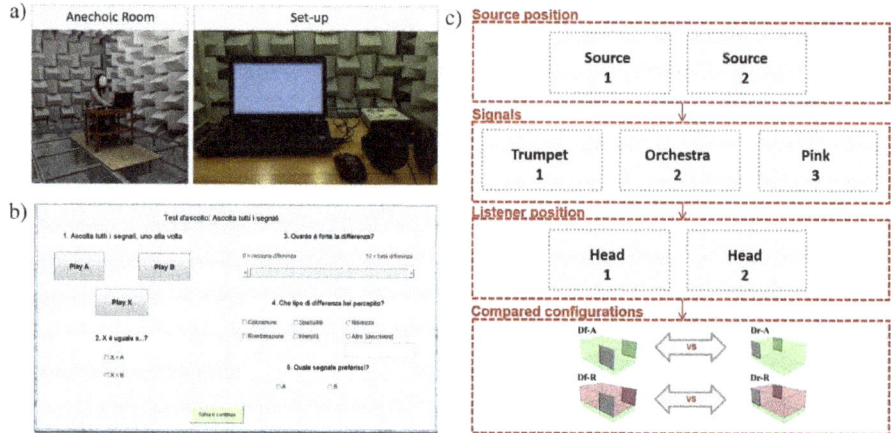

Figure 3. (a) Listening test anechoic room set-up, (b) user-interface in Italian, and (c) listening test scheme.

2.2.1. Test Subjects and Experimental Environment

A group of twenty-four professors, research assistants, and students aged between 25 to 50 years old with normal hearing ability have been involved in the test. All the listeners were volunteers interested in acoustic topics and no one of them could be considered as an expert listener, based on their musical experience. All of them provided written consent for the anonymized use of their test results. The normal hearing ability of each listener was tested by using the app "Loud Clear Hearing Test," developed by JPSB Software [32] and the same headphones (Sennheiser 600 HD) subsequently used in the listening test. This procedure is helpful for a more accurate screening compared to just self-reported hearing ability, which is often used in acoustic investigations.

The listening test sessions have been conducted in the anechoic room at Politecnico di Torino (Figure 3a), which has a background noise of L_{Aeq} = 17.3 dB. During the two days test, the room conditions, as well as the set-up, have been kept unvaried. The equipment consisted of one computer, a sound card (Tascam US-144 MKII), and headphones (Sennheiser 600 HD). The environment was made comfortable for the listeners and they were familiarized with the test procedure by an illustrated written and verbal explanation.

2.2.2. ABX Method

The ABX methodology [31] is a standard psychoacoustic test for the determination of audible differences between two signals. In this procedure, three stimuli are presented to the listener: stimulus

"A" and stimulus "B," which have a known difference, and stimulus "X", which regards the task of the listener who has to identify whether it is the same as "A" or the same as "B." If there is no audible difference between the two signals, the listener's responses should be binomially distributed such that the probability of replying "X = A" is equal to the probability of replying "X = B," i.e., 50%. This score is interpreted as indicating no perceptual difference between A and B. The minimum number of correct answers needed to indicate a perceptual difference can be given by the inverse cumulative probability of a binomial distribution, based on the number of trials, confidence level and probability of correct answer.

For the sake of this investigation, an ad-hoc routine in Matlab 2018b (MathWorks, Natick, MA, USA) with an intuitive user interface in Italian language has been implemented to present the test to each participant (Figure 3b).

2.2.3. Test Procedure

The listening test consisted of signals recorded in the same "head" position (Figure 1), i.e., Head 1 and Head 2 for the front-rear asymmetric configurations (Df-A, Dr-A, Df-R, and Dr-R). Figure 3c depicts the test structure. A pair of two different configurations are compared in each experiment (Df-A vs. Dr-A or Df-R vs. Dr-R), while the sources, the artificial head, and the music/signal passage remain unvaried within each pair of samples.

The auditory tests consisted of 48 stimuli (24 pairs), which were created by convolving the binaural impulse responses obtained from in-situ measurements with three anechoic music passages. The three music/signal passages were chosen based on different style, tempo, and spectral contents: an orchestra track ("Water Music Suite"—Handel/Harty, Osaka Philarmonic Orchestra, Anechoic Orchestral Music Recordings, Denon, Kawasaki, Japan), a solo instrument trumpet (MAHLER_tr1_21.wav, Mahler, Odeon anechoic signals database) and pink noise. The temporal and spectral contents of the first two samples are shown in Figure 4. The pink noise was included in the test for its objective and perceptual acoustic properties, although it is not a realistic signal for concert halls. Pink noise has a well-known spectral density that decreases at a rate of 6 dB per octave which leads, on average, to the same amount of power for every octave band. From a perceptual point of view, the signal sounds flatter to the ear. The orchestra and trumpet signals present some differences below 400 Hz, where the trumpet sample has less energy (Figure 4a). Figure 4b,c shows the temporal development and the characteristics of the transients in the signals. The trumpet sample is constituted by abrupt onsets and reasonably damped offsets, while the orchestra sample is a more sustained signal that has ramped onsets and damped offsets. The listening test samples are made available in an open-access repository [33].

A sample length of 5 s was chosen to be long enough in order to give the listener the necessary time to assess the full extent of their acoustic perception and, at the same time, short enough to avoid excessive fatigue. Given the comparative structure of the test, no equalization has been applied for the sound level between the conditions in each pair.

The test was structured as a double-blind test, i.e., the administrator did not know the answers either, in order to avoid any accidental cues to the listeners. Moreover, the test was based on a fully randomized order of presentation of A and B pairs, as well as a random distribution of the correct answers, i.e., X could be randomly A or B. After listening to A and B, the listeners were asked to answer to the question "Which one is X?" by choosing between one of three options, that is, "sample A" and "sample B."

Compulsorily, the listeners had to listen to all of the three samples (A, B, and X) in order to continue to the next step of the test. However, they could freely choose the listening order of the three samples (A, B, and X) and repeat the samples as many times as they judged necessary.

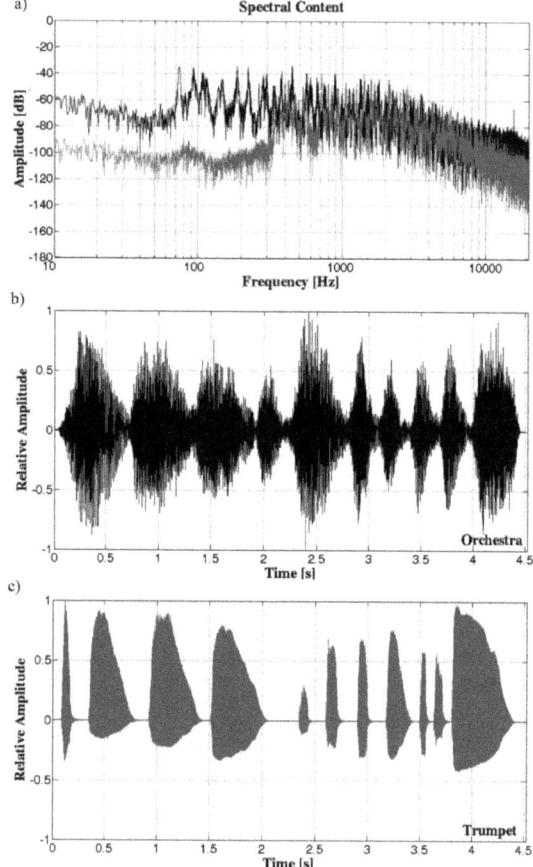

Figure 4. Orchestra and trumpet anechoic stimuli: (**a**) spectral content, (**b**) waveform of the orchestra sample, and (**c**) waveform of the trumpet sample.

The listeners did not receive any instructions on which features of the sound samples they should concentrate on. This aspect was investigated (Figure 3b) by asking them to give more details on their answers related to:

- "How strong is the difference?" The answer was given on a 0–10 scale.
- "What kind of difference could you perceive?" The answer was given by selecting the relevant attributes (coloration, spaciousness, clarity, reverberance, and loudness) that have been perceived as different. Listeners could choose more than one option or indicate other unincluded attributes.
- "Which signal do you prefer?" The answer was given by choosing between A and B.

The authors explained the case study and the purpose of the experiment at the end of the individual test. The listeners could not take breaks during the test, which lasted about 30 min. After the test, the listener's impressions and opinion were collected. Further information was gathered on their experience with previous listening tests, on their music skills, as well as on their age and general health conditions.

3. Results

3.1. Objective Results

Figures 5–9 show the results of each objective room acoustic parameter in all the considered hall conditions. Each parameter is given with respect to the source-to-receiver distance (S1 and S2). Moreover, the figures provide the objective acoustic parameter differences between the configurations Df-Dm, Df-Dr, Dm-Dr for an easier direct comparison to the JND values for the absorptive (-A) and reflective (-R) conditions, respectively. Differences within ±1 JND of the parameters are highlighted through a gray area. A summary of these differences has been given numerically in the tables in Appendix A.

The results of EDT (Figure 5) do not show a strong dependence on the source-to-receiver distance for both S1 and S2 in both the reflective (-R) and absorptive (-A) conditions. EDT values of the reflective conditions result higher for source location S2 compared to S1 for source-to-receiver distances between 8–12 m. The ΔEDT graph shows that there are a few significant differences between the configurations Df-Dm, Df-Dr, Dm-Dr in the reflective (-R) and absorptive (-A) conditions, i.e., >1 JND. These differences result higher for source location S2 and occur at a larger number of receivers in the absorptive (-A) conditions. However, no significant trend could be observed with respect to the source-to-receiver distance.

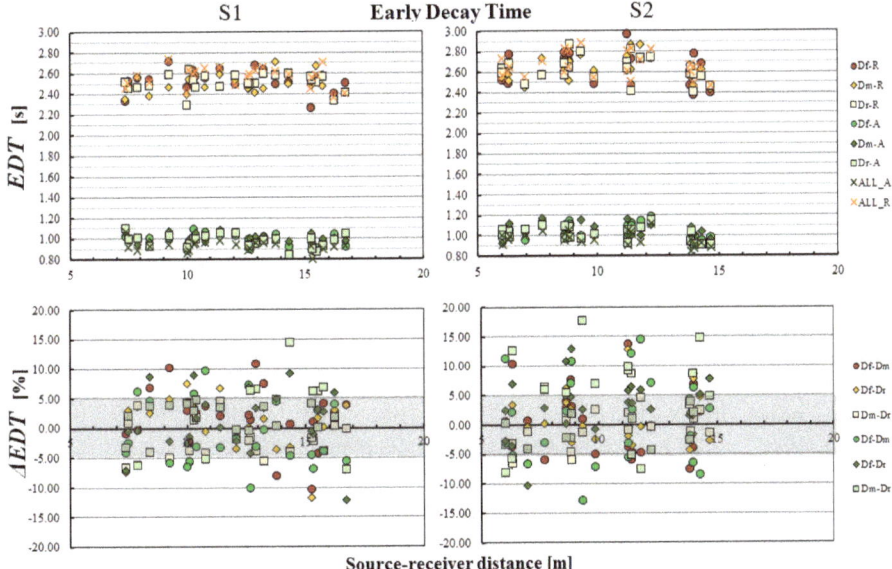

Figure 5. EDT parameter averaged over 500 Hz and 1000 Hz for S1 and S2 source-to-receiver distance. ΔEDT represents the parameter differences between the configurations Df-Dm, Df-Dr, Dm-Dr in the reflective (-R) and absorptive (-A) conditions. Differences equal ±1 JND of the parameters are highlighted through a gray area.

The results of T_{30} (Figure 6) show a decrease at the farthest positions for both S1 and S2 in the reflective conditions Df-R, Dm-R, Dr-R, and All-R. Conversely, there is no decreasing trend in the absorptive conditions Df-A, Dm-A, Dr-A, and All-A. T_{30} values of the reflective conditions result higher for source location S2 compared to S1 for the nearest receivers. Very few receiver locations seem to present differences (ΔT_{30}) higher than the JND between the different diffuser locations Df-Dm,

Df-Dr, Dm-Dr in the reflective (-R) and absorptive (-A) conditions. However, no significant trend can be detected considering the overall receivers and the source-to-receiver distance.

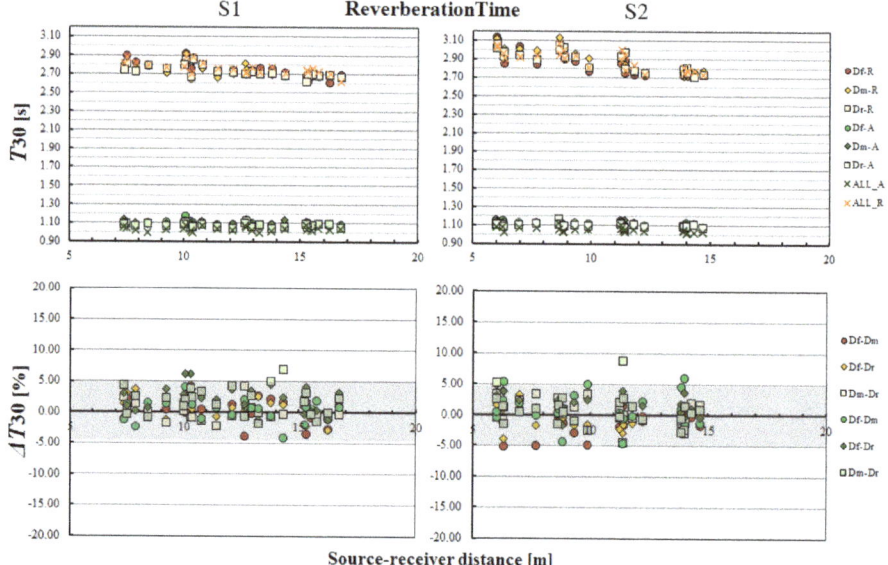

Figure 6. T_{30} parameter averaged over 500 Hz and 1000 Hz for S1 and S2 source-to-receiver distance. ΔT_{30} represent the parameter differences between the configurations Df-Dm, Df-Dr, Dm-Dr in the reflective (-R) and absorptive (-A) conditions. Differences equal ±1 JND of the parameters are highlighted through a gray area.

The results of C_{80} (Figure 7) present different trends for S1 and S2 with respect to the source-to-receiver distance in both the reflective (-R) and absorptive (-A) conditions. Generally, it can be noticed that ΔC_{80} values present a few differences higher than the JND between the different diffuser locations Df-Dm, Df-Dr, Dm-Dr in the reflective (-R) and absorptive (-A) conditions. However, it not possible to detect a significant general trend of differences due to the diffuser location when a comparison is made overall the source-to-receiver distances.

The results of D_{50} (Figure 8) show a decrease at the farthest positions both for S1 and S2 in the reflective (-R) and absorptive (-A) conditions. It can be noticed that D_{50} values present a higher variability in the absorptive conditions at each receiver position for both sources. Generally, it can be noticed that ΔD_{50} values present a few differences higher than the JND between the different diffuser locations Df-Dm, Df-Dr, Dm-Dr in the reflective (-R) and absorptive (-A) conditions. However, no significant trend can be detected considering the overall receivers.

The results of T_s (Figure 9) show an increase at the most distant positions both for S1 and S2 in the reflective (-R) and absorptive (-A) conditions. Only a very few receiver locations seem to present differences higher than the JND between the different diffuser locations. This is observed mainly for the reflective (-R) conditions.

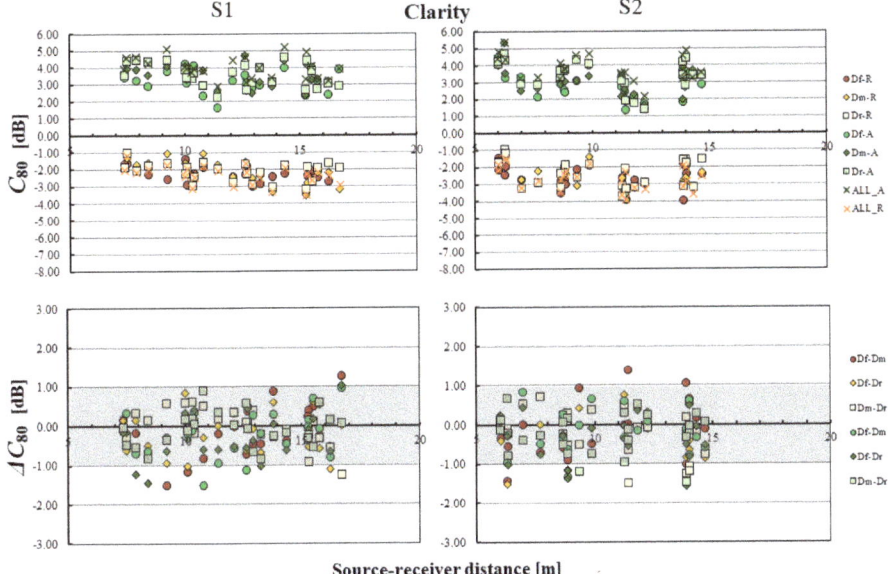

Figure 7. C_{80} parameter averaged over 500 Hz and 1000 Hz for S1 and S2 source-to-receiver distance. ΔC_{80} represents the parameter differences between the configurations Df-Dm, Df-Dr, Dm-Dr in the reflective (-R) and absorptive (-A) conditions. Differences equal ±1 JND of the parameters are highlighted through a gray area.

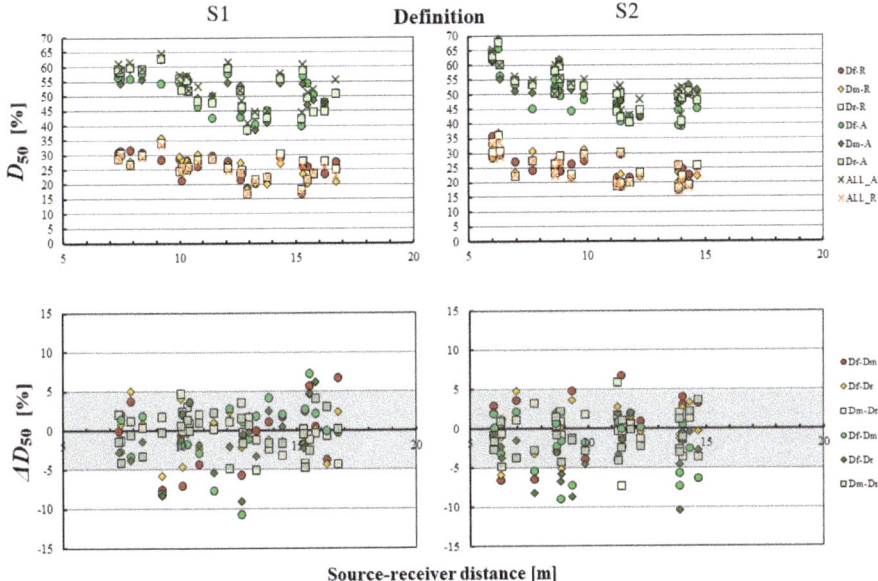

Figure 8. D_{50} parameter averaged over 500 Hz and 1000 Hz for S1 and S2 source-to-receiver distance. ΔD_{50} represents the parameter differences between the configurations Df-Dm, Df-Dr, Dm-Dr in the reflective (-R) and absorptive (-A) conditions. Differences equal ±1 JND of the parameters are highlighted through a gray area.

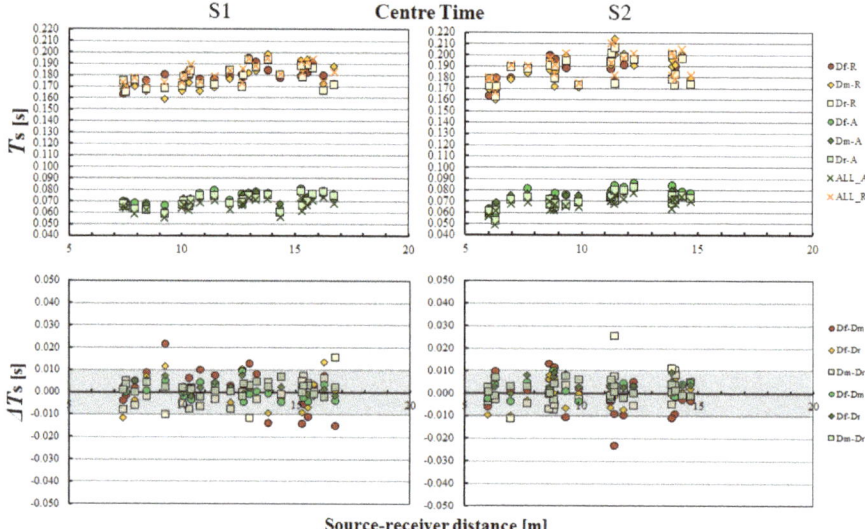

Figure 9. T_s averaged over 500 Hz and 1000 Hz for S1 and S2 source-to-receiver distance. ΔT_s represent the parameter differences between the configurations Df-Dm, Df-Dr, Dm-Dr in the reflective (-R) and absorptive (-A) conditions. Differences equal ±1 JND of the parameters are highlighted through a gray area.

A statistical analysis has been performed on the data shown in Appendix A to investigate the main factor (that is the absorptive/reflective conditions, source S1 and S2, source-to-receiver distance) effects on the variability of the objective acoustic parameters in the comparisons between the tested configurations (Df-Dm, Df-Dr, Dm-Dr). To this aim, only differences above the JND have been considered since it is not meaningful from an acoustic point of view to investigate data lower than the perceived ones. Thus, *EDT*, which resulted in the most affected parameter, was retained suitable for a statistical analysis given the relatively high number of receiver locations that showed differences above the JND. However, it was not possible to apply an ANOVA analysis since the assumptions of normality of data distribution and homogeneity of variance are violated. Given this result, the Kruskal–Wallis (KW) test, which is a non-parametric test and an extension of the Mann–Whitney U Test for more than two groups, has been applied [34]. The Kruskal–Wallis test did not show a statistically significant result ($p > 0.05$) for the differences due to the diffusers location variations.

Table 3 shows the differences in the spatial mean values of each parameter obtained in the conditions with the three different locations of the diffusive surfaces (Df, Dm, and Dr) with respect to the absorptive (All-A) and reflective (All-R) conditions. It can be noticed that the overall results show significant differences for the *EDT* in all the configurations (Df, Dm, and Dr) and also for T_{30} in the Df and Dm configurations with respect to All-A. However, this might be due to the variation of the equivalent absorption area, which decreases when one part of the lateral absorptive walls is set into a diffusive condition. This effect is not evident with respect to the reflective condition (All-R).

A more detailed analysis of the objective parameters has been performed at the head positions. Table 4 gathers the differences of the objective parameters between each compared pair for source position S1 and S2 in the subjective test. The objective parameters at the head position have been evaluated as the values of the parameters obtained at the nearest microphone positions, i.e., microphone position 18 for head 1 and average values of microphone positions 14 and 15 for head 2. The conditions Df and Dr, i.e., the subjectively compared conditions, that lead to differences between the objective parameters above the JND are highlighted in bold.

The combination of the listening position head 1 and source location S1 presents a greater number of parameters (EDT, D_{50}, and $IACC_L$) that reveal differences above the JND in the comparison of Df-A towards Dr-A. Conversely, in the reflective condition (-R), significant differences (>JND) are present only for T_s values. No significant differences can be observed for the combination of the listening position head 2 and source location S1 in both the reflective (-R) and absorptive (-A) conditions.

The combination of the listening position head 2 and source location S2 presents significant differences (>JND) for EDT only in both conditions (-R and -A). No significant differences can be observed for the combination of the listening position head 1 and source location S2 in both reflective (-R) and absorptive (-A) conditions.

Table 3. Spatial mean values and overall standard deviation of reverberation time (T_{30}), early decay time (EDT), clarity (C_{80}), definition (D_{50}), center time (T_s) in the eight conditions. Differences (Δ = All − D) with respect to the reference configurations All-A and All-R are given in brackets for each configuration (Df, Dm, and Dr). The differences above the JND have been highlighted in bold.

		EDT * [s]		T_{30} * [s]		C_{80} [dB]		D_{50} [%]		T_s [s]	
		Mean	SD	Mean	SD	Mean	SD	Mean	SD	Mean	SD
Source 1	All-A	0.92	0.06	1.04	0.02	4.0	0.6	53.9	6.4	0.066	0.005
	All-R	2.57	0.10	2.76	0.06	−2.3	0.6	25.2	3.8	0.182	0.008
	Df-R	2.53	0.10	2.75	0.08	−2.3	0.5	25.8	4.2	0.179	0.007
	Δ	(1.6)		(0.4)		(0.0)		(−0.6)		(0.003)	
	Dm-R	2.49	0.09	2.74	0.07	−2.1	0.7	26	4.5	0.178	0.01
	Δ	(3.1)		(0.7)		(−0.2)		(−0.8)		(0.004)	
	Dr-R	2.52	0.08	2.73	0.06	−2.1	0.6	25.7	4.0	0.179	0.009
	Δ	(1.9)		(1.1)		(−0.2)		(−0.5)		(0.003)	
	Df-A	0.99	0.06	1.10	0.02	3.3	0.7	50.7	6.4	0.072	0.005
	Δ	**(−7.6)**		**(−5.8)**		(0.7)		(3.2)		(−0.006)	
	Dm-A	1.00	0.05	1.10	0.02	3.5	0.6	51.0	6.4	0.072	0.005
	Δ	**(−8.7)**		**(−5.8)**		(0.5)		(2.9)		(−0.006)	
	Dr-A	0.98	0.06	1.08	0.02	3.6	0.7	51.7	6.8	0.07	0.006
	Δ	**(−6.5)**		(−3.8)		(0.4)		(2.2)		(−0.004)	
Source 2	All-A	0.97	0.07	1.05	0.03	3.8	0.8	53.8	7.0	0.067	0.006
	All-R	2.66	0.12	2.89	0.1	−2.6	0.7	24.2	4.8	0.191	0.012
	Df-R	2.64	0.15	2.87	0.13	−2.6	0.6	24.9	4.2	0.186	0.012
	Δ	(0.8)		(0.7)		(0.0)		(−0.7)		(0.005)	
	Dm-R	2.64	0.12	2.90	0.13	−2.6	0.7	24.7	5.0	0.187	0.013
	Δ	(0.8)		(−0.3)		(0.0)		(−0.5)		(0.004)	
	Dr-R	2.63	0.12	2.88	0.11	−2.4	0.8	25.1	5.0	0.186	0.012
	Δ	(1.2)		(0.3)		(−0.2)		(−0.9)		(0.005)	
	Df-A	1.05	0.07	1.12	0.02	3.0	0.8	49.2	7.3	0.075	0.007
	Δ	**(−8.2)**		**(−6.7)**		(0.8)		(4.6)		(−0.008)	
	Dm-A	1.04	0.07	1.11	0.03	3.2	0.9	51.3	7.1	0.072	0.007
	Δ	**(−7.2)**		**(−5.7)**		(0.6)		(2.5)		(−0.005)	
	Dr-A	1.02	0.07	1.10	0.02	3.4	0.9	52.1	7.3	0.070	0.007
	Δ	**(−5.2)**		(−4.8)		(0.4)		(1.7)		(−0.003)	

* EDT and T_{30} differences (Δ = (All − D) × 100/All) are given in [%].

Table 4. Objective acoustic parameters obtained at the head position for the compared pairs of configurations. The differences above the JND between Df and Dr values in the reflective (-R) and absorptive (-A) conditions have been highlighted in bold.

Parameters		EDT	T_{30}	C_{80}	D_{50}	Ts	$IACC_E$	$IACC_L$	IACC
Units		(s)	(s)	(dB)	(%)	(s)	(-)	(-)	(-)
JND		5% R ≈ 0.10 s A ≈ 0.05 s	5% R ≈ 0.10 s A ≈ 0.05 s	1 dB	5%	10 ms	0.075	0.075	0.075
S1—Head 1	Df-R	2.51	2.71	−2.3	22	**0.180**	0.436	0.117	0.223
	Dm-R	2.45	2.81	−1.6	27	0.170	0.439	0.149	0.220
	Dr-R	2.51	2.70	−1.7	24	**0.170**	0.448	0.104	0.196
	Df-A	**0.99**	1.12	3.6	**43**	0.077	0.611	0.168	0.428
	Dm-A	0.93	1.10	4.7	54	0.067	0.605	0.232	0.466
	Dr-A	**0.93**	1.10	4.1	**52**	0.069	0.617	0.251	0.458
S1—Head 2	Df-R	2.57	2.76	−2.6	21	0.188	0.350	0.158	0.178
	Dm-R	2.58	2.70	−2.9	20	0.191	0.367	0.125	0.188
	Dr-R	2.59	2.71	−2.6	22	0.190	0.408	0.158	0.210
	Df-A	1.03	1.08	3.1	43	0.078	0.478	0.153	0.349
	Dm-A	1.01	1.08	3.0	40	0.078	0.430	0.177	0.301
	Dr-A	0.99	1.06	3.6	43	0.074	0.545	0.140	0.407
S2—Head 1	Df-R	2.73	2.73	−2.8	23	0.196	0.408	0.120	0.190
	Dm-R	2.74	2.73	−3.0	22	0.191	0.414	0.124	0.215
	Dr-R	2.75	2.75	−2.9	23	0.196	0.358	0.127	0.172
	Df-A	1.18	1.10	1.7	43	0.086	0.426	0.214	0.315
	Dm-A	1.10	1.08	1.6	42	0.083	0.486	0.192	0.351
	Dr-A	1.15	1.08	1.4	45	0.083	0.435	0.140	0.319
S2—Head 2	Df-R	**2.81**	2.91	−3.0	21	0.195	0.464	0.162	0.250
	Dm-R	2.68	2.94	−3.0	20	0.196	0.437	0.190	0.257
	Dr-R	**2.67**	2.93	−3.4	19	0.198	0.450	0.193	0.264
	Df-A	**1.04**	1.13	2.9	46	0.078	0.546	0.225	0.393
	Dm-A	1.09	1.13	2.8	48	0.077	0.535	0.260	0.419
	Dr-A	**0.99**	1.12	3.1	47	0.074	0.595	0.295	0.457

3.2. Subjective Results

The subjective data gathered from the listening tests have been analyzed based on binomial distribution [35] in order to determine the statistical significance of the test results. The inverse cumulative probability is used to evaluate the minimum number of correct answers that are needed to indicate a perceptual difference. The inverse cumulative probability is given as a function of the trials (corresponding to the thirty-one listeners), probability of correct answers (50%), and confidence level (95%). Therefore, the minimum number of correct answers necessary to indicate a significant difference between pairs at a 95% confidence level was found to be 15, i.e., correct answers should result equal or higher than 15.

Figure 10 shows the correct answers for each music/signal passage, listening (head), and source position. The dashed horizontal line indicates the minimum number of correct answers necessary to detect a significant perceptual difference between configurations compared in one pair. No significant variations of the location of diffusive surfaces were significantly perceived in any of the compared pairs. Some of the listeners could still indicate a few differences relying on different attributes as presented in Figure 11, which shows the occurrences of each attribute given in the correct answers. Further, according to the feedback of the listeners, for each signal (trumpet, orchestra, pink noise), more than 75% of the correct answers were given by relying on two or more attributes (reverberance, coloration, and spaciousness). Among them, reverberance is the main attribute when the orchestra and pink noise samples are compared in the reflective condition.

Finally, given the small perceived differences, it was not possible to collect reliable results regarding the preference indicated by the listeners.

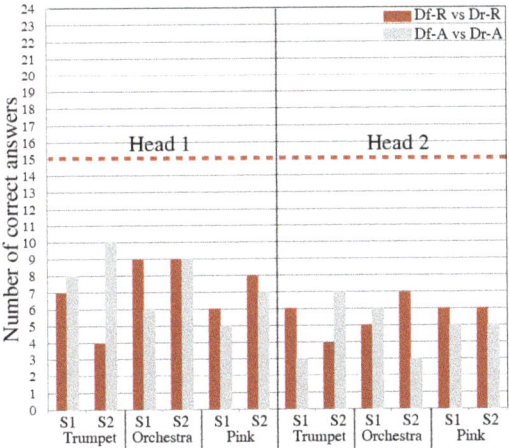

Figure 10. Listening test results. The dashed horizontal line indicates the minimum number of correct answers necessary to indicate a significant perceptual difference between configurations compared in one pair. *x*-axis indicates the pairs compared in each trial.

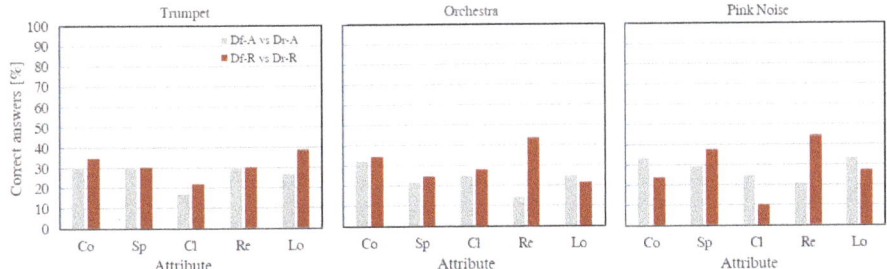

Figure 11. Listening test results. The listeners' subjective evaluations on the perceived differences between front and rear location of the diffusive surfaces in the absorptive and reflective conditions (Df-A vs. Dr-A and Df-R vs. Dr-R). The *y*-axis depicts the occurrences of each attribute given in the correct answers. The *x*-axis reports the attributes Co—coloration, Sp—spaciousness, Cl—clarity, Re—reverberance, Lo—loudness.

4. Discussion

This work aims to give more insight into the design aspects of concert halls related to the effects of diffusive surfaces location. Based on the results presented above, a few practically relevant comments can be made in order to achieve a more mindful design of concert halls and intervene in those areas that could lead to the required objective and perceived acoustic quality.

The objective analyses presented in Figures 5–9 and Appendix A showed that the objective parameters are not significantly influenced by the diffusive surface location. These results confirm the findings of previous investigations Jeon et al. [18] and Jeon et al. [12], i.e., the diffusers installation on lateral walls do not have any significant effect on the overall acoustic parameters. However, a few significant differences could be observed at single receiver positions. Generally, no clear trend can be observed for T_{30}, C_{80}, D_{50}, and T_s variations in the different configurations in both the absorptive (-A) and reflective (-R) conditions. EDT was shown to be the most affected parameter. The differences over the configurations show that this is more evident for source location S2 and occurs at a larger number of receivers in the absorptive (-A) conditions. However, no significant trend could be observed with respect to the source-to-receiver distance and the statistical analysis did not show a statistically significant difference between the different diffusers locations. Generally, when the different configurations have been compared to the reference conditions (Table 3), no significant differences resulted in the reflective condition while in the absorptive conditions EDT and T_{30} resulted in the most affected.

It was shown that in the absorptive conditions (-A), the combination of the listening position head 1 and source location S1 presented a greater number of parameters (EDT, D_{50}, and $IACC_L$) that reveal differences above 1 JND in comparison to Df-A with Dr-A (Table 4). Conversely, the listening position head 2 and source location S2 presented significant differences (>1 JND) for EDT only in both the reflective (-R) and absorptive (-A) conditions. Given these differences, as in previous studies [7,12], it was not possible to correlate the objective parameters differences in these two positions to the perceived differences.

The subjective test did not show significant perceived differences between the configurations Df-A and Dr-A or Df-R and Dr-R, i.e., front-rear asymmetric conditions of the diffusive surface location with respect to the listener position. Some of the listeners could still indicate a few differences by relying on different attributes as presented in Figure 10, which shows the occurrences of each attribute given in the correct answers. However, it was not possible to identify the preferred location of the diffusers due to the small perceived differences. It was observed that for each signal (trumpet, orchestra, pink noise), more than 75% of the correct answers were given relying on two or more attributes (reverberance, coloration, and spaciousness). These attributes have been also highlighted as the most affected in previous studies [4,7,20,21,23]. Reverberance seems to be the main attribute when the orchestra and pink noise samples are compared in the reflective condition. However, despite the differences between the samples typologies it was not possible to determine a significant difference between them, which is in line with the findings in [12–16].

The objective and subjective results highlight the need for further investigations on new parameters. More systematic investigations might focus on the number of peaks (N_p) proposed by [19], which correlates to the listener preference or the 'effective duration' of the autocorrelation function (τ_e), which correlates to the intimacy and reverberance [36] and has been proposed as key factor to 'preferred' values of several room criteria in relation to different kind of music signals [37].

It should be highlighted that this study focuses on perceptual differences within a shoebox hall only. Different results might be expected for different hall's shapes and volumes [12,16].

Further research could be performed, as indicated in Kim et al. [9] and Jeon et al. [19], also by taking into account the diffuser shape, size, and directivity of the polar distributions of diffuse reflections. In the ESPRO hall, the diffusers are alternatively vertically or horizontally oriented, i.e., a uniform directivity might be approximated. As it was shown in [9] large and sparse diffuser profiles might result in more effectivity over the acoustic parameters. Moreover, the extension over other surfaces might lead to more significant differences [16]. From the designers' perspective become more interesting the configurations that do not lead to any significant variation on the objective parameters and on the subjective perception. In this way, there might be more freedom on the aesthetical choices that can be applied to the design of a concert hall once that the acoustic optimal conditions have been obtained.

Limitations of the Study

Given the conditions studied in this paper, it should be noted that the receiver's area could be extended also at closer or further locations from the source positions. However, given the small spatial variability of the measured objective parameters, we would expect a limited effect also in the very rear part of the hall. The overall number of measurements in this project was made in an automatized way: the surfaces of the room were varied from a control room and the overall set-up of sources and microphone positions were set in the most representative locations in order to avoid entering the room with the risk of variation of its conditions. Therefore, in the attempt to reach the right tradeoff between the gathered acoustic information, room configurations, number of microphones and sources, and time limitations on the use of the room itself, it was concluded that the presented protocol was the most suitable one.

The results of this study have highlighted some important issues related to the relevance of the diffuser's location in a performance hall. However, it should be underlined that only two listening positions have been used in this investigation. Given the differences that might occur due to source-receiver locations, it might be useful to increase the number of listening positions in order to have clearer evidence of the diffusive surface effects on the overall sound field perception. It might be useful to investigate also more representative positions of the front and rear rows of listeners. However, given the time limitations of the use of the ESPRO for this project, it was not possible to extend the number of dummy head positions. It should be considered that the simplification introduced by an ensemble generated from a single source location on the stage might have influenced the spatial impression for the orchestra sample. When technical and budget availability may cover important experimental costs, multiple sources might be a more accurate representation for this case as shown for the orchestra of loudspeakers in [38,39]. Moreover, in each receiver position, a multi-microphone technique could be used to enable multichannel 3D sound reproduction. Therefore, a spatial sound reproduction could have led to a more realistic listening condition. It might have been easier to identify differences when head movements are allowed since they are naturally used when attending concerts [40].

One of the limitations of this study is related to the use of non-individual HRTF, which could have affected the performance of the subject by diminishing the effects of the different surface locations. Research on the use of individual HRTF data sets has shown that their use would allow for better performance of the subjects in localization tasks and lower front-back confusions [41]. It was not possible to apply individual HRTFs due to the amount of technical effort that should be put to measure these data sets [42]. However, since the same dummy head was used in all the measurements here, this could not have any influence on the relative differences between the compared conditions.

The reverberation time characteristics of around 1 s in the absorptive conditions might have influenced the perception and preference of music samples, which are usually played in rooms with longer reverberation times for optimal listening. However, since the test was based on relative comparisons the influence on the distinction of the differences. Based on the JND definition, it measures the sensitivity of the listeners to a change in a given parameter and is focused on acoustic conditions typically found in concert halls or auditoria [1]. In very large or very small rooms the relations between the different parameters may change and consequently, the perceived differences may also be affected [43]. Therefore, the effects investigated in this research should be considered valid for the room volume of the case study and related ranges of reverberation time.

These aspects remain open to future research where also investigations with experts might lead to a more detailed description of other attributes related to the acoustic quality [44]. Although, previous studies on diffusive-to-reflective surface discrimination have shown compatible results between experts and non-expert listeners [7]. Moreover, also the effects of the diffuser location over the stage area and musicians' perception could be investigated with specific protocols as in [8]. The effect of diffusers on a different type of performances and related preference remains a crucial point to be further investigated given the importance of the specific effects recreated by the artists' work [45].

Finally, this study is by no means comprehensive, many other diffusive surface locations strategies exist, and further investigations of additional strategies will be useful to refine and expand the findings presented here over a larger number of hall's shapes and volumes.

5. Conclusions

In situ measurements and perceptual listening tests have been used to investigate the influence of diffusive surface location on the acoustic parameters used in the design process of concert halls and on the perceived acoustic sound field. The case study involved a real concert hall with variable acoustics (ESPRO, IRCAM, Paris, France), where eight hall configurations have been generated by modifying the characteristics of the lateral walls. The objective evaluation has been carried out by analyzing the variation of the ISO 3382-1 [17] acoustic parameters T_{30}, EDT, C_{80}, D_{50}, and $IACC$ in each configuration, while the perceptual tests have been performed using the ABX method in order to determine whether listeners are sensitive to variations of diffusive surface location. This study gives further insight into the importance of the quantification of the trade-off between the design effort and objective and subjective efficacy of the diffusers application in shoebox halls.

The main conclusions can be summarized as follows:

- The objective parameters are not significantly influenced by the diffusive surface location. No clear trend can be observed for T_{30}, C_{80}, D_{50}, and T_s variations in the different configurations in both the absorptive (-A) and reflective (-R) conditions. EDT results as the most affected parameter.
- The perceived differences between the front-rear asymmetric conditions of the diffusive surface location with respect to the listener position do not show significant differences. However, some of the listeners could still indicate a few differences relying on two or more attributes (reverberance, coloration, and spaciousness). Reverberance seems to be the main attribute when the orchestra and pink noise samples are compared in the reflective condition.

Future work should include different hall shapes and volumes in order to have also a more generalized overview of the interaction between room shape and effects of diffusive surfaces. More effort should be put into the investigation of differently shaped surfaces, i.e., different diffusion patterns and scattering values, and different degrees of diffusive surface extensions. More adequate sound sources and reproduction systems might be used in order to have more accurate results although the technical and economical effort for these improvements seems to be important.

The findings of this study should be seen as a milestone based on in situ results towards the redaction of reliable guidelines, which could enable an easier design process for architects and practitioners alike. The limited effects of the diffusive surfaces give space to a broad field of design alternatives from the designers' perspective. In this way, there might be more freedom on the aesthetical choices that can be applied to the design of a concert hall once that the acoustic optimal conditions have been obtained. It might be useful to investigate the boundaries of this filed within which the dialog between designers and acousticians would promote further aspects related to creativity.

Author Contributions: L.S. and A.A. conceived and designed the data collection campaigns; L.S. collected data on site; L.S. and S.D.B. performed data analysis; L.S. drafted and curated the first version of the manuscript. All the authors revised the paper. All authors have read and agreed to the published version of the manuscript.

Funding: This research was funded through a Ph.D. scholarship awarded to the first author by the Politecnico di Torino (Turin, Italy).

Acknowledgments: The authors would like to thank all the participants in the listening test sessions for the collaboration. They would also like to express their gratitude to Michael Vorländer, Sönke Pelzer, Renzo Vitale, and Fabian Knauber from the Institute of Technical Acoustic in Aachen, and Olivier Warusfel and Markus Noisternig from IRCAM in Paris for their technical support and availability during the measurements.

Conflicts of Interest: The authors declare no conflicts of interest.

Appendix A

Table A1. Objective acoustic parameter differences between the configurations with different diffuser locations in the reflective (-R) condition for S1. Differences equal or higher than the JNDs of the parameters are given in bold font.

R	d-S1 [m]	ΔEDT [%]			ΔT_{30} [%]			ΔC_{80} [dB]			ΔD_{50} [%]			ΔT_s [s]		
		Df-Dm	Df-Dr	Dm-Dr	Df-Dm	Df-Dr	Dm-Dr	Df-Dm	Df-Dr	Dm-Dr	Df-Dm	Df-Dr	Dm-Dr	Df-Dm	Df-Dr	Dm-Dr
6	7.37	-0.8	**-7.4**	**-6.6**	-1.1	1.4	2.5	0.2	0.1	0.0	0.0	2.1	2.1	-0.004	**-0.012**	-0.008
5	7.48	2.0	2.9	0.9	2.3	3.1	0.8	-0.3	-0.6	-0.3	2.0	0.8	-1.3	-0.001	0.004	0.005
4	7.87	-0.2	3.7	3.9	2.3	3.7	1.3	-0.2	0.2	0.3	3.8	**5.0**	1.2	0.002	-0.004	-0.006
3	8.4	**6.8**	2.6	-3.9	-0.1	-0.1	0.0	-0.6	-0.5	0.2	0.5	0.9	0.4	0.009	0.007	-0.002
2	9.22	**10.3**	4.9	-4.8	0.5	-1.2	-1.7	**-1.5**	-0.9	0.6	**-7.6**	**-5.8**	1.8	**0.022**	**0.012**	**-0.010**
12	10	3.0	**7.5**	4.4	1.0	1.2	0.2	0.2	0.8	0.6	-0.7	4.1	4.8	0.002	-0.003	-0.005
11	10.08	-2.2	**-5.9**	-3.7	0.0	2.4	2.4	**-1.2**	**-1.0**	0.2	**-7.0**	**-4.6**	2.4	0.002	0.002	-0.001
1	10.3	2.0	3.5	1.5	4.3	3.5	-0.8	0.0	0.6	0.6	0.0	3.0	3.0	0.006	-0.001	-0.008
10	10.38	2.6	4.5	1.9	1.3	0.5	-0.8	0.4	0.2	-0.2	0.0	0.9	1.0	-0.002	0.000	0.002
9	10.79	3.8	-0.5	-4.1	0.5	-0.8	-1.3	-0.8	-0.3	0.5	-4.3	-2.2	2.1	**0.010**	0.004	-0.006
8	11.43	2.1	**6.7**	4.6	1.6	-0.7	-2.3	-0.2	0.0	0.2	0.9	1.1	0.2	0.008	0.004	-0.003
7	12.1	-1.6	-3.4	-1.8	1.3	0.7	-0.5	0.0	0.3	0.3	1.2	2.4	1.1	0.003	-0.005	-0.008
18	12.63	2.2	0.1	-2.0	-3.9	0.2	4.3	-0.7	-0.6	0.1	**-5.6**	-2.1	3.5	**0.010**	**0.010**	0.000
17	12.69	1.4	-0.2	-1.6	-0.5	1.2	1.7	0.4	0.5	0.1	-1.1	-1.4	-0.3	-0.002	-0.001	0.000
16	12.93	**10.9**	**6.7**	-3.9	-0.2	0.6	0.8	-0.5	-0.1	0.4	0.2	1.9	1.7	**0.013**	0.001	**-0.012**
15	13.26	**7.5**	1.5	**-5.6**	2.6	2.6	0.0	-0.5	-0.7	-0.2	0.0	-1.2	-1.2	0.008	0.005	-0.003
14	13.79	**-7.9**	-3.5	4.8	2.2	1.6	-0.6	0.9	0.6	-0.3	1.1	-1.2	-2.3	**-0.014**	-0.009	0.004
13	14.35	0.7	-3.2	-3.9	1.6	1.3	-0.3	-0.4	-0.5	-0.1	1.8	-1.4	-3.2	-0.002	-0.003	-0.001
24	15.26	**-10.3**	**-11.7**	-1.6	1.1	2.6	1.4	0.2	-0.1	-0.3	-1.7	-1.4	0.2	**-0.014**	-0.009	0.005
23	15.31	1.0	-1.2	-2.2	-3.5	-2.2	1.3	0.4	-0.5	-0.9	2.5	-1.7	-4.2	-0.005	0.002	0.007
22	15.51	-4.2	1.8	**6.3**	-0.8	-0.4	0.4	0.5	0.6	0.1	**5.7**	4.6	-1.1	**-0.011**	-0.007	0.004
21	15.76	4.2	0.2	-3.9	0.0	-0.2	-0.3	-0.3	-0.6	-0.3	0.5	0.2	-0.3	0.003	0.004	0.000
20	16.23	2.6	2.9	0.4	-2.7	-2.8	-0.1	-0.5	**-1.1**	-0.6	-3.7	-4.3	-0.7	0.007	**0.014**	0.006
19	16.71	3.9	3.7	-0.2	1.2	0.8	-0.3	**1.3**	0.0	**-1.3**	**6.7**	2.4	-4.3	**-0.015**	0.001	**0.016**

Table A2. Objective acoustic parameter differences between the configurations with different diffuser locations in the reflective (-R) condition for S2.

R	d-S2 [m]	ΔEDT [%]			ΔT₃₀ [%]			ΔC₈₀ [dB]			ΔD₅₀ [%]			ΔT_s [s]		
		Df-Dm	Df-Dr	Dm-Dr	Df-Dm	Df-Dr	Dm-Dr	Df-Dm	Df-Dr	Dm-Dr	Df-Dm	Df-Dr	Dm-Dr	Df-Dm	Df-Dr	Dm-Dr
2	6.01	0.6	-2.7	-3.2	0.4	3.8	3.4	0.2	0.2	0.1	2.9	2.1	-0.8	-0.006	-0.010	-0.004
3	6.02	-3.0	-2.7	0.3	2.0	1.6	-0.4	-0.3	-0.4	-0.1	0.1	-1.8	-1.9	0.002	0.000	-0.002
1	6.28	-3.7	-6.7	-3.1	1.7	0.8	-0.8	-1.4	-1.5	-0.1	-6.6	-6.0	0.7	0.007	0.005	-0.002
4	6.32	10.4	3.4	-6.4	-5.2	-4.0	1.3	-0.6	-0.8	-0.2	-0.5	-1.0	-0.6	0.010	0.007	-0.003
5	6.96	0.8	-0.4	-1.1	0.9	3.2	2.3	0.0	0.5	0.5	3.6	4.8	1.2	0.001	-0.010	-0.011
6	7.69	-5.9	0.2	6.5	-5.0	-1.8	3.4	-0.7	0.0	0.7	-6.5	-3.3	3.2	0.001	-0.003	-0.004
8	8.63	4.1	6.2	2.0	-1.5	1.2	2.7	-0.6	-0.3	0.2	-0.2	0.8	0.9	0.013	0.006	-0.007
9	8.64	3.0	3.8	0.8	-0.4	-0.2	0.2	-0.2	-0.5	-0.2	-2.9	-2.5	0.4	0.006	0.008	0.002
7	8.82	7.7	2.7	-4.6	-0.9	-0.7	0.2	-0.9	-0.6	0.3	-0.9	0.0	0.9	0.007	0.000	-0.007
10	8.85	3.1	-3.0	-5.9	-1.3	-0.4	0.9	-0.6	-1.2	-0.5	-4.3	-5.2	-0.9	0.012	0.006	-0.005
11	9.32	1.0	-0.2	-1.2	-2.9	-1.7	1.3	0.9	0.4	-0.5	4.8	3.6	-1.2	-0.010	-0.007	0.004
12	9.87	-4.9	-2.4	2.7	-4.9	-1.6	3.4	-0.5	-0.1	0.4	-3.9	-2.1	1.8	0.002	0.000	-0.002
14	11.25	-3.8	-1.9	2.0	-0.2	0.5	0.7	0.0	0.4	0.4	1.1	2.8	1.7	-0.003	-0.006	-0.003
15	11.26	13.7	12.7	-0.8	-1.8	-2.2	-0.4	0.3	0.8	0.5	0.2	-0.1	-0.3	0.003	0.001	-0.002
13	11.4	-5.9	2.3	8.7	1.6	-2.9	-4.5	1.4	-0.1	-1.5	6.7	-0.7	-7.4	-0.023	0.002	0.026
16	11.42	-4.8	-3.5	1.3	-1.7	-1.6	0.1	0.0	-0.6	-0.6	-1.2	-1.6	-0.4	-0.009	-0.001	0.008
17	11.79	-4.7	-0.3	4.6	-0.9	-1.3	-0.4	0.5	0.4	-0.1	1.0	1.5	0.5	-0.010	-0.007	0.002
18	12.23	-0.2	-0.5	-0.3	-0.2	-1.0	-0.8	0.0	-0.1	-0.1	0.9	-0.4	-1.3	0.005	0.000	-0.005
20	13.87	-7.5	-4.4	3.4	-1.1	-2.3	-1.3	1.0	-0.2	-1.3	1.3	-1.3	-2.6	-0.011	0.000	0.011
21	13.88	-1.9	-3.0	-1.1	0.4	-0.3	-0.8	-1.0	-0.8	0.2	-2.8	-0.8	1.9	0.004	-0.001	-0.005
19	14	-3.9	-1.6	2.4	1.6	-1.3	-2.9	0.7	-0.4	-1.1	3.1	-0.5	-3.6	-0.009	0.001	0.010
22	14.02	6.8	7.8	0.9	0.0	-0.9	-1.0	0.5	-0.6	-1.2	4.0	2.6	-1.4	-0.002	0.007	0.008
23	14.31	2.2	4.7	2.5	-0.3	1.6	1.9	0.2	-0.1	-0.2	2.0	3.4	1.3	-0.003	0.001	0.003
24	14.68	-1.3	-2.7	-1.4	-1.7	0.0	1.8	-0.1	-0.9	-0.8	3.2	-0.3	-3.6	-0.003	0.002	0.005

Table A3. Objective acoustic parameter differences between the configurations with different diffuser locations in the absorptive (-A) condition for S1.

R	d-S1 [m]	ΔEDT [%]			ΔT30 [%]			ΔC80 [dB]			ΔD50 [%]			ΔTs [s]		
		Df-Dm	Df-Dr	Dm-Dr	Df-Dm	Df-Dr	Dm-Dr	Df-Dm	Df-Dr	Dm-Dr	Df-Dm	Df-Dr	Dm-Dr	Df-Dm	Df-Dr	Dm-Dr
6	7.37	−4.1	−7.1	−3.1	−1.2	3.1	4.3	−0.1	−0.2	0	−1.3	−2.7	−1.4	0.001	0.002	0.001
5	7.48	−2.5	−0.5	2.1	1.2	1	−0.3	0.3	−0.1	−0.5	1.5	−2.6	−4.1	0.000	0.004	0.003
4	7.87	6.2	−0.4	−6.2	−2.4	0.2	2.6	−0.7	−1.2	−0.5	−3.3	−3.8	−0.5	0.005	0.005	0.000
3	8.4	4.6	8.6	3.8	1.5	0.8	−0.7	−0.6	−1.4	−0.8	1.9	−1.4	−3.3	0.001	0.005	0.005
2	9.22	−5.8	−2.1	3.9	1.4	3.7	2.3	−0.3	−0.6	−0.4	−8.1	−8.3	−0.2	0.005	0.007	0.002
12	10	−6.5	−2.5	4.2	1.8	2.9	1	0.2	0.3	0.2	−1.8	−1.2	0.6	0.000	0.001	0.002
11	10.08	−5.5	−1.4	4.3	4.1	6.2	2	−0.2	−0.3	−0.1	2.1	2.1	0	0.000	0.000	0.002
1	10.3	5.8	9	3	2.2	6.1	3.9	−0.1	0.1	0.2	−1.7	−0.3	1.3	−0.002	0.002	0.001
10	10.38	−3	1.7	4.8	1.3	3.7	2.4	0.4	0.4	0.1	3.6	3.6	0	0.000	0.000	0.001
9	10.79	9.7	4.1	−5.2	−1.1	2.3	3.4	−1.5	−0.6	0.9	−2.9	−1.8	1	−0.003	−0.002	−0.002
8	11.43	−3.1	0.2	3.5	0.7	2	1.3	−0.9	−0.6	0.3	−7.7	−5.4	2.3	0.004	0.002	0.001
7	12.1	−1.8	−2.1	−0.3	−0.4	3.8	4.2	−0.6	−0.5	0	2.8	−2	−4.8	0.004	0.004	0.000
18	12.63	7.2	6.8	−0.4	2	1.3	−0.7	−1.1	−0.5	0.6	−10.7	−9.1	1.6	0.002	0.007	−0.002
17	12.69	−10	−4.3	6.4	1.4	0.8	−0.6	0.5	0.4	−0.1	1.1	−0.5	−1.6	0.009	−0.001	0.004
16	12.93	−3.1	3.4	6.6	0.7	3.5	2.8	0.3	−0.4	−0.6	−1	−0.3	0.8	−0.004	0.002	0.003
15	13.26	−0.3	3.3	3.6	0.7	−1.2	−1.8	−0.2	−1	−0.8	1.8	−3.2	−5.1	−0.001	0.005	0.005
14	13.79	4.5	5	0.5	−0.5	4.5	5	0.3	0	−0.3	4.2	2.6	−1.6	0.000	0.002	0.002
13	14.35	−4.6	9.2	14.5	−4.2	2.5	7	−0.4	−0.6	−0.2	2.1	0.5	−1.6	−0.001	0.002	0.007
24	15.26	−4.5	−0.5	4.2	−2	−0.2	1.8	0	−0.3	−0.3	−1.8	−2.2	−0.4	−0.004	0.002	0.000
23	15.31	−6.8	−1.1	6.2	0.8	4.1	3.3	0.2	−0.1	−0.2	2.7	−2	−4.7	0.002	0.000	0.003
22	15.51	3	3.2	0.2	0.6	0.3	−0.3	0.7	0.2	−0.5	7.2	4.7	−2.5	−0.002	−0.001	0.003
21	15.76	−3.9	2.7	6.9	1.9	0.4	−1.5	−0.1	0.5	0.6	2.1	6.2	4.1	−0.004	−0.002	−0.001
20	16.23	4	5.9	1.8	−1	−1.1	−0.1	−0.8	−0.6	0.2	0	3	3.1	0.003	0.001	−0.002
19	16.71	−5.5	−12.1	−7	0.9	3.2	2.3	1	1	0.1	−0.3	0	0.2	−0.004	−0.002	0.002

Table A4. Objective acoustic parameter differences between the configurations with different diffuser locations in the absorptive (−A) condition for S2.

R	d-S2 [m]	ΔEDT [%]			ΔT_{30} [%]			ΔC_{80} [dB]			ΔD_{50} [%]			ΔT_s [s]	
		Df-Dm	Df-Dr	Dm-Dr	Df-Dm	Df-Dr	Dm-Dr	Df-Dm	Df-Dr	Dm-Dr	Df-Dm	Df-Dr	Dm-Dr	Df-Dm	Df-Dr	Dm-Dr
2	6.01	−3	−2.7	0.3	−0.4	4.8	5.3	0.2	0.1	−0.1	1.8	−0.8	−2.6	−0.002	0.000	0.003
3	6.02	11.3	2.5	−8	0.5	3.2	2.6	0.1	0.2	0.1	0.2	−1.4	−1.6	0.001	0.002	0.001
1	6.28	2.2	−3.5	−5.6	2.4	3.9	1.4	−1	−0.3	0.7	−3.1	−2.3	0.9	0.004	0.003	−0.001
4	6.32	−5.2	6.9	12.7	5.3	3.8	−1.4	−0.2	−1	−0.8	1.2	−3.7	−4.9	−0.001	0.006	0.007
5	6.96	−6.6	−10.3	−4	1.9	2.4	0.5	0.8	0.4	−0.4	2.2	−1.5	−3.7	−0.003	0.000	0.003
6	7.69	−3	2.9	6.1	0	1.1	1.1	−0.5	−0.8	−0.3	−5.5	−8.2	−2.7	0.005	0.008	0.003
8	8.63	5.1	10.9	5.5	0.4	−1.1	−1.5	0.3	−0.2	−0.5	0.6	2.1	1.5	0.001	0.001	0.000
9	8.64	2.4	0.2	−2.2	1	2.5	1.5	−0.4	−0.3	0.1	−2.2	−3	−0.8	0.002	0.002	0.000
7	8.82	7.1	4.7	−2.2	−4.3	−1.7	2.8	−1.3	−1.2	0.2	−9	−6.8	2.2	0.002	0.007	−0.003
10	8.85	10.9	13	1.9	0.6	1.2	0.5	−0.6	−1.4	−0.8	−1.5	−5.8	−4.2	0.005	0.009	0.004
11	9.32	−12.8	2.6	17.6	3.1	2.1	−1	0	−1.2	−1.2	−7.3	−8.7	−1.4	0.001	0.008	0.008
12	9.87	−7.1	−0.7	6.9	5	2.4	−2.5	0.6	−0.1	−0.7	−1.7	−4.5	−2.8	−0.003	0.003	0.006
14	11.25	−3.1	5.8	9.2	0	2.2	2.1	−0.3	0	0.3	−4	1.8	5.8	0.004	0.003	−0.001
15	11.26	−5.5	3.8	9.9	0.3	0.1	−0.3	0.6	−0.4	−1	0.8	−3.3	−4.1	−0.002	0.005	0.006
13	11.4	12.1	6.4	−5.1	−4.6	3.9	8.8	−0.1	−0.5	−0.4	0	−2.5	−2.4	0.001	0.004	0.003
16	11.42	2.6	3.7	1	−0.6	2.1	2.7	−0.4	−0.5	−0.1	−2.4	−1.4	1	0.004	0.005	0.001
17	11.79	14.5	6	−7.5	−0.2	1.2	1.4	−0.2	0.4	0.5	2	1.9	−0.2	0.005	0.003	−0.002
18	12.23	7.1	2.5	−4.3	2.2	1.5	−0.6	0.1	0.3	0.2	0.2	−2.1	−2.3	0.003	0.003	0.000
20	13.87	4.3	1.6	−2.6	1.1	0.5	−0.5	−0.2	−0.5	−0.3	−5.8	−4.6	1.2	0.003	0.004	0.001
21	13.88	−2	1.9	3.9	4.6	1.8	−2.7	−0.1	−1.6	−1.5	−7.4	−10.4	−3	0.005	0.009	0.004
19	14	−6.5	1.6	8.6	6.1	3.6	−2.3	0.6	0.5	−0.1	−0.7	−1.6	−0.9	−0.002	0.002	0.004
22	14.02	6.3	8.6	2.1	−1.7	−0.3	1.5	−0.8	−0.8	0	0.5	−0.6	−1.1	0.003	0.004	0.001
23	14.31	−8.4	5.1	14.7	0.7	1.1	0.4	−0.3	−0.1	0.3	−2.6	−0.4	2.1	0.004	0.003	−0.001
24	14.68	2.7	7.7	4.9	−1.3	−0.4	0.9	−0.6	−0.5	0.1	−6.4	−2.7	3.6	0.005	0.004	−0.001

References

1. Beranek, L. *Concert Halls and Opera Houses: How They Sound*; Acoustical Society of America: New York, NY, USA, 1996.
2. Haan, C.; Fricke, F.R. An evaluation of the importance of surface diffusivity in concert halls. *Appl. Acoust.* **1997**, *51*, 53–69. [CrossRef]
3. Cox, T.; D'Antonio, P. *Acoustic Absorbers and Diffusers. Theory, Design and Application*; CRC Press: Boca Raton, FL, USA, 2007.
4. Ryu, J.K.; Jeon, J.Y. Subjective and objective evaluations of a scattered sound field in a scale model opera house. *J. Acoust. Soc. Am.* **2008**, *124*, 1538. [CrossRef]
5. Hanyu, T. A theoretical framework for quantitatively characterizing sound field diffusion based on scattering coefficient and absorption coefficient of walls. *J. Acoust. Soc. Am.* **2010**, *128*, 1140. [CrossRef] [PubMed]
6. Jeon, J.Y.; Kim, J.H.; Seo, C.K. Acoustical remodeling of a large fan-type auditorium to enhance sound strength and spatial responsiveness for symphonic music. *Appl. Acoust.* **2012**, *73*, 1104–1111. [CrossRef]
7. Shtrepi, L.; Astolfi, A.; D'Antonio, G.; Guski, M. Objective and perceptual evaluation of distance-dependent scattered sound effects in a small variable-acoustics hall. *J. Acoust. Soc. Am.* **2016**, *140*, 3651–3662. [CrossRef] [PubMed]
8. Jeon, J.Y.; Kim, Y.S.; Lim, H.; Cabrera, D. Preferred positions for solo, duet, and quartet performers on stage in concert halls: In situ experiment with acoustic measurements. *Build. Environ.* **2015**, *93*, 267–277. [CrossRef]
9. Kim, Y.H.; Kim, J.H.; Jeon, J.Y. Scale Model Investigations of Diffuser Application Strategies for Acoustical Design of Performance Venues. *Acta Acust. United Acust.* **2011**, *97*, 791–799. [CrossRef]
10. Lokki, T.; Pätynen, J.; Tervo, S.; Siltanen, S.; Savioja, L. Engaging concert hall acoustics is made up of temporal envelope preserving reflections. *J. Acoust. Soc. Am.* **2011**, *129*, EL223–EL228. [CrossRef]
11. Robinson, P.W.; Pätynen, J.; Lokki, T.; Jang, H.S.; Jeon, J.Y.; Xiang, N. The role of diffusive architectural surfaces on auditory spatial discrimination in performance venues. *J. Acoust. Soc. Am.* **2013**, *133*, 3940–3950. [CrossRef] [PubMed]
12. Jeon, J.Y.; Jo, H.I.; Seo, R.; Kwak, K.H. Objective and subjective assessment of sound diffuseness in musical venues via computer simulations and a scale model. *Build. Environ.* **2020**, *173*. [CrossRef]
13. Shtrepi, L. Investigation on the diffusive surface modeling detail in geometrical acoustics based simulations. *J. Acoust. Soc. Am.* **2019**, *145*, EL215–EL221. [CrossRef] [PubMed]
14. Shtrepi, L.; Astolfi, A.; Puglisi, G.E.; Masoero, M. Effects of the Distance from a Diffusive Surface on the Objective and Perceptual Evaluation of the Sound Field in a Small Simulated Variable-Acoustics Hall. *Appl. Sci.* **2017**, *7*, 224. [CrossRef]
15. Shtrepi, L.; Astolfi, A.; Pelzer, S.; Vitale, R.; Rychtáriková, M. Objective and perceptual assessment of the scattered sound field in a simulated concert hall. *J. Acoust. Soc. Am.* **2015**, *138*, 1485–1497. [CrossRef]
16. Shtrepi, L.; Rychtáriková, M.; Astolfi, A. The Influence of a Volume Scale-Factor on Scattering Coefficient Effects in Room Acoustics. *Build. Acoust.* **2014**, *21*, 153–166. [CrossRef]
17. ISO3382-1. *Acoustics-Measurement of Room Acoustic Parameters-Part 1: Performance of Spaces*; International Standards Organization: Geneva, Switzerland, 2009.
18. Jeon, J.Y.; Seo, C.K.; Kim, Y.H.; Lee, P.J. Wall diffuser designs for acoustical renovation of small performing spaces. *Appl. Acoust.* **2012**, *73*, 828–835. [CrossRef]
19. Jeon, J.Y.; Jang, H.S.; Kim, Y.H.; Vorlander, M. Influence of wall scattering on the early fine structures of measured room impulse responses. *J. Acoust. Soc. Am.* **2015**, *137*, 1108–1116. [CrossRef]
20. Torres, R.; Kleiner, M.; Dalenback, B.-I. Audibility of "Diffusion" in room acoustics auralization: An initial investigation. *Acust. Acta Acust.* **2000**, *86*, 919–927.
21. Takahashi, D.; Takahashi, R. Sound Fields and Subjective Effects of Scattering by Periodic-Type Diffusers. *J. Sound Vib.* **2002**, *258*, 487–497. [CrossRef]
22. Singh, P.K.; Ando, Y.; Kurihara, Y. Individual subjective diffuseness responses of filtered noise sound fields. *Acustica* **1994**, *80*, 471–477.
23. Vitale, R. Perceptual Aspects of Sound Scattering in Concert Halls. Ph.D. Thesis, Logos Verlag Berlin GmbH, Aachen, Germany, 2015.

24. ISO17497-1. *Acoustics-Measurement of Sound Scattering Properties of Surfaces. Part 1: Measurement of the Random-Incidence Scattering Coefficient in a Reverberation Room*; International Standards Organization: Geneva, Switzerland, 2004.
25. ISO17497-2. *Acoustics-Sound-Scattering Properties of Surfaces. Part 2: Measurement of the Directional Diffusion Coefficient in a Free Field*; International Standards Organization: Geneva, Switzerland, 2012.
26. Shtrepi, L.; Astolfi, A.; D'Antonio, G.; Vannelli, G.; Barbato, G.; Mauro, S.; Prato, A. Accuracy of the random-incidence scattering coefficient measurement. *Appl. Acoust.* **2016**, *106*, 23–35. [CrossRef]
27. Peutz, V. Nouvelle examen des theories de reverberation. *Rev. D'acoust.* **1981**, *57*, 99–109.
28. Peutz, V. The Variable Acoustics of the Espace de Projection of Ircam (Paris). In Proceedings of the 59th International Audio Engineering Society Convention, Hamburg, Germany, 28 March 1978.
29. ITA TOOLBOX. Available online: http://www.ita-toolbox.org/ (accessed on 26 May 2020).
30. Farina, A. Simultaneous measurement of impulse response and distortion with a swept-sine technique. In Proceedings of the 108th International Audio Engineering Society Convention, Paris, France, 19–22 February 2000.
31. Boley, J.; Lester, M. Statistical analysis of ABX results using signal detection theory. In Proceedings of the 127th International Audio Engineering Society Convention, New York, NY, USA, 9–12 October 2009.
32. JPSB Software. Available online: https://download.cnet.com/developer/jpsb-software/i-10273146 (accessed on 26 May 2020).
33. Listening Test Database. Available online: https://zenodo.org/record/3893723#.XvLSjcARWMo (accessed on 16 June 2020).
34. Sigel, S.; Castellan, N.J. *Non Parametric Statistics for the Behavioral Sciences*, 2nd ed.; McGraw-Hill: New York, NY, USA, 1988; pp. 116–126, 184–194. ISBN 978-0070573574.
35. Clark, D. High-resolution subjective testing using a double-blind comparator. *J. Audio Eng. Soc.* **1982**, *30*, 330–338.
36. Ando, Y. *Concert Hall Acoustics*; Springer: New York, NY, USA, 1985.
37. D'Orazio, D.; Garai, M. The autocorrelation-based analysis as a tool of sound perception in a reverberant field. *Riv. Estet.* **2017**, *1*, 133–147. [CrossRef]
38. Pätynen, J.; Tervo, S.; Lokki, T. A loudspeaker orchestra for concert hall studies. In Proceedings of the Seventh International Conference on Auditorium Acoustics, Oslo, Norway, 3–5 October 2008.
39. D'Orazio, D. Anechoic recordings of Italian opera played by orchestra, choir, and soloists. *J. Acoust. Soc. Am.* **2020**, *147*, EL157–EL163. [CrossRef]
40. Kim, C.; Mason, R.; Brookes, T. Head movements made by listeners in experimental and real-life listening activities. *J. Audio Eng. Soc.* **2003**, *61*, 425–438.
41. Wenzel, E.M.; Arruda, M.; Kistler, D.J.; Wightman, F.L. Localization using non individualized head–related transfer functions. *J. Acoust. Soc. Am.* **1993**, *94*, 111–123. [CrossRef]
42. Richter, J.-G.; Behler, G.; Fels, J. Evaluation of a Fast HRTF measurement system. In Proceedings of the 140th International Audio Engineering Society Convention, Paris, France, 4–7 June 2016.
43. Martellota, F. The just noticeable difference of center time and clarity index in large reverberant spaces. *J. Acoust. Soc. Am.* **2010**, *128*, 654–663. [CrossRef]
44. Galiana, M.; Llinares, C.; Page, Á. Subjective evaluation of music hall acoustics: Response of expert and non-expert users. *Build. Environ.* **2012**, *58*, 1–13. [CrossRef]
45. D'Orazio, D.; De Cesaris, S.; Morandi, F.; Garai, M. The aesthetics of the Bayreuth Festspielhaus explained by means of acoustic measurements and simulations. *J. Cult. Heritage* **2018**, *34*, 151–158. [CrossRef]

© 2020 by the authors. Licensee MDPI, Basel, Switzerland. This article is an open access article distributed under the terms and conditions of the Creative Commons Attribution (CC BY) license (http://creativecommons.org/licenses/by/4.0/).

Article

A Machine Learning Based Prediction Model for the Sound Absorption Coefficient of Micro-Expanded Metal Mesh (MEMM)

Yaw-Shyan TSAY * and Chiu-Yu YEH

Department of Architecture, National Cheng Kung University, Tainan 701, Taiwan; arch54100@apps.arch.ncku.edu.tw
* Correspondence: tsayys@mail.ncku.edu.tw; Tel.: +886-6-2757575#54155

Received: 17 September 2020; Accepted: 22 October 2020; Published: 28 October 2020

Abstract: Recently, micro-perforated panels (MPP) have become a popular sound absorbing material in the field of architectural acoustics. However, the cost of MPP is still high for the commercial market in Taiwan, and MPP is still not very popular compared to other sound absorbing materials and devices. The objective of this study is to develop a prediction model for MEMM via a machine learning approach. An experiment including 14 types of MEMM was first carried out in a reverberation room based on ISO 354. To predict the sound absorption coefficient of the MEMM, the capability of three conventional models and three machine learning (ML) models of the supervised learning method were studied for the development of the prediction model. The results showed that in most conventional models, the sound absorption coefficient of using an equivalent perimeter had the best agreement compared with other parameters, and the root mean square error (RMSE) between prediction models and experimental data were around 0.2~0.3. However, the RMSE of all ML models was less than 0.1, and the RMSE of the gradient boost model was 0.033 in the training sets and 0.062 in the testing sets, which showed the best agreement with the experiment data.

Keywords: building acoustics; sound absorption coefficient; prediction models; supervised learning method

1. Introduction

The micro-perforated panel (MPP) has recently become a popular sound absorber in the field of noise control and building acoustics. The attractive appearance, durability, and environmental friendliness of MPP relative to the conventional porous absorbing materials made from minerals and synthetics, which present problems of indoor air quality [1]. However, the cost of MPP is still high for the commercial market in Taiwan, so MPP is not yet popular in public buildings like public transportation stations because the costs are prohibitive to the government. In this study, we developed a micro-expanded metal mesh (MEMM) absorber with a lower cost and high sound-absorbing quality.

Many studies have proposed prediction models for perforated panels and MPPs, but most of them have assumed that the shape of the perforations are circular [1–4]. However, the MEMM expanding process makes the holes on both sides uneven (Figure 1). Therefore, before using these theoretical models to predict the absorption coefficient of MEMM, we have to make assumptions regarding transforming the geometric conditions of the perforation.

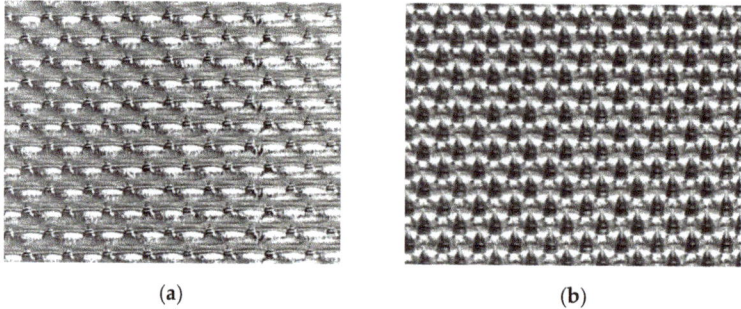

Figure 1. Picture of the MEMM: (**a**) Small side; (**b**) Large side.

Furthermore, machine learning (ML) can solve specific problems or perform certain tasks through relevant information and experience and has been widely used in various studies of prediction models, including image and speech recognition [5,6], market analysis [7], etc. In the ML approach, a prediction model can be trained with input data to achieve a goal without solving theoretic equations. In the field of building acoustics, ML has been adopted to predict the reverberation time (RT) of the auditorium, thus demonstrating that using neural networks has a higher correlation coefficient (R^2) than traditional prediction formulas, such as Sabine and Eyring [8]. Falcon Perez [9] developed the prediction model of indoor acoustic parameters in a single room; compared with the Sabine formula, the neural network had a higher mean accuracy in predicting RT. In addition, using the speech recognition method to predict RT also provided accurate prediction results [10]. Most of such ML models have achieved higher accuracy than conventional models.

The objective of this study is to develop a MEMM absorber with a lower cost that provides high sound-absorbing quality and propose a prediction model via the machine learning approach.

2. Materials and Methods

An experiment including 14 types of MEMM was first carried out in a reverberation room based on ISO 354. To predict the sound absorption coefficient of the MEMM, the capability of three conventional models and three machine learning models of supervised learning method were studied for the development of the prediction model.

With regard to practical use, the conventional model and the machine learning model need to input different variables (Table 1). Since we can determine the input variables of machine learning, the four variables used in this study are all simple geometric conditions that can be easily obtained. In contrast, the conventional model needs to use the microphotograph of the MEMM to convert the orifice diameter.

Table 1. Input variables of prediction models (O: necessary; X: unnecessary).

	Conventional Models	Machine Learning Models
Panel thickness	O	O
Airspace depth	O	O
Perforation ratio	O	X
Orifice diameter	O	X
Coefficient of viscosity of air	O	X
Density of air	O	X
Velocity of air	O	X
Horizontal center distance of the hole	X	O
Vertical center distance of the hole	X	O

2.1. Experiment

The sound absorption performance of the MEMM was measured in the reverberation room of the architectural acoustics lab at National Cheng Kung University, Taiwan (Figures 2 and 3). The volume of the reverberation room is 171.3 m^3, the surface area is 184.3 m^2, and the floor area is 32.8 m^2. The laboratory uses a floating structure to reduce the outside interference of the experiment. The experiment was based on ISO 354:2003 [11], and the rating of sound absorption was based on ISO 11654 [12]. Measurements were analyzed in 1/3-octave bands with the center frequencies of 125~4000 Hz.

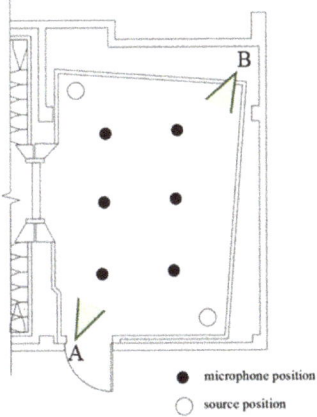

Figure 2. Reverberation room plan.

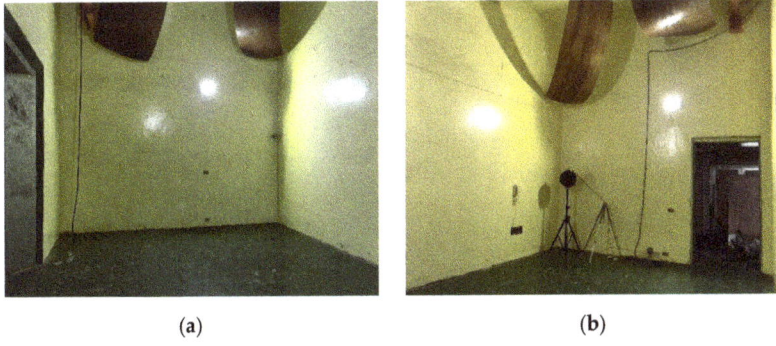

Figure 3. Picture of reverberation room: (**a**) from position A; (**b**) from position B.

The test specimen is composed of a MEMM and a closed air space layer, which together represent the structure of a common ceiling construction in Taiwan. The total area of the test specimen was 10.8 m^2 (3 m × 3.6 m), and each unit was 600 mm × 600 mm, and the 18 mm thick lumber core plywood is used for edge sealing (Figure 4). In the experiment, five kinds of MEMM with different hole distances and panel thicknesses were utilized with different air space depths, as shown in Table 2.

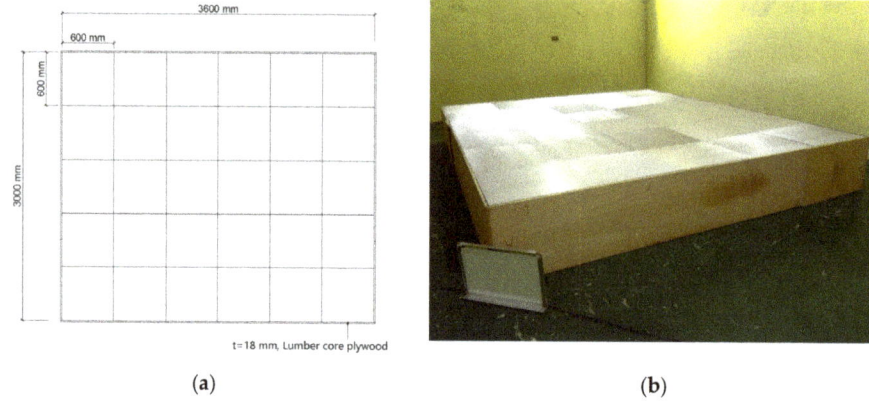

Figure 4. Specimen of MEMM absorber: (**a**) Structure of specimen; (**b**) Picture of specimen.

Table 2. Experiment cases.

Case Number	Horizontal Center Distance of the Hole (mm)	Vertical Center Distance of the Hole (mm)	Thickness of the Panel (mm)	Air Space Depth (mm)
A1	1	2	0.5	210
A2				260
A3				460
B1	2	4	0.5	210
B2				260
B3				460
C1	1	2	0.6	210
C2				260
C3				460
D1	2	4	0.6	210
D2				260
D3				460
E1	1	2	0.8	200
E2				450

2.2. The Conventional Models

In this study, two prediction models for the perforated panel and one for MPP were adapted for the prediction models of MEMM. Note that in these models, the assumptions of the holes are circular perforations, so the parameters of the geometric conditions of the MEMM have to be transformed to fit the model. Therefore, we adapted the circle diameters and perforation ratio obtained via equivalent area, equivalent perimeter, and the circumcircle on different sides of the panel (Figure 5, Table 3).

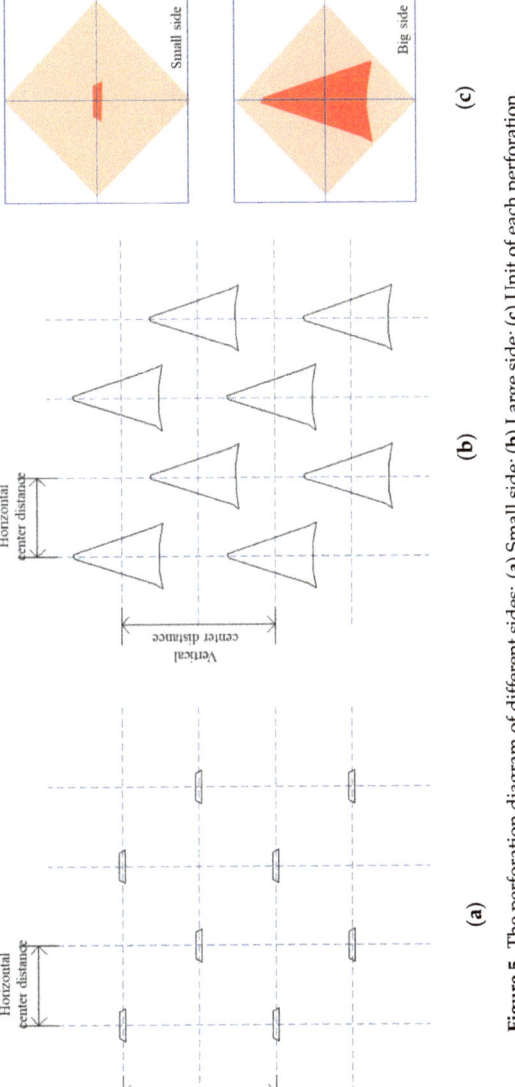

Figure 5. The perforation diagram of different sides: (a) Small side; (b) Large side; (c) Unit of each perforation.

Table 3. Circle orifice diameter transforming method.

	Transforming Method	Diagram
Equivalent perimeter	perimeter of the hole = L $2\pi r = L$	
Equivalent area	cross-sectional area of the hole = S $\pi r^2 = S$	
Circumcircle	$r = R$	

The detailed information of perforation in the microphotograph was converted to obtain different diameters and perforation ratios. For example, when considering case E on the condition of the small side, the results are as follows: circle diameters = 0.198 mm and perforation ratio = 0.014 obtained by equivalent area; circle diameters = 0.301 mm and perforation ratio = 0.033 obtained by equivalent perimeter; circle diameters = 0.482 mm and perforation ratio = 0.084 obtained by circumcircle. Furthermore, the basic parameter settings used in these models, such as viscosity coefficient of air, density of air, and velocity of air, were set at a temperature of 25 °C.

2.2.1. The Semi-Theoretical Model of a Perforated Panel

The semi-theoretical model of a perforated panel assumes that each hole in the perforated panel and the air space layer behind it are considered Helmholtz resonances [2,13,14]. The panel thickness, orifice diameter, and distance of each hole, perforation ratio, and air space depth are represented in Equations (1)–(9).

Equations (1)–(3) show the acoustic impedance of hole Z_h, acoustic resistance R, which is based on the viscosity and heat conduction of the inner wall of the hole, and correction factor δ for the orifice diameter. Equations (4) and (5) represent the characteristic impedance z_a and propagation constant γ_a of air. Equations (6)–(8) obtain the transfer matrix of the air space layer, the transfer matrix with the perforated panel, and the acoustic impedance Z of the entire sound absorber, respectively. Finally, the sound absorption coefficient α can be obtained from Equation (9).

$$Z_h = R + j\omega\rho_0 \frac{t + 2\delta d}{S}, \tag{1}$$

$$R \cong 4 \frac{0.83 \times 10^{-2} \sqrt{f}}{S} \frac{t+d}{d}, \tag{2}$$

$$\delta = 0.4\left(1 - 1.47\sigma^{1/2} + 0.47\sigma^{3/2}\right), \tag{3}$$

$$z_a = \rho_0 c_0, \tag{4}$$

$$\gamma_a = jk, \tag{5}$$

$$\begin{bmatrix} t_{11} & t_{12} \\ t_{21} & t_{22} \end{bmatrix} = \begin{bmatrix} \cosh\gamma D & z\sinh\gamma D \\ \frac{1}{z}\sinh\gamma D & \cosh\gamma D \end{bmatrix}, \tag{6}$$

$$\begin{bmatrix} T_{11} & T_{12} \\ T_{21} & T_{22} \end{bmatrix} = \begin{bmatrix} 1 & Z_h \\ 0 & 1 \end{bmatrix}\begin{bmatrix} t_{11} & t_{12} \\ t_{21} & t_{22} \end{bmatrix}, \tag{7}$$

$$Z = \frac{T_{11}}{T_{21}}, \tag{8}$$

$$\alpha = 1 - \left|\frac{Z-1}{Z+1}\right|^2, \tag{9}$$

2.2.2. Lee & Kwon's Model

For the prediction model of perforated panels, a transfer matrix method is used instead of an equivalent circuit to calculate the effect of sound through the perforated panel and the air space layer [3]. The calculation considered the panel thickness, orifice diameter, perforation ratio, and air space depth.

In this model, the empirical formula of Rao and Munjal [15] was corrected via Equation (10) to calculate the normalized acoustic impedance ξ of the perforated panel. Equations (11) and (12) represent the transfer matrix of the perforated panel and the air space layer, respectively. The transfer matrix of the whole sound absorber can be obtained by Equation (13). Equations (14) and (15) calculate the reflection coefficient γ' of sound and obtain the sound absorption coefficient α.

$$\xi = \left[7.337 \times 10^{-3} + j \times 1.3 \times 2.2245 \times 10^{-5}(1+51t)(1+204d)f\right]/\sigma, \tag{10}$$

$$\begin{bmatrix} P_{11} & P_{12} \\ P_{21} & P_{22} \end{bmatrix} = \begin{bmatrix} 1 & \rho_0 c_0 \xi \\ 0 & 1 \end{bmatrix}, \tag{11}$$

$$\begin{bmatrix} S_{11} & S_{12} \\ S_{21} & S_{22} \end{bmatrix} = \begin{bmatrix} \cos kD & (j\rho_0 c_0)\sin kD \\ (j/\rho_0 c_0)\sin kD & \cos kD \end{bmatrix}, \tag{12}$$

$$[T] = [P][S], \tag{13}$$

$$\gamma' = \frac{T_{11} - \rho_0 c_0 T_{21}}{T_{11} + \rho_0 c_0 T_{21}}, \tag{14}$$

$$\alpha = \frac{4Re(1+\gamma'/1-\gamma')}{[1+Re(1+\gamma'/1-\gamma')]^2 + [Im(1+\gamma'/1-\gamma')]^2}, \tag{15}$$

2.2.3. Maa's Model for MPP

In Maa's model, the perforation of MPP was limited to less than 1 mm in diameter and 1% in perforation ratio [4,16]. The panel thickness, orifice diameter, and distance of each hole, the perforation ratio, and the air space depth are all taken into consideration in this model, as shown in Equations (16)–(20).

Equation (16) calculates the acoustic impedance Z_{MPP} of the MPP; the acoustic resistance r and mass reactance ωm are calculated by Equations (17) and (18), respectively. Equation (19) shows the perforation constant k' for the holes of the MMP, and the sound absorption coefficient α can be obtained using Equation (20) through the previously calculated sound resistance and reactance.

$$Z_{MPP} = r + j\omega m \tag{16}$$

$$r = \frac{32\eta t}{\sigma \rho_0 c_0 d^2} k_r, \; k_r = \left[1 + \frac{k'^2}{32}\right]^{1/2} + \frac{\sqrt{2}}{32} k' \frac{d}{t}, \tag{17}$$

$$\omega m = \frac{\omega t}{\sigma c} k_m, \; k_m = 1 + \left[1 + \frac{k'^2}{2}\right]^{-1/2} + 0.85 \frac{d}{t}, \tag{18}$$

$$k' = d\sqrt{\omega \rho_0 / 4\eta}, \tag{19}$$

$$\alpha = \frac{4r}{(1+r)^2 + (\omega m - \cot(\omega D/c))^2},\quad (20)$$

2.3. The Machine Learning Model

In the ML model, a supervised learning method was used to train and obtain the prediction model in this study. In the supervised learning method, the MEMM characteristics of panel thickness, hole-to-hole distance, and air space depth were defined as input objects, and the output value was the sound absorption coefficient.

The sound absorption coefficient of each frequency band was predicted via the ML process, and the results were then compared with the experiment data to verify the applicability. The root-mean-square error (RMSE) value between the prediction models and experiment data were used to evaluate applicability.

The following describes the basic process of ML. First, the data were sorted and divided into training, validation, and testing sets. In this study, the proportion was 65% training and validation set and 35% testing sets. Furthermore, the k-fold cross validation was used (with k = 5), so that the training and validation set would be shuffled.

The second step involves selecting models and adjusting the parameters of the model. In this study, we used three kinds of ensemble learning methods, including the gradient boosting (Gboost) model, the average model, and the stacking model, to obtain the prediction model. The last two models are the combination method of the Gboost model and the three linear models. The algorithms of the linear model are shown in Table 4 [17]. Note that the advantage of the ensemble learning method is that it combines multiple learners to produce more accurate results than individual learners [18].

Table 4. Algorithm of linear regression models.

Model	Objective Function
Lasso	$\min_{w} \frac{1}{2n_{samples}} \|Xw - y\|_2^2 + \alpha\|w\|_1$
Elastic net (ENet)	$\min_{w} \frac{1}{2n_{samples}} \|Xw - y\|_2^2 + \alpha\rho\|w\|_1 + \frac{\alpha(1-\rho)}{2}\|w\|_2^2$
Kernel ridge (KRR)	combines Ridge regression with the kernel trick Ridge: $\min_{w}\|Xw - y\|_2^2 + \alpha\|w\|_2^2$

1. The Gboost model

The Gboost model is a combination of the gradient descending model and the boost model. A boost model can generate a strong learner from an ensemble of weak learners, each of which can barely do better than random guessing [19]. The process of the Gboost model is to build a model, then increase the weight of data that is incorrectly predicted on this model to create a second model, and repeat the same steps to obtain a better-performing model.

2. The average model

The average model constructs and trains different models, combines the models to reduce errors and overfitting, and averages the output values of each model. In this study, the average model is a combination of the above models (three linear models and Gboost), and the average value of the four models was used as the final output value.

3. The stacking model

The problem that an averaging method may encounter is that not every model is good, and a bad model will cause the averaged result to be worse. Therefore, the stacking model is an improvement of the average model by assigning weight to the contribution of each model (in the average model, each model contributes the same amount). The specific method uses trainable combiners, which develop a learner (another meta-model) in order to first combine the models. Doing so allows researchers to determine which learners are likely to be successful in which part of the feature space and then combine them accordingly [19]. The prediction results of the first-level model are used as the input variables of the meta-model (Figure 6).

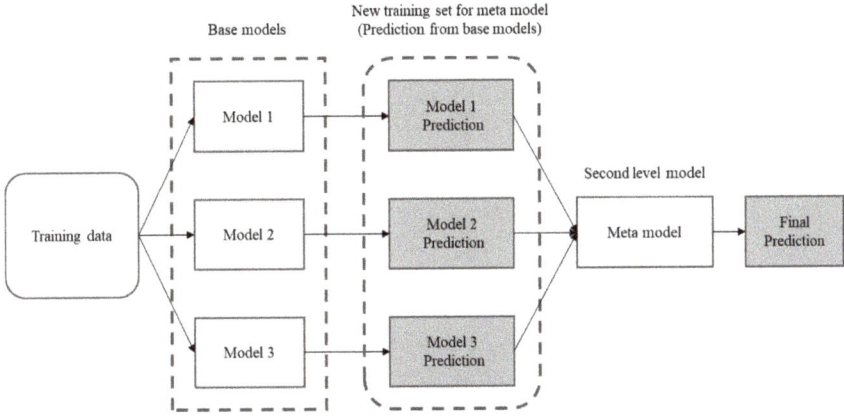

Figure 6. The stacking model.

3. Results

3.1. The Conventional Model

The equivalent area, equivalent perimeter, and circumcircle of the holes are used to obtain the value of the orifice diameter. Comparing the small and big sides, we found that using the geometric condition of the small side was more accurate in any transform method; therefore, the following discussion uses the results of the small side. The results of different transform methods are shown in Figure 7. Furthermore, Figure 8 shows the results of different models with the equivalent perimeter condition.

The orifice diameter converted by equivalent perimeter shows that Maa's formula had the best agreement with the experiment. However, the three models are more accurate in the low frequency band, and the error is higher in the high frequency.

Figure 9 shows that the predicted value tends to under-estimate the sound absorption coefficient of the MEMM. The root mean square error (RMSE) of the different models are listed in Table 5.

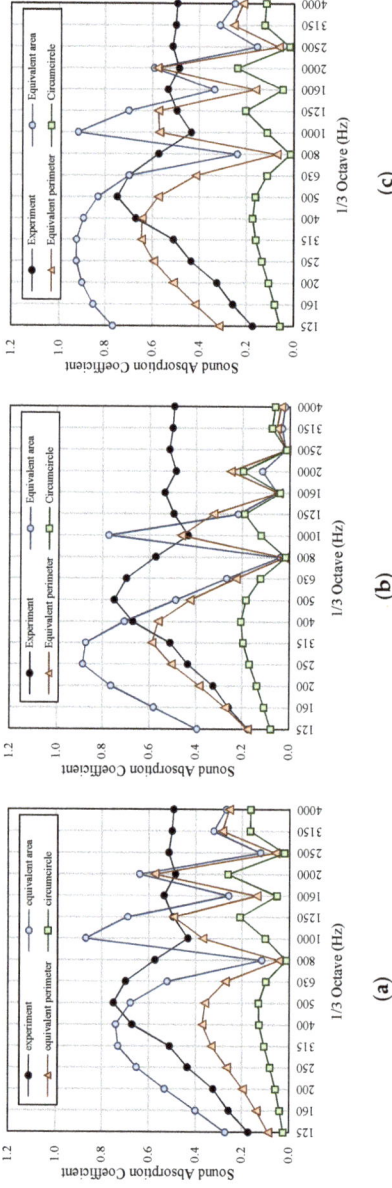

Figure 7. Comparison of the different methods to the converted orifice diameter (using the small side condition): case E1. (**a**) Semi-theoretical model; (**b**) Lee & Kwon's model; (**c**) Maa's model.

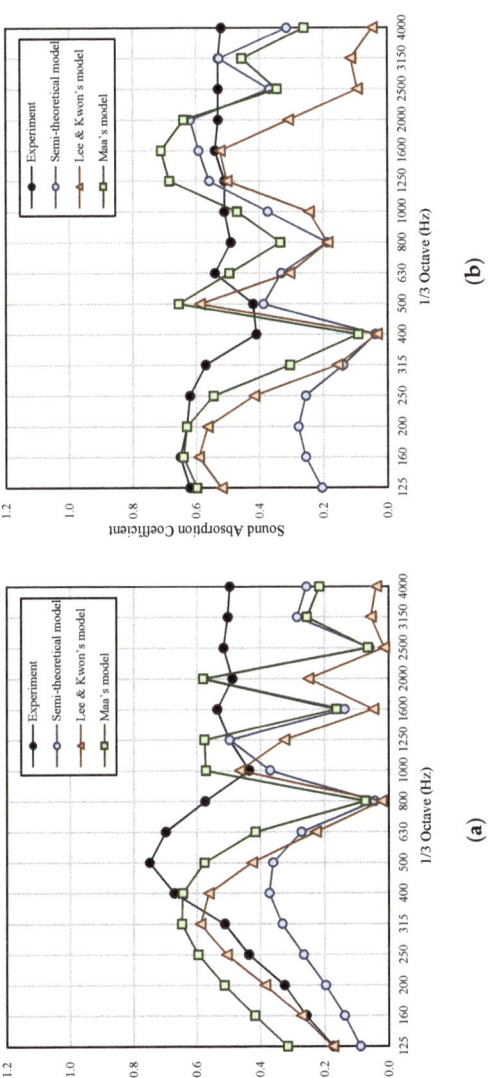

Figure 8. Results of conventional models with small side equivalent perimeter condition: (**a**) case E1; (**b**) case E2.

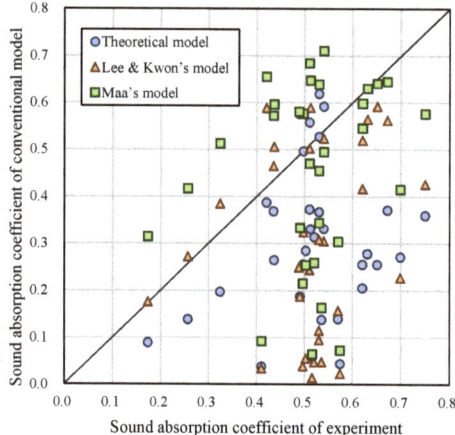

Figure 9. Scatter plot of conventional models.

Table 5. RMSE of conventional models.

Model	Converted Method	RMSE
Semi-theoretical model	Equivalent area	0.248
	Equivalent perimeter	0.276
	Circumcircle	0.413
Lee & Kwon's model	Equivalent area	0.394
	Equivalent perimeter	0.301
	Circumcircle	0.400
Maa's model	Equivalent area	0.368
	Equivalent perimeter	0.212
	Circumcircle	0.401

3.2. The ML Model

In the machine learning part (Figures 10 and 11), the prediction of the basic linear regression models was poor and failed to express the characteristics of such resonance absorption structure performing better in a specific frequency band (resonance frequency). For the ensemble model, the Gboost model and the stacking model were more accurate. The Gboost and stacking models are shown to express better sound absorption at resonance frequency. Compared to conventional models, ML models had better predictive ability for medium and high frequencies. Nevertheless, the average model could not fit the sound absorption trend. Furthermore, the prediction performance of the training set was better than that of the test set. The nomenclature and prediction results of other cases are attached in Appendix A.

Table 6 shows the RMSE of each model's prediction value and the experimented value. The RMSE of all ML models was less than 0.1. The Gradient boost model had a RMSE of 0.033 in the training set and 0.062 in the testing set, which is superior to the conventional theoretical model and shows the best agreement with experiment data.

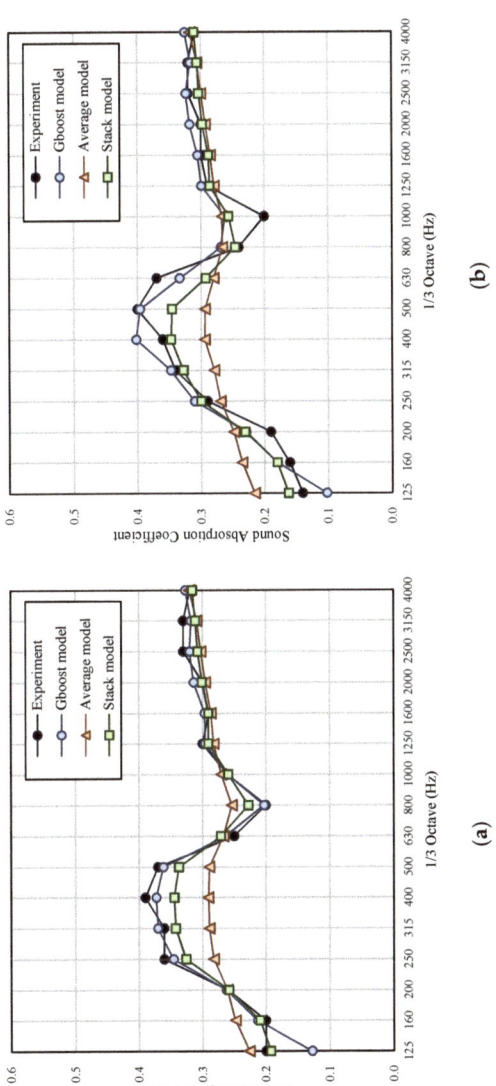

Figure 10. Results of machine learning models: (**a**) training set: case A2; (**b**) testing set: case A1.

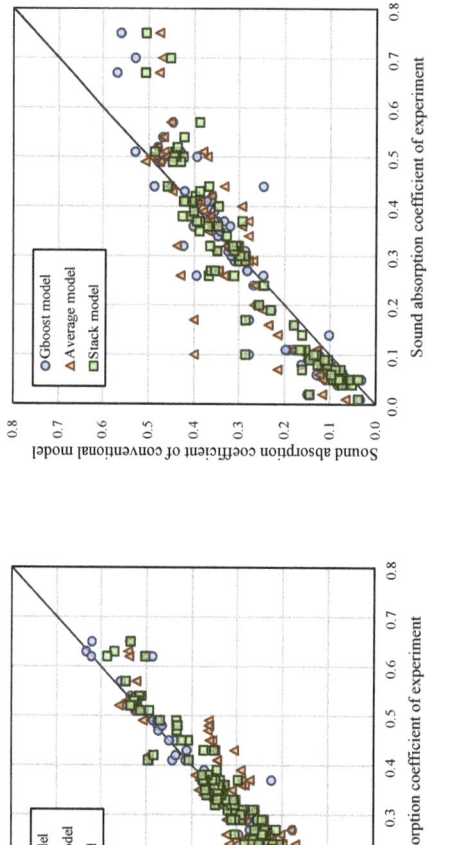

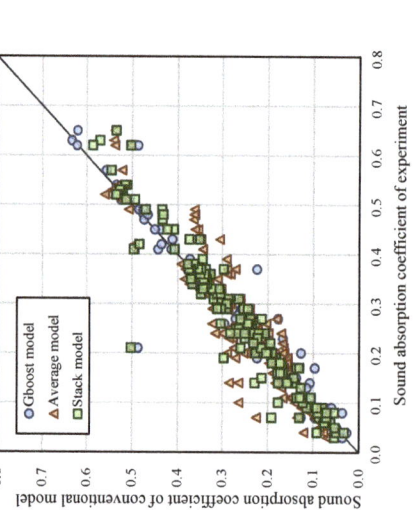

Figure 11. Scatter plot of machine learning models: (**a**) training set; (**b**) testing set.

Table 6. RMSE of ML models.

	Model	RMSE
Training set	Lasso	0.069
	ENet	0.069
	KRR	0.070
	Gboost	0.033
	Average	0.056
	Stack	0.040
Testing set	Lasso	0.092
	ENet	0.092
	KRR	0.095
	Gboost	0.062
	Average	0.081
	Stack	0.067

4. Discussion and Conclusions

This study attempted to use different models to predict the sound absorption coefficient of the MEMM and discuss its applicability. After comparing the conventional model and the machine learning model, the machine learning model is more accurate with regard to predictive ability. As for use, the geometric conditions of materials used in machine learning are simple to obtain, while traditional models require microphotograph and conversion methods for orifice diameter.

In the first resonance frequency band, the predictive ability of the conventional model was not poor, but regardless of the model, there was several resonance frequency bands (several resonance frequencies), which are not show in the experimental results. The assumptions of the conventional model were all speculated to be ideal physical environments. In the literature, the model was compared with the measurement of impedance tubes to demonstrate the results of several resonance frequency bands [3]. However, the results of this study, whose experiment was conducted in a reverberation room, has no such situation. The measurement method of sound absorption consists of small-size—impedance tubes [20,21] and large-size—reverberation room [11]. Further research is warranted to clarify the difference and connection between the two.

The ML model did not have this problem because the sound absorption coefficient of the training set was obtained from the reverberation room experiment. In the comparison of each ML model, the average model was usually found to underestimate the sound absorption capacity of the MEMM at the resonance frequency. It was speculated that during the average process, the result was affected by the poorly performing linear model. Furthermore, the predictive ability still differed between the training set and the testing set. If the number of the entire data set is increased, the model could be further improved to decrease the performance gap between the two.

The generalization of machine learning models primarily depends on the data set. Therefore, its predictive ability is more credible within the scope of the experiment of this study. For other types of MEMM, their applicability remains to be studied. Furthermore, if the number and diversity of experimental specimen are increased in the future, the scope of application of the prediction models can be continuously increased.

Author Contributions: Conceptualization, Y.-S.T.; methodology, Y.-S.T. and C.-Y.Y.; software, C.-Y.Y.; validation, Y.-S.T. and C.-Y.Y.; data curation, Y.-S.T. and C.-Y.Y.; writing—original draft preparation, C.-Y.Y.; writing—review and editing, Y.-S.T.; visualization, C.-Y.Y.; supervision, Y.-S.T. All authors have read and agreed to the published version of the manuscript.

Funding: This research received no external funding.

Conflicts of Interest: The authors declare no conflict of interest.

Nomenclature

Z_h	acoustic impedance of the hole (Pa·s/m^3)
R	acoustic resistance based on the viscosity and heat conduction of the inner wall of the hole (Pa·s/m^3)
f	the frequency (Hz)
ω	the angular frequency (rad/s)
ρ_0	the density of air (kg/m^3)
t	the panel thickness (m)
δ	correction factor (-)
d	the orifice diameter (m)
$S = \pi d^2/4$	cross section area of the hole (m^2)
σ	the perforation ratio (-)
z_a	characteristic impedance of air (Pa·s/m^3)
γ_a	propagation constant of air (rad/m)
c_0	the velocity of air (m/s)
k	the wave number (rad/m)
D	cavity thickness/airspace depth (m)
Z	acoustic impedance of the absorber (Pa·s/m^3)
α	the absorption coefficient (-)
ξ	the normalized acoustic impedance of the panel (Pa·s/m^3)
$[]$	The overall transfer matrix for perforated panel system
γ'	the pressure reflection coefficient (-)
Z_{MPP}	acoustic impedance of the MPP (Pa·s/m^3)
r	relative acoustic resistance (Pa·s/m^3)
$x_m = \omega m$	mass reactance (Pa·s/m^3)
k'	the perforate constant
k_r	the resistance coefficient
k_m	mass reactance coefficient
η	coefficient of viscosity (Pa·s)

Appendix A

Training Set

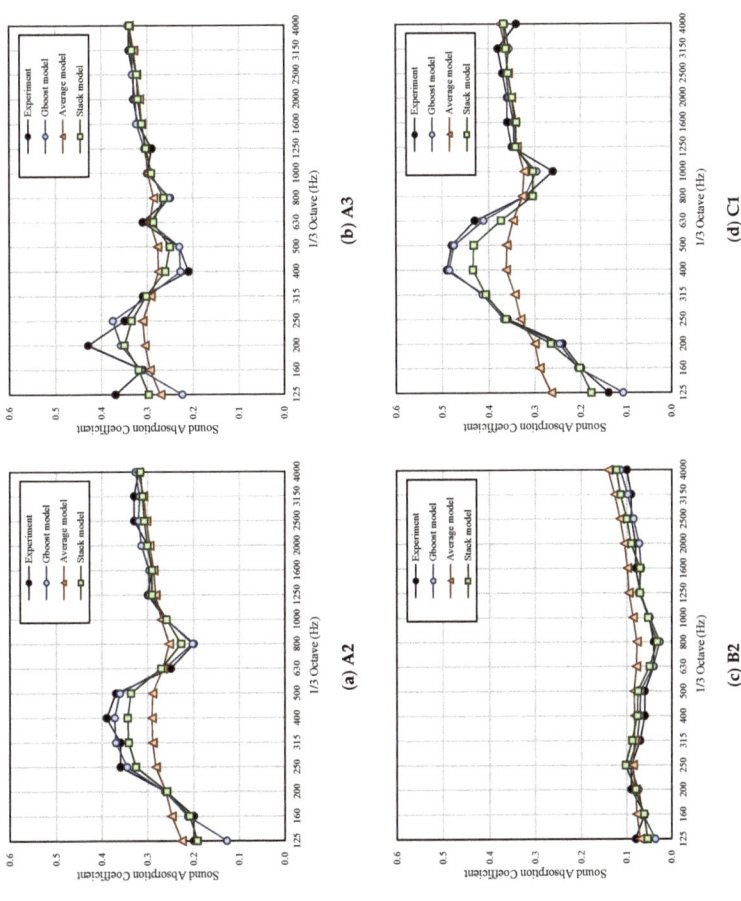

Figure A1. *Cont.*

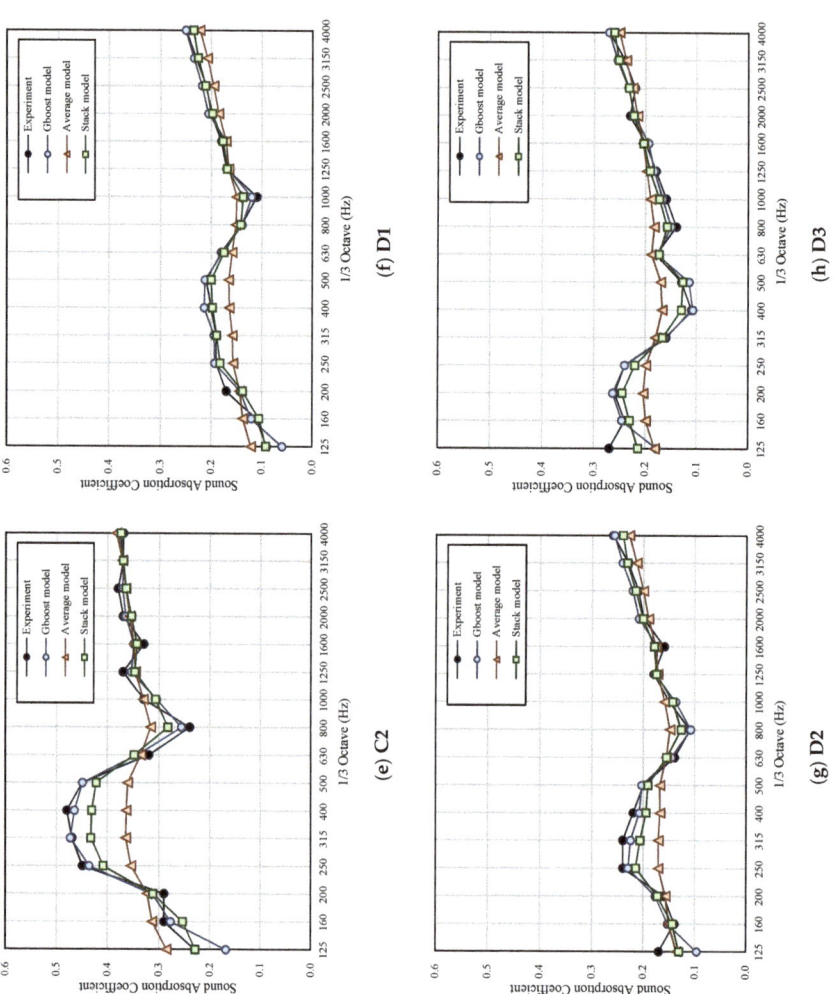

Figure A1. *Cont.*

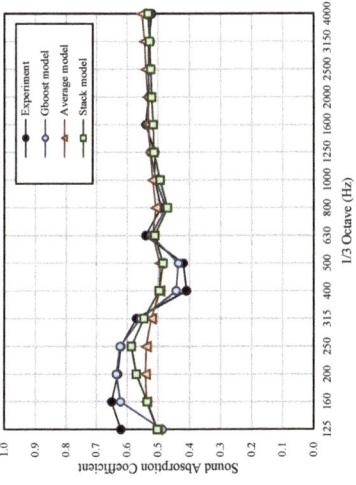

(j) E2

Figure A1. Prediction Results of the ML Model.

Testing Set

Figure A2. *Cont.*

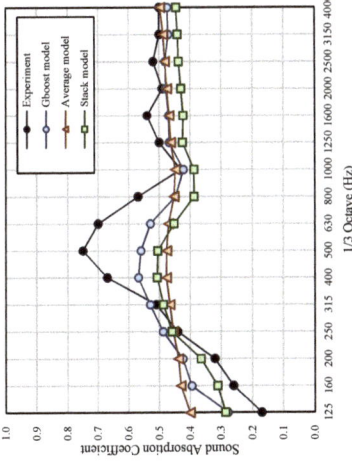

Figure A2. Prediction Results of the ML Model.

References

1. Mosa, A.I.; Putra, A.; Ramlan, R.; Esraa, A.A. Micro-Perforated Panel Absorber Arrangement Technique: A Review. *J. Adv. Res. Dyn. Control. Syst.* **2018**, *10*, 372–381.
2. Iwase, T. Chapter 3.4 Sound reflection, absorption and insulation. In *Handbook of Noise Control*, 1st ed.; The Institute of Noise Control Engineering of Japan, Ed.; Gihodobooks: Tokyo, Japan, 2001; pp. 138–148. (In Japanese)
3. Lee, D.H.; Kwon, Y.P. Estimation of the absorption performance of multiple layer perforated panel systems by transfer matrix method. *J. Sound Vib.* **2004**, *278*, 847–860. [CrossRef]
4. Maa, D.Y. Theory and design of microperforated panel sound-absorbing constructions. *Sci. Sin.* **1975**, *18*, 55–71.
5. Ciregan, D.; Meier, U.; Schmidhuber, J. Multi-Column Deep Neural Networks for Image Classification. In Proceedings of the 2012 IEEE Conference on Computer Vision and Pattern Recognition, Providence, RI, USA, 16–21 June 2012; pp. 3642–3649.
6. Hinton, G.; Deng, L.; Yu, D.; Dahl, G.E.; Mohamed, A.R.; Jaitly, N.; Senior, A.; Vanhoucke, V.; Nguyen, P.; Sainath, T.A.; et al. Deep neural networks for acoustic modeling in speech recognition: The shared views of four research groups. *IEEE Signal. Process. Mag.* **2012**, *29*, 82–97. [CrossRef]
7. Patel, J.; Shah, S.; Thakkar, P.; Kotecha, K. Predicting stock and stock price index movement using trend deterministic data preparation and machine learning techniques. *Expert Syst. Appl.* **2015**, *42*, 259–268. [CrossRef]
8. Nannariello, J.; Fricke, F. The prediction of reverberation time using neural network analysis. *Appl. Acoust.* **1999**, *58*, 305–325. [CrossRef]
9. Falcon Perez, R. Machine-Learning-Based Estimation of Room Acoustic Parameters. Master's Thesis, Aalto University, Espoo, Finland, 2018.
10. Gamper, H.; Tashev, I.J. Blind reverberation time estimation using a convolutional neural network. In Proceedings of the 2018 16th International Workshop on Acoustic Signal Enhancement (IWAENC), Tokyo, Japan, 17–20 September 2018; pp. 136–140.
11. ISO 354. *Acoustics—Measurement of Sound Absorption in a Reverberation Room*; ISO: Geneva, Switzerland, 2003.
12. ISO 11654. *Acoustics—Sound Absorbers for Use in Buildings—Rating of Sound Absorption*; ISO: Geneva, Switzerland, 1997.
13. Koyasu, M. *Sound Absorbing Material*, 1st ed.; Gihodobooks: Tokyo, Japan, 1976. (In Japanese)
14. Hiroyuki, S. Acoustic analysis of acoustic structural features using Excel (2) Analysis of sound absorption structure. *Soc. Heat. Air Cond. Sanit. Eng. Jpn.* **2006**, *80*, 919–927. (In Japanese)
15. Rao, K.N.; Munjal, M.L. Experimental evaluation of impedance of perforates with grazing flow. *J. Sound Vib.* **1986**, *108*, 283–295.
16. Maa, D.Y. Potential of microperforated panel absorber. *J. Acoust. Soc. Am.* **1998**, *104*, 2861–2866. [CrossRef]
17. Documentation of Scikit-Learn 0.16.1. Available online: https://scikit-learn.org/0.16/documentation.html (accessed on 5 September 2020).
18. Kuncheva, L.I.; Bezdek, J.C.; Duin, R.P. Decision templates for multiple classifier fusion: An experimental comparison. *Pattern Recognit.* **2001**, *34*, 299–314. [CrossRef]
19. Zhang, C.; Ma, Y. *Ensemble Machine Learning: Methods and Applications*; Zhang, C., Ma, Y., Eds.; Springer Science & Business Media: New York, NY, USA, 2012.
20. ISO 10534-1. *Acoustics—Determination of Sound Absorption Coefficient and Impedance in Impedance Tubes—Part. 1: Method Using Standing Wave Ratio*; ISO: Geneva, Switzerland, 1996.
21. ISO 10534-2. *Acoustics—Determination of Sound Absorption Coefficient and Impedance in Impedance Tubes—Part. 2: Transfer-Function Method*; ISO: Geneva, Switzerland, 1998.

Publisher's Note: MDPI stays neutral with regard to jurisdictional claims in published maps and institutional affiliations.

© 2020 by the authors. Licensee MDPI, Basel, Switzerland. This article is an open access article distributed under the terms and conditions of the Creative Commons Attribution (CC BY) license (http://creativecommons.org/licenses/by/4.0/).

MDPI
St. Alban-Anlage 66
4052 Basel
Switzerland
Tel. +41 61 683 77 34
Fax +41 61 302 89 18
www.mdpi.com

Applied Sciences Editorial Office
E-mail: applsci@mdpi.com
www.mdpi.com/journal/applsci

www.ingramcontent.com/pod-product-compliance
Lightning Source LLC
LaVergne TN
LVHW070229100526
838202LV00015B/2111